XVI Workshop on High Energy Spin Physics (D-SPIN 2015)

AA002747

Journal of Physics: Conference Series Volume 678

Dubna, Russia
8 - 12 September 2015

ISBN: 978-1-5108-2089-0
ISSN: 1742-6588

XVI Workshop on High Energy Spin Physics (D-SPIN 2015)

Journal of Physics: Conference Series Volume 678

Dubna, Russia
8 - 12 September 2015

IOP
Institute of Physics

Printed from e-media with permission by:

Curran Associates, Inc.
57 Morehouse Lane
Red Hook, NY 12571

Some format issues inherent in the e-media version may also appear in this print version.

Copyright© (2015) by the Institute of Physics
All rights reserved. The material featured in this book is subject to
IOP copyright protection, unless otherwise indicated.

Printed by Curran Associates, Inc. (2016)

For permission requests, please contact the Institute of Physics
at the address below.

Institute of Physics
Dirac House, Temple Back
Bristol BS1 6BE UK

Phone: 44 1 17 929 7481
Fax: 44 1 17 920 0979

techtracking@iop.org

Additional copies of this publication are available from:

Curran Associates, Inc.
57 Morehouse Lane
Red Hook, NY 12571 USA
Phone: 845-758-0400
Fax: 845-758-2634
Email: curran@proceedings.com
Web: www.proceedings.com

Table of contents

Volume 678

XVI Workshop on High Energy Spin Physics (D-SPIN2015)
8–12 September 2015, Dubna, Russia,

Accepted papers received: 10 December 2015
Published online: 5 February 2016

Preface

Papers

Journal of Physics: Conference Series **678** (2016) 011001 doi:10.1088/1742-6596/678/1/011001

Welcome address

Richard Lednicky
Vice-Director, JINR, 141980, Dubna, Moscow Region, Russia
E-mail: lednicky@fzu.cz

Dear Colleagues,
Ladies and Gentlemen, on behalf of the Directorate of Joint Institute for Nuclear Research (JINR) it is a pleasure for me to welcome you here to Dubna for the 16th International Workshop on High Energy Spin Physics. It provides an opportunity to present and discuss the news accumulated during last year. Another important feature of this series of workshops has always been the participation of a large number of physicists from the former Soviet Union and Eastern European countries, for which long trips have previously been limited by financial (and earlier also by bureaucratic) reasons. It thus represents an important addition to the series of large International Symposia on spin physics held in even-numbered years in different countries, including the Symposium held in Dubna in 2012.

JINR has a long-lasting tradition of experimental and theoretical studies of spin phenomena. The workshops on high energy spin physics started in Dubna in 1981 due to the initiative of L. Lapidus, an outstanding theoretical physicist. Since then, these meetings have been held in Dubna in every odd year and have become regular thanks to Anatoly Vasilievich Efremov, the chairman for many years.

Recent years have brought a lot of new experimental results, and above all the discovery and determination of quantum characteristics of the Higgs boson at the Large Hadron Collider.

The spin structure of the nucleon is studied in high energy scattering of leptons on polarized nucleons (COMPASS, Jlab, HERMES), and in the collisions of polarized hadrons (RHIC, IHEP, JINR). The joint description of so many different high-energy processes is made possible by the application of the fundamental theories of strong interactions - quantum chromodynamics (QCD) due to its remarkable properties of factorization, local quark-hadron duality and asymptotic freedom, which allows the calculations of characteristics of hard sub-processes in the framework of perturbation theory. At the same time, the theory doesn't predict the universal (process independent) parton distribution functions (PDF's) and fragmentation functions, thus requiring the use of modeling techniques. A number of reports at the meeting are devoted to the development and application of such models.

The theoretical description of processes involved besides spin including the internal transverse motion turns out to be rather complex - the number of PDF's taking into account the transverse momentum dependence (TMD) increases and the simplicity of the parton model with its probabilistic interpretation is lost. One of the problems to be solved is the evolution of these functions with the change of the wavelength of the probe. A possible approach to its solution will be presented at this Workshop.

New data on the spin distribution of the "sea" $\bar{u}$ and $\bar{d}$ quarks from the production process of W^+ and W^- bosons in collisions of polarized protons will be presented.

All hope for overcoming the so-called spin crisis rests now on possible contributions to the nucleon spin of orbital angular momenta of quarks and gluons, which can be determined by

Content from this work may be used under the terms of the Creative Commons Attribution 3.0 licence. Any further distribution of this work must maintain attribution to the author(s) and the title of the work, journal citation and DOI.
Published under licence by IOP Publishing Ltd

Journal of Physics: Conference Series **678** (2016) 011001 doi:10.1088/1742-6596/678/1/011001

measuring the Generalized Parton Distributions (GPD's). Theoretical and experimental aspects of GPDs will be presented as well.

Although the overall description of single spin asymmetries (SSA) is achieved in the existing theory, its development continues. Here, T-odd PDF's lose the key properties of universality and become "effective", depending on the processes in which they are observed. In particular, the most fundamental prediction of QCD is the change in their opposite sign in Sivers electroproduction of pions and Drell-Yan pair production on a transversely polarized target. This conclusion is supposed to be checked in the experiment COMPASS at SPS CERN and in the collider experiments at RHIC, NICA and PANDA-PAX.

Considerable attention at the workshop will be paid to further development of the projects of polarization studies in FNAL, plans of further research on the modified Jlab accelerator, as well as plans to create the world's nuclear electron colliders eRHIC, LHeC, MEIC/EIC and EIC@HIAF (China).

The program of getting polarized proton and antiproton beams from the Lambda particle decays at the U-70 IHEP accelerator in Protvino for the SPASCHARM project will be presented. The importance of comparative study of spin effects induced by particles and antiparticles will be particularly stressed.

Especially interesting are plans to create in the IKP (Juelich, Germany), the unique European complex for the measurement of the electric dipole moment (EDM) of protons and nuclei. Since the dipole moment of fundamental particles violates the conservation laws of spatial and temporal parity, its detection would indicate a violation of the Standard Model and, in particular, contribute to understanding of the baryon asymmetry of the Universe.

Our Institute is also well-known for building powerful accelerators, such as the famous Synchrophasotron and the Nuclotron, the first superconducting relativistic heavy ion synchrotron. The current workshop is taking place in an important period for the development of the accelerator complex NICA (Nuclotron based Ion Collider fAcility) - the start of the civil construction of its collider part.

The main physics goal of NICA is a study of hot and dense QCD matter. The important properties of this matter, like critical point and mixed phase, are expected to show up just in the NICA energy range. The NICA program is complementary to the program of FAIR@GSI and will continue the program of BES@RHIC with much higher luminosity.

Another important direction of experiments at NICA, represented by Spin Physics, will explore both fixed target and collider modes, using polarized deuteron and proton beams. There will be dedicated session devoted to spin physics at the NICA complex at the workshop. Thus, the spin physics program will be an important part of the sophisticated fixed target experiment BM@N (Baryonic Matter at Nuclotron), which is under preparation. The spin physics program in the collider mode will be realized with the help of the main Multi Purpose Detector (MPD), as well as with the help of the dedicated Spin Physics Detector (SPD), the proposal of which is under preparation. The main goal is the study of Single Spin Asymmetries in Drell-Yan processes and an investigation of new PDFs. There are also plans to study spin-related signals of collective effects in heavy ion collisions; particularly, the vorticity may serve as a new source of hyperon polarization. The NICA spin physics program was recently presented and discussed in International Workshop on Hadron Structure and Spectroscopy organized by JINR in the historic Russian town of Suzdal.

As for the NICA status:

- the civil construction of the collider part of the NICA complex started by preparatory works this year;
- the magnet assembly workshop is in operation, the series supercoducting magnets production has started;
- the MPD sub-detectors are close to the stage of fabrication;

- the SPD Letter of Intent was prepared and the SPD working group has been formed;
- the international cooperation around NICA is steadily growing.

The Workshop is supported by the Russian Foundation for Basic Research, the International Committee of Spin Physics Symposia, MePHI, EPS and, of course, by JINR (in particular, by the Heisenberg-Landau, Bogolyubov-Infeld and Votruba-Blokhintsev programs). I would like to thank the organizers and wish you a nice stay here in Dubna and JINR full of interesting talks and illuminating discussions. The workshop is opened.

Peer review statement

All papers published in this volume of *Journal of Physics: Conference Series* have been peer reviewed through processes administered by the proceedings Editors. Reviews were conducted by expert referees to the professional and scientific standards expected of a proceedings journal published by IOP Publishing.

Content from this work may be used under the terms of the Creative Commons Attribution 3.0 licence. Any further distribution of this work must maintain attribution to the author(s) and the title of the work, journal citation and DOI.
Published under licence by IOP Publishing Ltd

Journal of Physics: Conference Series **678** (2016) 012001 doi:10.1088/1742-6596/678/1/012001

The QCD evolution of TMD in the covariant approach

A.V. Efremov[1], **O.V. Teryaev**[1] **and P. Zavada**[2]

[1]Bogoliubov Laboratory of Theoretical Physics, JINR, 141980 Dubna, Russia
[2]Institute of Physics AS CR, Na Slovance 2, CZ-182 21 Prague 8, Czech Republic

E-mail: zavada@fzu.cz

Abstract. The procedure for calculation of the QCD evolution of transverse momentum dependent distributions within the covariant approach is suggested. The standard collinear QCD evolution together with the requirements of relativistic invariance and rotational symmetry of the nucleon in its rest frame represent the basic ingredients of our approach. The obtained results are compared with the predictions of some other approaches.

1. Introduction

In our previous study we discussed various aspects of the covariant quark-parton model, see [1, 2, 3, 4, 5, 6] and citations therein. The discussion included distribution and structure functions (PDF-parton distribution function, TMD-transverse momentum distributions, unpolarized and polarized structure functions) in the leading order. For a fixed Q^2 we obtained the set of relations and rules, which interrelate some of them, for example PDF withTMD:

$$q(x) \rightleftarrows q(x, p_T), \tag{1}$$

as we have studied in [2, 3].

The aim of the present talk is to discuss a possible procedure for the Q^2- evolution in the covariant approach. The starting point will be the use of a standard algorithm for the collinear QCD evolution of integrated distribution: $q(x) \to q(x, Q^2)$. Then, the evolved distribution $q(x, Q^2)$ is inserted into the original relation (1). In this way we have defined the procedure for the evolution of TMD:

$$q(x) \to q(x, Q^2) \to q(x, p_T, Q^2). \tag{2}$$

In the present study we will discuss unpolarized distributions only. In the next section we collect corresponding relations defining the procedure (1) in detail. The last section involves the numerical results and its discussion.

2. Evolution of TMD

The well known algorithm for evolution of the unpolarized PDF reads:

$$\frac{d}{d\ln Q^2} q(x, Q^2) = \int_x^1 \frac{dy}{y} P\left(\frac{x}{y}\right) q(x, Q^2), \tag{3}$$

Content from this work may be used under the terms of the Creative Commons Attribution 3.0 licence. Any further distribution of this work must maintain attribution to the author(s) and the title of the work, journal citation and DOI.
Published under licence by IOP Publishing Ltd

Journal of Physics: Conference Series **678** (2016) 012001 doi:10.1088/1742-6596/678/1/012001

where P is the corresponding splitting function. We have shown [5] that in the leading order approach the relativistic invariance and spherical symmetry in the nucleon rest frame imply

$$\rho_q(p,Q^2) = 4\pi p^2 M G_q(p,Q^2) = -x^2\left(\frac{q(x,Q^2)}{x}\right)' = q\left(x,Q^2\right) - xq'\left(x,Q^2\right); \qquad p\left(x\right) = \frac{Mx}{2}. \tag{4}$$

where $\rho_q(p)$ is the probability distribution of the quark momentum $p = |\mathbf{p}|$ in the rest frame, $f'\left(x,Q^2\right)$ denotes df/dx. This relation is valid for any sufficiently large Q^2, when the quark can be considered effectively free in any frame in the sense of Ref. [1]. The relation (3) implies

$$\frac{d}{d\ln Q^2} q'\left(x,Q^2\right) = -\frac{1}{x}P\left(\frac{x}{x}\right) q\left(x,Q^2\right) + \int_x^1 \frac{dy}{y}\frac{d}{dx}P\left(\frac{x}{y}\right) q\left(y,Q^2\right). \tag{5}$$

Since

$$\frac{d}{dx}P\left(\frac{x}{y}\right) = -\frac{y}{x}\frac{d}{dy}P\left(\frac{x}{y}\right), \tag{6}$$

integration by parts gives

$$\frac{d}{d\ln Q^2}\left(xq'\left(x,Q^2\right)\right) = \int_x^1 \frac{dy}{y}\frac{d}{dx}P\left(\frac{x}{y}\right)\left(yq'\left(y,Q^2\right)\right). \tag{7}$$

This equality together with Eqs. (3) and (4) gives

$$\frac{d}{d\ln Q^2}\rho_q\left(p\left(x\right),Q^2\right) = \int_x^1 \frac{dy}{y}P\left(\frac{x}{y}\right)\rho_q\left(p\left(y\right),Q^2\right) \tag{8}$$

or equivalently

$$\frac{d}{d\ln Q^2}G_q\left(p,Q^2\right) = \frac{1}{4\pi p^2 M}\int_p^{M/2}\frac{dp'}{p'}P\left(\frac{p}{p'}\right)\rho_q\left(p',Q^2\right). \tag{9}$$

Further, in Ref. [4, 5] (see also [7]) we proved the relation for the unpolarized TMD:

$$f_1^q(x,p_T,Q^2) = MG_q(\tilde{p},Q^2) = \frac{1}{4\pi\tilde{p}^2}\left(q\left(\xi,Q^2\right) - \xi q'\left(\xi,Q^2\right)\right), \tag{10}$$

where

$$\tilde{p}(x,p_T) = \frac{M\xi}{2}, \qquad \xi = x\left(1+\left(\frac{p_T}{Mx}\right)^2\right). \tag{11}$$

Relation (10) with the use of (9) give the TMD evolution:

$$\frac{d}{d\ln Q^2}f_1^q(x,p_T,Q^2) = \frac{1}{4\pi\tilde{p}^2}\int_{\tilde{p}}^{M/2}\frac{dp'}{p'}P\left(\frac{\tilde{p}}{p'}\right)\rho_q\left(p',Q^2\right) \tag{12}$$

Note the same splitting function in convolutions (3), (8), (9) and (12), which follows from the fact the distributions $q\left(x,Q^2\right)$, $\rho_q(p,Q^2)$ and $f_1^q(x,p_T,Q^2)$ are equivalent, as a result of relativistic invariance and rotational symmetry.

At the same time, in an accordance with Eq. (10), the evoluted TMD can be expressed directly by means of the evoluted PDF. In the next this relation will be used for numerical calculations of TMD evolution.

Journal of Physics: Conference Series **678** (2016) 012001 doi:10.1088/1742-6596/678/1/012001

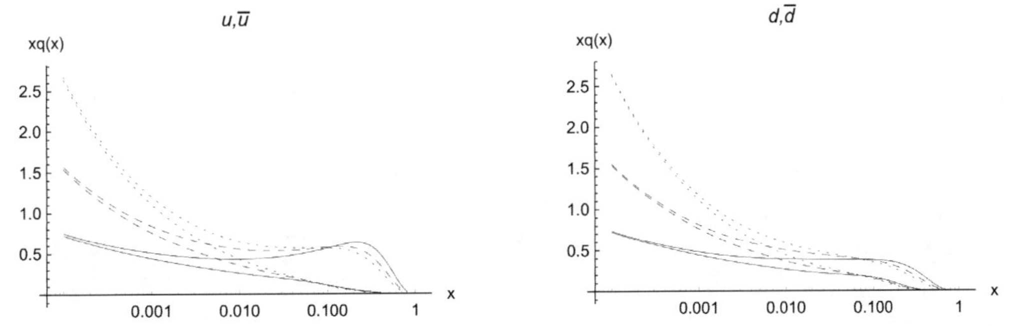

Figure 1. Input PDF of $u, \bar{u}$ (left) and $d, \bar{d}$ (right) quarks at different scales: $Q^2 = 4, 40, 400GeV$ (solid, dashed, dotted curves)

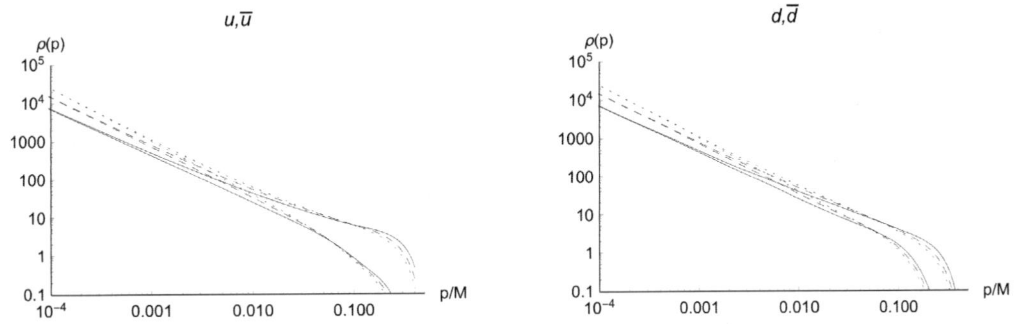

Figure 2. Distribution of the momentum in the nucleon rest frame for the quarks $u, \bar{u}$ (left) and $d, \bar{d}$ (right) at different scales: $Q^2 = 4, 40, 400GeV$ (solid, dashed, dotted curves)

3. Results

For numerical calculation we have used the PDF set MSTW2008(LO) [8] at the three different scales, which are displayed in Fig.1. In the next figure (Fig.2) we have shown their representation in terms of the distributions $\rho_q(p, Q^2)$ calculated from the relation (4). Finally, in Fig.3 we have displayed the corresponding TMDs. One can observe two important features:

i) The transverse moments of quarks satisfy the condition $p_T < M/2$. This condition follows from the constraint $0 < x_B < 1$ and the condition of relativistic invariance and rotational symmetry as we have shown in Ref. [6].

ii) Dependence of TMD on the scale Q^2 is rather weak.

These results are well compatible with the predictions obtained within the statistical approach and presented in Ref. [9]. On the other hand our results on TMD differs rather substantially e.g. from the results of the QCD evolution in Ref. [10]. The possible resolution of this contradiction is that evolution considered here may be attributed to the 'soft' non-perturbative component of TMDs and even can be used to disentangle it from the 'hard' perturbative component. Apparently these discrepancies require further study.

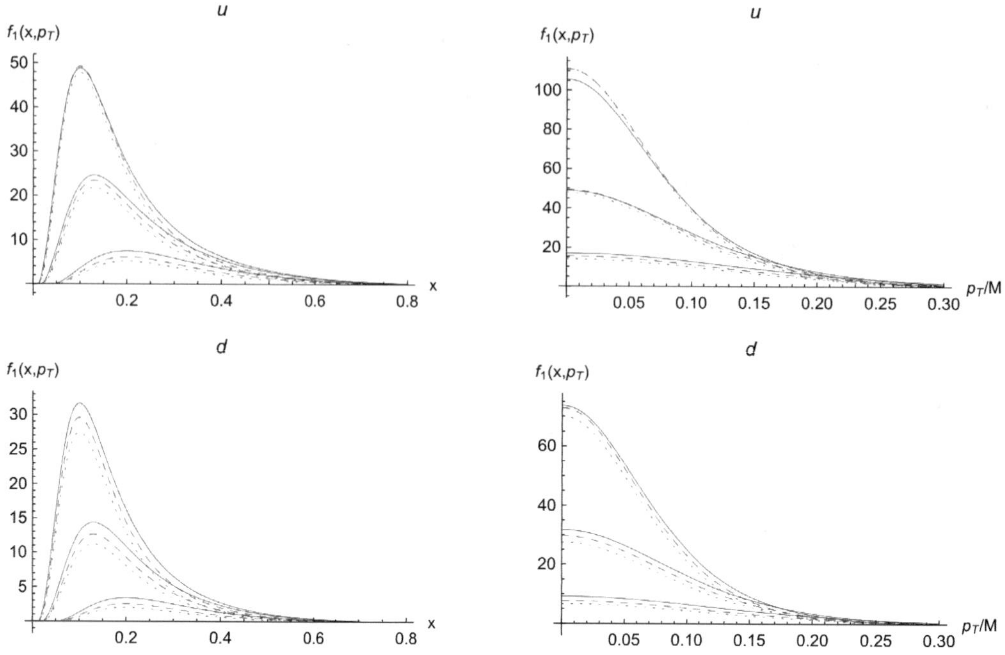

Figure 3. TMD at different scales: $Q^2 = 4, 40, 400GeV$ (solid, dashed, dotted curves) for u and d quarks. Sets of curves in left panels (from top) correspond to fixed $p_T/M = 0.1, 0.13, 0.20$. The curves in left panels (from top) correspod to fixed $x = 0.18, 0.22, 0.30$.

Acknowledgments
Authors of this work were supported by the Votruba-Blokhitsev Program of JINR Dubna. P.Z. was supported also by the project LG130131 of the MEYS (Czech Republic).

References
[1] P. Zavada, Phys. Rev. D **89**, no. 1, 014012 (2014) [arXiv:1307.0699 [hep-ph]].
[2] A. V. Efremov, P. Schweitzer, O. V. Teryaev and P. Zavada, Phys. Rev. D **83**, 054025 (2011) [arXiv:1012.5296 [hep-ph]].
[3] A. V. Efremov, P. Schweitzer, O. V. Teryaev and P. Zavada, Phys. Rev. D **80**, 014021 (2009) [arXiv:0903.3490 [hep-ph]].
[4] P. Zavada, Phys. Rev. D **83**, 014022 (2011) [arXiv:0908.2316 [hep-ph]].
[5] P. Zavada, Eur. Phys. J. C **52**, 121 (2007) [arXiv:0706.2988 [hep-ph]].
[6] P. Zavada, Phys. Rev. D **85**, 037501 (2012) [arXiv:1106.5607 [hep-ph]].
[7] U. D'Alesio, E. Leader and F. Murgia, Phys. Rev. D **81**, 036010 (2010) [arXiv:0909.5650 [hep-ph]].
[8] A. D. Martin, W. J. Stirling, R. S. Thorne and G. Watt, Eur. Phys. J. C **63**, 189 (2009) [arXiv:0901.0002 [hep-ph]].
[9] C. Bourrely, F. Buccella and J. Soffer, Int. J. Mod. Phys. A **28**, 1350026 (2013) [arXiv:1302.4281 [hep-ph]].
[10] S. M. Aybat and T. C. Rogers, Phys. Rev. D **83**, 114042 (2011) [arXiv:1101.5057 [hep-ph]].

Journal of Physics: Conference Series **678** (2016) 012002 doi:10.1088/1742-6596/678/1/012002

About measurements of Higgs boson parity

I. F. Ginzburg[1,2]

[1] Sobolev Inst. Mathem., Av. Ac. Koptyug, 4, 630090 Novosibirsk, Russia,

[2] Novosibirsk State University, ul. Pirogova, 2, 630090 Novosibirsk, Russia

E-mail: ginzburg@math.nsc.ru

Abstract. Recently CMS and ATLAS announced that they had measured the Higgs boson parity. In this note we show that their approach can determine this parity only under the additional assumption that an extension of Standard Model of some special type is realized in Nature.

We show that the used approach gives no information about the Higgs boson parity when assuming most other extensions of the Standard Model.

In recent papers the CMS [1] and ATLAS [2] collaborations announced that they had measured the parity of Higgs boson with mass of 125 GeV. Based on the proposals [3], these collaborations study correlations in the momentum distributions of leptons produced in decays $h \to ZZ^* \to (\ell_1\bar{\ell}_1)(\ell_2\bar{\ell}_2)$ or $h \to WW^* \to (\ell_1\nu)(\ell_2\nu)$ with leptons $\ell = e,\ \mu$.

The SM Higgs boson is definitely P-even. The problem of measuring its parity appears only in the extended models of the Higgs sector (beyond Standard Model – BSM). We subdivide such models into two groups. The approach of refs. [1]-[3] allows us to determine parity of observed Higgs boson for the models of the first group and gives no information on the parity of observed Higgs boson for the wider family of the models of the second group.

In the first group of models the Higgs boson is single scalar, and the BSM contains some new interactions, perhaps with CP violation [3]. To explain the main statement, I start with a brief description of these models in terms useful for me.

In the SM the interaction of Higgs boson with gauge bosons comes from a kinetic term of the Lagrangian

$$D^\mu\phi^* D_\mu\phi\,. \tag{1}$$

The electroweak symmetry breaking (EWSB) gives expansion of ϕ in the form $\phi = \dfrac{1}{\sqrt{2}}\begin{pmatrix} G^+ \\ v+h+iG^0 \end{pmatrix}$, where $v = 246$ GeV is v.e.v. of the Higgs field. Besides, $G^0, G^\pm$ are components of the Goldstone mode accumulated in the longitudinal component of Z, W and omitted below. After substitution of this expansion in (1) terms $\propto v^2$ give masses of W and Z and terms $\propto v$ give the interactions hWW and hZZ. For brevity we write down only the interaction with Z (the interaction with W differs by coefficients only)

$$(\bar{g}v)\, h\; Z^\mu Z_\mu\,. \tag{2}$$

The BSM interactions in this model can produce new terms in the effective Lagrangian of the form (*dimension 5 operator*)

$$\Delta L = \frac{h}{v}(A\, Z^{\mu\nu}Z_{\mu\nu} + B\,\varepsilon_{\mu\nu\alpha\beta}Z^{\alpha\beta}Z^{\mu\nu})\,. \tag{3}$$

Content from this work may be used under the terms of the Creative Commons Attribution 3.0 licence. Any further distribution of this work must maintain attribution to the author(s) and the title of the work, journal citation and DOI.
Published under licence by IOP Publishing Ltd

Journal of Physics: Conference Series **678** (2016) 012002 doi:10.1088/1742-6596/678/1/012002

(the factor v in the denominator is written to have dimensionless parameters A and B). The term with B describes a pseudoscalar contribution in the observed process. The data [1], [2] show that this contribution is small. Together with the measured SM-similar cross sections for Higgs observation, it allows to write that the pseudoscalar component of Higgs is absent or small.

The second group consists of many other models, discussed now as possible candidates for BSM physics. Their common feature is the existence of additional particles similar to a Higgs boson – both scalars and pseudoscalars with their possible mixing. For example, we consider the simplest variant of such models – the widely discussed two-Higgs-doublet model, 2HDM (see, e.g., [4]) *(the Higgs sector of MSSM is its particular case)*.

In the 2HDM the basic Higgs doublet ϕ_1 is supplemented by a second scalar doublet ϕ_2. The kinetic term is a sum of two terms, similar to (1). The EWSB with standard decomposition for neutral components like the SM case $\phi_i^0 = (v_i e^{i\xi_i} + \zeta_i + i\eta_i)/\sqrt{2}$ gives four neutral fields $\zeta_{1,2}$, $\eta_{1,2}$ (where $v_{1,2}$ are v.e.v.'s of the fields ϕ_1 and $|v_1|^2 + |v_2|^2 = v^2$). One linear combination of $\eta_{1,2}$ gives a neutral component of the Goldstone field G^0, the orthogonal linear combination of η_i is denoted by $\tilde{\eta}$. In the CP-conserving case a linear combination of the fields ζ_i forms two P-even Higgses h and H, while $\tilde{\eta}$ describes a P-odd Higgs A.

In the CP-violating case the fields ζ_i and $\tilde{\eta}$ are mixed, forming three Higgs fields h_a, having no definite P-parity. We denote the observed Higgs boson as h_1. The value of scalar–pseudoscalar mixing for h_1 can be either small or large (with limitations that appear beyond the Higgs sector). To determine Higgs P-parity, one needs to measure this mixing.

The interaction of h_a with Z comes from a kinetic term in precisely the same way as in the SM and can be written as

$$g_{SM}^Z \sum_a \chi_a^V h_a Z^\mu Z_\mu \,. \tag{4}$$

In this main approximation the form of $h_a ZZ$ interaction does not depend on the P-parity of h_a (only the "P-even part of h_a" interacts with Z), the terms like $B\, \varepsilon_{\mu\nu\alpha\beta} Z^{\alpha\beta} Z^{\mu\nu}$) (3) don't appear. Therefore, **the experiments [1], [2] do not allow us to draw any conclusions about the Higgs boson parity** if this model (and many other models) is realized.

In the radiative corrections small terms like (3) can appear. The CMS–ATLAS data support this smallness.

In order to finish our discussion, we show that – within 2HDM – even a big admixture of P-odd components in the observed Higgs boson does not contradict the modern data. For this goal we show that, with a suitable choice of parameters, the same values of the cross sections $gg \to h \to \gamma\gamma$ and $gg \to h \to ZZ$ can be obtained both in the SM and in the strongly CP-violating case of 2HDM with $M_1 = M_h = 125$ GeV,

$$\frac{\sigma(gg \to h_1 \to \gamma\gamma)}{\sigma(gg \to h \to \gamma\gamma)_{SM}} = 1\,, \qquad \frac{\sigma(gg \to h_1 \to ZZ)}{\sigma(gg \to h \to ZZ)_{SM}} = 1\,. \tag{5}$$

Some sets of parameter values which satisfy these equations (with taking into account some other limitations) are presented in [5] for the cases when the $t\bar{t}h_1$ production cross section will differ significantly from its SM value.

We are interested in the case when, in addition to (5), future measurements of the $t\bar{t}h_1$ production cross section will give results which are very close to predictions of the SM. In our calculations we use relative couplings, determined for the neutral Higgs bosons h_a with mass M_a and for the charged Higgs bosons $H^\pm$ with mass $M_\pm$:

$$\chi_a^P = \frac{g_a^P}{g_{\rm SM}^P} \;\; (P = V\,(W, Z), q = (t, b, ...))\,, \quad \chi_a^\pm = \frac{g(H^+ H^- h_a)}{2M_\pm^2/v}\,. \tag{6}$$

(The ratio $\chi_3^V/(\chi_1^V\sqrt{1-(\chi_2^V)^2}\,)$ describes the admixture of the P odd state in h_1 [6].)

Using the well-known equations for the two-photon and two-gluon widths, collected e.g. in [4], [7], we determine two benchmark sets of parameters, giving ratios (5) at $|\chi_1^t|=1$:

$$\begin{aligned}&(I)\quad \chi_1^V=0.9,\quad \chi_1^\pm=0.4,\quad Re(\chi_1^t)=0.9,\;\; Im(\chi_1^t)=0.43\,;\\ &(II)\quad \chi_1^V=0.8,\quad \chi_1^\pm=1.4,\quad Re(\chi_1^t)=0.74,\;\; Im(\chi_1^t)=0.67\,.\end{aligned}\tag{7}$$

(We neglected all fermion contributions except t-quarks.)

The couplings χ_a^V obey a sum rule $\sum(\chi_a^V)^2=1$. Therefore, in case (I) the sum $(\chi_2^V)^2+(\chi_3^V)^2=0.19$, which allows us to have $\chi_2^V\approx\chi_3^V\approx 0.3$ (the admixture of the P-odd to P-even components of the h_1 about 0.3). In case (II) the sum $(\chi_2^V)^2+(\chi_3^V)^2=0.36$, which allows us to have $\chi_2^V\approx\chi_3^V\approx 0.4$ (the admixture of the P-odd to P-even components of the h_1 about 0.5).

This simple analysis shows that a big P-odd admixture in the observed Higgs boson is compatible with SM-like values for many observed rates.

Certainly, detailed analysis of data with taking into account full equations is necessary.

Summary. The results [1], [2] show definitely that all observations at the LHC are consistent with the expectations for the standard model Higgs boson with the quantum numbers $J^{PC}=0^{++}$. Nevertheless, these data give no model independent information about the parity of the observed Higgs boson.

The signal of indefinite P-parity of the Higgs boson can be obtained at the LHC relatively soon in the observations $h_1\to\tau\bar\tau$ [8] and, perhaps, in the study of the process $pp\to t\bar t h+...$ [9]. Unfortunately, these experiments cannot provide information about the shares of the P-odd and P-even components in the observed Higgs boson (since even in the 2HDM the Yukawa sector is almost arbitrary for a given scalar sector). The value of h_1WW coupling, measurable in the experiments with W fusion, will be very important for the problem considered. If it is found that $\varepsilon^V=(1-(\chi_1^V)^2)\ll 1$, then the P-odd fraction of the observed Higgs will be $\leqslant\sqrt{\varepsilon^V}$.

To measure the shares of the P-odd and P-even components in the Higgs boson, one should observe two other neutral Higgs bosons $h_{2,3}$ and their couplings to gauge bosons (a very complicated experiment). Direct measurement of this mixing is possible at a photon collider (see, e.g., [10]) when it is constructed.

The discussions with S. Eidelman, I. Ivanov, D. Kazakov, K. Melnikov and M. Vysotsky were useful. This work was supported by the grants RFBR 15-02-05868, NSh-3802.2012.2 and NCN OPUS 2012/05/B/ST2/03306 (2012-2016).

[1] The CMS Collaboration. *Phys. Rev. Lett.* **110**, 081803 (2013); *Phys. Rev.* **D 92**, 012004 (2015).
[2] The ATLAS Collaboration. *Eur. Phys. J.* **C75**, 231 (2015); *Eur. Phys. J.* **C**; arXiv:1506.05669 [hep-ex].
[3] S.Y. Choi et al., *Phys. Lett.* **B 553**, 61 (2003); S. Bolognesi et al., *Phys. Rev.* **D 86**, 095031 (2012); C. P. Buszello et al., *Eur. Phys. J.* **C32**, 209 (2004).
[4] T.D. Lee, *Phys. Rev.* **D8**, 1226 (1973); J.F. Gunion, H.E. Haber, G. Kane and S. Dawson, *The Higgs Hunter's Guide* (Addison-Wesley, Reading, 1990); G. C. Branco, P. M. Ferreira, L. Lavoura, M. N. Rebelo, M. Sher and J. P. Silva, Phys. Rep. **516**, 1 (2012).
[5] D. Fontes, J.C. Romao, J.P. Silva, R. Santos, arXiv:1506.00860 [hep-ph].
[6] I.F. Ginzburg, K.A. Kanishev. *Phys. Rev.* **D 92** (2015) 015024.
[7] I.F. Ginzburg, arXiv:1505.01984 [hep-ph].
[8] A. Askew, P. Jaiswa, T. Okui, H.B. Prosper, N. Sato, *Phys. Rev.* **D 91**, 075014 (2015); arXiv:1501.03156 [hep-ph]; S. Berge, W. Bernreuther, S. Kirchner, arXiv:1510.03850 [hep-ph].
[9] D. Fontes, J.C. Romao, J.P. Silva, arXiv:1506.00860 [hep-ph]; H.-L. Li, P.-C. Lu, Z.-G. Si, Y. Wang, arXiv:1508.06416 [hep-ph].
[10] I.F. Ginzburg, I.P. Ivanov, *Eur. Phys. J.* **C 22**, 411 (2001).

Journal of Physics: Conference Series **678** (2016) 012003 doi:10.1088/1742-6596/678/1/012003

Transition of polarized ions through the system of ring permanent magnets

V V Fimushkin, I V Gapienko, Yu A Plis and Yu V Prokofichev

Joint Institute for Nuclear Research, Joliot-Curie 6, Dubna, Moscow region, Russia

E-mail: yuriy@jinr.ru

Abstract. The task of the transportation of the polarized proton and deuteron beam of an energy 10–20 keV in the given magnetic field is considered. Specifically the magnetic field is produced by the system of a solenoid and some ring permanent magnets with contrary directed field. The Lorentz equations together with classic equations for vector and tensor polarizations have been solved. The results of the numerical simulation for a simple two-ring system are presented, which shows there is no essential depolarization in agrement with theoretical estimation.

The problem of transporting a polarized deuteron beam [1] with an energy of 10–20 keV in the magnetic field altering in direction and magnitude is considered. In the particular case of the polarized ion source, we have the system of a solenoid and two ring permanent magnets with counter fields (Fig. 1).

We choose the coordinate system with the origin at the center of the ring system, the axis x is along the direction of motion of the deuteron beam and the axis z is in the vertical direction. For paraxial trajectories the radial magnetic field is calculated as usual:

$B_r(x) = -0,5r(dB_x(x)/dx)$, then $B_y(x) = -0,5y(dB_x(x)/dx)$, $B_z(x) = -0,5z(dB_x(x)/dx)$.

So, the field on the axis of the solenoid is

$$B_x(x) = \frac{\mu_0 In}{2(r_2 - r_1)} \left[(x+L+l)\ln\frac{\sqrt{r_2^2 + (x+L+l)^2} + r_2}{\sqrt{r_1^2 + (x+L+l)^2} + r_1} - (x+L)\ln\frac{\sqrt{r_2^2 + (x+L)^2} + r_2}{\sqrt{r_1^2 + (x+L)^2} + r_1} \right], \tag{1}$$

where $\mu_0 = 4\pi 10^{-7}$ N/A^2, $x_1 = x + L$ - the distance from the point of measuring to the proximal end of the solenoid, L - the distance from the origin to the proximal end of the solenoid, $x_2 = x + L + l$ - the distance from the point of measuring to the distal end of the solenoid, l - the length of the solenoid, r_1 - the inner radius of the solenoid, r_2 - the outer radius, n - the number of turns per unit of length of the solenoid, I - the solenoid current. The magnetic field on the axis of the annular permanent magnet is calculated as the difference of the field values of two solid cylinders. For a cylinder with magnetization $M = B_0/\mu_0$ (B_0 - the residual field) the axial field in the function of the distance t from the center of the magnet can be calculated by:

$$B_x(t) = \frac{1}{2}\mu_0 M[(b-t)\sqrt{a^2 + (b-t)^2} + (b+t)\sqrt{a^2 + (b+t)^2}], \tag{2}$$

where a and $2b$ - the radius and length of the cylinder.

Content from this work may be used under the terms of the Creative Commons Attribution 3.0 licence. Any further distribution of this work must maintain attribution to the author(s) and the title of the work, journal citation and DOI.
Published under licence by IOP Publishing Ltd

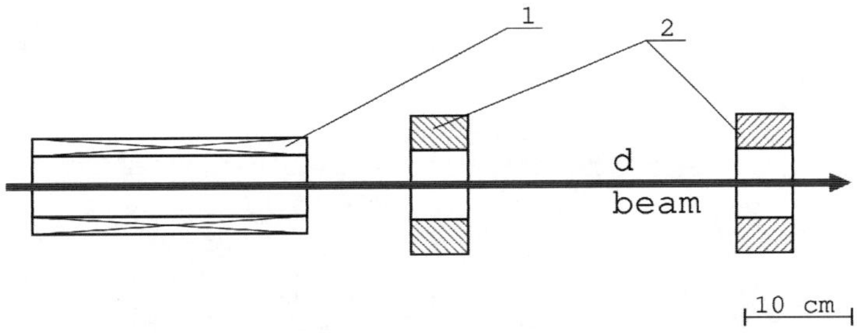

Figure 1. The magnetic system; 1 - a solenoid, 2 - permanent magnets

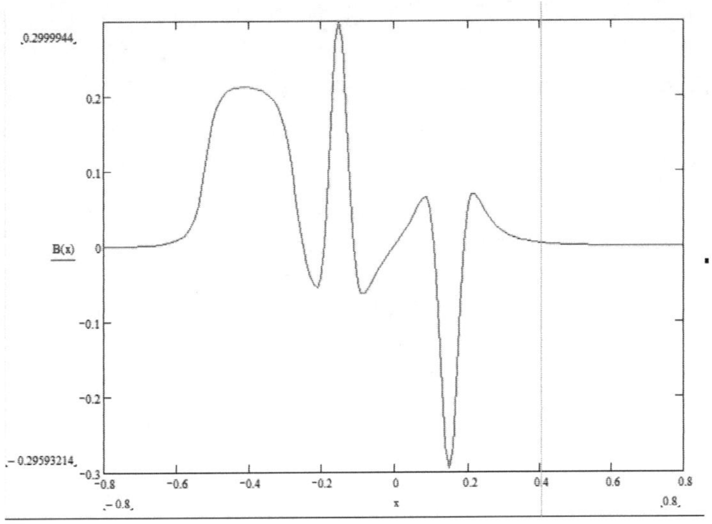

Figure 2. Distribution of the magnetic field along the axis , the distance is measured in m, the field is in T

Let the inner and outer radii of the ring be a_1 and a_2, respectively, and $2d$ - the distance between the rings. Then, shifting the origin to the center of the two rings, we obtain the total field of the two ring system:

$$\begin{aligned} B_x(x) = \frac{1}{2}\mu_0 M \Big[&[(-x-d)\sqrt{a_2^2+(-x-d)^2} + (x+d+2b)\sqrt{a_2^2+(x+d+2b)^2}] \\ &-[(-x-d)\sqrt{a_1^2+(-x-d)^2} + (x+d+2b)\sqrt{a_1^2+(x+d+2b)^2}] \\ &-[(d+2b-x)\sqrt{a_2^2+(d+2b-x)^2} + (x-d)\sqrt{a_2^2+(x-d)^2}] \\ &+[(d+2b-x)\sqrt{a_1^2+(d+2b-x)^2} + (x-d)\sqrt{a_1^2+(x-d)^2}]\Big]. \end{aligned} \tag{3}$$

The distribution of the magnetic field of the entire system is shown in Fig. 2.

To check the validity of the paraxial approximation we find an exact solution. For a cylinder with magnetization M the radial field equals

$$B_r = \frac{\mu_0 M}{2\pi}\sqrt{\frac{a}{r}}\left[\frac{m-2}{\sqrt{m}}K(m) + \frac{2}{\sqrt{m}}E(m)\right]_{\zeta_-}^{\zeta_+}, \tag{4}$$

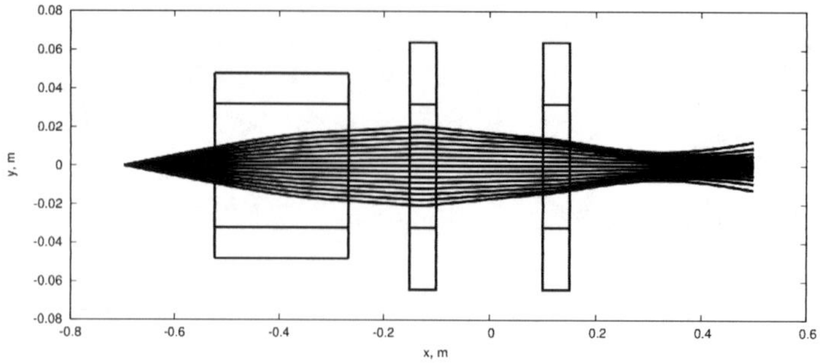

Figure 3. Trajectories of the deuterons in the 20 keV beam

where a and L are the radius and the length of the cylinder, respectively, $\zeta_{\pm} = x \pm L/2$,

$$m = \frac{4ar}{(a+r)^2 + \zeta^2}, \quad K(m) = \int_0^{\pi/2} \frac{d\theta}{\sqrt{1 - m\sin^2\theta}}, \quad E(m) = \int_0^{\pi/2} \sqrt{1 - m\sin^2\theta}\, d\theta. \tag{5}$$

Calculation of the radial magnetic field from the exact formula shows that to a radius of 15 mm the difference between the exact value and the approximate one is less than 1%.

Consider the situation with spin rotation in a radial magnetic field in the center of the ring system. The condition of "fast transition" [2] is

$$\gamma B_r \ll \frac{d\varphi}{dt}, \tag{6}$$

where

$$\gamma = \frac{\mu_d}{\hbar}, \quad \cot\varphi = \frac{B_x}{B_r}, \quad \frac{1}{\sin^2\varphi}\frac{d\varphi}{dt} \approx \frac{1}{B_r}\frac{dB_x}{dt}, \quad \frac{d\varphi}{dt} \approx v\frac{dB_x}{dx}\frac{B_r}{B_x^2 + B_r^2}.$$

As $B_x(0) = 0$, $B_r = -(r/2)dB_x/dx$, so

$$\frac{d\varphi}{dt} \approx \frac{v}{B_r}\frac{dB_x}{dx}. \tag{7}$$

In this case, for $E = 20$ keV ($v = 1.38 \times 10^6$ m/s), $dB_x(0)/dx = -0.32$ T/m, $r = 0.015$ m

$$\gamma B_r = \frac{\mu_d B_r}{\hbar} = 10^5\,\text{rad/s}, \quad \frac{d\varphi}{dt} \approx \frac{v}{B_r}\frac{dB_x}{dx} = 1.8 \times 10^8\,\text{rad/s},$$

Thus, the condition of fast passage without changing the polarization is performed.

With use of the Lorentz equations the trajectories of a deuteron beam in the magnet system were calculated (Fig. 3). Further we consider the system of equations including the motion of deuterons in the magnetic field

$$\frac{d^2\vec{r}}{dt^2} = \frac{e}{m_d}\left[\vec{v} \times \vec{B}\right], \quad \frac{d^2\vec{r}}{dx^2} = \frac{e}{m_d v^2}\left[\vec{v} \times \vec{B}\right] \tag{8}$$

and the classic equations for vector and tensor polarizations [2]. For the vector polarization we have

$$\frac{d\vec{P}}{dt} = \frac{\mu_d}{\hbar}\left[\vec{P} \times \vec{B}\right], \quad \frac{d\vec{P}}{dx} = \frac{\mu_d}{\hbar v}\left[\vec{P} \times \vec{B}\right], \tag{9}$$

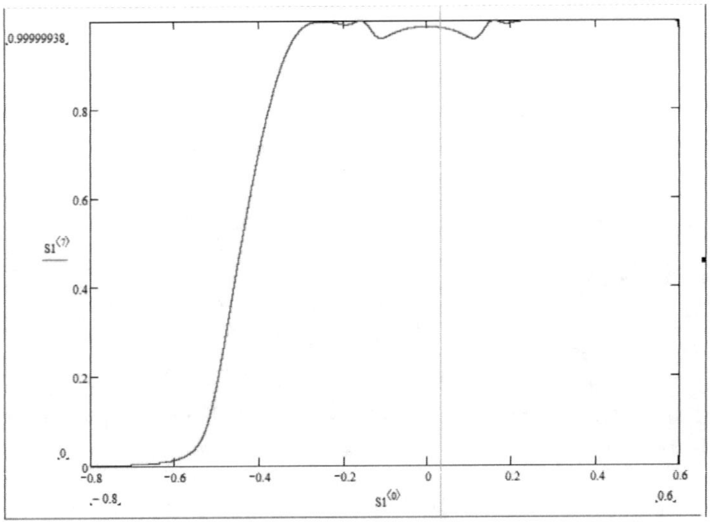

Figure 4. Changing the vertical polarization P_z along the way , the energy $E = 20$ keV, the initial polarization value $P_x = 0$, $P_y = -1$, $P_z = 0$, the initial value of $y = 15$ mm, the rest values are zero

where $\mu_d = 0,433073489 \times 10^{-26}$ JT^{-1} is the magnetic moment of the deuteron. For the tensor polarization

$$\frac{dP_{ik}}{dx} = -\frac{\mu_d}{\hbar v}\left(\varepsilon_{jkl}P_{ij}B_l + \varepsilon_{jil}P_{kj}B_l\right). \tag{10}$$

We assume that the residual induction of permanent magnets is $B_0 = 1.124$ T. The length of each magnet is $2b = 0.05$ m, the internal radius is $a_1 = 0.032$ m and the outer radius is $a_2 = 0.064$ m. The distance between the rings $2d = 0.25$ m. The initial polarization of the deuteron beam is horizontal, for injection into the accelerator after the magnetic system it must be vertical.

The parameters of the solenoid are determined by the requirement that the polarization rotates to 90°. For a deuteron energy of 20 keV with $L = 0.275$ m, $l = 0.245$ m, $r_1 = 0.028$ m, $r_2 = 0.043$ m, $n = 588$ we need $I = 74$ A.

The calculation results are shown in Fig. 4. With all diversity of initial conditions the loss of polarization is $\Delta P_z \leq 0.02$. The calculation shows also that at the output of the magnetic system there is a nonzero $P_{xy} \simeq 0.1$.

Conclusion

For a polarized deuteron beam of an energy 20 keV and a normalized emittance of 1.2 π mm rad in the magnetic system consisting of the 90° spin-rotating solenoid and 0.4 T permanent magnet rings the depolarization is expected less then 2%. The beam diameter is up to 30 mm (half of rings aperture).

References

[1] Fimushkin V V, Kovalenko A D, Kutuzova L V, Prokofichev Yu V, Belov A S, Turbabin A V and Zubets V N 2012 *Proc. 20th Int. Symp. on Spin Physics* (Dubna: PEPAN vol 45 part 1) pp 297-300
[2] Schieck Hans Paetz gen. 2012 *Nuclear Physics with Polarized Particles* (Berlin: Springer)

Journal of Physics: Conference Series **678** (2016) 012004 doi:10.1088/1742-6596/678/1/012004

Study for a Design of Magnet System for the SPD Detector NICA LHEP JINR

Ivan P. Yudin
Joint Institute for Nuclear Research, 141980 Dubna Moscow region, Russia
E-mail: yudin@jinr.ru

Abstract. The choice of magnet system for the Spin Physics Detector of the NICA Collider of LHEP JINR is given. The inverse problem of magnetostatics is solved for a magnetic field of 0.5 tesla in the aperture a) ϕ 3 m x 5 m and b) ϕ 3 m x 6 m. We also discuss the design of the magnet with a field of 0.3 T.

The paper presents the results obtained for the "warm" and SC versions of the magnetic system: currents (ampere-turns), the geometry (size) of the coil and the iron yoke, weight (on the whole and the individual elements), the magnet transportation and assembly.

1. Introduction

The spin physics detector (SPD) is a 4π spectrometer to be used for studying of the nucleon structure in heavy ion collisions at energies provided by the NICA collider of the Joint Institute for Nuclear Research (Dubna) [1, 2, 3].

A constituent part of the SPD is a solenoid magnet with a coil and a steel flux return yoke. The coil we have considered in the variant as "warm", and in other variant - as a superconducting (SC).

The inverse problem of magnetostatics is solved for a magnetic field of 0.5 tesla (and than for B=0.3 T) in the aperture a) ϕ 3 m x 5 m and b) ϕ 3 m x 6 m. We solved the inverse problem as a sequence of direct problems of magnetostatics [4, 5].

Figure 1 shows the designations of the basic sizes to perform calculations.

2. Calculating of "warm" magnet

2.1. Initial parameters are here as follows: the "warm" coil is with the Cu cable 4.5 x 4.5 mm^2, the induction B = 0.5 T; aperture ϕ 3m x 5m.

The resulting magnetic system configuration in this case is shown in Figure 2. Main characteristics of the SPD magnet are shown in the Table 1.1. Weights of the main magnet components are shown in the Table 1.2.

2.2. Initial parameters are here as follows: the "warm" coil is with the Cu cable 4.5 x 4.5 mm^2, the induction B = 0.5 T; aperture ϕ 3m x 6m.

Main characteristics of the SPD magnet in this case are shown in the Table 1.1. Weights of the main magnet components are shown in the Table 1.2.

Content from this work may be used under the terms of the Creative Commons Attribution 3.0 licence. Any further distribution of this work must maintain attribution to the author(s) and the title of the work, journal citation and DOI.
Published under licence by IOP Publishing Ltd

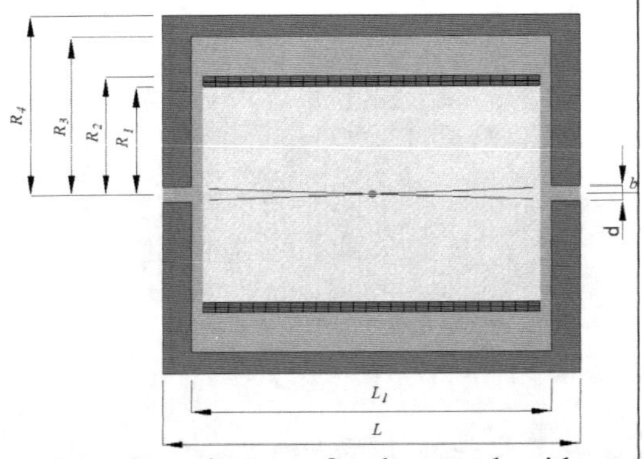

Fig.1 Magnet sizes are for the search with an calculation

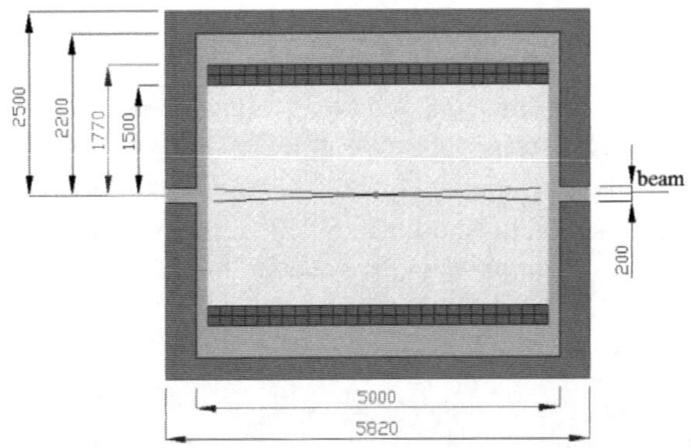

F
Fig.2 "Warm" magnet sizes are after the calculation

Table 1.1. Main characteristics of the SPD "warm" magnet

	L_1 =5 m	L_1 =6 m
Rated induction, T	0.5	0.5
Rated current density in the coil, A/мм2	1.4736	1.5052
Rated current, A	29.84	30.48
Total current (ampere-turns) at the rated induction, kA	1987	2383
Insulated Cu cable cross section, mm2	4.5 x 4.5	4.5 x 4.5
Conductor weight, kg	117693	138157

Table 1.2. Weights of the main "warm" magnet components

	L_1 =5 m	L_1 =6 m
Weight of the yoke without poles, tons	172.76	207.31
Weight of the 2 poles, tons	124.78	124.78
Total (weight of the yoke with the poles), tons	297.54	332.09

3. Calculating of a superconducting (SC) magnet

3.1. Initial parameters are here as follows: the SC coil is with the NbTi cable 4.5 x 20.4 mm^2, the induction B = 0.5 T; aperture ϕ 3m x 5m.

The resulting magnetic system configuration in this case is shown in Figure 3. Main characteristics of the SPD magnet are shown in the Table 2.1. Weights of the main magnet components are shown in the Table 2.2.

3.2. Initial parameters are here as follows: the SC coil is with the NbTi cable 4.5 x 20.4 mm^2, the induction B = 0.5 T; aperture ϕ 3m x 6m.

Journal of Physics: Conference Series **678** (2016) 012004 doi:10.1088/1742-6596/678/1/012004

Main characteristics of the SPD magnet in this case are shown in the Table 2.1. Weights of the main magnet components are shown in the Table 2.2.

3.3. Initial parameters are here as follows: the SC coil is with the NbTi cable 4.5 x 20.4 mm^2, the induction B = 0.5 T; aperture ϕ 3m x 7m.

Main characteristics of the SPD magnet in this case are shown in the Table 2.1. Weights of the main magnet components are shown in the Table 2.2.

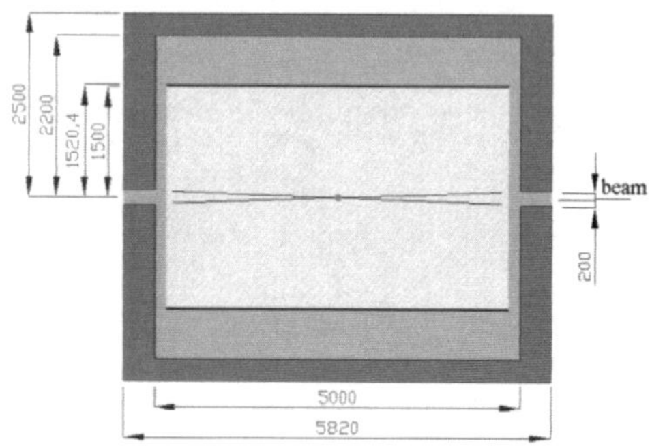

Fig.3. SC magnet sizes are after the calculation

Fig.4. 3D view of "warm" magnet for the assembly

Table 2.1. Main characteristics of the SPD SC magnet

	L_l =5 m	L_l =6 m	L_l =7 m
Rated induction, T	0.5	0.5	0.5
Rated current, A	1790	1791	1792
Total current (ampere-turns) at the rated induction, kA	1987	2386	2787
Insulated SC cable cross section, mm2	4.5 x 20.4	4.5 x 20.4	4.5 x 20.4
Conductor weight, kg	5700	6840	8208

Table 2.2. Weights of the main SC magnet components

	L_l =5 m	L_l =6 m	L_l =7 m
Weight of the yoke without poles, tons	172.76	207.31	248.89
Weight of the 2 poles, tons	124.78	124.78	124.78
Total (weight of the yoke with the poles), tons	297.54	332.09	373.67

Figure 4 shows the magnetic system when SP detector assembly. The yoke consists of a barrel part and two end caps. In addition, two rings support provides overall structural rigidity.

4. Conclusion

The study for a design of the Spin Physics Detector of the NICA Collider of LHEP JINR is present. The results for the "warm" and SC versions of the magnetic system: currents (ampere-turns), the geometry (size) of the coil and the iron yoke, weight (on the whole and the individual elements), the magnet transportation and assembly, were obtained.

When discussing the magnetic system SPD detector with B=0.3 T, it should be said to reduce by 60% the weight of the magnetic yoke, and 60% of the weight of the warm conductor. For SC magnet the weight of the SC conductor remains the same as in the variant B= 0.5 T, but the number of ampere-turns is reduced by the same 60%.

Acknowledgements

I am grateful to Professor Savin I.A. and Dr. Kukhtin V.V.

References

[1] http://nica.jinr.ru/files/SPD_LOI.pdf
http://nica.jinr.ru/files/Spin_program/SPIN-Praha-2013/
http://theor.jinr.ru/~spin/2013/

[2] I.A. Savin and A.P. Nagaytsev, Phys. Part. Nucl., 2004, t. 35, iss. 1 and references therein.

[3] A. Guskov Spin physics with direct photons at NICA SPD, pp.363-366.
V.I. Sharov Possibilities for the spin-dependent observables measurement in elastic NN scattering at NICA, pp.367-370 // Proc. of XV Advanced Research Workshop on High Energy Spin Physics (DSPIN-13)(Dubna, October 8–12, 2013). — Dubna, JINR. — 423p. ISBN 978-5-9530-0315-3.

[4] Yudin I.P., Izvestia Saratov State Univ. Seria Physics. 2010, Vol.10, No2, P.25-30.

[5] I.P. Yudin The method of assembly functions in the problem of a momentum restoring of charged particles, pp.389-392 // Proc. of XV Advanced Research Workshop on High Energy Spin Physics (DSPIN-13)(Dubna, October 8–12, 2013). — Dubna, JINR. — 423p. ISBN 978-5-9530-0315-3.

Journal of Physics: Conference Series **678** (2016) 012005 doi:10.1088/1742-6596/678/1/012005

Upgraded polarized target for polarized Drell-Yan measurement at COMPASS

Michael Pešek

Charles University Prague, Faculty of Mathematics and Physics, Prague, Czech Republic University of Torino, Department of Physics, Torino, Italy

On behalf of the COMPASS Collaboration

Abstract

The first ever polarized Drell-Yan measurement, presently being done by the COMPASS collaboration at CERN is described. Emphasis is put on the polarized target and its upgrade.

1. Polarized Drell-Yan process at COMPASS

When parton transverse momentum is taken into account then the proton structure can be described at leading twist by eight so called Transverse Momentum Dependent parton distribution functions (TMDs). They are summarized in Tab.1. They can be measured in a variety of processes like (polarized) Semi-Inclusive Deep Inelastic Scattering (SIDIS) or Drell-Yan (DY) process. In Fig.1. the leading order Feynman graphs for these two processes are shown. There are two TMDs which are of great interest: the Sivers and Boer-Mulder functions. These are naïve T-odd and thus process dependent and are predicted to change sign when measured in SIDIS or DY process. The measurement of the sign change is the main goal of COMPASS DY programme, see Ref. [1] for details.

		Parent hadron polarization		
		Unpolarized	Longitudinal	Transverse
Quark polarization	Unpolarized	$f_1(x,k_T^2)$ Number density		$f_{1T}^{\perp}(x,k_T^2)$ Sivers
	Longitudinal		$g_1(x,k_T^2)$ Helicity	$f_1(x,k_T^2)$
	Transverse	$h_1^{\perp}(x,k_T^2)$ Boer-Mulders	$h_{1L}^{\perp}(x,k_T^2)$	$h_1(x,k_T^2)$ Transversity $h_{1T}^{\perp}(x,k_T^2)$

Table 1: Transverse Momentum Dependent parton distribution functions

The COMPASS(COmmon Muon and Proton Apparatus for Structure and Spectroscopy) collaboration runs a fixed target experiment located at the CERN North Area. It uses a secondary hadron or tertiary muon beam from SPS accelerator with momentum up to 280 GeV/*c*. The apparatus is two staged magnetic spectrometer (for measurements of

Content from this work may be used under the terms of the Creative Commons Attribution 3.0 licence. Any further distribution of this work must maintain attribution to the author(s) and the title of the work, journal citation and DOI.
Published under licence by IOP Publishing Ltd

particles scattered at large and small angles) with both hadronic and electromagnetic calorimetry and particle identification in both stages.

For the DY measurement the original COMPASS apparatus [2] required several important modifications. These include modification of the polarized target, building a hadron absorber with beam dump, scintillating fibers based vertex detector, new large drift chamber and improved calorimetry. The modifications mainly result from using high intensity negative hadron beam (10^8 hadrons/s, 97 % pions) to maximise the number of DY events as the corresponding cross-sections are low, of the order of a tenth of nanobarn. The use of hadron absorber is necessary to avoid saturating the detectors with the hadronic background. This then limits the experiment to measurement of muon pairs only.

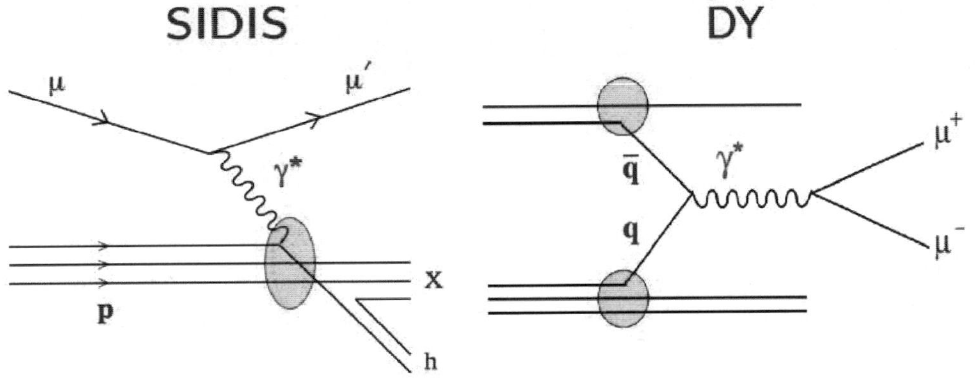

Figure 1: Leading order Feynman graphs for Semi Inclusive Deep Inelastic Scattering ($\mu+p\rightarrow\mu'+h+X$) and for Drell-Yan processes ($q+\bar{q}\rightarrow\mu^++\mu^-$)

2. COMPASS polarized target

The polarized target [3] is the key component of the experiment. Building of the polarization needs microwave system (microwave cavity and a pair of microwave generators) for inducing the polarization via the Dynamic Nuclear Polarization mechanism [4] while sustaining the polarization needs low temperature and high magnetic field.

The target has two cells of 55 cm length and 4 cm diameter containing about 700 ml of solid ammonia each. The cells are separated by a 20 cm gap to ensure a proper vertex assignment to each cell. After reaching the high polarization, about 75-90%, in the solenoid field of 2.5 T and temperature about 300 mK the material is cooled down to about 60 mK, and the field is switched to that of a dipole of 0.64 T. The low temperature is obtained by a powerful dilution refrigerator with a cooling power of 5 mW at 75 mK. At these low temperatures the spin-lattice relaxation time for protons is of the order of 1 000 hours which is sufficient for the measurement.

During the buildup the polarization is measured through a continuous Nuclear Magnetic Resonance by five coils in each cell where three of them are mounted outside a cell and two inside it. Cells are polarized in opposite direction to minimise the systematic errors.
The target original design in 2010 and 2011 consisted of three cells of 30-60-30 cm lengths and 4 cm diameter and of ten NMR coils which were all mounted outside the cells. A different design was chosen for DY to cope with higher thermal input form the high intensity hadron beam and to allow a proper assignment of interaction vertex to a target cell. This leads to a

modification of the microwave cavity i.e. a change of microwave stoppers which prevent microwaves to leak from one cell to another. A new layout is shown in Fig.2.

The target requires constant monitoring of all its parameters; it is done by about 30 thermometers, 10 pressure gauges and flow meters. Because of the high radiation dose expected in the experimental hall, it was decided to change the set up the remote control of the whole experiment. This mean abbandoning the LabView system for target monitoring and to develop a new one. The new system called *ptread* has been developed using the C++ language. It is an open-source modular software which allows for many connected measuring devices and several outputs. Most important is a possibility to communicate with COMPASS centralized Detector Control System. Fig.3 shows the structure diagram of the *ptread*.

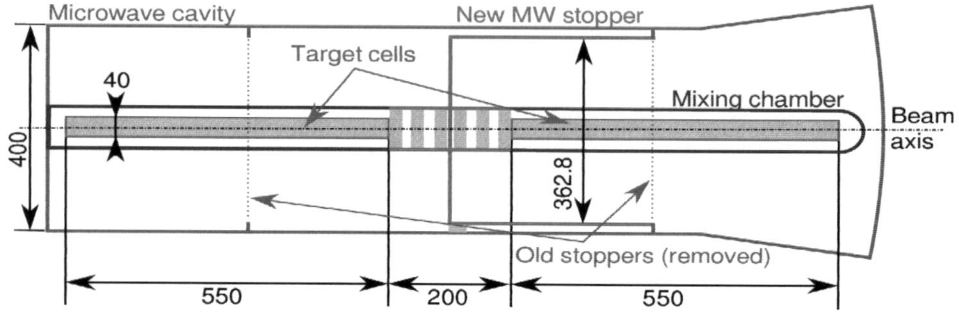

Figure 2: Modified microwave cavity. Numbers denote dimensions in mm.

The target superconducting magnet underwent heavy refurbishment in 2012-2014. The cryogenic part (thermal shielding) was greatly improved. Also huge improvement to superconducting circuitry, mainly the trim coils which homogenize the magnetic field. A new Magnet Safety System and Magnet Control System were designed by the PH-DT group [6] at CERN.

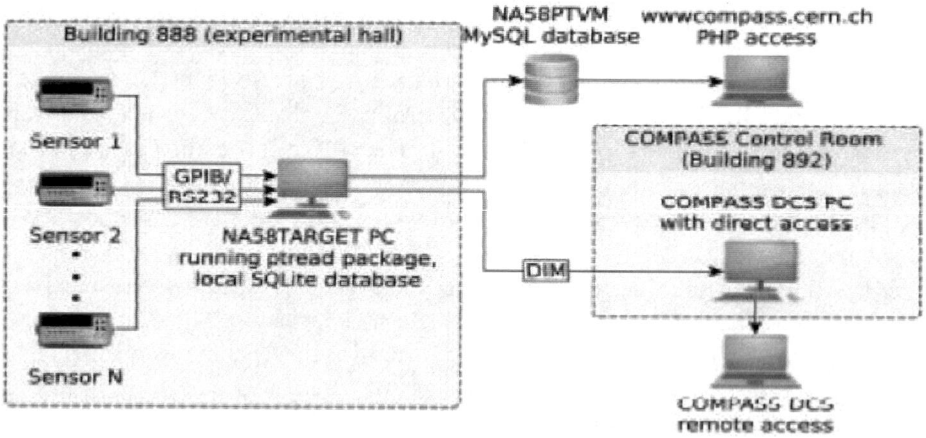

Figure 3: Diagram of *ptread* communication. The Linux computer with *ptread* package reads sensors. Data can be stored locally in SQLite database, sent to MySQL database and published by a DIM server [5] for the COMPASS DCS.

3. Present status

The year 2014 allowed for a short two-months pilot DY run. The majority of the target systems were commissioned except the superconducting magnet. The magnet was fully commissioned in April 2015 and the target is fully operational since then. Data-taking is ongoing.

4. Conclusion

Year 2015 is devoted to the first ever polarized Drell-Yan measurement using negative pion beam and transversely polarised proton target at COMPASS. Data should allow to verify the predicted sign change of the Sivers and Boer-Mulders TMDs. This goal required heavy modifications of the apparatus, mainly the polarized target which underwent modifications concerning the target cells, microwave cavity and superconducting magnet. The new design of the target cells allows for better distribution of the heat load caused by the high intensity hadron beam, thus he target provides high degree of proton polarization (around 80 %) with sufficient relaxation time (>1000 h). The new design also allow to separate well enough the vertices of interaction originating from the different cells. This would be very difficult with the old design as the hadron absorber does not allow any tracking for more than 2 m after the target.

5. Acknowledgements

The author acknowledges the support by the LG13031 MEYS Czech Republic grant.

6. References

[1] COMPASS Collaboration, COMPASS-II Proposal, CERN-SPSC, CERN-SPSC 2010-014, SPSC-P-340, http://wwwcompass.cern.ch/compass/publications/#proposal

[2] COMPASS Collaboration, P. Abbon et al., Nucl. Instr. and Meth. A577 (2007) 455.

[3] J. Koivuniemi et al., Journal of Physics: Conference Series 150 (2009) 012023

[4] A. Abragam, M. Goldman, Rep. Prog. Phys. 41 (1978) 395.

[5] C. Gaspar et al., Distributed information management system, http://dim.web.cern.ch/dim/

[6] CERN Detector Technologies group, http://ph-dep-dt.web.cern.ch

XVI Workshop on High Energy Spin Physics (D-SPIN2015) IOP Publishing
Journal of Physics: Conference Series **678** (2016) 012006 doi:10.1088/1742-6596/678/1/012006

Deformed nucleon, vorticity and proton spin

G. Musulmanbekov

Abstract. In the framework of the proposed by the author Strongly Correlated Quark Model the valence quarks are treated as vortical solitons, and the proton spin arises from the additive sum of the valence quark vorticities. The spin "crisis" is a manifestation of the irrotational character of vortical fields quarks.

JINR, Dubna, Russia

E-mail: genis@jinr.ru

One the most challenging key questions in hadronic physics is understanding the spin structure of the proton. From non-relativistic quark model we know that the proton spin is composed of spins of constituent quarks additive sum of which gives one half. However, deep inelastic scattering experiments have found out that only $\sim 20\%$ of proton spin is carried out by quark spins. It is natural to describe the proton spin according to the sum rule

$$\frac{1}{2} = J_q + J_g, \tag{1}$$

where J_q and J_g are total angular momenta of quarks and gluons, respectively. There are two important tasks emerging from this sum rule: to separate total angular momenta both quarks and gluons into intrinsic spins and orbital angular momenta and to define observable quantities corresponing them. So, question is: where does the proton spin come from? In this paper we describe the spin of the proton as arising from the hadronic flow circulating around valence quarks. Considerations are performed in the framework of so–called Strongly Correlated Quark Model (SCQM) [1]. The ingredients of the model are the following. Suppose a single quark of definite color embedded in vacuum. Because of its color charge the surrounding vacuum becomes polarized that in turn results in a creation of quark and gluon condensate. The quark, as a defect in vacuum, experiences a pressure from the surrounding vacuum due to the zero point radiation or quantum fluctuations. If we place the corresponding antiquark in the vicinity of the first one their color fields of opposite sign interfere destructively in the overlapped regions eliminating each other, mostly in space between them. This effect leads to decreasing of condensates density in the same regions and overbalancing of the vacuum pressure acting on quark (antiquark) from all directions. As a result the attractive force between quark and antiquark emerges and quark and antiquark start to move towards each other. The density of the remaining condensate around quark (antiquark) is identified with hadronic matter distribution. At maximum displacement in $\overline{q}q-$ system, that corresponds to small overlapping of polarization fields, hadronic matter distributions have maximum extent and values. The more close quarks come to each other, the more is the destructive interference of their fields in between them, and the less hadronic matter distributions remain around the quarks. In that way quark and antiquark start to oscillate near their center of mass. For such interacting $\overline{q}q-$ pair located on X axis at distance $2x$ from each other the total Hamiltonian is

Content from this work may be used under the terms of the Creative Commons Attribution 3.0 licence. Any further distribution of this work must maintain attribution to the author(s) and the title of the work, journal citation and DOI.
Published under licence by IOP Publishing Ltd

$$H = \frac{m_{\overline{q}}}{(1-\beta^2)^{1/2}} + \frac{m_q}{(1-\beta^2)^{1/2}} + V_{\overline{q}q}(2x), \tag{2}$$

were $m_{\overline{q}}$, m_q- current masses of valence antiquark and quark, $\beta = \beta(x)-$ their velocity depending on displacement x and $V_{\overline{q}q}-$ quark–antiquark potential energy with separation $2x$. It can be rewritten as

$$H = \left[\frac{m_{\overline{q}}}{(1-\beta^2)^{1/2}} + U(x)\right] + \left[\frac{m_q}{(1-\beta^2)^{1/2}} + U(x)\right] = H_{\overline{q}} + H_q, \tag{3}$$

were $U(x) = \frac{1}{2}V_{\overline{q}q}(2x)$ is potential energy of quark or antiquark. Quark (antiquark) with the surrounding quark and gluon condensate forms the constituent quark. It is natural to assume that the potential energy of quark (antiquark), $U(x)$, is equivalent to the mass M_Q of constituent quark:

$$2U(x) = 2c\int_{-\infty}^{\infty}dz'\int_{-\infty}^{\infty}dy'\int_{-\infty}^{\infty}dx'\rho(x,\mathbf{r}') \approx 2M_Q(x) \tag{4}$$

where c corresponds to the constituent mass at maximal displacement, $M_Q(x_{\max})$ and hadronic matter density distribution, $\rho(x,\mathbf{r}')$, is defined as

$$\rho(x,\mathbf{r}') = [\varphi(x,\mathbf{r}')]^2 = \left[\varphi_Q(x'+x,y',z') - \varphi_{\overline{Q}}(x'-x,y',z')\right]^2. \tag{5}$$

Here φ_Q and $\varphi_{\overline{Q}}$ are color fields around quark and antiquark located at distance $2x$ from each other. Owing to their oposite color "charges" they possess opposite phases and consequently interfere destructively eliminating each other in overlap region.

Our quark – antiquark system is identical to the periodic, so called, "breather" solution of (nonlinear) sine–Gordon (SG) equation [2]. The SG equation in (1+1) dimensions in a reduced form for the scalar function $\phi(x,t)$ reads

$$\Box\phi(x,t) + \sin\phi(x,t) = 0, \tag{6}$$

where x, $t-$ dimensionless. A breather is a periodic solution representing the bound state of soliton–antisoliton pair which oscillate near their center of mass. During the oscillations of soliton–antisoliton pair their density profile

$$\varphi_{s-as}(x,t) = \frac{d\phi_{br}(x,t)}{dx} \tag{7}$$

evolves like our quark–antiquark system, i.e. at maximal displacement soliton and antisoliton are maximal and at minimum displacement they "annihilate". This similarity is not surprising because our quark–antiquark system was formulated in close analogy with the model of dislocation–antidislocation [3], which in its continuous limit is described by breather solution of SG equation. It can be shown that soliton, antisoliton and breather obey relativistic kinemaics, i.e. their energies, momenta and shapes are transformed according to Lorentz transformations. Thus the dynamics of our strongly correlated quark system can be described by sine–Gordon equation. Since the above consideration of quarks as solitons is classical the problem is to construct quantum states around them. Although soliton solution of SG equation looks like extended (quantum) particle the relation of classical solitons to quantum particles is not so trivial. Technique of quantization of classical solitons with usage of various methods has been developed by many authors. The most known of them is semiclassical method of quantization (WKB) which allows one to relate classical periodic orbits with the quantum energy levels [6].

Generalization to three–quark system in baryons is performed according to $SU(3)_{color}$ symmetry: a pair of quarks have coupled representations $3 \otimes 3 = 6 \oplus \overline{3}$ in $SU(3)_{color}$ and for quarks within the same baryon only the $\overline{3}$ (antisymmetric) representation is realized. Hence, the antiquark can be replaced by two correspondingly colored quarks to get color singlet baryon and destructive interference takes place between color fields of three valence quarks (VQs). Putting aside the mass and charge differences of valence quarks we may say that inside baryon three quarks oscillate along the bisectors of equilateral triangle. Therefore, keeping in mind that quark and antiquark in mesons and three quarks in baryons are strongly correlated, we can consider each of them separately as undergoing oscillatory motion in 1+1 dimension. Hereinafter we consider VQ oscillating along $X-$ axis and $Z-$ axis is perpendicular to the plane of oscillation XY. The spacial distribution of color fields of quarks in Eq. 5 is approximated by a gaussian function. It has been shown in papers [5] that the wave packet solutions of time dependent Schrodinger equation for harmonic oscillator move in exactly the same way as corresponding classical oscillators. These solutions are called "coherent states". This relationship justifies (partly) our semiclassical treatment of quantum objects.

The parameters of the model, namely, maximum displacement, $x_{\max}$, parameters of 3–D gaussian function, $\sigma_{x,y,z}$, for hadronic matter distribution around VQ, and the maximal constituent mass, c, are chosen to be

$$x_{\max} = 0.64\ fm,\ \sigma_{x,y} = 0.24\ fm,\ \sigma_z = 0.12\ fm,\ c = M_{Q(\overline{Q})}(x_{\max}) \approx 360\ MeV, \tag{8}$$

where m_Δ and m_N are masses delta–isobar and nucleon respectively. They are adjusted by comparison of calculated and experimental values of inelastic cross sections, $\sigma_{in}(s)$, and inelastic overlap function $G_{in}(s,b)$ for pp and $\overline{p}p-$ collisions [4]. The current mass of valence quark is taken to be 5 MeV. The model demonstrates the relationship between constituent and current quark states inside a hadron. At maximum displacement quark is nonrelativistic, constituent one (VQ surrounded by condensate), since the influence of polarization fields of other quarks becomes minimal and the VQ possesses the maximal potential energy corresponding to the mass of constituent quark. At the origin of oscillation, $x = 0$, antiquark and quark in mesons and three quarks in baryons, being close to each other, have maximum kinetic energy and correspondingly minimum potential energy and mass: they are relativistic, current quarks (bare VQs). This configuration corresponds to so called "asymptotic freedom". In the intermediate region there is increasing (decreasing) of constituent quark mass by dressing (undressing) of VQs due to decreasing (increasing) of the destructive interference effect. On the other hand this dynamical mechanism of VQ dressing (undressing) corresponds to the chiral symmetry breaking (restoration).The proposed model has some important consequences for the nucleon structure: i) all three quarks and accompanying them gluons are strongly correlated; ii) inasmuch as quarks oscillate along the bisectors of equilateral triangle they don't possess an angular orbital momentum; iii) nucleons are non-spherical, oblate objects, they are flattened along the axis perpendicular to the plane of quarks oscillations.

So far we dealt only with the scalar color field φ around VQs. How can one include spin in the framework of these classical considerations? According to the prevailing belief, the spin is a quantum feature of microparticle and has no classical analog. However, Belinfante as early as in 1939 demonstrated that the spin of electron may be regarded as an angular momentum generated by a circulating flow of electromagnetic energy in the wave field of electron [7]

$$\mathbf{s} = \int_a^\infty \mathbf{r} \times (\mathbf{E} \times \mathbf{B}) d^3\mathbf{r}, \tag{9}$$

where $\mathbf{E} \times \mathbf{B}$ is the well–known Poynting vector. Now we should proceed to four-potential color field $A^\mu = (\varphi, \mathbf{A})$ including into consideration, besides appearing above the scalar part,

Journal of Physics: Conference Series **678** (2016) 012006 doi:10.1088/1742-6596/678/1/012006

φ, three–component magnetic vector potential $\mathbf{A}$ (color indices are implicit). By analogy with the electron we assume that intersecting chromoelectric, $\mathbf{E}_{ch} = -\,\mathrm{grad}\,\mathbf{A}$, and chromomagnetic, $\mathbf{B}_{ch} = \mathbf{rot A}$, fields create around quark (antiquark) the circulating flow $\mathbf{S}_{ch} = \mathbf{E}_{ch} \times \mathbf{B}_{ch}$, the momentum density of color field. Generated by this chromomagnetic flow the total angular momentum is defined by the integral

$$\mathbf{J} = \int_a^\infty \mathbf{r} \times (\mathbf{E}_{ch} \times \mathbf{B}_{ch}) d^3\mathbf{r}, \tag{10}$$

where the lower limit of integration, a, corresponds ot the size of the quark. It is natural to beleive that the angular momentum vector, $\mathbf{J}$, of each quark points perpendicular to the plane of quark oscillation. During quark oscillation the extent of the color potential, A^μ, around the quark changes from maximum at the maximal quark displacement to minimum at its position near the center of mass of a nucleon. As to the circulating flow, $\mathbf{S}_{ch}$, the requirement of conservation of the angular momentum, $\mathbf{J}$, make it to be concentrated in a narrower shell around the quark at its approach to the origin of oscillation. Thus, according to our approach the quark is a vortex, and a nucleon consists of three oscillating quark–vortices turning into constituent states at maximum displacement and current (relativistic) states – at minimum displacement.

With aim to analyse the "null" result of DIS measurements of the proton spin we proceed to the hydrodynamical analogy. In hydrodynamics a key concept in the dynamics of vortices is the vorticity, a vector that describes circulation around unit area in the fluid; it is defined as the curl of the flow velocity: $\xi = \nabla \times \mathbf{v}$, where $\mathbf{v}$ is the vector of the flow velocity and ξ is its vorticity. In ideal liquids the flow velocity is inversely proportional to the distance from the center of the vortex, $v \sim 1/r$ and the velocity field and the vortex are called irrotational. For irrotational vortices the line integral in the velocity field, called circulation, $\mathbf{\Gamma} = \oint_\Gamma \mathbf{v}{\cdot}d\mathbf{r}$ is zero along any closed contour that does not encloses the vortex axis and has a fixed value, Γ, for any contour that encloses the axis. Since circulation relates to vorticity by Stock's theorem

$$\oint_\Gamma \mathbf{v}{\cdot}d\mathbf{r} = \int_S (\nabla \times \mathbf{v}){\cdot}d\mathbf{S} = \int_S \xi{\cdot}d\mathbf{S}, \tag{11}$$

the same effect takes place for the vorticity, as well. Therefore, according to our approach the proton spin is combined as additive sum of the spins of vortical quarks oscillating inside the proton without orbiting. Hence, the proton spin can not be decomposed to the intrinsic and orbital parts. In DIS the transverse size (wavelength), d, of the virtual photon is inversely proportional to the momentum transfer by incident lepton as $d^2 \propto 1/Q^2$. At small $Q^2$the wavelength of virtual photon is comparable with the sizes of quarks, and the photon probes the total vorticity (spin) of the proton. At large Q^2some areas inside the proton covered by the short wavelength photon does not include any quark–vortex axis. It is just the reason of spin "crisis". Qualitatively other known spin phenomena can be explained by this approach: single spin asymmetry, cross section differences in elastic scattering of protons with parallel and anti–parallel spins.

[1] G. Musulmanbekov 2001 in *Frontiers of Fundamental Physics 4* ed B G Sidharth and M V Altaisky (New York: Kluwer Acad./Plenum Pub.) p.109

[2] Rajaraman R *Phys. Rep.* C **21** 229

[3] Kontorova T A and Frenkel Ya I 1938 *Zh. Eksp. Teor. Fiz.* **8**, 1340 (in russian)

[4] Musulmanbekov G 2004 *Phys. Atom. Nucl.*, **67** 89

[5] Schrodinger E 1926 *Naturwissenschaften*, **14** 664

[6] Dashen R Hasslacher B and Neveu **A** 1974 *Phys. Rev.* D **10** 4114

[7] Belinfante F J 1939 *Physica* **6** 887

Journal of Physics: Conference Series **678** (2016) 012007 doi:10.1088/1742-6596/678/1/012007

A gravitating electroweak bag model

Alexander Burinskii

Theor. Phys. Lab. NSI, Russian Academy of Sciences, B. Tulskaya 52, Moscow 115191 Russia

E-mail: burinskii@mail.ru

Abstract.

Gravitational and electromagnetic (EM) field of electron is described by the Kerr-Newman (KN) black hole solution with a topological defect. Regularization of this defect by the Higgs field leads to the smooth source which shares much in common with the known MIT- and SLAC-bag models, but has the advantage, of matching gravitational and electromagnetic fields of the electron. This model is flexible, and the rotating KN bag takes the shape of a thin disk with a circular string positioned on the sharp border of the disk. We consider the lowest excitations of the KN solution and the corresponding deformations of the bag surface, setting a preliminary correspondence with electroweak sector of the SM.

1. Introduction

The Kerr-Newman rotating black hole (BH) solution has gyromagnetic ratio $g = 2$, as that of the Dirac electron. As a result, the gravitational and EM fields of electron are described by the Kerr-Newman (KN) solution [1]. Extreme high spin of electron, $a = J/m >> m$, corresponds to the over-rotating KN solution without horizons, and the naked Kerr singular ring forms the Einstein-Rosen bridge, or a wormhole to another sheet of the Kerr space-time, where the effective charge and mass change the signs. The Kerr-Schild metric has the form [1]

$$g_{\mu\nu} = \eta_{\mu\nu} + 2Hk_\mu k_\nu, \tag{1}$$

where $\eta_{\mu\nu}$ is metric of auxiliary Minkowski space M^4, and k_μ is a null vector field, $k_\mu k^\mu = 0$, forming the Principal Null Congruence (PNC) $\mathcal{K}$. In the Kerr oblate spheroidal coordinates, the retarded and advanced sheets of metric are related by a smooth transfer of the Kerr PNC via disk $r = 0$ spanned by the Kerr singular ring $r = 0, \ \cos\theta = 0$, where r is the Kerr ellipsoidal radial coordinate (see for example [2, 3]). The surface $r = 0$ represents a disklike "door" from negative sheet $r < 0$ to positive one $r > 0$. The null vector fields $k^{\mu\pm}(x)$ differs on these sheets, and two different null congruences $\mathcal{K}^\pm$, create two different metrics $g^\pm_{\mu\nu} = \eta_{\mu\nu} + 2Hk^\pm_\mu k^\pm_\nu$ on the same Minkowski background M^4. This feature of the Kerr geometry caused the long-term search for different models of the source of KN solution, which could remove this mystery, retaining the external KN field. For the KN solution with parameters of the electron, the size of the Kerr singular ring (wormhole) $a = \hbar/2m$ corresponds to the Compton wavelength. Therefore, *the KN gravity turns out to be not weak in the Compton zone of the electron*, that spoils the flat space required by quantum theory. The singular space-time of the KN geometry should be "regularized" by some source. For nearly five decades, many authors have made important contribution to study of this problem, starting from H. Keres (1966), and then by W. Israel

Content from this work may be used under the terms of the Creative Commons Attribution 3.0 licence. Any further distribution of this work must maintain attribution to the author(s) and the title of the work, journal citation and DOI.
Published under licence by IOP Publishing Ltd

Journal of Physics: Conference Series **678** (2016) 012007 doi:10.1088/1742-6596/678/1/012007

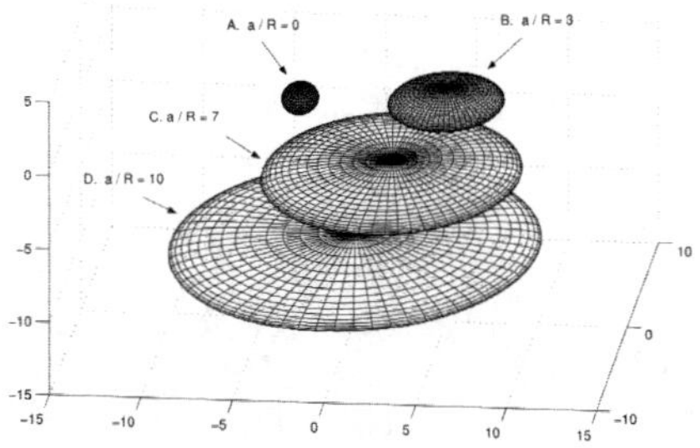

Figure 1. (A): Spherical bad with zero rotation, $a/R = 0$, and the rotating disk-like bags for different ratios a/R, (B): $a/R = 3$;(C): $a/R = 7$, (D):$a/R = 10$.

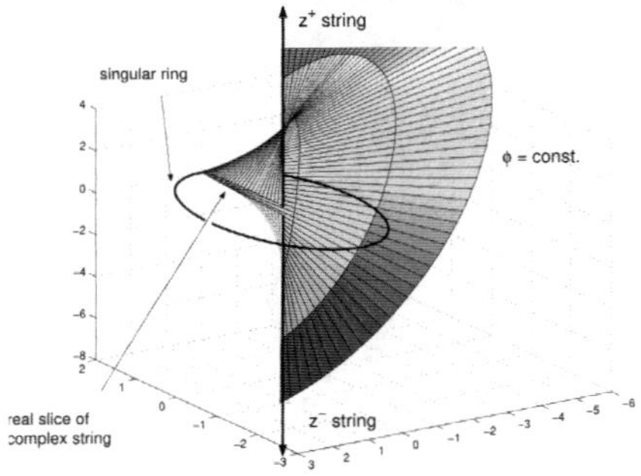

Figure 2. The Kerr surface $\phi = const.$ Vector potential is dragged by the Kerr singular ring, forming the closed loop along the bag border.

(I970), V. Hamity (1976), C. López (1984) at al. The most advanced model of the bubble-source was suggested by López [4]. The singular curved region has to be excised and replaced by the regular flat metric $\eta_{\mu\nu}$, which is to be matched with external KN solution along the surface of "zero-gravity potential"

$$H = H_{KN} \equiv \frac{mr - e^2/2}{r^2 + a^2 \cos^2 \theta} = 0. \tag{2}$$

It yields $r = r_e = e^2/2m$, where r is the Kerr oblate spheroidal coordinate. The López model was generalized in [2, 3] to the spinning soliton model, and then in [5, 12] to the spinning bag model. As is known, the bags are deformable and may take the stringy form. The determined by (2) bag for parameters of an electron takes the form of oblate disk of the Compton radius and has the thickness of the classical radius r_e. It shows that the KN bag model should be associated with the *dressed electron*.

2. Domain Wall bubble and the supersymmetric source of the KN solution

In [2, 3] this bag was formed as a domain-wall (DW) interpolating between the supersymmetric flat interior of the bubble and the external KN solution. The superpotential W depends on three chiral fields Φ^i, $i = 1, 2, 3$, one of which is considered as the Higgs field, $H(x) \equiv \Phi^1$. Supersymmetry provides concentration of the Higgs condensate in a pseudo-vacuum state inside the bag [5], where the effective field model coincides with the known Nielsen-Olesen models of the vortex string. Interaction of the Higgs field $H(x) = |H|e^{i\chi(x)}$ with the KN vector potential

Journal of Physics: Conference Series **678** (2016) 012007 doi:10.1088/1742-6596/678/1/012007

inside the disk is determined by the equation $\nabla_\nu \nabla^\nu A_\mu = I_\mu = \frac{1}{2}e|H|^2(\chi,_\mu + eA_\mu)$. Rhs of this eq. shows that phase $\chi,_\mu$ compensates potential A_μ and the current I_μ is expelled from interior of bubble to boundary of the disk. It indicates superconducting nature of the internal vacuum and leads to very important peculiarity [2, 5]: so far as the flux of the vector potential is dragged by the Kerr ring in ϕ-direction (see Fig.2), it forms a quantum Wilson loop, $\oint eA_\phi d\phi = -4\pi ma$, and **the spin of this soliton is to be quantized**, $J = ma = n\hbar/2,\ n = 1, 2, 3...$ Note also that the Higgs condensate is oscillating with the frequency $\omega = 2m$.

The Dirac field is consistently embedded into KN bag and gets mass from the Higgs field [6, 5], which is one of the characteristic features of the bag models [7, 8]. Another important feature of the bag models is pliancy to deformations leading to formation of the stringy structure. The KN bag forms a circular string positioned along the sharp border of the disk, see Fig.1. This string appears as the result of regularization the closed singular string formed by the Kerr ring [9, 10]. Strings of this sort were considered in the low energy string theory as fundamental solitonic strings [11]. Excitations of this string create deformations of the bag boundary. Minimum of the energy density $H^{(ch-r)}$ is achieved on the solutions satisfying the first order Bogomolnyi equations [12], which determine structure and features of the bag, in particular, quantization of the angular momentum. When the Bogomolnyi equations are saturated, the mass-energy of DW turns into full differential, and the total mass is determined by $\Delta W = W(R+\delta) - W(R-\delta) = -\mu\eta^2$,

$$M_{ch} = 2\pi\Delta W \int_{-1}^{1} dX(R^2 + a^2X^2) = 4\pi(R^2 + \frac{1}{3}a^2)\Delta W. \quad (3)$$

All the fields may be extended to $r < 0$, where we obtain an "anti-DW" which gives the same contribution to total mass, but with opposite sign. The DW bag-like source of the KN solution forms a bubble-antibubble system similar to the known kink-antikink solutions. The front side of the anti-DW wall 'looks' on the nonphysical ("advanced") sheet of the KN solution, and its negative contribution to mass should not be considered in the case of stationary KN solution.

3. The lowest excitations

The stationary KN bag solution may be excited by the circular EM traveling waves propagating along the sharp border of the disk. The lowest excitations of the KN solution were considered in [12]. The general solution for vector potential has the form [1]

$$A_\mu dx^\mu = -Re\ [(\frac{\psi}{r + ia\cos\theta})k_\mu dx^\mu + \chi d\bar{Y}], \quad \chi = 2\int (1 + Y\bar{Y})^{-2}\psi dY \quad (4)$$

and defined by analytic function $\psi(Y,\tau)$, where $Y = e^{i\phi}\tan\frac{\theta}{2}$ is complex projective angular variable, $\tau = t - r - ia\cos\theta$ is complex retarded-time parameter. According [1]

$$H = \frac{mr - |\psi|^2/2}{r^2 + a^2\cos^2\theta}, \quad (5)$$

and function ψ determines the surface of "zero gravity" from $H(r) = 0$, (2). The function $\psi = -e$ corresponds to stationary KN solution (2) leading to the constant cut-off $R = r_e$, see Fig.3A. The lowest nontrivial excitations

$$\psi = \frac{e}{Y}e^{i\omega\tau} \quad (6)$$

yields also the constant cut-off $R = r_e$. Another excitation, considered in [12],

$$\psi = e(1 + \frac{1}{Y}e^{i\omega\tau}) \quad (7)$$

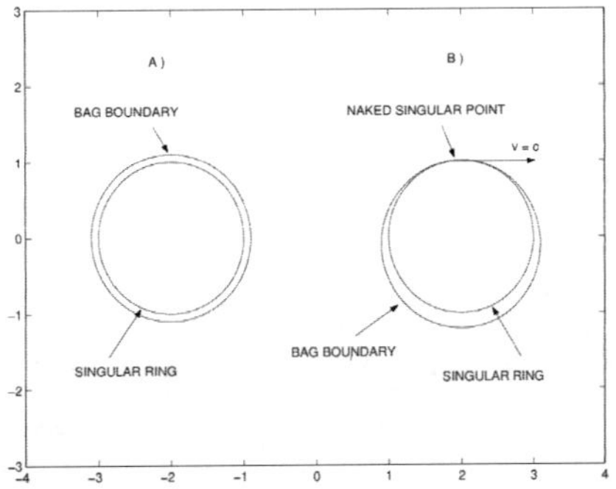

Figure 3. Section of the bag in equatorial plane. Distance between bag boundary and singular ring acts as a cut-off parameter R. (A)Axially symmetric solution yields the constant cut-off $R = r_e$. (B) Traveling wave deforms the bag boundary, creating singular point.

deforms the bag boundary in the equatorial plane as shown in Fig.3B.

One sees that in the equatorial plane $\cos\theta = 0$ the solution (7) takes the form $\psi = e(1+e^{-i(\phi-\omega t)})$, and the cut-off parameter $R = |\psi|^2/2m = \frac{e^2}{m}(1+\cos(\phi-\omega t)$ becomes depending on $\phi - \omega t$. Vanishing R at $\phi = \omega t$ creates the singular pole which circulates along the ring-string together with traveling wave, reproducing the known effect of zitterbewegung of the Dirac electron. This singular pole may be interpreted as a point-like bare electron, and therefore, the KN bag model unites the naked and dressed electron in the unique gravitating bag-string-quark system [12] generating the external gravitational and electromagnetic fields of the electron. This excitation changes topology of the bag surface, combining the DW and anti-DW boundaries in the single nonorientable surface. The masses of the anti-DW and DW are mutually cancelled, and mass of the confined fermion is determined only by the energy of the external EM field. Dual rotations of the EM field in the solution (7) allows to get another solution with opposite charge and one more neutral solution which can be associated with positron and neutrino. One can assume that muon can belong to the next very high excitation. The topologically more simple DW-antiDW solutions of type (6) and their dual analogs may correspond to intermediate bosons. These solutions are initially massless and acquire mass from the Yukawa coupling with the Higgs field in accord with basic conception of the Standard Model, but with the essential advantage of consistency with gravity.

Acknowledgments
This work was supported by the RFBR grant N 13-01-00602.

References

[1] Debney G C, Kerr R P and Schild A 1969 *J. Math. Phys.* **10** 1842
[2] Burinskii A 2010 *J. Phys. A: Math. Theor.* **43** 392001
[3] Burinskii A 2014 *Int. J. of Mod. Phys.* **A 29** 1450133
[4] López C A 1984 *Phys. Rev. D* **30** 313
[5] Burinskii A 2015 *JETP (Zh. Eksp. Teor. Phys.)* **148** 228
[6] Burinskii A 2015, *Grav. and Cosmology* **21**, 28
[7] Chodos A et al. 1974 *Phys. Rev.* **D 9** 3471
[8] Bardeen W. A. at al. 1974 *Phys. Rev.* **D 11**, 1094
[9] Burinskii A Ya 1974 *Sov. Phys. JETP* **39** 193
[10] Ivanenko D D and Burinskii A Ya 1974 *Izv. Vuz. Fiz.*, **5**, 135
[11] Burinskii A 1995 *Phys. Rev.* **D 52** 5826
[12] Burinskii A 2015 *JETP (Zh. Eksp. Teor. Fiz.)* **148** 937

Journal of Physics: Conference Series **678** (2016) 012008 doi:10.1088/1742-6596/678/1/012008

Model for non-perturbative contributions to the quark distributions

B I Ermolaev[1], **S I Troyan**[2]

[1] Ioffe Physico-Technical Institute, 194021 St.Petersburg, Russia
[2] St.Petersburg Institute of Nuclear Physics, 188300 Gatchina, Russia

E-mail: ermolaev@mail.cern.ch

Abstract. In the present talk we briefly discuss the most general form of QCD Factorization, which we call Basic Factorization, and show how it is related to the conventional forms of factorization: K_T-Factorization and Collinear Factorization. When Basic Factorization is applied to hadronic scattering amplitudes, integration over momenta of the partons connecting perturbative and non-perturbative blobs runs over the whole phase space but result of the integration must be finite. This obvious requirement allows us to obtain theoretical restrictions on the fits for the involved non-perturbative contributions. In order to model non-perturbative contributions in Basic Factorization, we suggest the Resonance Model which can universally be used to describe hadronic reactions with unpolarized or polarized hadrons.

1. Introduction

QCD factorization is actually the only available means to apply results of Perturbative QCD to description of hadronic reactions at high energies. It is supposed to compensate, though approximative, lack of knowledge on QCD in the infra-red region. This method proved to be quite efficient for description of all kinds of hadronic processes and such important their ingredients as parton distributions in the hadrons. In essence, factorization works as follows: First, appropriate non-perturbative inputs are introduced and then Evolution Equations (or fixed-order perturbative calculations) evolve them to perturbative domain. However, actual situation is more involved and because of that different forms of factorization were invented. Choice of a particular form depends mostly on perturbative methods used in the calculations. In order to illustrate it, let us consider factorization for the DIS non-singlet structure function $f_{NS}(x, Q^2)$ as an example. We consider it under the approximation of single-parton collisions, when only one parton with momentum k in the initial hadron with momentum p interacts with the virtual photon with momentum q while other partons are spectators. In this case, QCD factorization for each structure function in the hadronic tensor $W_{\mu\nu}$ is conventionally depicted as shown in Fig. 1, where the upper blob corresponds to the parton-photon scattering and the lowest blob corresponds to emission of a single active parton off the hadron[1].

When the DGLAP[2] or its generalizations to the small-x, or to the small-Q^2 regions (see e.g. Ref. [3] and Refs therein for detail) are used to describe $f_{NS}(x, Q^2)$, an appropriate non-perturbative input $\phi(x_0, \mu^2)$ is supposed to be on-shell. In this case Collinear Factorization[4] is usually used, Using the DGLAP equations to evolve the scale μ^2 up to Q^2 (with $Q^2 > \mu^2$) and

[1] For an overview of double-parton collisions see recent Ref. [1].

Content from this work may be used under the terms of the Creative Commons Attribution 3.0 licence. Any further distribution of this work must maintain attribution to the author(s) and the title of the work, journal citation and DOI.
Published under licence by IOP Publishing Ltd

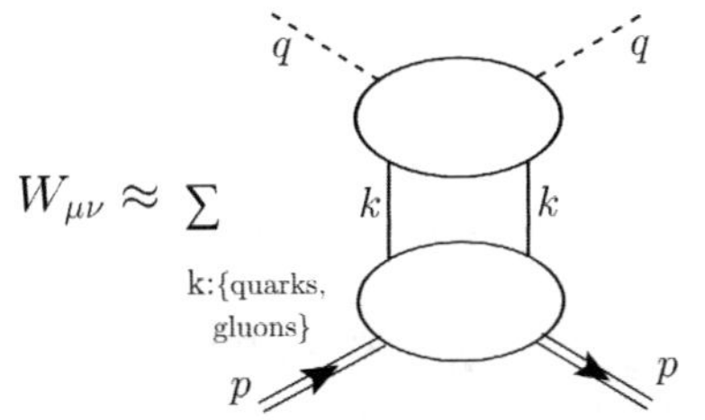

Figure 1. QCD factorization for the DIS hadronic tensor under the approximation of single-parton collisions. The s-cut of the graph is implied.

complementing the issue by the x-evolution, one arrives at the result which can be written as follows:

$$f_{NS}(x, Q^2) = \int_x^1 \frac{d\beta}{\beta} f_{NS}^{(pert)}(x/\beta, Q^2/\mu^2)\phi(\beta, \mu^2), \tag{1}$$

where $f_{NS}^{(pert)}$ is the perturbative component. It includes both the coefficient function and anomalous dimension. The factorization scale μ^2, i.e. the starting point of the perturbative Q^2 -evolution in the literature is usually chosen in the perturbative domain ($x_0 \sim 1$ and $\mu \sim$ few GeV), so the input (integrated quark distribution) $\phi(x_0, \mu^2)$ contains both perturbative and non-perturbative contributions.

Obviously, Collinear Factorization cannot be applied when BFKL[5] is used to account for perturbative contributions, because the initial partons (gluons) here are essentially off-shell, with virtuality $k^2 \approx -k_\perp^2$, and integration over $k_\perp$ runs down to zero. In this case another form of QCD factorization, K_T -Factorization[6] named also High-Energy Factorization[7], can be applied. K_T -Factorization can also be used beyond the BFKL context. In this case the expression for $f_{NS}(x, Q^2)$ is

$$f_{NS}(x, Q^2) = \int_{\beta_0}^1 \frac{d\beta}{\beta} \int_0^{w(1-x)} \frac{dk_\perp^2}{k_\perp^2} f_{NS}^{(pert)}(x/\beta, Q^2/k_\perp^2)\Phi(\beta, k_\perp^2), \tag{2}$$

where $\Phi(\beta, k_\perp^2)$ denotes the non-perturbative input (unintegrated quark distribution) in K_T-Factorization, $\beta_0 = x + k_\perp^2/w$ and $w = 2pq$. Obviously, K_T -Factorization is more general than Collinear Factorization. Momentum k of the active parton in Fig. 1 is parameterized in K_T-Factorization as follows:

$$k = \beta p + k_\perp, \tag{3}$$

which operates with three parameters: $\beta, |k_\perp|$ and the azimuthal angle[2]. Nevertheless, momentum k is four-component, so one of its longitudinal components is left unaccounted by the parametrization of Eq. (3).

Recently in Ref. [9] we have derived a new and most general form of QCD factorization: Basic Factorization where the parametrization of k accounts for all four components of k and therefore is most general:

$$k_\lambda = -\alpha\left(q_\lambda - (q^2/w)p_\lambda\right) + \beta\left(p_\lambda - (p^2/w)q_\lambda\right) + k_\lambda^\perp, \tag{4}$$

i.e. it is the standard Sudakov parametrization[8]. In Basic Factorization

$$f_{NS}(x, Q^2) = \int_{\beta_0}^1 \frac{d\beta}{\beta} \int_0^{w(1-x)} dk_\perp^2 \int_0^w d\alpha f_{NS}^{(pert)}(x/\beta, Q^2/k^2) \frac{B}{k^2 k^2} \Psi(\alpha, k^2), \tag{5}$$

[2] Dependence on the azimuthal angle is often neglected as unessential.

Journal of Physics: Conference Series **678** (2016) 012008 doi:10.1088/1742-6596/678/1/012008

where the factor B is

$$B = k_\perp^2 + w\left(\alpha^2 + \beta^2\right)/2. \tag{6}$$

It arises from handling the spinors, while the factors k^2 correspond to the propagators of the intermediate (active) quarks. In practical implementations, one can use the approximation $b \approx k_\perp^2$. $\Psi(\alpha, k^2)$ is a new non-perturbative input (totally unintegrated quark distribution). Let us remind that $k^2 = -w\alpha\beta - k_\perp^2$. Optical theorem relates f_{NS} to the non-singlet part of the Compton amplitude A_{NS} in the forward kinematics. In Basic Factorization A_{NS} is given by the following expression:

$$A_{NS}(x, Q^2) = \int \frac{d\beta}{\beta} dk_\perp^2 d\alpha A_{NS}^{(pert)}(x/\beta, Q^2/k^2) \frac{B}{k^2 k^2} T(\alpha, k^2), \tag{7}$$

where A_{NS} is the perturbative and T is the non-perturbative input (altogether non-perturbative quark-hadron scattering amplitude). In contrast to Eq. (5), integration over k in Eq. (7) runs over the whole phase space. Basic Factorization approximately can be reduced to K_T-Factorization providing the most essential values of α are outlined by the following restriction:

$$w|\alpha\beta| \ll k_\perp^2, \tag{8}$$

which is the standard restriction for the ladder momenta in all available perturbative Evolution Eqs, including DGLAP and BFKL. On its turn, K_T-Factorization can approximately be reduced to Collinear Factorization providing the dependence of $\Phi(\beta, k_\perp^2)$ on $k_\perp^2$ has a specific form: it should be one or several sharp peaks.

Another important result obtained in Ref. [9] is theoretical restrictions on the fits for non-perturbative inputs in all forms of QCD factorization. Integration in Eq. (7) runs over the whole phase space and it should yield a finite result despite that the integrand can have both infra-red (IR) and ultra-violet (UV) divergences. They were regulated in Ref. [9] with imposing restrictions on the non-perturbative inputs. As non-perturbative QCD is known very poor, the only way to specify the non-perturbative inputs is to model them. In Ref. [10] we suggested the Resonance Model for non-perturbative contributions to the quark-hadron scattering amplitudes. It embraces the cases of non-polarized and polarized hadrons with arbitrary orientation of the hadron spin. Applying Optical theorem to the amplitudes makes possible to obtain expressions for the quark distributions in the hadrons. Below we describe an improved version of this model.

2. Resonance Model for non–perturbative inputs

First of all, let us note that the non-perturbative inputs to the non-singlet quark-hadron scattering amplitudes and to the Compton scattering amplitudes off hadrons (the lowest blob in Fig. 1, with no cut) are identical. Fig. 1 explicitly demonstrates that this blob includes both the quark density matrix

$$\hat{\rho}_q(k, S_q) = \frac{1}{2}\left(\hat{k} + m_q\right)\left(1 - \gamma_5 \hat{S}_q\right), \tag{9}$$

where k, m_q and S_q are the quark momentum, mass and spin respectively. We suggest that the hadron spinor structure can be described similarly:

$$\hat{T} = (\hat{p} + M_H) T_U - (\hat{p} + M_H) \gamma_5 \hat{S}_H T_S, \tag{10}$$

with p, M_H and S_H being the hadron momentum, mass and spin respectively while T_U and T_S are non-perturbative invariant amplitudes, each depends on k^2 and the invariant energy s_1, with

$$s_1 = (p-k)^2 \approx k^2 + w\alpha + M_H^2 = w\alpha - w\alpha\beta - k_\perp^2 + M_H^2. \tag{11}$$

Eq. (10) satisfies the conformity: when the hadron is replaced by a quark, $\hat{T}$ converts into the quark density matrix. Now we have to specify $T_{U,S}$. As the both of them can be handled identically, we skip the subscripts U and S in what follows and will use the generic notation T for them both, arriving therefore to T participating in Eq. (7). We also acknowledge that it may be possible that $T_U = T_S$. Any model for T should guarantee integrability of the factorization convolutions. For example, the convolution in Eq. (7) can potentially have both infra-red (IR) divergence at $k^2 = 0$ and ultra-violet (UV) divergence at large $|\alpha|$ even with $A^{(pert)}$ being free of such divergences. As shown in Ref. [10], IR singularities in the convolutions are absent if

$$T \sim \left(k^2\right)^{1+\eta} \tag{12}$$

at $k^2 \to 0$, with $\eta > 0$, while UV divergences are killed when

$$T \sim (\alpha)^{-\chi} = (\alpha)^{1+\eta} (\alpha)^{-1-\eta-\chi}, \tag{13}$$

at large $|\alpha|$, with $\chi > 0$. Then, any model for T should guarantee applicability of the Optical theorem, so T should have a non-zero imaginary part with respect to the invariant energy s_1. Finally, after reducing to K_T-factorization, a dependence of T on $k_\perp$ should have a peaked form. We choose the following expression for T in Basic Factorization:

$$T = R\left(k^2\right) Z\left(s_1\right), \tag{14}$$

with $R \sim (k^2)^{1+\eta}$ at small k^2 and $Z \sim (\alpha)^{-1-\eta-\chi}$ at large $|\alpha|$. We suggest the following general formula for Z:

$$Z = Z_n = \prod_1^n \frac{1}{(s_1 - M_r^2 + \imath\Gamma_r)}, \tag{15}$$

with $n = 2, 3, \ldots$. The case $n = 1$ is excluded by Eq. (13). In what follows we focus on the simplest option $n = 2$, where Z_2 can easily be represented as an interference of two resonances:

$$Z_2 = \frac{1}{(\Delta M^2 + \imath\Delta\Gamma)} \left[\frac{1}{s_1 - M_1^2 + \imath\Gamma_1} - \frac{1}{s_1 - M_1^2 + \imath\Gamma_1}\right], \tag{16}$$

where $\Delta M^2 = M_1^2 - M_2^2, \Delta\Gamma = \Gamma_1 - \Gamma_2$. Applying the Optical theorem to Eq. (14) allows us to obtain the non-perturbative contribution Ψ to the quark distributions in the hadrons:

$$\Psi = \Im_{s_1} T = \widetilde{R}(k^2) \left[\frac{\Gamma_2}{(s_1 - M_2^2)^2 + \Gamma_2^2} - \frac{\Gamma_1}{(s_1 - M_1^2)^2 + \Gamma_1^2}\right], \tag{17}$$

with $\widetilde{R}(k^2) = R(k^2)/\left(\Delta M^2 + \imath\Delta\Gamma\right)$. Obviously, the expression Ψ is of the Breit-Wigner type. Because of that we name our model the Resonance Model. The guiding idea for our suggesting the Resonance Model is based on simple physical grounds: after emitting an active quark off the hadron, the remaining set of partons becomes unstable and therefore it can be described through resonances. Unfortunately, we could not contrive a similar argumentation for specifying R. It is absolutely arbitrary save the only restriction: $R \sim (k^2)^{1+\eta}$ at small k^2. For instance, it could be such an option:

$$R = \left(\frac{k^2}{k^2 + M^2}\right)^{1+\eta}. \tag{18}$$

Journal of Physics: Conference Series **678** (2016) 012008 doi:10.1088/1742-6596/678/1/012008

3. Transition from Basic Factorization to conventional forms of QCD factorization

The expression for the totally unintegrated quark distributions $\Psi(\alpha,\beta,k_\perp^2)$ in Eq. (17) is obtained for Basic factorization. In order to obtain the unintegrated quark distributions $\Phi\left(\beta,k_\perp^2\right)$ from $\Psi\left(\alpha,\beta,k^2\right)$, the α -dependence of $\Psi(\alpha,\beta,k^2)$ should be integrated out. The integration region is given by Eq. (8), so we can use the approximation $k^2 \approx -k_\perp^2$ in expressions for $\Psi(\alpha,\beta,k_\perp^2)$. Therefore (see Ref. [10] for detail),

$$\Phi(\beta,k_\perp^2) = \int_0^\zeta d\alpha \Psi(\alpha,\beta,k_\perp^2), \tag{19}$$

with $\zeta = k_\perp^2/(w\beta)$. Using Eq. (17), we easily arrive at the following expression for Φ:

$$\Phi = R\left(k_\perp^2\right)\left[\frac{\Gamma_1}{(\zeta-\mu_1^2)^2+\Gamma_1^2} + \frac{\Gamma_2}{(\zeta-\mu_2^2)^2+\Gamma_2^2}\right], \tag{20}$$

where we have skipped terms independent of ζ. In Eq. (20) we denoted $\mu_1^2 = M_H^2 - M_1^2$, $\mu_2^2 = M_H^2 - M_2^2$. Obviously, Φ in Eq. (20) is also given by expressions of the Breit-Wigner type. Let us note that in Eq. (20) $\zeta > 0$ while signs of μ_1^2 and μ_2^2 cannot be fixed a priory. It means that in the case of positive μ_1^2 and μ_2^2, the both resonances are within their resonant region. On the contrary, if $\mu_1^2 > 0$ and $\mu_2^2 < 0$ (or vice versa), contributions to Φ come from one sharp resonance and a tail of the other resonance. Such a scenario has a certain resemblance to the Duality conception. Transition from $\Phi(\beta,k_\perp^2)$ in K_T-Factorization to the integrated quark distribution $\phi(\beta)$ in Collinear Factorization (both Φ and φ are non-perturbative) can be done with integration of $\Phi(\beta,k_\perp^2)$ over $k_\perp$, leading to the following expressions, depending on the signs of μ_1^2 and μ_2^2: If $\mu_1^2,\ \mu_2^2 > 0$,

$$\varphi = R(\mu_1^2,\beta) + R(\mu_2^2,\beta) \tag{21}$$

and if $\mu_1^2 > 0,\ \mu_2^2 < 0$, then

$$\varphi = R(\mu_1^2,\beta). \tag{22}$$

We remind that, being altogether non-perturbative, φ does not coincide with the integrated distributions ϕ conventionally used in Collinear Factorization (see e.g. Eq. (1)), where both perturbative and non-perturbative contributions are present. Transition from φ to ϕ is done with applying perturbative evolution means to φ (see REf. [10] for detail).

4. Conclusion

In the present talk we have given a brief description of Basic Factorization which is the most general form of QCD factorization and demonstrated its relations to the conventional forms of QCD factorization: K_T-Factorization and Collinear Factorization. We presented the Resonance Model which is supposed to universally describe non-perturbative contributions to the quark-hadron scattering amplitudes and quark distributions in hadrons, where the hadrons can be either polarized or unpolarized.

5. Acknowledgements

B.I. Ermolaev is grateful to Organizing Committee of DSPIN 2015 for financial support.

References

[1] Szczurek A 2015 *Acta Phys. Polon. Supp.* **8** 483

[2] Altarelli G and Parisi G 1977 *Nucl. Phys.* B **126** 297;
Gribov V N and Lipatov L N 1972 *Sov. J. Nucl. Phys.* **15** 438;
Lipatov L N 1972 *Sov. J. Nucl. Phys.* **20** 95;
Dokshitzer Yu L 1977 *Sov. Phys. JETP* **46** 641
[3] Ermolaev B I, Greco M and Troyan S I 2010 *Riv. Nuovo Cim.* **33** 57
[4] Amati D, Petronzio R and Veneziano G 1978 *Nucl. Phys.* B **140** 54;
Efremov A V and Radyushkin A V 1980 *Teor. Mat. Fiz.* **42** 147; 1980 *Theor. Math. Phys.* **44** 573; 1980 *Teor. Mat. Fiz.* bf 44 17; 1976 *Phys. Lett.* B **63** 449; 1977 *Lett. Nuovo Cim.* **19** 83;
Libby S and Sterman G. 1978 *Phys. Rev.* D **18** 3252;
Brodsky S J and Lepage G P 1979 *Phys. Lett.* B **87** 359; 1980 *Phys. Rev.* D **22** 2157;
Collins J C and Soper D E 1981 *Nucl. Phys.* B **193** 381;
Collins J C and Soper D E 1982]it Nucl. Phys. B **194** 445;
Collins J C , Soper D E and Sterman G 1985 *Nucl. Phys.* B **250** 199;
Efremov A V and Radyushkin A V *Report JINR E2-80-521*; 2009 *Mod. Phys. Lett.* A **24** 2803
[5] Kuraev E A, Lipatov L N and Fadin V S 1976 *Sov. Phys. JETP* **44** 443; 1977 *Sov. Phys. JETP* **45** 199;
Balitsky I I and Lipatov L N 1978 *Sov. J. Nucl. Phys.* **28** 822
[6] Catani S, Ciafaloni M and Hautmann F 1990 *Phys. Lett.* B **242** 97; 1991 *Nucl. Phys.* B **366** 135
[7] Collins J C and Ellis R K 1991 *Nucl. Phys.* B **360** 3
[8] Sudakov V V 1956 *Sov. Phys. JETP* **3** 65
[9] Ermolaev B I , Greco M and Troyan S I 2011 *Eur. Phys. J.* C **71** 1750; 2012 *Eur. Phys. J.* C **72** 1953
[10] Ermolaev B I, Greco M and Troyan S I 2015 *Eur. Phys. J.* C **75** 306.

Journal of Physics: Conference Series **678** (2016) 012009 doi:10.1088/1742-6596/678/1/012009

Photo-absorption sum rules σ_{-1} in different environments

S.B. Gerasimov

BLTPh, Joint Institute for Nuclear Research, Dubna, Russia

E-mail: gerasb@theor.jinr.ru

Abstract. Combining the spin-dependent dispersion GDH-sum rule, the iso-vector,photo-excitation states projecting Cabibbo-Radicati sum rule, and the relativistic dipole-moment-fluctuation (or generalized Gottfried) sum rule with the three valence quark configuration of nucleons taken into account for the composition of the ground and the excited states of the nucleon, the relevant moments of the distribution and correlation functions of the quark electric dipole moment operators in the nucleon ground state are expressed via the experimentally measurable nucleon resonance photo-excitation amplitudes.These functions are of interest for checking detailed quark-configuration structure of the nucleon state vector.

1. Introduction

As is known, the non-relativistic dipole sum rules continue to be a useful tool in the theory of the atomic and nuclear photo-effect.

$$\sigma_n(E1) = \int_{thr}^{\infty} d\omega\ \omega^n\ \sigma_{E1}(\omega)$$

Examples: $n = -2$ $\rightarrow$ Kramers-Heisenberg sum rule (SR) for static electric-dipole polarizability of a given quantum system;
$n = -1$ $\rightarrow$ the bremsstrahlung-weighted SR, dependent of charged-"parton" correlation in a given system;
$n = 0$ $\rightarrow$ the famous Thomas-Reiche-Kuhn SR, known as a precusor of the Quantum Mechanics.
As far as the nucleon is the relativistic 3q system, the relativistic generalization of sum rules is needed, including the spin degrees of freedom and spin-dependent interactions and correlations of partons.

2. QED and Atoms

In what follows we consider relativistic dipole moment fluctuation sum rules in the "valence-parton" approximation, that is neglecting virtual particle-antiparticle configurations in the ground state of the considered systems or diffractively produced in the final states of photo-absorption reactions. The anomalous magnetic moment sum rules express a model-independent correspondence between static properties of a particle (or bound system of particles) and

Content from this work may be used under the terms of the Creative Commons Attribution 3.0 licence. Any further distribution of this work must maintain attribution to the author(s) and the title of the work, journal citation and DOI.
Published under licence by IOP Publishing Ltd

Journal of Physics: Conference Series **678** (2016) 012009 doi:10.1088/1742-6596/678/1/012009

integrals over the photo-absorption spectrum. For particles with the spin $S = 1/2$ the sum rule for the anomalous magnetic moment κ reads [1, 2, 3, 4]

$$\frac{2\pi^2\alpha\kappa^2}{m^2} = \int_{thr}^{\infty} \frac{d\nu}{\nu}(\sigma_p(\nu) - \sigma_a(\nu)) \tag{1}$$

In QED,the validity of the sum rule for free electron was checked in two lowest orders of the perturbation theory [5, 6]. Later on, for the physical reasons, we shall replace κ^2 entering different sum rules just by its integral expression in the GDH sum rule. In particular, for the relativistic dipole moment fluctuation sum rule

$$4\pi^2\alpha[\frac{1}{3} < D^2 > -\frac{\kappa^2}{4m^2}] = \int_{thr}^{\infty} \frac{d\nu}{\nu}\sigma_{tot}(\nu) \tag{2}$$

we get another form to be used later

$$4\pi^2\alpha[\frac{1}{3} < D^2 >] = \int_{thr}^{\infty} \frac{d\nu}{\nu}(\sigma_p(\nu) \tag{3}$$

We apply first derived sum rule to the system of the highly ionized atom Pb^{81+},initiated about half-century ago by J.S. Levinger and co-workers [7, 8, 9]. Using the form of the sum rule with our included term κ_{atom} we reduced deviation between left- and right-hand sides of the sum rule, discovered in earlier works, to one-half percent. Numerically:

$$4\pi^2\alpha\frac{1}{3} < D^2 > [937.2b] - 4\pi^2\alpha(\frac{\kappa}{2M})^2[67.9b] = \int_{thr}^{\infty} \frac{d\nu}{\nu}\sigma_{tot}(\nu)[874b] \tag{4}$$

where for the amomalous magnetic moment of the atom we used the total magnetic moment of electron bound in the lowest S-wave orbit [10]

3. Relativistic Constituent Quark Model and Nucleons

Following formally to the $p_z \to \infty$ techniques derivation of the Cabibbo-Radicati [11]or GDH sum rule [4], we can obtain the relation

$$4\pi^2\alpha(\frac{1}{3} < \vec{D}^2 > -(\frac{\kappa_N}{2m_N})^2) = \int \frac{d\nu}{\nu}\sigma_{tot}^{res}(\nu) \tag{5}$$

We use the definitions

$$\hat{D} = \int \vec{x}\hat{\rho}(\vec{x})d^3x = \sum_{j=1}^{3} Q_q(j)\vec{d_j}, \tag{6}$$

$$\hat{r_1^2} = \int \vec{x}^2\hat{\rho}(\vec{x})d^3x = \sum_{j=1}^{3} Q_q(j)\vec{d_j}^2 \tag{7}$$

The defined operators $Q_q(j)$ and $\vec{d_j}$ are the electric charges and configuration and spin variables of point-like interacting quarks in the infinite-momentum frame of the bound system.

Finally, we relate the electric dipole moment operator correlators sucessively for the proton,the neutron, the pure "isovector-nucleon" and the isovector part of the mean-squared radii operators, which all are sandwiched by the nucleon state vectors in the "infinite - momentum frame", with experimentally measurable data on the resonance parts of the

Journal of Physics: Conference Series **678** (2016) 012009 doi:10.1088/1742-6596/678/1/012009

photoabsorption cross sections on the proton and neutron presently known below ~ 2 GeV [12]. The listed operator mean values are parametrized as follows

$$R_V = \frac{1}{2}(< r_1^2 >_P - < r_1^2 >_N) = \alpha - \frac{1}{2}\beta \tag{8}$$

$$J_P = \frac{1}{3} < \hat{D}^2 >_P = \frac{8}{27}\alpha + \frac{1}{27}\beta + \frac{8}{27}\gamma - \frac{8}{27}\delta \tag{9}$$

$$J_N = \frac{1}{3} < \hat{D}^2 >_N = \frac{2}{27}\alpha + \frac{4}{27}\beta + \frac{2}{27}\gamma - \frac{8}{27}\delta \tag{10}$$

$$J_V = \frac{1}{3} < \hat{D}^2 >_V = \frac{2}{3}\alpha + \frac{1}{3}\beta + \frac{2}{3}\gamma - \frac{4}{3}\delta \tag{11}$$

where $< \vec{d_1}^2 >=< \vec{d_2}^2 >= \alpha$,$< \vec{d_3}^2 >= \beta$, $< \vec{d_1} \cdot \vec{d_2} >= \gamma$, $< \vec{d_1} \cdot \vec{d_3} >=< \vec{d_2} \cdot \vec{d_3} >= \delta$ indices "1" and "2" refer to the like quarks (i.e. to the $u(d)$- quarks inside the proton (neutron)), and "3" to the odd quark. Evaluation of the relativistic electric dipole moment fluctuation and the isovector charge radius sum rules for the nucleon was carried out with the available compilation [12] of the resonance pion-photoproduction data on the proton and neutron with the helicity $\lambda = 1/2(3/2)$ $A_{1/2}^{P(N)}$ and $A_{3/2}^{P(N)}$ and all integrals over photoexcited nucleon resonances were taken in the narrow resonance approximation:

$$J_{p(a)}^{res} \simeq \frac{4\pi m_n |A_{3/2(1/2)}^{res}|^2}{m_{res}^2 - m_n^2}, \tag{12}$$

where $m_{n(res)}$ is the nucleon (or resonance) mass. Solving the system of the linear equations and evaluating the $R_V, J_{P,N,V}$ with the help of experimentally known partial amplitudes of main photo-excited resonances, we find our final results for the numerical values α, β and the opening angle θ_{12} and θ_{13} between vectors $\vec{d_1}$ and $\vec{d_2}$ and vectors $\vec{d}_{1(2)}$ and $\vec{d_3}$:

$$\alpha^{1/2} = 0.75 \pm 0.06 fm \tag{13}$$

$$\beta^{1/2} = 0.77 \pm 0.12 fm \tag{14}$$

$$\theta_{12} \simeq 120^0 \tag{15}$$

$$\theta_{13} \simeq \theta_{23} \sim 120^0 \tag{16}$$

$$< r_1{}^2 >_V = 0.25 \pm 0.02 fm^2 (exp : .29 fm^2) \tag{17}$$

Having got the values of the "global" parameters α, β, etc. defined in the infinite momentum frame, we put all three vectors with the pertinent numerical parameters into the plane transverse to $p_z \to \infty$ momentum and define parameters of the closed contour circumscribing their ends. The resulting curve is either ellipse with very small excentricitet or the circle. This fact is, in our opinion, at variance with rather popular "quark-diquark" models of the nucleon with, presumably, more noticeable asymmetry of the spatial correlations of the uu- and ud - quark pairs.

4. Concluding Remarks

1. The properness was demonstrated of using the current-algebra and dispersion relation based relativistic sum rules for establishing new relations between different characteristics of the electron bound in the hydrogen-like, highly-charged ions.

2. With the accuracy about 15%, the isovector charge radius of the nucleon was calculated via the Cabibbo-Radicati sum rule through the integral of total resonance photo-production of

hadrons on the nucleons. The mentioned deficit should be referred to the "pion-sea" presence in the nucleon state vector.

3. According to our suggested Ansatz for the including of the interference of photo-absorption amplitudes on different quarks, their effect is very significant and it is strongly diminishes the difference of $J_P - J_N$ calculated for the real photo-absorption. For the virtual photon absorption in the reactions of electro-production hadrons on nucleons, the more strong suppression of the interference amplitudes in the "deep-inelastic" region will result in the enhancement of mentioned difference, approaching the result to the deep-inelastic version of the generalized Gottfried sum rule [13, 14].

Acknowledgments *The author would like to thank Dr. O.V. Selyugin for fruitful discussions and help in preparation of the paper.*

[1] S.B. Gerasimov, Yad. Phys. **2** (1965) 598; [Sov. J. Nucl. Phys. (1966) 430;
[2] S.B. Gerasimov, Yad. Phys. **5** (1967) 1263;
[3] S.D. Drell and A.C. Hearn, Phys. Rev. Letters, **16** (1966) 908;
[4] M. Hosoda and K. Yamamoto, Prog. Theo. Phys., **36** (1966) 425;
[5] S.B. Gerasimov and J. Moulin, Nucl. Phys. B **98** (1975) 349;
[6] D.A. Dicus and R. Vega, Phys. Lett. B **501** (2001) 44;
[7] W.B. Payne and J.S. Levinger, Phys. Rev. **101** (1956) 1020;
[8] J. S. Levinger and M.L. Rustgi, Phys. Rev. **103** (1956) 439;
[9] J.S. Levinger, M.L. Rustgi and K. Okamoto, Phys. Rev. **103** (1957) 1191;
[10] S.B. Gerasimov, Czech. J. Phys. **55** (2005) A-2009;
[11] N. Cabibbo and L.A. Radicati, Phys. Lett. **19** (1966) 687;
[12] Particle Data Group (Olive, K.A. et al.) Chin. Phys. C**38** (2014) 090001;
[13] K. Gottfried, Phys. Rev. Lett. **18** (1967) 1174;
[14] A.L. Kataev, arXiv: 0311091;

Journal of Physics: Conference Series **678** (2016) 012010 doi:10.1088/1742-6596/678/1/012010

Gravitational lensing and polarization in astrophysics

Alexander F. Zakharov[1,2,3,4]

[1]Institute of Theoretical and Experimental Physics, B. Cheremushkinskaya 25, 117218 Moscow, Russia
[2]Bogoliubov Laboratory of Theoretical Physics, Joint Institute for Nuclear Research, 141980 Dubna, Russia
[3]National Research Nuclear University MEPhI (Moscow Engineering Physics Institute), 115409, Moscow, Russia
[4]North Carolina Central University, Durham, NC 27707, USA

E-mail: zakharov@itep.ru

Abstract. We discuss astrophysical problems connected with gravitational lensing where current and future polarimetric observations are very important to clarify theoretical models. Namely, we consider polarization observations for exoplanet searches with gravitational microlensing, optical polarization for gravitational lens systems, polarization signatures of cosmological gravitational waves.

1. Basics of gravitational lensing

There are a number of books and reviews on gravitational lensing and microlensing [1, 2, 3, 4, 5, 6, 7, 8], however we have to remind basics of gravitational lensing.

A science in a modern sense started since I. Newton investigations and corresponding monographs. In particular, in his book [9], he described a number of experiments to establish laws of optics and he listed queries which have to be confirmed (or disprove) with future experiments. His first query (hypothesis) was the following one [9]. "Query 1. Do not not bodies act upon light at a distance and by their action bend its rays, and is not this action strongest at the least distance?" Since in the Newton's times scientists (and people) do not use a fundamental system of units, sometimes each county (or country) used their own system of units. After reading a classical book [10], one can understand that based on his corpuscular theory of light, Newton proved a correctness of Query 1 in a mathematical way.[1] However, Newton thought that physics is an experimental science and physical laws have to be confirmed with experiments. Therefore, clearly Newton could derive the following relation for a deflection angle θ

$$\delta\theta = \frac{2GM}{c^2 p}, \tag{1}$$

where M, p, c and G are the gravitating body mass, the impact parameter, the speed of light and the Newton constant, respectively. If $M = M_\odot, p = R_\odot$, then $\theta_\odot = 0.''87$.

[1] The relation was derived but not published by Cavendish [11, 12] and at the first time it was published by J. Soldner [13, 14].

Content from this work may be used under the terms of the Creative Commons Attribution 3.0 licence. Any further distribution of this work must maintain attribution to the author(s) and the title of the work, journal citation and DOI.
Published under licence by IOP Publishing Ltd

In 1915 Einstein found that in the framework of general relativity the deflection angle of light has to be in two times greater in comparison with Newtonian gravity [15],[2] namely

$$\delta\theta = \frac{4GM}{c^2 p}. \tag{2}$$

As it was mentioned earlier [1], in Newton's times the estimate for speed of light was much smaller $c = 2 \times 10^{10}$ cm/c, therefore, assuming that Newton had an opportunity to check Eq. (1), he could confirm the relation, in spite of the fact that the factor 2 has been lost because now we know that the prediction of general relativity Eq. (2) has been confirmed at the first time in 1919 [17, 18]. One of the first astrophysical applications of light deflection or gravitational lensing has been considered by O. Chwolson [19].[3]

2. Caustics and magnification

As is shown [2], the gravitational lens equation may be written in the following form: when the distance between a source and an observer is D_s, the distance between gravitational lens and an observer is D_d , D_{ds} is the distance between the gravitational lens and a source. If we suppose the small angle of deflection then we have the following simple expression for lens equation

$$\boldsymbol{\eta} = D_s\boldsymbol{\xi}/D_d - D_{ds}\boldsymbol{\beta}(\boldsymbol{\xi}), \tag{3}$$

where the vectors $(\boldsymbol{\eta}, \boldsymbol{\xi})$ define the coordinates in lens plane and in source plane respectively,

$$\boldsymbol{\beta}(\boldsymbol{\xi}) = \frac{4G}{c^2}\int \rho(\boldsymbol{R})\frac{(\boldsymbol{r}-\boldsymbol{R})}{|\boldsymbol{r}-\boldsymbol{R}|^2}dXdY, \tag{4}$$

where $\boldsymbol{R} = \{X, Y\}$ is the point vector in lens plane, $\rho(\boldsymbol{R})$ is the surface mass density of the gravitational lens. We introduce the following variables [2]

$$\boldsymbol{x} = \boldsymbol{\xi}/R_0, \quad \boldsymbol{y} = D_s\boldsymbol{\eta}/(R_0 D_d), \tag{5}$$

where $R_0 = \sqrt{2r_g D_d D_{ds}/D_s}$ is the Einstein - Chwolson radius. We also introduce the following notation for scaled [2] angle

$$\boldsymbol{\alpha} = \boldsymbol{\beta} D_{ds} D_d/(D_s R_0). \tag{6}$$

In gravitational lensing, the surface mass density is normalized with the critical surface mass density [23]

$$\rho_{cr} = \frac{c^2 D_s}{4\pi G D_d D_{ds}}.$$

For typical lensing situations the critical surface mass density is of the order $\rho_{cr} = 10^4 M_\odot pc^{-2}$ [23]. Therefore, if we define the scaled surface mass density by the following expression

$$\sigma = \rho/\rho_{cr},$$

[2] Earlier, Einstein derived Eq.(1) and asked his assistant E. Freundlich to check the prediction in 1914 [2]. The Soldner's result has been practically forgotten for a century.

[3] Chwolson described circular images [19] and Einstein obtained basic expressions for gravitational lensing [20]. Moreover, it was found that Einstein analyzed gravitational lens phenomenon in his unpublished notes in 1912 [21, 22].

Journal of Physics: Conference Series **678** (2016) 012010 doi:10.1088/1742-6596/678/1/012010

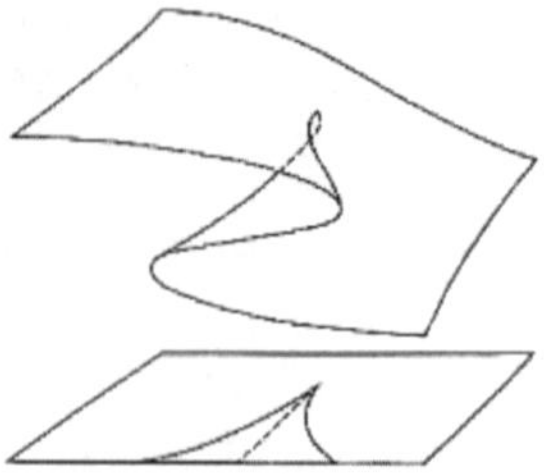

Figure 1. Cusp type singularity as a result of projection of two-dimensional surface onto plane. A touching the two folds on the plane gives the cusp.

then we have the expression for the angle $\boldsymbol{\alpha}$

$$\boldsymbol{\alpha}(\boldsymbol{x}) = \int \frac{\sigma(\boldsymbol{x'})(\boldsymbol{x} - \boldsymbol{x'})}{|\boldsymbol{x} - \boldsymbol{x}'|^2} d^2\boldsymbol{x}. \tag{7}$$

As it was shown [24], we may introduce the scalar potential ψ, such, that

$$\boldsymbol{\alpha}(x) = \nabla\psi(\boldsymbol{x}), \tag{8}$$

where

$$\psi(\boldsymbol{\xi}) = \int \sigma(\boldsymbol{x}') \ln |\boldsymbol{x} - \boldsymbol{x}'| d^2\boldsymbol{x}. \tag{9}$$

Then

$$\boldsymbol{y} = \boldsymbol{x} - \boldsymbol{\alpha}(\boldsymbol{x}). \tag{10}$$

If we introduce

$$\phi(\boldsymbol{x}, \boldsymbol{y}) = \frac{(\boldsymbol{x} - \boldsymbol{y})^2}{2} - \psi(\boldsymbol{x}). \tag{11}$$

Then we can write the lens equation in the expression [24]

$$\nabla\phi(\boldsymbol{x}, \boldsymbol{y}) = 0. \tag{12}$$

It is easy to see that mapping $x \mapsto y$ is Langrange's mapping [25], since that is gradient mapping [26]. Really, if we consider the function

$$S = \frac{\boldsymbol{x} * \boldsymbol{x}}{2} - \psi(\boldsymbol{x}), \tag{13}$$

then $\boldsymbol{y} = \nabla S$ Singularities of Lagrange's mappings are described [27, 26], and their bifurcations are described [28]. Stable singularities for two-dimensional Lagrange mappings are only folds and cusps [27, 29]. Therefore, a point singularity arisen in point mass model of gravitational lens (the Schwarzschild gravitational lens) is unstable and disappeared under small perturbations [5]. A formation of a cusp singularity as a projection of two dimensional surface on plane is shown in Fig. 1.

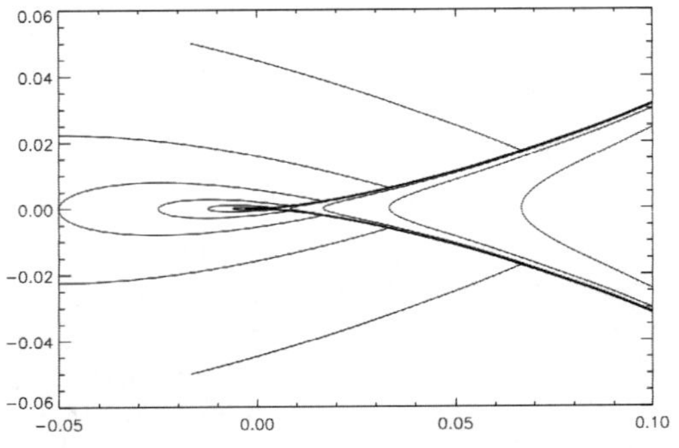

[t]

Figure 2. Contours for magnification near cusp type singularities. Contours are $10 \times 2^i, \forall i \in [-1, 14], i \in Z$ [30].

Eq. (10) defined the mapping of points on the lens plane into points on the source plane. Using the Jacobian matrix we define the local mapping $\boldsymbol{y} = A\delta\boldsymbol{x}$, where

$$A = \begin{pmatrix} \dfrac{\partial y_1}{\partial x_1} & \dfrac{\partial y_1}{\partial x_2} \\ \dfrac{\partial y_2}{\partial x_1} & \dfrac{\partial y_2}{\partial x_2} \end{pmatrix} = \begin{pmatrix} 1 + \psi_{11} & \psi_{12} \\ \psi_{21} & 1 + \psi_{22} \end{pmatrix}, \tag{14}$$

where

$$\psi_{ij} = \frac{\partial^2 \psi}{\partial x_i \partial x_j}(i, j = 1, 2). \tag{15}$$

The magnification of an image at $\boldsymbol{x}$ is

$$\mu(\boldsymbol{x}) = 1/\det A(\boldsymbol{x}) \tag{16}$$

since of square for transformation is determined by the Jacobian matrix. The singular points of mapping are characterized by $\det A(\boldsymbol{x}) = 0$ (where the mapping is not one-to-one). The set is formed by so-called critical curves [2]. Images of critical curves are caustics. The most amplified images are for sources located on caustics because an amplification for these cases is infinite, however, since sources have finite sizes but not infinitesimal ones amplifications for finite sources located on (or near) caustics is high but not infinite (corresponding integrals from singular functions are convergent). One can find asymptotic relations for magnification near folds [2] and cusps [30].

3. Searches of exoplanets with gravitational microlensing

3.1. Light curve analysis

In October 1995 the first extrasolar planet has been found [31]. In October 2015 almost 2000 exoplanets (1969 planets in 1249 planetary systems including 490 multiple planetary systems) have been discovered.[4] Majority of exoplanets have been found searching transits with space borne telescope Kepler. Exoplanet search are one of the hottest topic of modern astronomy because here we have an intersection of many different branches of science including not only astronomy and astrophysics but also astrobiology including abundance and origin of life [32].

[4] http://exoplanet.eu/catalog/.

Journal of Physics: Conference Series **678** (2016) 012010 doi:10.1088/1742-6596/678/1/012010

Already before the discovery of first exoplanets Mao and Paczynski showed how efficient is gravitational microlensing as a tool to search for extrasolar planets, including the low mass ones, even at relatively large distances from their host stars [33]. Since there is a proper motion in a system consisting of a source, a lens and an observer, a source is passing areas with different amplification of a gravitational lens (sometimes a source is passing areas with high amplifications near caustics), therefore there are variation of light curves. Later on, observations and simulations gave the opportunity to confirm the robustness of these conclusions. Exoplanets near the snow line may be also detected with this technique as it was shown, for instance, in Fig. 8 presented in [34]. In 2006 one of the lightest exoplanet with mass around 5.5 $M_\oplus$ has been found [35] with gravitational microlensing and later it has been proven that in average each star has one or more exoplanet [36] (practically such an assumption has been used for simulations of possible detection of exoplanets with pixel lensing [37].

3.2. Polarization curve analysis

For extended sources, the importance of polarization measurements was pointed out for point-like lens in [39] and for binary lens in [40]. For point-like lens polarization could reach 0.1% while for binary lens it could reach a few percent since the magnification gradient is much greater near caustics. It has been shown that polarization measurements could resolve degeneracies in theoretical models of microlensing events [40]. Calculations of polarization curves for microlensing events with features in the light curves induced by the presence of an exoplanet and observed towards the Galactic bulge have been done [41, 42, 43, 44, 45]. Here we emphasize that measurements of then polarization angle could give additional information about the gravitational microlensing model.[5] If polarization measurements are possible, in principle, one could measure polarization as a function of direction for an orientation of polarimeter and an instant for microlensing event. If a polarimeter is fixed one has an additional function of time to explain observational data, but if a polarimeter could be rotated, polarization is an additional function of direction at each instant. Such an information could help to resolve degeneracies and confirm (or disprove) microlensing models for observed phenomena.

4. Polarization in CMB and gravitational waves

CMB (cosmic microwave background) radiation is polarized. The CMB radiation is polarized because it was scattered off of free electrons during decoupling. The quadrupole anisotropy (which produced CMB polarization) could arise from three types of perturbations: Scalar (due to density fluctuations), Vector (due to vorticity induced by defects/strings), Tensor (due to gravity waves). The polarization pattern in the sky can be decomposed into two components: Curl-free component, called "E-mode" (electric-field like) or "gradient-mode", with no handedness Grad-free component, called "B-mode" (magnetic-field like) or "curl-mode", with handedness. The E-mode may be due to both the scalar and tensor perturbations, but the B-mode is due to only vector or tensor perturbations because of their handedness. One should mention that gravitational lensing can transform E-mode into B-mode. In 2014 BICEP2 collaboration claimed the discovery of primordial gravitational waves (or initial tensor perturbations) at a level $r \approx 0.2$ [46], however, joint analysis of BICEP2 and Planck collaborations showed that $r < 0.12$ and the initial result of BICEP2 collaboration has been caused by cosmic dust [47]. The dust originates B-mode of CMB and initially this systematic effect was underestimated.

References

[1] Bliokh P V and Minakov A A 1989 *Gravitational Lenses* (Kiev: Naukova Dumka) (in Russian)
[2] Schneider P, Ehlers J and Falco E E 1992 *Gravitational Lenses* (Berlin: Springer)

[5] We call polarization angle the angle which corresponds to a direction with the maximal polarization.

[3] Blandford R and Narayan R 1992 *Ann. Rev Astron. Astrophys.* **30** 311
[4] Zakharov A F 1997 *Gravitational Lenses and Microlenses* (Moscow: Yanus) (in Russian)
[5] Zakharov A F and Sazhin M V 1998 *Physics – Uspekhi* **41** 945
[6] Petters A O, Levine H and Wambsganss J 2001 *Singularity theory and gravitational lensing* (Boston: Birkhäuser)
[7] Mollerach S and Roulet E 2002 *Gravitational lensing and microlensing* (Singapore: World Scientific)
[8] Zakharov A F 2008 *Phys. Part. Nucl.* **39** 1176
[9] Newton I 1717 *Opticks or A Treatise of the Reflections, Refractions, Inflections & Colors of Light* (London).
[10] Newton I 1687 *Philosophiae Naturalis Principia Mathematica (Mathematical Principles of Natural Philosophy)* (London).
[11] Will C 1988 *Am. J. Phys.* **56** 413
[12] Trimble V 2001 *Gravitational Lensing: Recent Progress and Future Goals, ASP Conference Proceedings* vol 237, ed T G Brainerd and C S Kochanek (San Francisco, Astronomical Society of the Pacific)
[13] Soldner J G 1804 *Astronomisehes Jahrbuch für das Jahr 1804* 161
[14] Jaki S 1978 *Found. Phys.* **8** 927
[15] Einstein A 1915 *Sitzungsber. preuss. Akad. Wiss.* **47** 831
[16] Einstein A 1911 *Annalen der Physik* **340** 898
[17] Eddington A S 1919 *The Observatory* **42** 119
[18] Dyson F W, Eddington A S and Davidson C 1920 *Phil. Trans. Roy. Soc. London. Series A* **220** 291
[19] Chwolson O 1924 *Astronom. Nachrichten* **221** 329
[20] Einstein A 1936 *Science* **84** 506
[21] Renn J, Sauer T and Stachel J 1997 *Science* **275** 184
[22] Sauer T 2008 *Arch. Hist. Exact Sci.* **62** 1
[23] Wambsganss J 1990 *Gravitational Microlensing* Dissertation der Fakultät für Physik der Ludwig - Maximilians - Universität, München
[24] Schneider P 1985 *Astron. & Astrophys.* **143** 413
[25] Arnold V I 1978 *Mathematical Methods of Classical Mechanics* (Berlin: Springer)
[26] Arnold V I 1983 *Russian Math. Surveys* **38** 87
[27] Arnold V I 1972 *Funct. Anal. Appl.* **6** 254
[28] Arnold V I 1976 *Comm. Pure Appl. Math.* **29** 557
[29] Arnold V I 1992 *Catastrophe theory* (Berlin: Springer)
[30] Zakharov A F 1995 *Astron. & Astrophys.* **293** 1
[31] Mayor M and Queloz D 1995 *Nature* **378** 355
[32] Perryman M 2014 *The Exoplanet Handbook* (Cambridge: Cambridge University Press)
[33] Mao S, Paczynski B 1991 *Astrophys. J.* **374** L37
[34] Mao S 2012 *Research Astron. Astrophys.* **12** 947
[35] Beaulieu J-P, Bennett D P, Fouqué P, Williams A *et al* 2006 *Nature* **439** 437
[36] Cassan A, Kubas D, Beaulieu J-P, Dominik M *et al* 2012 *Nature* **481** 167
[37] Ingrosso G, De Paolis F, Nucita A A, Strafella F, Calchi Novati S, Jetzer Ph and Zakharov A 2009 *Mon. Not. R. Astron. Soc.*, **399** 219
[38] Ingrosso G, De Paolis F, Nucita A A, Strafella F, Calchi Novati S, Jetzer Ph and Zakharov A 2010 *Gen. Rel. Grav.* **43** 1047
[39] Bogdanov M B, Cherepashchuk A M, Sazhin M V 1996 *Astrophys. & Space Science* **235** 219
[40] Agol E 1996 *Mon. Not. R. Astron. Soc.* **279** 571
[41] Ingrosso G, De Paolis F, Nucita A A, Strafella F, Calchi Novati S, Jetzer Ph and Zakharov A 2012 *Mon. Not. R. Astron. Soc.*, **426** 1496
[42] Ingrosso G, De Paolis F, Nucita A A, Strafella F, Calchi Novati S, Jetzer Ph, Liuzzi G and Zakharov A, 2014 *Physica Scripta* **89** 084001
[43] Zakharov A, Ingrosso G, De Paolis F, Nucita A A, Strafella F, Calchi Novati S and Jetzer Ph 2014 *Adv. Space Res.* **54** 1319
[44] Zakharov A, Ingrosso G, De Paolis F, Nucita A A, Strafella F, Calchi Novati S and Jetzer Ph 2014 *Proc. of the International Astronomical Union, Symposium S293 (Formation, Detection, and Characterization of Extrasolar Habitable Planets)* **8** (Cambridge: Cambridge University Press) p. 36
[45] Zakharov A, Ingrosso G, De Paolis F, Nucita A A, Strafella F, Calchi Novati S and Jetzer Ph 2014 *proceedings of the XV Advanced Research Workshop on High Energy Spin Physics (DSPIN-13)*, ed. A.V. Efremov and S.V. Goloskokov, p. 155 (Dubna:JINR)
[46] Ade P A R, Aikin R W, Barkats D *et al.* 2014 *Phys. Rev. Lett.* **112** 241101
[47] BICEP2/Keck and Planck Collaborations, Ade P A R, Aikin R W, Barkats D *et al.* 2015 *Phys. Rev. Lett.* **114** 101301

Journal of Physics: Conference Series **678** (2016) 012011 doi:10.1088/1742-6596/678/1/012011

Possible measurements of the spin one observables in elastic dN, dd collisions at the NICA deuteron beams

V I Sharov

Veksler and Baldin Laboratory of High Energy Physics, Joint Institute for Nuclear Research, Dubna, Russia

E-mail: vsharov41@mail.ru

Abstract. The report shows the possibilities of studying the spin one observables in the elastic dN and dd interactions at the $NICA$ collider of the $VBLHEP$ $JINR$. The use of the colliding deuteron beams would allow us to carry out the measurements of the differential cross sections $I^0(dN, dd)$ of the elastic scattering of unpolarized deuterons and the differential cross sections $I^{pol}(dN, dd)$ and the vector $A_y(E_d, \theta)$ and tensor $A_{yy}(E_d, \theta)$ and $A_{xx}(E_d.\theta)$ analyzing powers in elastic collisions of the vector and tensor polarized deuterons. The planned luminosity of the colliding polarized deuteron beams will provide sufficiently high elastic events counting rate. The use of the colliding beams of the polarized deuterons for the spin one dN and dd observables research has a number of significant advantages in comparison with the experiments with the "fixed" target. The angular acceptance of the collider detector covers the full solid angle 4π radians while the wide ranges of the energies of the dN, dd interactions and the 4-momentum transfer squared are available.

1. Introduction.

Spin physics has been a traditional experimental and theoretical field of fundamental research at $JINR$. Therefore, one of the main physical tasks for the NICA collider will be research in the particle spin physics [1] as a continuation of the JINR research programme in this field. Obtaining new data on the spin-dependent observables in the elastic NN, dN and dd scattering at the $T_{N,d\,Lab}$ energies above 1 GeV could be one of the priority areas of the $VBLHEP$ research at the collider.

In our reports on previous conferences [2, 3] we briefly showed the possibilities of measuring the spin-dependent NN observables with the colliding beams of polarized protons and deuterons at the $NICA$ collider. The requirements to the parameters of colliding beams, detectors and appropriate infrastructure at $NICA$ had been listed. In this report, we will show the opportunities for the spin one dN and dd observables research using the oncoming bunches deuterons of the $NICA$ collider. The use of the colliding beams of the polarized deuterons for the spin one observables research has a number of significant advantages in comparison with the experiments with the "fixed" target.

The angular acceptance of the collider detectors covers the full solid angle 4π radians while the wide ranges of the energies of the dN, dd interactions and the 4-momentum transfer squared are available. These advantages allow us to obtain simultaneously a complete set of the energy and angular dependencies of the vector $A_y(E_d, \theta)$ and tensor $A_{yy}(E_d, \theta)$, $A_{xx}(E_d, \theta)$ components

Content from this work may be used under the terms of the Creative Commons Attribution 3.0 licence. Any further distribution of this work must maintain attribution to the author(s) and the title of the work, journal citation and DOI.
Published under licence by IOP Publishing Ltd

of the analyzing powers for the dN, dd reactions in the energy region of the colliding deuteron beams of $NICA$.

A series of unpolarized experiments on deuteron-deuteron and deuteron-proton elastic scattering have been performed at *Dubna*, ANL, *Serpukhov*, $CERN$, *Fermilab* and *Brookhaven*. The investigated kinematical range covers colliding energy from $\sim 1\,GeV$ to $400\,GeV$ and 4-momentum transfer squared region from $\sim 0.08\ (GeV/c)^2$ to $1.4\ (GeV/c)^2$. Data from dp and dd elastic reactions at energy above a few GeV may be used to study aspects of deuteron structure and to probe features of the $2N$ and $3N$ interactions. The most widely used relativistic nuclear scattering model, the relativistic impulse approximation (RIA) and the Glauber multiple-scattering theory have been successfully applied to calculate the differential cross sections up to the ISR energies and spin one observables in dd, dp elastic scattering up to a few GeV.

Because of the comparatively simple structure of deuteron, a relatively complete theoretical treatment of the dp and dd scattering processes is feasible. Due to the spin $1-1/2$ structure of the dp system and availability of both polarized proton and polarized deuteron beams and targets, a large number of spin observables may be measured in elastic scattering experiments. A prime objective of the proposed measurements is to obtain a sufficiently large number of independent dp spin observables, measured with sufficient statistical accuracy. These data allow, in a model independent manner, reconstruction of the dp elastic scattering amplitude. At this time, we are still a long way from reaching this objective. Therefore, comparison between the experiment and the theoretical models has to be made on the level of observables rather than the scattering amplitude.

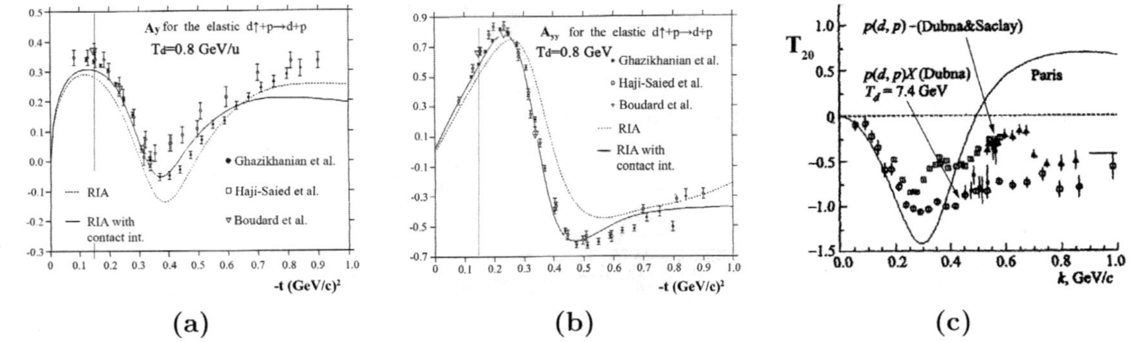

Figure 1. **(a)** Vector analyzing power A_y versus $t_{1,3}$ for the elastic dp scattering at $T_d = 0,8\,GeV$ [4, 5]. **(b)** Tensor analyzing power A_{yy} versus $t_{1,3}$ for the elastic dp scattering at $T_d = 0,8\,GeV$ [4, 5]. **(c)** Tensor analyzing power T_{20} for the elastic dp backward scattering in comparison with the same for $dp- > pX$ break-up [6, 7, 8].

Experiments on the measurement of the dp and dd spin one observables have been planned and carried out using polarized deuteron beams at Dubna accelerators (Synchrophasotron, Nuclotron), Saclay (SATURNE II) and ANL ZGS. As an example, below we show some experimental results of measurements of the vector and tensor analyzing powers in the elastic dp scattering.

2. Formalism of the elastic spin one observables.

For a two-body reaction characterized by the scattering matrix M and initial ρ_{in} and outgoing ρ_{out} spin density matrixes, the intensity I^{pol} of outgoing particles is given by the Cartesian spin tensors as [9, 10]:

$$I^{pol} = Tr\rho_{out} = TrM\rho_{in}M^{\dagger} = I^0[1 + \frac{3}{2}p_yA_y + \frac{1}{2}p_{zz}A_{zz} + \frac{2}{3}p_{xz}A_{xz} + \\ + \frac{1}{6}(p_{xx} - p_{yy})(A_{xx} - A_{yy})]. \quad (1)$$

In this expression both the polarizations (p_i, p_{ij}) and analyzing powers (A_i, A_{ij}) are referred to the right-handed coordinate system having z-axis along the momentum vector $\vec{k}_{in}$ direction of the input particle, and y-axis along the vector product $\vec{k}_{in} \times \vec{k}_{out}$. In the following, this coordinate system will be referred to as S.

The colliding deuteron beams are polarized in a polarized ion source where the quantization axis is oriented vertically along the z axis of the source system. Denoting the tensor moments which express the beam polarization referred to the spin symmetry axis z at the polarized ion source by (p_z, p_{zz}), the analyzing power moments (A_i, A_{ij}) in S are given by the transformation expression as: $(A_i, A_{ij}) = (p_z, p_{zz})D^i_j(\Phi, \Theta, \Psi)$, where Φ, Θ and Ψ are the Euler angles rotating coordinate system of the (p_z, p_{zz}) into S. The angle between z and vector $\vec{k}_{in}$ is β, and φ denotes the angle between the projection of z on the xy-plane of S and $\vec{k}_{in} \times \vec{k}_{out}$. In terms of β and φ, the Euler angles which carry the polarized ion source system into S are $\Phi = 0$, $\Theta = -\beta$ and $\Psi = -(90^\circ + \varphi)$.

Using these Euler angles values we finally obtain

$$I^{pol} = I^0[1 + 1.5p_zA_ysin\beta cos\varphi + 0.25p_{zz}A_{zz}(3cos^2\beta - 1) - \\ - p_{zz}A_{xz}sin\beta cos\beta sin\varphi - 0.25p_{zz}(A_{xx} - A_{yy})sin^2\beta cos2\varphi]. \quad (2)$$

We write now the expressions of the particle outputs for specific cases of scattering at the angle θ to the left (L direction $+x$), right (R direction $-x$), up (U direction $+y$) and down (D direction $-y$) in the coordinate system S. Using the relationship $A_{xx} + A_{yy} + A_{zz} = 0$ for the Cartesian tensor analyzing forces, we can get the following expressions for the outputs to the right, left, up and down:

$$I^{Lpol}(\theta_H) = I^{L0}(\theta_H)[1 + 1.5p_zA_y(\theta_H) + 0.5p_{zz}A_{yy}(\theta_H)], \quad (3)$$

$$I^{Rpol}(\theta_H) = I^{R0}(\theta_H)[1 - 1.5p_zA_y(\theta_H) + 0.5p_{zz}A_{yy}(\theta_H)], \quad (4)$$

$$I^{Upol}(\theta_V) = I^{U0}(\theta_V)[1 + 0.5p_{zz}A_{xx}(\theta_V)], \quad (5)$$

$$I^{Dpol}(\theta_V) = I^{D0}(\theta_V)[1 + 0.5p_{zz}A_{xx}(\theta_V)], \quad (6)$$

where θ_H is the scattering angle in xz-plane for right, left cases, and θ_V is the scattering angle in yz-plane for up, down cases.

For the left-right scattering case we get the asymmetry expression $\mathbf{a} = 3p_zA_y(\theta_H) = I^{Lpol}/I^{L0} - I^{Rpol}/I^{R0}$ and can obtain value of the vector analyzing power $A_y(\theta_H) = \mathbf{a}/(3p_z)$ or the vector component of the deuteron beam polarization $p_z = \mathbf{a}/(3A_y(\theta_H))$. For the sum of the left-right yields we get $I^{Lpol}/I^{L0} + I^{Rpol}/I^{R0} = 2 + p_{zz}A_{yy}(\theta_H)$, and can obtain value of the tensor component of analyzing power $A_{yy}(\theta_H)$ or value of the tensor component of the deuteron beam polarization p_{zz}. For the case of up-down scattering the sum of the yields gives $I^{Upol}/I^{U0} + I^{Dpol}/I^{D0} - 2 = p_{zz}A_{xx}(\theta_V)$, and we can obtain value of the tensor component $A_{xx}(\theta_V)$ of the analyzing power.

Assuming the deuteron is a loosely coupled system, one can apply the above formalism to describe the elastic dd and quasi-elastic dN - deuteron scattering on individual nucleon in the deuteron (See Fig.2a). To measure the elastic dp, dd differential cross section we have to select these elastic events among all inelastic ones. The parameters of the $NICA$ detectors have to provide selection of the elastic dp and dd events.

The polarimetry of the polarized colliding dd beams at the $NICA$ was previously discussed in [2, 3]. Here we only note that the detection of the elastic dp, dd scattering events by the $NICA$ detectors provide the ability to determine the left-right asymmetries of the reaction yields. This will enable us to estimate the polarization values of the colliding beams directly at the interaction point. This could be done both for proton and deuteron colliding beams.

3. Kinematics of the dp and dd elastic collisions.

Before proceeding to the yield evaluations for the elastic dp, dd collisions at the $NICA$ deuteron beams we briefly recall features of the elastic scattering kinematics:

- for experiments with the "fixed" target, in Lab. system ($p_{2\,Lab} = 0$, $E_{2\,Lab} = m_2$),
- for the system of the colliding beams ($m_1 \neq m_2, \beta_1 = \beta_2, p_p = 0.5 \cdot p_d, E_p = 0.5 \cdot E_d$ when neutron in deuteron is spectator), and
- for the colliding particles center-of-mass ($\vec{p}_p + \vec{p}_d = 0$).

Figure 2b shows the dependencies of the proton and deuteron Lab. kinetic energies $T_{p,d\,Lab}$ from the kinetic energy $T_{p,d\,Col}$ in the collider.

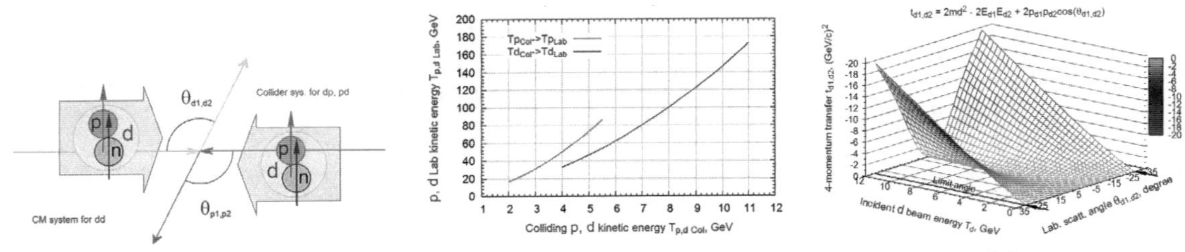

(a) (b) (c)

Figure 2. (a) Schematic view of the quasi-elastic dp scattering in the colliding beams of deuterons. (b) Dependencies of $T_{p,d\,Lab}$ from $T_{p,d\,col}$. (c) Elastic $d_1 + N_1 \rightarrow d_2 + N_2$ collision at Lab. system.

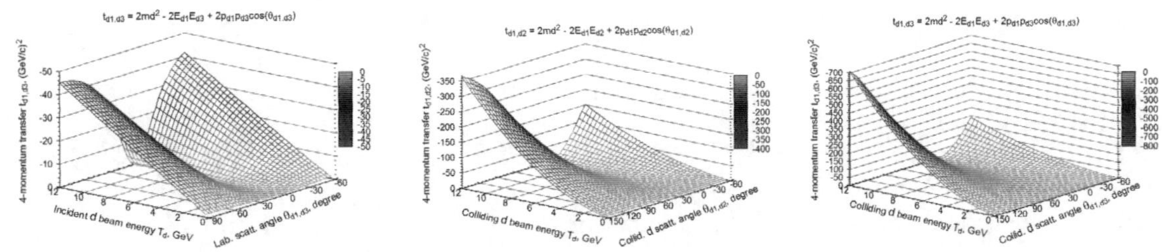

(a) (b) (c)

Figure 3. (a) Elastic $d_1 + d_2 \rightarrow d_3 + d_4$ collision at Lab. system. (b) Elastic $d_1 + N_1 \rightarrow d_2 + N_2$ collision at the $NICA$ collider. (c) Elastic $d_1 + d_2 \rightarrow d_3 + d_4$ collision at the $NICA$ collider (CM system).

The kinematic characteristic ranges for the elastic dp and dd collisions at the Lab, SM, and Col. systems are shown in figures 2c, 3$a \div$ 3c. The dependencies of the four-momentum transfer squared $t_{1,3}$ from the beam deuteron energy $T_{d\,Lab,CM}$ and from deuteron scattering angle $\theta_{d\,lab,CM}$ in the Lab, CM, and Col coordinate systems are shown by the three-dimensional figures.

Note a significant benefit of the collisions research in center-of-mass in comparison with experiments with "fixed" targets.

Journal of Physics: Conference Series **678** (2016) 012011 doi:10.1088/1742-6596/678/1/012011

4. The elastic event yields in the *dd* and *dp* collisions.

Unfortunately, the existing *dN* and *dd* data are quite poor, however they will provide a realistic assessment of the outputs of the elastic *dp* and *dd* scattering processes. In Fig. 4 we present the existing data on differential *dp*, *pd*, and *dd* cross-sections at the Center-of-Mass system or/and at the Collider system.

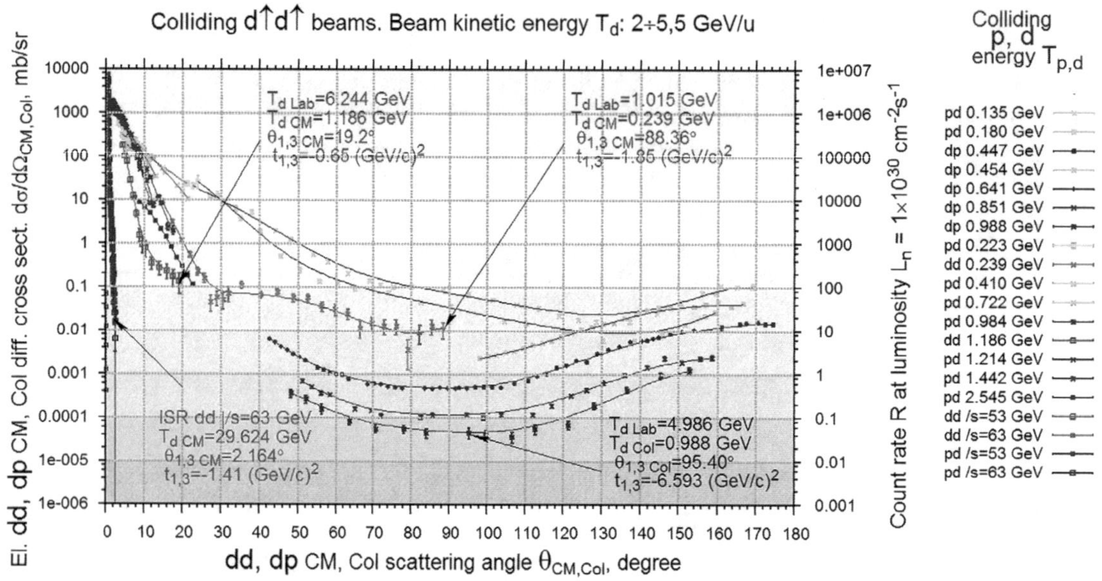

Figure 4. Available experimental data set on the differential cross sections for the elastic *dp* [11–13] (blue symbols), *dd* [14, 15] (magenta symbols), and *pd* [16–21] (green symbols) collisions in the *CM* and Collider systems.

The x-axis represents the deuteron scattering angle θ_d in the CM or $Col.$ systems of the colliding deuterons. The left logarithmic scale represents values of the differential cross sections $d\sigma/\Omega_{CM,Col}(dd.dp, pd)$ in the CM or $Col.$ systems of the colliding deuterons. The legend indicating the symbols representing the $d\sigma/\Omega_{CM,Col}(dd.dp, pd)$ values obtained at a given kinetic energy $T_d\,CM,\,Col$ of the colliding deuterons is shown on the right edge of the Fig. 4. The presented dp, dd, pd data sets are approximated with the Bezier curves of degree n (the number of data points) that connect the endpoints.

The kinematic characteristics of the boundary points of multiple data sets are shown in Fig. 4 by the arrows. There the values of the deuteron kinetic energy $T_{d\,CM\,Col}$, the deuteron scattering angle $\theta_{d\,CM\,Col}$, and the squared four-momentum transfer $t_{d1,d3}$ are specified. These data will facilitate the estimations of the outputs of the elastic *dp* and *dd* processes in the CM or Col systems of the interacting particles.

Right logarithmic scale in Fig. 4 shows the values of the count rate R of the elastic *dp*, *dd* events expected for appropriate values of the cross sections $d\sigma/d\Omega_{CM,Col}$ when the luminosity of the colliding deuteron beams is equal to $L_d = 1 \times 10^{30}\,cm^{-2}s^{-1}$. The use of this scale allows us to quickly get reasonable estimation of the expected counting rates for the elastic events.

5. Conclusion.

The possibilities of measuring the spin-one observables in the elastic *dd* and quasi-elastic *dp* interactions at the colliding $\vec{d}\vec{d}$ beams of the $NICA$ collider were considered. The formalism [9, 10] that specifies the spin-one observables in the elastic *dp* and *dd* scattering was briefly considered.

Journal of Physics: Conference Series **678** (2016) 012011 doi:10.1088/1742-6596/678/1/012011

The kinematic characteristics of the dp and dd scattering processes at the Laboratory system and in the Center of mass system of the colliding particles were analyzed. It is shown that in the elastic dp and dd scattering at the colliding d beams the broader ranges both the energy values of the interacting particles and the square of the four-momentum transfer are achieved in comparison with the experiment using the "fixed" target. Other significant advantages of the measurements at the collider compared with the experiments with the "fixed" targets are the following: the angular acceptance of the collider detector extends to the full solid angle 4π rad and the so-called "target" in such measurements does not contain background impurities, which complicate the processing of the accumulated data. These advantages allow as to obtain simultaneously a complete set of the energy and angular dependencies of the vector and tensor analyzing power $A_y(E_d,\theta)$, $A_{yy}(E_d,\theta)$ and $A_{xx}(E_d,\theta)$ components for the dN, dd reactions in the energy region of the colliding deuteron beams of $NICA$.

Available existing data set [11–21] on the differential elastic dp, dd, and pd cross sections in the collider system or in the CM of the colliding particles were shown. The output estimations of the elastic dp, dd events were presented

The spin one observables data planned to be obtained in the measurements of the elastic dN and dd scattering will promote a creation of an adequate phenomenological and theoretical description of the dN and dd interaction over the $T_{d\,col}$ energy region of $4 \div 11\,GeV$ ($T_{d\,lab}$ $\sim 30 \div 170\,GeV$).

Acknowledgment
The author is grateful to his colleague E.V. Chernykh for the help during the talk and this contribution preparing.

References

[1] *NICA - Nuclotron - based Ion Collider fAcility - a Base for Fundavental Research and Innovation Development*, JINR, Dubna 2014, http://nica.jinr.ru/files/Booklet-NICA.pdf
[2] Sharov V I 2014 *Proc. of the XV Advanced ResearchWorkshop on High Energy Spin Physics* DSPIN-13 JINR E1,2-2014-12 367.
[3] Sharov V I 2014 *Proc. of the XXII International Baldin Seminar on High Energy Physics Problems* PoS(Baldin ISHEPP XXII) 102.
[4] Haji-Saied M et al. 1987 *Phys. Rev.* C **36** No. 5 2010.
[5] Ghazikhanian V et al. 1991 *Phys. Rev.* C **43** No. 4 1532.
[6] Punjabi V et al. 1995 *Phys. Lett.* B **350** 178.
[7] Azhgirey L S et al. 1997 *Phys. Lett.* B **391** 22.
[8] Azhgirey L S et al. 1986 *Phys. Lett.* B **387** 37.
[9] Darden S E 1970 *Proc. of the Third International Symposium, Madison 1970* Ed. by H.H. Barschall and W. Haeberli. The University of Wisconsin Press. 39.
[10] Ohlsen G G 1972 Polarization transfer and spin correlation experiments in nuclear physics *Rep. Prog. Phys.* **35** 717-801.
[11] Glagolev V V et al. 2012 *Eur. Phys. J.* A **48** 182.
[12] Dubal L et al. 1974 *Phys. Rev.* D **9** No.3 597.
[13] Sekiguchi K et al. 2002 *Phys. Rev.* C **65** 034003.
[14] Goggi G et al. 1979 *Nucl. Phys.* B **149** 381.
[15] Goshaw A T et al. 1970 *Phys. Rev. Lett.* **25** No. 4 249.
[16] Dalkhazhav N et al. 1968 *Jour. of Nucl. Phys.* **8** is. 2 342.
[17] Winkelmann E et al. 1980 *Phys. Rev.* **21** No. 6 2535.
[18] Booth N E et al. 1971 *Phys. Rev.* D **4** No. 5 1261.
[19] Coleman E et al. 1967 *Phys. Rev.* **164** No. 5 1655.
[20] Amaldi U et al. 1972 *Nucl. Phys.* B **39** 39.
[21] Korbel Z 1964 *Czech. J Phys.* B **14** 741.

Journal of Physics: Conference Series **678** (2016) 012012 doi:10.1088/1742-6596/678/1/012012

Sivers, Boer-Mulders and transversity distributions in the difference cross sections in SIDIS

Ekaterina Christova

Institute for Nuclear Research and Nuclear Energy, Bulgarian Academy of Sciences, Sofia, Bulgaria

E-mail: echristo@inrne.bas.bg

Elliot Leader

Imperial College, London, UK

E-mail: e.leader@imperial.ac.uk

Abstract. A major experimental program is presently underway to determine the Sivers, Böer-Mulders and transversity distributions, vital for understanding the internal structure of the nucleon. To this end we consider the Sivers, Böer-Mulders and transversity azimuthal asymmetries of the difference cross sections of hadrons with opposite charges in SIDIS reactions with unpolarized and transversely polarized target $l + N \to l' + h + X$, $h = \pi^\pm, K^\pm, h^\pm$. We show that on deuteron target these asymmetries are particularly simple and determine the sum of the valence-quark $Q_V = u_V + d_V$ transverse momentum dependent distributions without any contributions from the strange or other sea-quark functions.

At present, data on these asymmetries are presented for the integrated asymmetries i.e. the x_B- and z_h-dependent asymmetries. If data are available in small bins in Q^2, so that Q^2-dependence can be neglected, these expressions simplify dramatically leading to remarkably simple and powerful tests of the simplifying assumptions used in extracting these functions from the data.

1. Introduction

We consider the transverse momentum dependent (TMD) Sivers, Boer-Mulders (BM) and transversity functions, measured in transversely polarized or unpolarized semi-inclusive deep inelastic scattering (SIDIS) $l + N \to l' + h + X$.

At present, the extraction of these transverse momentum dependent (TMD) functions has been relatively simplistic, involving two key conventions:

a) The analysis is in leading order in perturbative QCD

b) Both the parton distributions (PDFs) and fragmentation functions (FFs) which depend on parton intrinsic momentum $k_\perp$, are typically parametrized under the following conventions: 1) a factorized form for the x_B/z_h and $k_\perp$-dependences:

$$\Delta f(x_B \,\text{or}\, z_h, k_\perp^2) = \Delta f(x_B \,\text{or}\, z_h)\, e^{-k_\perp^2/\langle k_\perp^2\rangle}\,. \tag{1}$$

2) the $x_B(z_h)$-dependence is proportional to the collinear PDFs (FFs), 3) the Q^2-evolution is in the collinear PDFs (FFs) and 4) the $k_\perp$–dependence is Gaussian type and flavour independent.

Content from this work may be used under the terms of the Creative Commons Attribution 3.0 licence. Any further distribution of this work must maintain attribution to the author(s) and the title of the work, journal citation and DOI.
Published under licence by IOP Publishing Ltd

The functional form of the TMD functions is still under discussion and these are the simplest, commonly used at present *standard* parametrizations. Additional assumptions about the sea-quark contributions are made as well. For BM functions it is assumed that they are proportional to Sivers functions.

The goal of our studies is to show that if one uses the so called "difference" asymmetries one can 1) obtain information about the valence-quark TMDs without contributions from the sea-quarks, 2) test the standard parametrization and the assumption about BM function using relations between measurable quantities only. These tests are crutial as they would verify to what extend the used standard parametrizations provide a reliable information on the TMD distributions for the present set of data.

The "difference" cross section asymmetries are combinations of the type:

$$A^{h-\bar{h}} \equiv \frac{\Delta\sigma^h - \Delta\sigma^{\bar{h}}}{\sigma^h - \sigma^{\bar{h}}} \qquad (2)$$

where σ^h and $\Delta\sigma^h = \sigma^{h\uparrow} - \sigma^{h\downarrow}$ are the unpolarized and polarized cross sections respectively. The arrows indicate the polarization of the target, and h and $\bar{h}$ are hadrons with opposite charges.

Previously the difference cross section asymmetries [1, 2] have appeared rather useful in the simple collinear picture and the valence quark helicity parton densities and fragmentation functions were determined directly [3]. Here we extend our studies to the non-collinear picture when transverse momentum dependence is included.

Presently, data on Sivers, Boer-Mulders (BM) and transversity asymmetries are presented as functions of only one of the kinematic variables x_B or z_h, sometimes with, in addition, the Q^2 and P_T dependence, with integration over the measured intervals of the other variables. We obtain analytic expressions for the x_B and z_h dependence of the asymmetries. These expression strongly simplify if in the measured kinematic intervals the binning in Q^2 is small enough to allow the neglect of the Q^2-dependence of the parton densities and FFs. Here we shall present the general formulae for Sivers asymmetries only, and the formulae when this simplification is valid for all asymmetries. Most of the presented results can be found in more details in ref. [4].

We present the asymmetries on deuteron target, when the asymmetries provide information on the sum of the valence-quark TMDs Δf_{Q_V}, $Q_V = u_V + d_V$ and the results appear especially simple. The results on proton targets are obtained with the simple replacement $\Delta f_{Q_V} \to e_u^2 \Delta f_{u_V} - e_d^2 \Delta f_{d_V}$ for final pions and $\Delta f_{Q_V} \to e_u^2 \Delta f_{u_V}$ for final kaons. The explicit expressions for them are given in ref.[4].

2. Sivers functions

The Sivers distribution function $\Delta^N f_{q/S_T}(x_B, k_\perp)$ appears in the expression for the number density of unpolarized quarks q with intrinsic transverse momentum $\mathbf{k}_\perp$ in a transversely polarized proton $p^\uparrow$ with 3-momentum $\mathbf{P}$ and transverse spin $\mathbf{S_T}$ [5]:

$$f_{q/p^\uparrow}(x_B, \mathbf{k}_\perp) = f_{q/p}(x_B, k_\perp) + \frac{1}{2}\,\Delta^N f_{q/S_T}(x_B, k_\perp)\; \mathbf{S_T} \cdot (\hat{\mathbf{P}} \times \hat{\mathbf{k}}_\perp) \qquad (3)$$

where $f_{q/p}(x_B, k_\perp)$ are the unpolarized x_B and $k_\perp$-dependent parton densities and the triple product induces a definite azimuthal $\sin(\phi_h - \phi_S)$-dependence, ϕ_h and ϕ_S are the azimuthal angles of the final hadron and the spin of the target, $x_B = Q^2/2(P.q)$ is the usual Bjorken variable.

To access the Sivers TMDs one considers the $\sin(\phi_h - \phi_S)$ azimuthal moment $A_{UT}^{\sin(\phi_h - \phi_S)}$ of the transverse single-spin asymmetry in SIDIS [6]. In accordance with this we define Sivers asymmetry for the difference cross section analogously:

$$A_{UT}^{Siv,h-\bar{h}}(x_B,Q^2,z_h,P_T^2) = \frac{1}{S_T}\frac{\int d\phi_h d\phi_S \left[d^6\sigma^{\uparrow} - d^6\sigma^{\downarrow}\right]^{h-\bar{h}} \sin(\phi_h-\phi_S)}{\int d\phi_h d\phi_S \left[d^6\sigma^{\uparrow} + d^6\sigma^{\downarrow}\right]^{h-\bar{h}}} \tag{4}$$

where $d^6\sigma^{\uparrow,\downarrow} = d\sigma^{\ell p^{\uparrow,\downarrow}\to\ell' hX}/(dx_B\, dQ^2\, dz_h\, dP_T^2\, d\phi_h\, d\phi_S)$ stands for the differential cross section of SIDIS with an unpolarized lepton beam on a transversely polarized target in the kinematic region $P_T \simeq k_\perp \ll Q$ at order $(k_\perp/Q)$, P_T is the transverse momentum of the final hadron in the $\gamma * -p$ c.m. frame, and z_h, Q^2 and y are the usual measurable SIDIS quantities:

$$z_h = \frac{(P.P_h)}{(P.q)}, \quad Q^2 = -q^2, \quad q = l - l', \quad y = \frac{(P.q)}{(P.l)}, \quad Q^2 = 2MEx_B y \tag{5}$$

with l and l', P and P_h the 4-momenta of the initial and final leptons, and hadrons. Throughout the paper we follow the notation and kinematics of ref. [6].

In general, only the valence-quark TMD functions enter the difference cross sections. When a deuteron target is used, a further simplification occurs – independently of the final hadrons, only one parton density enters – the sum of the valence quarks:

$$\Delta f_{Q_V/S_T}(x_B,k_\perp) \equiv \Delta f_{u_V/S_T}(x_B,k_\perp) + \Delta f_{d_V/S_T}(x_B,k_\perp) \tag{6}$$

Following the convention for the standard parametrizations, we adopt the following form for valence-quark Sivers function:

$$\Delta f_{Q_V/S_T}(x_B,k_\perp) = \Delta f_{Q_V/S_T}(x_B)\,\sqrt{2e}\,\frac{k_\perp}{M_S}\,\frac{e^{-k_\perp^2/\langle k_\perp^2\rangle_S}}{\pi\langle k_\perp^2\rangle} \tag{7}$$

where

$$\Delta f_{Q_V/S_T}(x_B,Q^2) = 2\mathcal{N}_{Siv}^{Q_V}(x_B)\,Q_V(x_B,Q^2) \tag{8}$$

and $Q_V = u_V + d_V$ is the sum of the collinear valence PDFs, and

$$\langle k_\perp^2\rangle_S = \frac{\langle k_\perp^2\rangle\,M_S^2}{\langle k_\perp^2\rangle + M_S^2}\,. \tag{9}$$

The unknowns are M_S (involved in the definition of $\langle k_\perp^2\rangle_S$) and $\mathcal{N}_{Siv}^{Q_V}(x_B)$. They are to be determined in the integrated z_h- and x_B- Sivers difference asymmetries, for which we obtain:

$$A_{UT}^{Siv,h-\bar{h}}(z_h) = \mathcal{B}_{Siv}^h(z_h)\frac{z_h}{\sqrt{\langle p_\perp^2\rangle + z_h^2\langle k_\perp^2\rangle_S}}, \qquad h = \pi^+, K^+, h^+ \tag{10}$$

$$\mathcal{B}_{Siv}^h(z_h) = A_{Siv}\frac{\int dx_B \int dQ^2\, \frac{1+(1-y)^2}{Q^4}\, \Delta f_{Q_V/S_T}(x_B,Q^2)\, D_{u_V}^h(z_h,Q^2)}{2\int dx_B \int dQ^2\, \frac{1+(1-y)^2}{Q^4}\, Q_V(x_B,Q^2)\, D_{u_V}^h(z_h,Q^2)} \tag{11}$$

and

$$A_{UT}^{Siv,h-\bar{h}}(x_B) = \mathcal{C}_{Siv}^h(x_B)\,\mathcal{N}_{Siv}^{Q_V}(x_B), \qquad h = \pi^+, K^+, h^+ \tag{12}$$

$$\mathcal{C}_{Siv}^h(x_B) = A_{Siv}\frac{\int dQ^2\, \frac{1+(1-y)^2}{Q^4}\, Q_V(x_B,Q^2) \int dz_h\, z_h \left[D_{u_V}^h(z_h,Q^2)\right]/\sqrt{\langle P_T^2\rangle_S}}{\int dQ^2\, \frac{1+(1-y)^2}{Q^4}\, Q_V(x_B,Q^2) \int dz_h\, D_{u_V}^h(z_h,Q^2)}\,. \tag{13}$$

Eqs. (10) and (12) determine $\mathcal{N}^{Siv}(x_B)$ and $\langle k_\perp^2\rangle_S$ without any contributions from the sea quarks.

For bins corresponding to a reasonably small interval ΔQ^2 in Q^2, we replace the integral over Q^2 by ΔQ^2 times the Q^2-dependent functions evaluated at the mean value $\bar{Q}^2$ for the bin. Then (10) and (12) become particularly simple:

$$A_{UT}^{Siv,h-\bar{h}}(z_h,\bar{Q}^2) = \bar{B}_{Siv}(\bar{Q}^2)\,\frac{z_h}{\sqrt{\langle p_\perp^2\rangle + z_h^2\langle k_\perp^2\rangle_S}}, \tag{14}$$

$$A_{UT}^{Siv,h-\bar{h}}(x_B,\bar{Q}^2) = \bar{C}_{Siv}^h(\bar{Q}^2)\,\mathcal{N}_{Siv}^{Q_V}(x_B), \tag{15}$$

as $\bar{B}_{Siv}$ and $\bar{C}_{Siv}^h$ are constant factors:

$$\bar{B}_{Siv}(\bar{Q^2}) = A_{Siv}\,\frac{\int dx_B\,[1+(1-\bar{y})^2]\,\Delta f_{Q_V/S_T}(x_B,\bar{Q^2})}{2\int dx_B\,[1+(1-\bar{y})^2]\,Q_V(x_B,\bar{Q^2})}, \quad \forall h \tag{16}$$

$$\bar{C}_{Siv}^h(\bar{Q^2}) = A_{Siv}\,\frac{\int dz_h\,z_h\left[D_{u_V}^h(z_h,\bar{Q^2})\right]/\sqrt{\langle P_T^2\rangle_S}}{\int dz_h\,D_{u_V}^h(z_h,\bar{Q^2})}, \quad h=\pi^+,K^+ \tag{17}$$

Here

$$\bar{y} = \frac{\bar{Q}^2}{2MEx_B}, \quad A_{Siv} = \frac{\sqrt{e\pi}}{2\sqrt{2}}\,\frac{\langle k_\perp^2\rangle_S^2}{M_S\,\langle k_\perp^2\rangle}, \quad \langle P_T^2\rangle_S = \langle p_\perp^2\rangle + z_h^2\langle k_\perp^2\rangle_S \tag{18}$$

Eq. (14) is remarkably strong prediction both for its explicit z_h behaviour, determined solely by the Gaussian dependence on $k_\perp$, and for its independence of h. Eq. (16) implies that Sivers x_B-asymmetries should have the same x_B-behaviour for all final hadrons.

3. Boer-Mulders distributions

The extraction of Boer-Mulders and transversity distributions is more complicated as compared to Sivers parton densities. The reason is that Sivers functions enter the cross section in convolution with the unpolarized TMD fragmentation functions, *known* from multiplicities, while the BM and transversity functions enter the cross sections in convolution with the transversely polarized TMD FFs, the so called Collins functions $\Delta^N D_{h/q^\uparrow}(z_h,p_\perp)$. The latter can, in principle, be extracted from $e^+e^-\to h_1h_2X$, but at present are rather poorly known.

The Boer-Mulders function determines the distribution of transversely polarized quarks $q^\uparrow$ in an unpolarized proton p [7]:

$$\Delta^N f_{q^\uparrow/p}(x_B,k_\perp) \equiv \Delta f_{s_y/p}^q(x_B,k_\perp) = -\frac{k_\perp}{m_p}\,h_1^\perp(x_B,k_\perp)\,\cdot \tag{19}$$

It is accessed measuring the $\cos 2\phi_h$-momentum in unpolarized SIDIS differential cross section. The difference BM asymmetry is defined respectively:

$$A_{BM}^{h-\bar{h}} = \frac{\int d\phi_h\,\cos 2\phi_h\,d^5\sigma^{h-\bar{h}}}{\int d\phi_h\,d^5\sigma^{h-\bar{h}}}\,. \tag{20}$$

On deuteron target, independently of the final hadrons $h-\bar{h}$, it determines only the sum of the valence quark BM distribution:

$$\Delta f_{s_y/p}^{Q_V}(x_B,k_\perp) \equiv \Delta f_{s_y/p}^{u_V}(x_B,k_\perp) + \Delta f_{s_y/p}^{d_V}(x_B,k_\perp) \tag{21}$$

that is parametrized accordingly:

$$\begin{aligned}\Delta f^{Q_V}_{s_y/p}(x_B,k_\perp,Q^2) &= \Delta f^{Q_V}_{s_y/p}(x_B,Q^2)\,\sqrt{2e}\,\frac{k_\perp}{M_{BM}}\,\frac{e^{-k_\perp^2/\langle k_\perp^2\rangle_{BM}}}{\pi\langle k_\perp^2\rangle_{BM}}\\ \Delta f^{Q_V}_{s_y/p}(x_B,Q^2) &= 2\mathcal{N}^{Q_V}_{BM}(x_B)\,Q_V(x_B,Q^2),\end{aligned} \tag{22}$$

where $\mathcal{N}^{Q_V}_{BM}(x_B)$ and M_{BM} are to be determined from the integrated BM asymmetries. Here we present only the x_B-dependent asymmetry. For all final h it determines the same $\mathcal{N}^{Q_V}_{BM}$:

$$A_{BM}(x_B)^{h-\bar{h}} = \mathcal{C}^h_{BM}(x_B)\,\mathcal{N}^{Q_V}_{BM}(x_B), \qquad \forall h \tag{23}$$

where for $h=\pi^+,K^+$ we have:

$$\mathcal{C}^h_{BM}(x_B) \;=\; -4e\,A_{BM}\,A_{Coll}\,\frac{\int\int dQ^2\,dz_h\,\frac{1-y}{Q^4}\,Q_V(x_B)\,z_h\,\Delta^N D_{h/u_V\uparrow}(z_h)\,/\langle P_T^2\rangle_{BM}}{\int\int dQ^2\,dz_h\,\frac{1+(1-y)^2}{Q^4}\,Q_V(x_B)\,D^h_{u_V}(z_h)}\,. \tag{24}$$

For $h=h^\pm$ the relative expression is given in [4].

If Q^2-evolution can be neglected, then $A_{BM}(x_B)^{h-\bar{h}}$ is determined solely by $\mathcal{N}^{Q_V}_{BM}$:

$$A^{h-\bar{h}}_{BM}(x_B,\bar{Q}^2) = \frac{1-\bar{y}}{1+(1-\bar{y})^2}\,\bar{\mathcal{C}}^h_{BM}(\bar{Q}^2)\,\mathcal{N}^{Q_V}_{BM}(x_B), \quad h=\pi^+,K^+,h^+ \tag{25}$$

where $\bar{\mathcal{C}}^h_{BM}$ are constant factors. This implies, in particular, that $A_{BM}(x_B)^{h-\bar{h}}$ have the same x_B-behaviour:

$$A^{\pi^+-\pi^-}_{BM}(x_B) \simeq A^{K^+-K^-}_{BM}(x_B) \simeq A^{h^+-h^-}_{BM}(x_B) \quad \text{or} \quad \frac{A^{\pi^+-\pi^-}_{BM}(x_B)}{A^{h^+-h^-}_{BM}(x_B)} = \ldots = \text{const} \tag{26}$$

4. Relations between BM and Sivers asymmetries

In current analysis an additional simplifying assumption is made, namely the BM function is assumed proportional to its chiral-even partner – the Sivers function [8]. In our case this reads:

$$\Delta f^{Q_V}_{s_y/p}(x,k_\perp) = \frac{\lambda_{Q_V}}{2}\,\Delta f_{Q_V/S_T}(x,k_\perp), \tag{27}$$

where λ_{Q_V} is a fitting parameter. This implies that BM and Sivers x_B-asymmetries measure the same $\mathcal{N}^{Q_V}_{Siv}$ and they are related:

$$\frac{A^{h-\bar{h}}_{BM}(x_B)}{A^{Siv,h-\bar{h}}_{UT}(x_B)} = \frac{\lambda_{Q_V}}{2}\,\mathcal{C}^h(x_B), \quad h=\pi^+,K^+,h^+ \tag{28}$$

where $\mathcal{C}^h(x_B)$ is independent of $\mathcal{N}_{Siv}$:

$$\mathcal{C}^h(x_B) = \frac{-4\sqrt{2e}}{\sqrt{\pi}}\,\frac{\int\int dQ^2\,dz_h\,\frac{1-y}{Q^4}\,Q_V(x_B,Q^2)z_h\,\Delta^N D_{h/u_V\uparrow}(z_h,Q^2)/\langle P_T^2\rangle_{\widetilde{BM}}}{\int\int dQ^2\,dz_h\,\frac{1+(1-y)^2}{Q^4}\,Q_V(x_B,Q^2)z_h\,D^h_{u_V}(z_h,Q^2)/\sqrt{\langle P_T^2\rangle_{Siv}}}, \quad h=\pi^+,K^+ \tag{29}$$

If Q^2-evolution can be neglected the ratio of the asymmetries is completely fixed:

$$\frac{A^{h-\bar{h}}_{BM}(x_B)}{A^{Siv,h-\bar{h}}_{UT}(x_B)} = \frac{\lambda_{Q_V}}{2}\,\frac{1-\bar{y}}{1+(1-\bar{y})^2}\,\bar{\mathcal{C}}^h(\bar{Q}^2), \tag{30}$$

where $\bar{\mathcal{C}}^h$ is a constant factor.

Thus, if data exist for a range of x_B values at the same $\bar{Q}$, eq. (30) should allow a test of the used connection between BM and Sivers functions without requiring knowledge of BM, Sivers or even Collins functions, involving only measurable quantities.

5. Relations between BM and Collins asymmetries

The distribution of transversely polarized quarks $q^\uparrow$ in a transversely polarized proton $p^\uparrow$ defines the transversity distributions $h_{1q}(x_B)$ or $\Delta_T q(x_B)$:

$$\Delta_T q(x_B) = h_{1q}(x_B) = \int d^2\mathbf{k}_\perp \, h_{1q}(x_B, k_\perp), \tag{31}$$

where $h_{1q}(x_B, k_\perp)$ is the transversity distribution depending on the parton transverse momentum. It is selected by the Collins asymmetry, which for the difference cross sections is:

$$A_{UT}^{\sin(\phi_h+\phi_S),h-\bar{h}} = \frac{1}{S_T} \frac{\int d\phi_h d\phi_S \left[d^6\sigma^\uparrow - d^6\sigma^\downarrow\right]^{h-\bar{h}} \sin(\phi_h+\phi_S)}{\int d\phi_h d\phi_S \left[d^6\sigma^\uparrow + d^6\sigma^\downarrow\right]^{h-\bar{h}}}. \tag{32}$$

Again, on deuteron target it measures the sum of the valence-quarks transverse distributions:

$$h_{1Q_V}(x_B, k_\perp) \equiv h_{1u_V}(x_B, k_\perp) + h_{1d_V}(x_B, k_\perp) \tag{33}$$

$$h_{1Q_V}(x_B, k_\perp) = \frac{1}{2}\mathcal{N}_T^{Q_V}(x_B)\,[Q_V(x_B) + \Delta Q_V(x_B)]\,\frac{e^{-k_\perp^2/\langle k_\perp^2\rangle_T}}{\pi\langle k_\perp^2\rangle_T}, \tag{34}$$

where $\langle k_\perp^2\rangle_T$ and $\mathcal{N}_T^{Q_V}(x_B)$ are the unknown quantities.

Common for BM and Collins difference asymmetries is that they are convoluted with the same Collins FF, $\Delta^N D_{h^+/u_V^\uparrow}(z_h, p_\perp)$, and that the x_B and z_h dependencies factorize. This allows to relate the z_h-dependent BM and Collins asymmetries. This relation is especially simple for small enough bins in Q^2 with a completely fixed z_h-dependence and the same for all final hadrons:

$$\frac{A_{BM}^{h-\bar{h}}(z_h)}{A_{UT}^{Coll,h-\bar{h}}(z_h)} = \frac{\lambda_{Q_V}}{2}\,\frac{z_h\sqrt{\langle P_T^2\rangle_{\widetilde{T}}}}{\langle P_T^2\rangle_{\widetilde{BM}}}\,\bar{B}, \quad h = \pi^+, K^+ \tag{35}$$

where $\langle P_T^2\rangle_{\widetilde{BM}}$ and $\langle P_T^2\rangle_{\widetilde{T}}$ are fixed by the Gaussian form:

$$\langle P_T^2\rangle_{\widetilde{BM}} = \langle p_\perp^2\rangle_C + z_h^2\langle k_\perp^2\rangle_S, \quad \langle P_T^2\rangle_{\widetilde{T}} = \langle p_\perp^2\rangle_C + z_h^2\langle k_\perp^2\rangle \tag{36}$$

and $\bar{B}$ is independent of both z_h and the final hadron. Eq. (35) will hold only if $\Delta^N D_{h^+/u_V^\uparrow}$ enters both $A_{BM}^{h-\bar{h}}$ and $A^{Coll,h-\bar{h}}$ and thus, it presents a remarkably simple test, involving only measurable quantities, not only of the standard parametrization, but of the whole QCD picture.

Acknowledgments

E.C. is grateful to Oleg Teryaev for helpful discussions, to Bakur Parsamyan for stressing the experimental importance of the proton data, and for the hospitality of JINR where the paper was finalized. E.L is grateful to The Leverhulme Trust for an Emeritus Fellowship.

References

[1] E. Christova and E.Leader, Nucl. Phys. **B607** (2001) 369 (arXiv:hep-ph/0007303); Phys. Rev. D **79** (2009) 014019 (arXiv:0809.0191)
[2] E. Christova, Phys. Rev. **D 90** (2014) 054005 (arXiv:1407.5872);
[3] M. Alekseev et al., COMPASS collab., Phys. Lett. **B660** (2008) 458; E. Leader, A. Sidorov, D. Stamenov, arXiv:1406.4678
[4] E. Christova and E.Leader, arXiv:1507.01399
[5] D.Sivers, Phys. Rev. **D41** (1990) 83; **43**(1991) 261
[6] M. Anselmino, M. Boglione, U. D'Alesio, S. Melis, F. Murgia, E.R. Nocera and A. Prokudin, Phys. Rev. D **83** (2011) 114019 (arXiv:1101.1011)
[7] D. Boer and P.J.Mulders, Phys. Rev. **D 57** 5780 (1998)
[8] Vincenzo Barone, Alexei Prokudin and Bo-Qiang Ma, Phys. Rev. **D 78**, 045022 (2008)

Journal of Physics: Conference Series **678** (2016) 012013 doi:10.1088/1742-6596/678/1/012013

Sneutrino identification in tau pair production at ILC with polarized beams

A A Pankov and A V Tsytrinov

The Abdus Salam ICTP Affiliated Centre, Technical University of Gomel, 246746 Gomel, Belarus

E-mail: pankov@ictp.it

Abstract. Many scenarios of new physics can lead to deviations of the observables (cross sections, asymmetries) from the Standard Model predictions in e^+e^- collision. The possibility of uniquely identifying the indirect effects of s-channel sneutrino exchange, as predicted by supersymmetric theories with R-parity violation, against other new physics scenarios in high-energy electron-positron annihilation into tau pairs at the International Linear Collider has been studied.

1. Introduction

A wide range of new physics (NP) scenarios, beyond the Standard Model (SM), are characterized by novel interactions mediated by exchanges of very heavy states with mass scales significantly greater than the electroweak scale. Theoretical considerations and current experimental constraints indicate that the new objects may be too heavy to be directly produced at the CERN LHC and at the planned e^+e^- International Linear Collider (ILC). Thus, the non-standard interactions would only be revealed by indirect effects manifesting themselves as deviations from the predictions of the SM. In the case of indirect discovery many different NP scenarios may lead to very similar experimental signatures. There are many very different NP scenarios that predict new particle exchanges which can lead to contact interactions (CI) which may show up below direct production thresholds. These are compositeness, a Z' boson from models with an extended gauge sector, scalar or vector leptoquarks, R-parity violating sneutrino ($\tilde{\nu}$) exchange, bi-lepton boson exchanges, anomalous gauge boson couplings (AGC), virtual Kaluza–Klein (KK) graviton exchange in the context of gravity propagating in large extra dimensions, exchange of KK gauge boson towers or string excitations, *etc.*[1]. Of course, this list is not exhaustive, because other kinds of contact interactions may be at play. The purpose of the present analysis is to develop the tools that can be used to differentiate these scenarios at high energy e^+e^- colliders.

If R-parity is violated it is possible that the exchange of sparticles can contribute significantly to SM processes and may even produce peaks or bumps in cross sections if they are kinematically accessible. Below threshold, these new spin-0 exchanges may make their manifestation known via indirect effects on observables. Here we will investigate the question of whether the effects of the exchange of scalar (spin-0) sparticles can be differentiated at linear colliders from other contact interactions mentioned above [2].

Content from this work may be used under the terms of the Creative Commons Attribution 3.0 licence. Any further distribution of this work must maintain attribution to the author(s) and the title of the work, journal citation and DOI.
Published under licence by IOP Publishing Ltd

For a sneutrino in an R-parity-violating theory, we take the basic couplings to leptons to be given by $\lambda_{ijk}L_iL_j\bar{E}_k$. Here, L are the left-handed lepton doublet superfields, and $\bar{E}$ are the corresponding left-handed singlet fields. In this case observables associated with leptonic process

$$e^+ + e^- \to \gamma, Z, \tilde{\nu}_\mu \to \tau^+ + \tau^-, \tag{1}$$

will be affected by the exchange of $\tilde{\nu}$'s in the s-channel. If only the product of Yukawa coupling $\lambda_{121}\lambda_{323}$ is nonzero the s-channel $\tilde{\nu}_\mu$ exchange would contribute to the $\tau^+\tau^-$ pair final state.

2. Observables and numerical analysis

2.1. Polarized e^+, e^- beams

With P^- and P^+ denoting the longitudinal polarizations of the electrons and positrons the doubly polarized total cross section for the process (1) in the presence of contact interactions can be expressed as

$$\begin{aligned}\sigma^{\rm CI} = \sigma_+^{\rm CI} + \sigma_-^{\rm CI} &= \frac{1}{4}\,[(1-P^-)(1+P^+)\,(\sigma_{\rm LL}^{\rm CI} + \sigma_{\rm LR}^{\rm CI}) \\ &+ (1+P^-)(1-P^+)\,(\sigma_{\rm RR}^{\rm CI} + \sigma_{\rm RL}^{\rm CI})].\end{aligned} \tag{2}$$

For the SM it has the same form where one should replace the superscript CI $\to$ SM. In Eq. (2) the helicity cross sections $\sigma_{\alpha\beta}^{\rm CI} \propto |\mathcal{M}_{\alpha\beta}^{\rm CI}|^2$ $(\alpha, \beta = {\rm L}, {\rm R})$ are directly related to the individual CI couplings $\Delta_{\alpha\beta}$. Here the helicity amplitudes $\mathcal{M}_{\alpha\beta}^{\rm CI}$ can be written as $\mathcal{M}_{\alpha\beta}^{\rm CI} = \mathcal{M}_{\alpha\beta}^{\rm SM} + \Delta_{\alpha\beta}$. The $\Delta_{\alpha\beta}$ functions represent the contact interaction contributions coming from TeV-scale physics mentioned above [2].

Since the $\tilde{\nu}$ exchanged in the s-channel does not interfere with the s-channel SM γ and Z exchanges, the corresponding total cross section can be written as

$$\begin{aligned}\sigma^{\tilde{\nu}} &= \frac{1}{4}\,(1-P^-)(1+P^+)\,(\sigma_{\rm LL}^{\rm SM} + \sigma_{\rm LR}^{\rm SM}) + \frac{1}{4}\,(1+P^-)(1-P^+)\,(\sigma_{\rm RR}^{\rm SM} + \sigma_{\rm RL}^{\rm SM}) \\ &+ \frac{3}{2}\,\frac{1+P^-P^+}{2}\,(\sigma_{\rm RL}^{\tilde{\nu}} + \sigma_{\rm LR}^{\tilde{\nu}}).\end{aligned} \tag{3}$$

Here, $\sigma_{\rm RL}^{\tilde{\nu}} = \sigma_{\rm LR}^{\tilde{\nu}} \propto |\mathcal{M}_{\rm RL}^{\tilde{\nu}}|^2$, $\mathcal{M}_{\rm RL}^{\tilde{\nu}} = \mathcal{M}_{\rm LR}^{\tilde{\nu}} = \frac{1}{2}\,C_{\tilde{\nu}}^s\chi_{\tilde{\nu}}^s$, and $C_{\tilde{\nu}}^s$ and $\chi_{\tilde{\nu}}^s$ denote the product of the R-parity violating couplings and the propagator of the exchanged sneutrino. For the s-channel $\tilde{\nu}_\tau$ sneutrino exchange: $C_{\tilde{\nu}}^s\,\chi_{\tilde{\nu}}^s = (\lambda^2/(4\pi\alpha_{\rm em}))\,(s/(s - M_{\tilde{\nu}_\tau}^2 + iM_{\tilde{\nu}_\tau}\Gamma_{\tilde{\nu}_\tau}))$, where $\lambda^2 = \lambda_{121}\lambda_{323}$.

To uniquely identify the effect of the s-channel sneutrino exchange the double beam polarization asymmetry can be used:

$$\begin{aligned}A_{\rm double} &= \frac{\sigma(P_1,-P_2) + \sigma(-P_1,P_2) - \sigma(P_1,P_2) - \sigma(-P_1,-P_2)}{\sigma(P_1,-P_2) + \sigma(-P_1,P_2) + \sigma(P_1,P_2) + \sigma(-P_1,-P_2)} \\ &= P_1P_2\cdot\frac{(\sigma_{\rm LL}^{\rm SM} + \sigma_{\rm LR}^{\rm SM} + \sigma_{\rm RR}^{\rm SM} + \sigma_{\rm RL}^{\rm SM}) - 3(\sigma_{\rm LR}^{\tilde{\nu}} + \sigma_{\rm RL}^{\tilde{\nu}})}{(\sigma_{\rm LL}^{\rm SM} + \sigma_{\rm LR}^{\rm SM} + \sigma_{\rm RR}^{\rm SM} + \sigma_{\rm RL}^{\rm SM}) + 3(\sigma_{\rm LR}^{\tilde{\nu}} + \sigma_{\rm RL}^{\tilde{\nu}})},\end{aligned} \tag{4}$$

where $P_1 = |P^-|$, $P_2 = |P^+|$. For the whole set of contact interactions, with the exception of the s-channel sneutrino exchange, $A_{\rm double}^{\rm SM} = A_{\rm double}^{\rm CI} = P_1P_2 = 0.48$, where $P_1 = 0.8$, $P_2 = 0.6$. This is because these contact interactions contribute to the same amplitudes as SM. Thus, $A_{\rm double}^{\rm SM}$ and $A_{\rm double}^{\rm CI}$ are indistinguishable for any values of the contact interaction parameters. On the contrary, the $\tilde{\nu}$ exchange in the s-channel will force this observable to a smaller value, $\Delta A_{\rm double} = A_{\rm double}^{\tilde{\nu}} - A_{\rm double}^{\rm SM} \propto -P_1P_2\,|C_{\tilde{\nu}}^s\chi_{\tilde{\nu}}^s|^2 < 0$. The value of $A_{\rm double}$ below P_1P_2 can provide a signature of scalar exchange in the s-channel.

2.2. Unpolarized e^+, e^- beams

Alternatively, if the tau spins can be analyzed [3, 4], a spin-spin correlation can be formed which is sensitive to the spin of any new particle exchanged in s-channel of the process

$$e^+ + e^- \to \tau^+(h') + \tau^-(h), \tag{5}$$

where h (h') denotes the helicity of τ^- (τ^+). The spin-spin correlation, B_{ZZ}, as introduced in [3], can be written as:

$$\begin{aligned} B_{ZZ} &= \frac{\sigma(+,-)+\sigma(-,+)-\sigma(+,+)-\sigma(-,-)}{\sigma(+,-)+\sigma(-,+)+\sigma(+,+)+\sigma(-,-)} \\ &= \frac{(\sigma^{\rm SM}_{\rm LL}+\sigma^{\rm SM}_{\rm LR}+\sigma^{\rm SM}_{\rm RR}+\sigma^{\rm SM}_{\rm RL})-3(\sigma^{\tilde\nu}_{\rm LR}+\sigma^{\tilde\nu}_{\rm RL})}{(\sigma^{\rm SM}_{\rm LL}+\sigma^{\rm SM}_{\rm LR}+\sigma^{\rm SM}_{\rm RR}+\sigma^{\rm SM}_{\rm RL})+3(\sigma^{\tilde\nu}_{\rm LR}+\sigma^{\tilde\nu}_{\rm RL})}. \end{aligned} \tag{6}$$

This quantity is unity in the SM as well as in CI models but can be substantially smaller when s-channel scalars are present as shown in Fig. 1.

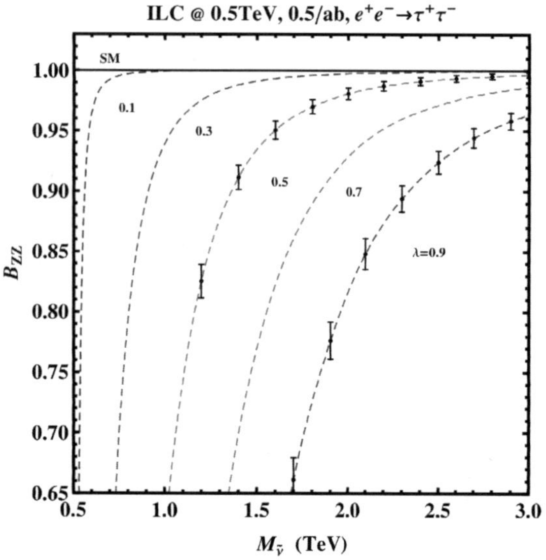

Figure 1. B_{ZZ} asymmetry as a function of sneutrino mass $M_{\tilde\nu}$ for different choices of λ. The error bars indicate the expected uncertainty (2σ-level) for ILC with $\sqrt{s} = 0.5$ TeV and $\mathcal{L}_{\rm int} = 0.5$ ab^{-1}.

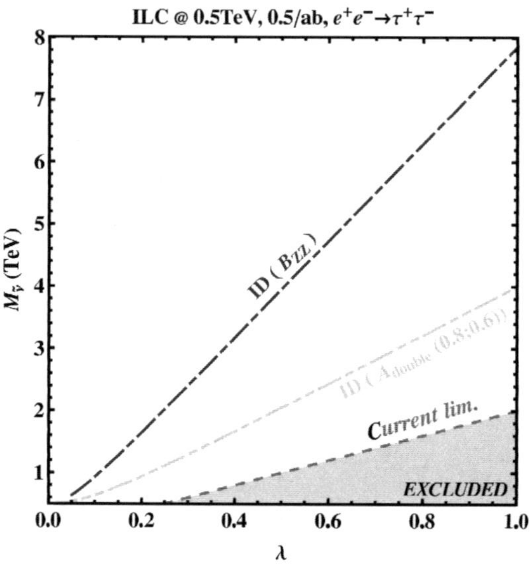

Figure 2. Identification reach on $M_{\tilde\nu}$ (95% C.L.) as a function of λ for the process $e^+e^- \to \tau^+\tau^-$ at the ILC with $\sqrt{s} = 0.5$ TeV and $\mathcal{L}_{\rm int} = 0.5$ ab^{-1}. Radiative corrections were included.

Identification reaches on the sneutrino mass $M_{\tilde\nu}$ (95% C.L.) plotted in Fig. 2 are obtained from χ^2 analysis of $A_{\rm double}$ and B_{ZZ} asymmetries. For comparison, current limits from low-energy data are also shown. The reach obtained using spin-spin correlation technique discussed above is superior to that found by examination of the double spin asymmetry $A_{\rm double}$.

3. Concluding remarks

We studied the possibility of uniquely identifying the indirect effects of s-channel sneutrino exchange, as predicted by supersymmetric theories with R-parity violation, against other new physics scenarios in high-energy electron-positron annihilation into tau pairs at the International

Linear Collider. New physics outlined above can lead to qualitatively similar alterations in SM cross sections and angular distributions but differ in detail. These detailed differences provide the key to the two major tools that are useful in accomplishing our task: (i) an asymmetry formed by polarizing both initial beams, A_{double}, and (ii) spin-spin correlation, B_{ZZ}.

Acknowledgments
Authors would like to thank P. Osland and G. Moortgat-Pick for the enjoyable collaboration on the subject matter covered here. This research has been partially supported by the Abdus Salam ICTP (TRIL and Associate Programmes) and the Collaborative Research Center SFB676/1-2006 of the DFG at the Department of Physics, University of Hamburg.

References
[1] Rizzo T G 2003 Contact interaction searches at the linear collider: Energy, luminosity and positron polarization dependencies *Preprint* hep-ph/0303056
[2] Tsytrinov A V Kalinowski J Osland P and Pankov A A 2012 Sneutrino identification in lepton pair production at ILC with polarized beams *Phys. Lett.* B **718** 94
[3] Bar-Shalom S Eilam G and Soni A 1998 R-parity violation and CP violating and CP conserving spin asymmetries in $l^+l^- \to \tilde{\nu} \to \tau^+\tau^-$: Probing sneutrino mixing at LEP-2, NLC and $\mu\mu$ colliders *Phys. Rev. Lett.* **80** 4629
[4] Augustin J E and Renard F M 1980 How to measure quark helicities in $e^+e^- \to$ hadrons *Nucl. Phys.* B **162** 341

Recent results from the NN-interaction studies with polarized beams and targets at ANKE-COSY

Sergey Dymov for the ANKE collaboration
Laboratory of Nuclear Problems, JINR, RU-141980 Dubna, Russia
Institut für Kernphysik, Forschungszentrum Jülich, D-52425 Jülich, Germany
E-mail: s.dymov@fz-juelich.de

Abstract. Adding to the nucleon-nucleon scattering database is one of the major priorities of the ANKE collaboration. Such data are necessary ingredients, not only for the understanding of nuclear forces, but also for the description of meson production and other nuclear reactions at intermediate energies.

By measuring the cross section, deuteron analysing powers, and spin-correlation parameters in the $dp \to \{pp\}_s n$ reaction, where $\{pp\}_s$ represents the $^1S_0$ state, information has been obtained on small-angle neutron-proton spin-flip charge-exchange amplitudes.

The measurements of *pp* elastic scattering by the COSY-EDDA have had a major impact on the partial wave analysis of this reaction above 1 GeV. However, these experiments only extended over the central region of c.m. angles, $30^0 < \theta_{\rm cm} < 150^0$, that has left major ambiguities in the phase shift analysis by the SAID group. In contrast, the small angle region is accessible at ANKE-COSY, that allowed measurement of the differential cross section and the analysing power at $5^0 < \theta_{\rm cm} < 30^0$ in the $0.8 - 2.8$ GeV energy range.

The data on the *pn* elastic scattering are much more scarce than those of *pp*, especially in the region above 1.15 GeV. The study of the $dp \to \{pp\}_s n$ reaction provides the information about the *pn* elastic scattering at large angles. The small angle scattering was studied with the polarized proton COSY beam and an unpolarised deuterium gas target. The detection the spectator proton in the ANKE vertex silicon detector allowed to use the deuterium target as an effective neutron one. The analysing powers of the process were obtained at six beam energies from 0.8 to 2.4 GeV.

1. Introduction

The nucleon-nucleon interaction is fundamental to the whole of nuclear physics and hence to the composition of matter as we know it. Apart from its intrinsic importance, it is also a necessary ingredient in the description of meson production and other processes. The meticulous investigation of the nucleon-nucleon interaction must be a communal activity across laboratories, with no single facility providing the final breakthrough.

The complete description of the NN interaction requires data as input to phase-shift analysies (PSA), from which the scattering amplitudes can be reconstructed. The PSA generally requires experiments with both beam and target particles polarised in the initial state, as well as polarisation determination of the final particles [1].

The existing database of elastic pp and pn differential cross sections is shown in the abundance plots of Fig. 1 (taken from Ref. [2]). Note that in the pp case the cross section is symmetric about 90° and so few points are plotted in the backward hemisphere. There is a heavy concentration

Content from this work may be used under the terms of the Creative Commons Attribution 3.0 licence. Any further distribution of this work must maintain attribution to the author(s) and the title of the work, journal citation and DOI.
Published under licence by IOP Publishing Ltd

Journal of Physics: Conference Series **678** (2016) 012014 doi:10.1088/1742-6596/678/1/012014

of pp points for $40° < \theta_{\rm cm} < 90°$, corresponding to the EDDA measurements [3], but there are very few at smaller angles, though this will be changed when the ANKE data are finalised [4]. On the other hand, it is seen that the pn database is even more sparse above 1.15 GeV (the maximum energy of the SATURNE "monochromatic" neutron beam).

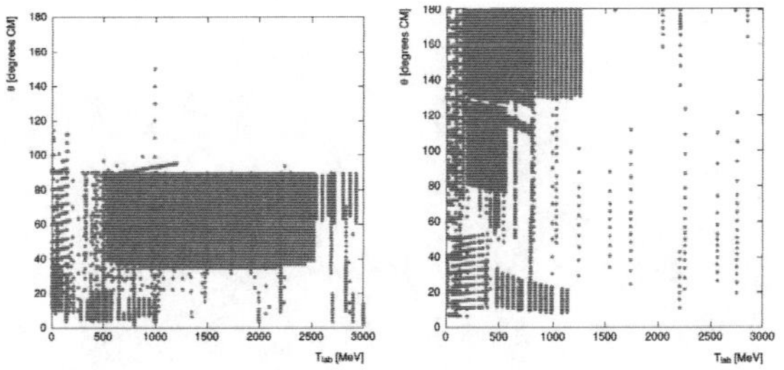

Figure 1. Abundance plots for elastic differential cross section experiments for pp (Left) and np (Right), taken from the SAID compilation [2].

The situation is broadly similar in the A_y measurements summarised in Fig. 2. For *pp* elastic scattering this parameter is antisymmetric around 90° but is seen from the figure that there have been relatively few measurements for $\theta_{\rm cm} < 30°$ for beam energies above 1.0 GeV. In the case of np scattering, there is no longer this antisymmetry but there are even fewer measurements above 1.15 GeV. It is important to note that isospin invariance requires the analysing powers for polarised protons and neutrons shown in the middle and right panels, correspondingly should be the same. It is in the context of these gaping holes in the nucleon-nucleon database that the ANKE NN programme should be judged.

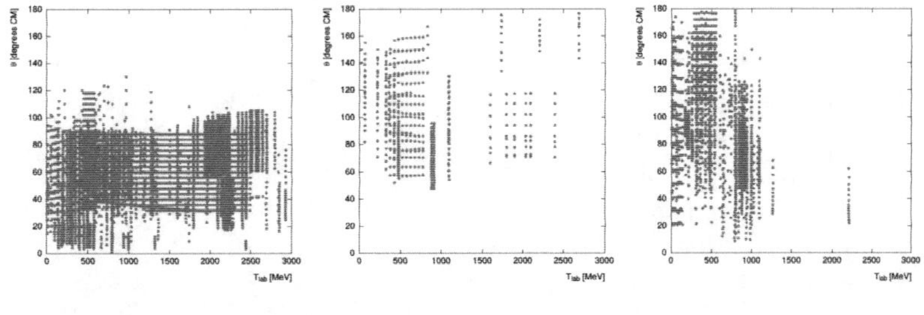

Figure 2. Abundance plots for experiments on the analysing power of (left) pp elastic scattering, (middle) the proton in pn elastic scattering, and (right) the neutron in pn elastic scattering.

2. The neutron-proton charge-exchange amplitudes

A great effort has been made in the investigation of the spin-dependent terms in large angle neutron-proton scattering. It was pointed out many years ago that the $dp \to \{pp\}_s n$ charge exchange at small angles is very sensitive to the spin-spin terms in the $np \to pn$ amplitude provided the excitation energy E_{pp} in the final *pp* system was kept low [5]. Under such conditions the $\{pp\}_s$ is in a $^1S_0$ state and the charge exchange necessarily involves a spin flip from the initial *np* spin-triplet of the deuteron. Furthermore, measurements of the deuteron tensor analysing powers A_{xx} and A_{yy} allow one to distinguish between the contributions from the three *np* spin-spin amplitudes.

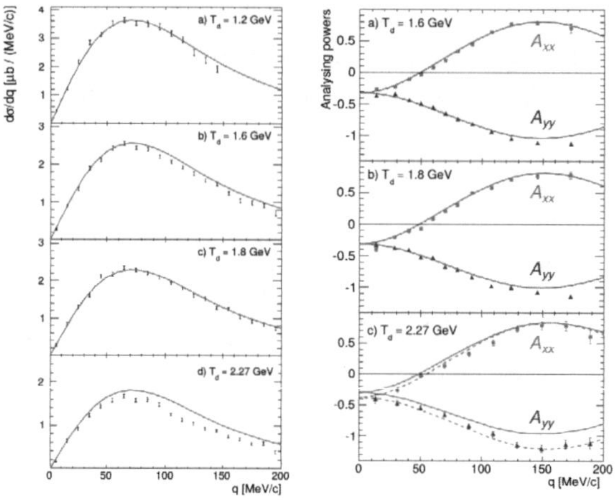

Figure 3. Differential cross sections (left panel) and cartesian deuteron analysing powers (right panel) for the $\vec{d}p \to \{pp\}_s n$ reaction for $E_{pp} < 3$ MeV at $T_d = 1.2$, 1.6, 1.8, and 2.27 GeV [9, 10, 11]. The impulse approximation predictions [8] have been evaluated with the SAID amplitudes [2] (solid curves) and also, at the highest energy, when the longitudinal spin-spin amplitude is scaled by a factor of 0.75 (dashed curves).

Measurements were carried out at Saclay [6, 7] but only in regions where the NN amplitudes were reasonably well known. These have been extended to higher energy at ANKE in fine steps in momentum transfer q. A cut of $E_{pp} < 3$ MeV was typically imposed but any contamination from triplet P-waves was taken into account in the theoretical modelling [8]. The ANKE differential cross section (including the 1.2 GeV data) and analysing power results at $T_d = 1.6$, 1.8, and 2.27 GeV [9, 10, 11] are compared in Fig. 3 to these impulse approximation predictions using as input up-to-date np amplitudes [2]. The satisfactory agreement at lower energies, and also in the values of the differential cross sections, shows that the theoretical description is adequate here. Above about 1 GeV, neutron-proton data become rather sparse. It comes therefore as no surprise that, when the same approach is employed on the higher energy data shown in Fig. 3, the current SAID amplitudes [2] give a poor overall description of the results. However, if the longitudinal spin-spin amplitude is reduced by a global factor of 0.75, the agreement is much more satisfactory. The charge exchange data can have a useful impact on the NN database.

Confirmation of these conclusions is to be found in the measurements of the deuteron-proton spin correlation parameters studied at ANKE with the polarised hydrogen gas cell [11]. Results on this are shown in Fig. 4 (left panel). In impulse approximation, these are sensitive to the interference between the longitudinal spin-spin amplitude and the two transverse ones. Whereas there is a satisfactory agreement with the theoretical predictions at 1.2 GeV, the model is much more satisfactory at 2.27 GeV if the longitudinal input is scaled by the 0.75 factor. In addition to measuring the spin correlations with the polarised cell, data were also obtained in parallel on the proton analysing power in the $d\vec{p} \to \{pp\}_s n$ reaction and the results are shown in right panel of Fig. 2. The message here is very similar to that for the other observables. At 600 MeV per nucleon the SAID input reproduces the experimental points very well but it seems that at 1135 MeV the SAID description of the spin-orbit amplitude has serious deficiencies.

Journal of Physics: Conference Series **678** (2016) 012014 doi:10.1088/1742-6596/678/1/012014

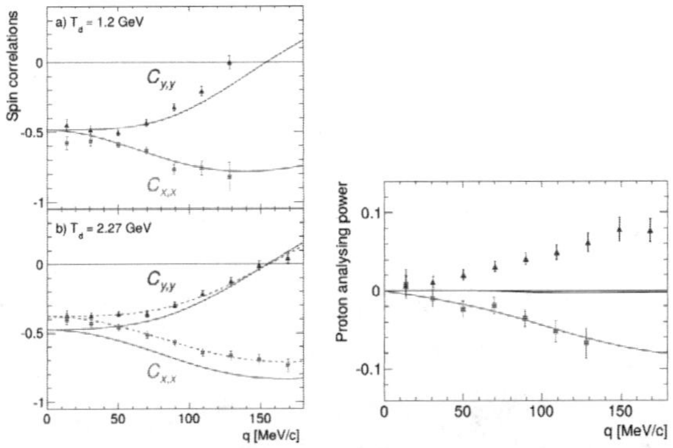

Figure 4. Left panel: Transverse spin correlation parameters in the $\vec{d}\vec{p} \to \{pp\}_s n$ reaction [11] at (a) 1.2 and (b) 2.27 GeV compared to the predictions of an impulse approximation model (solid curves). Better agreement is found at the higher energy if the longitudinal input is scaled by a factor of 0.75 (dashed curves). Right panel: Proton analysing power in the $d\vec{p} \to \{pp\}_s n$ reaction at 1.2 GeV (red squares) and 2.27 GeV (blue triangles) compared to impulse approximation predictions. Note that, with the current SAID input [2], the latter almost vanish at the higher energy.

2.1. Recent results

The vector and tensor analysing powers, A_y and A_{yy} , of the $\vec{p}\vec{d} \to \{pp\}_s n$ charge-exchange reaction have been measured at a beam energy of 600 MeV at ANKE by using an unpolarised proton beam incident on an internal storage cell target filled with polarised deuterium gas [12]. The low energy recoiling protons were measured in a pair of silicon tracking telescopes placed on either side of the target. By analysing events where both protons entered the same telescope, the charge-exchange reaction was measured for momentum transfers $q \geq 160$ MeV/c. These data provide a good continuation of the earlier results at $q \leq 140$ MeV/c obtained with a polarised deuteron beam [11]. They are also consistent with impulse approximation predictions with little sign evident for any modifications due to multiple scatterings.

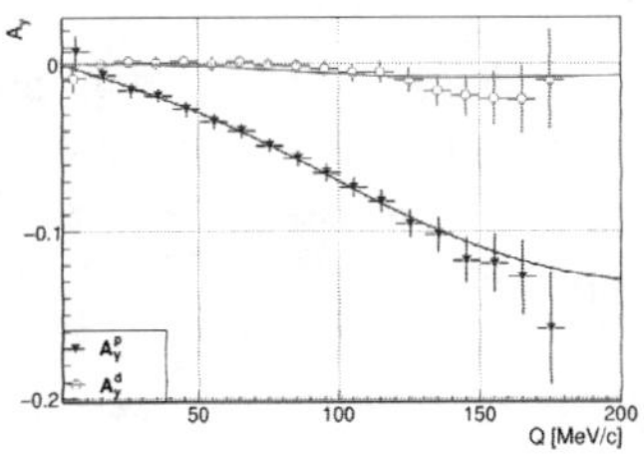

Figure 5. Values of the deuteron (vector) analysing power A_y^d (red circles) and proton analysing power A_y^p (blue triangles) for the $\vec{d}\vec{p} \to \{pp\}_s n$ reaction at 726 MeV. The curves are impulse approximation estimates [8].

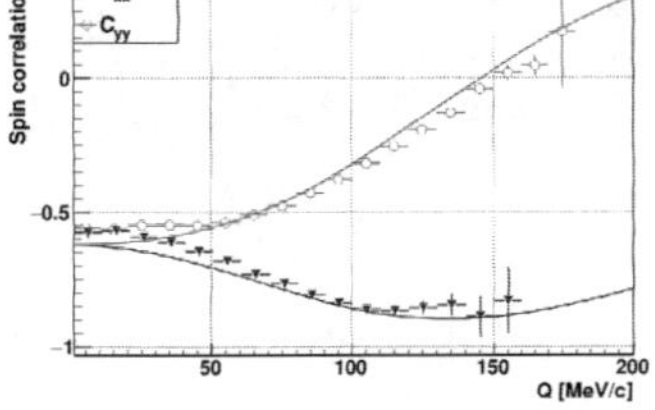

Figure 6. Values of the spin correlations $C_{y,y}$ (red circles) and $C_{x,x}$ (blue triangles) for the $dp \to \{pp\}_s n$ reaction at 726 MeV.

The charge exchange of vector polarised deuterons on a polarised hydrogen target has been

Journal of Physics: Conference Series **678** (2016) 012014 doi:10.1088/1742-6596/678/1/012014

studied at ANKE in a high statistics experiment at a deuteron beam energy of $T_d = 726$ MeV [13]. The measured analysing powers and spin correlations are sensitive to interference terms between specific neutron-proton charge-exchange amplitudes at a neutron kinetic energy of $T_n \approx \frac{1}{2}T_d = 363$ MeV. An impulse approximation calculation, which takes into account corrections due to the angular distribution in the diproton, describes reasonably the dependence of the data on both E_{pp} and the momentum transfer, as shown in Figs. 5, 6.

3. Analysing power in proton-proton elastic scattering at small angles

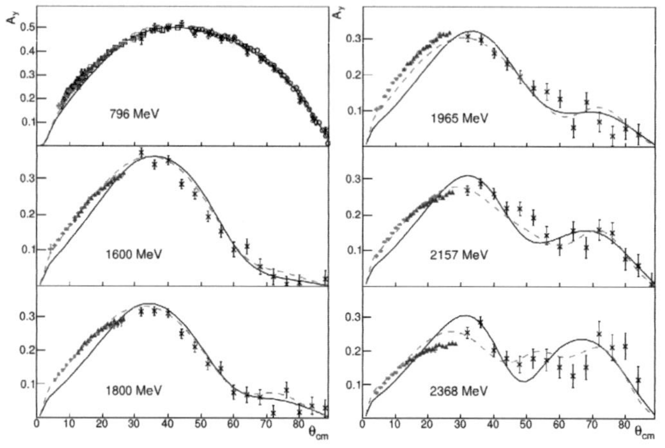

Figure 7. Comparison of the ANKE measurements of the proton analysing power in *pp* elastic scattering using the STT (red filled circles) and FD (blue filled triangles) systems with the curves corresponding to the SAID SP07 (solid black line) and the revised fit (dashed red) solutions [2]. Only statistical errors are shown. Also shown are selected results from EDDA (black crosses) [3] at the energies different by no more than 7 MeV and, at 796 MeV, LAMPF [17, 18, 19], and SATURNE [20] (black open symbols).

The proton analysing power in *pp* elastic scattering has been measured at small angles at COSY-ANKE at 796 MeV and five other beam energies between 1.6 and 2.4 GeV using a polarised proton beam [14]. In a strong-focusing synchrotron, such as COSY, resonances can lead to losses of polarisation of a proton beam during acceleration. In order to compensate for these effects, adiabatic spin-flip was used to overcome the imperfection resonances and tune-jumping to deal with the intrinsic ones [15]. The stripped-down version of the EDDA detector used as a polarimeter at COSY was calibrated during the EDDA data-taking periods against the full detector setup. The systematic uncertainty of the measurements of the polarisation was estimated to be 3% at each energy [16].

The asymmetries obtained by detecting the fast proton in the ANKE forward detector or the slow recoil proton in a silicon tracking telescope are completely consistent (see Fig. 7). Although the analysing power results agree well with the many published data at 796 MeV, and also with the most recent partial wave solution at this energy [2], the ANKE data at the higher energies lie well above the predictions of this solution at small angles. An updated phase shift analysis that uses the ANKE results together with the World data leads to a much better description of these new measurements.

4. Absolute differential cross section of proton-proton elastic scattering at small angles

The differential cross section for proton-proton elastic scattering has been measured at a beam energy of 1.0 GeV and in 200 MeV steps from 1.6 to 2.8 GeV for centre-of-mass angles in the range from $12^\circ - 16^\circ$ to $25^\circ - 30^\circ$, depending on the energy [4, 24].

The biggest challenge that has to be faced when measuring the absolute value of a cross section in a storage ring experiment is to establish the beam-target luminosity at the few percent level. It has been shown that this can be achieved by studying the energy loss through electromagnetic processes as the coasting uncooled beam passes repeatedly through the target chamber. There is a resulting change in the frequency of the machine that can be determined with high accuracy by studying the Schottky power spectrum of the beam [25], leading to a precise determination of the effective target density. The measurement of the beam intensity is a routine procedure for any accelerator and is performed at COSY using the high precision Beam Current Transformer device. In result, absolute normalisations of typically 3% were achieved.

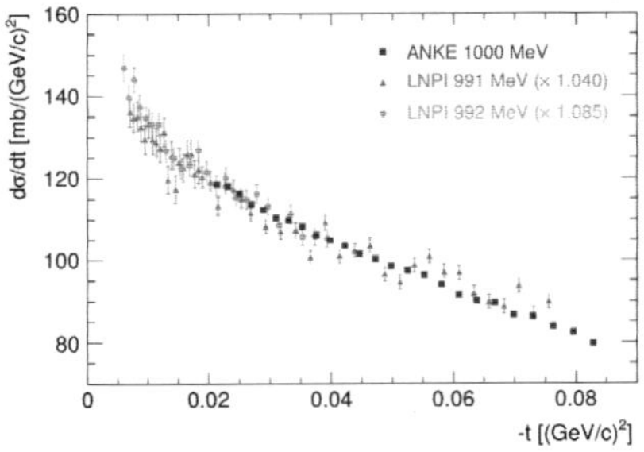

Figure 8. Differential cross section for pp elastic scattering. The ANKE data at 1000 MeV (blue squares) are compared to the Gatchina hydrogen data at 992 MeV (green circles) [21] scaled by a factor of 1.085 and methane results at 991 MeV (red triangles) [22] scaled by a factor of 1.04.

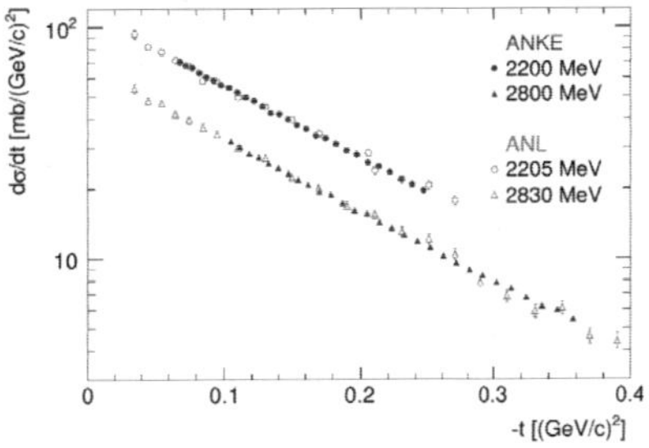

Figure 9. The ANKE pp differential cross section data at 2.2 GeV (closed blue circles) and 2.8 GeV (closed blue triangles) compared to the ANL results [23] at 2.2 GeV (open red circles) and 2.83 GeV (open red triangles). For presentational purposes, both higher energy data sets have been scaled downwards by 1.5.

There are very few data sets to which the ANKE results can be compared and these data are plotted together with our measurements in Figs. 8 and 9. In the vicinity of 1 GeV there are two measurements by the Gatchina group that were made with the IKAR recoil detector [21, 22]. Data are also available from the Argonne National Laboratory at 2.2 and 2.83 GeV [23].

The ANKE results shown for all energies in Fig. 10 could clearly have an impact on the current partial wave solutions. This is demonstrated in Fig. 11, where they are compared to both the SAID SP07 solution [2] and a modified one that takes the present data at all eight energies into account. Scaling factors in the partial wave analysis consistent with the overall uncertainties, have been included in the figure. The major changes introduced by the new partial wave solution are in the $^1S_0$ and $^1D_2$ waves at high energy. It should be noted that the modified solution does not weaken the description of the ANKE proton analysing powers presented in [14].

After extrapolating the values of the Coulomb-corrected differential cross section to the forward direction, the results are broadly compatible with the predictions of forward dispersion relations.

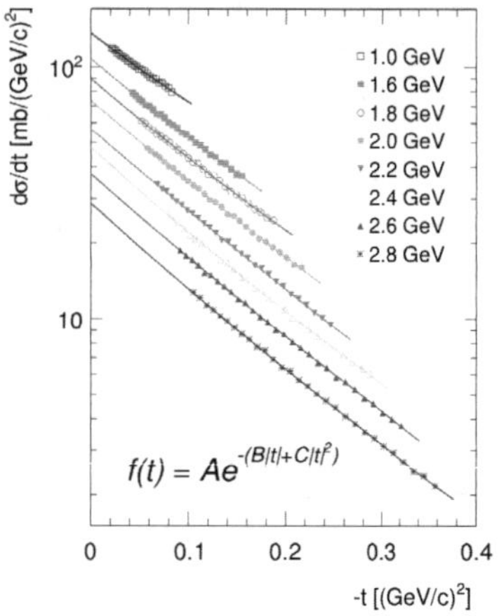

Figure 10. Combined ANKE data set of differential cross sections compared to fits of the $A\exp(-B|t|+C|t|^2)$ dependence. The data are scaled down sequentially in energy by factors of 1.2.

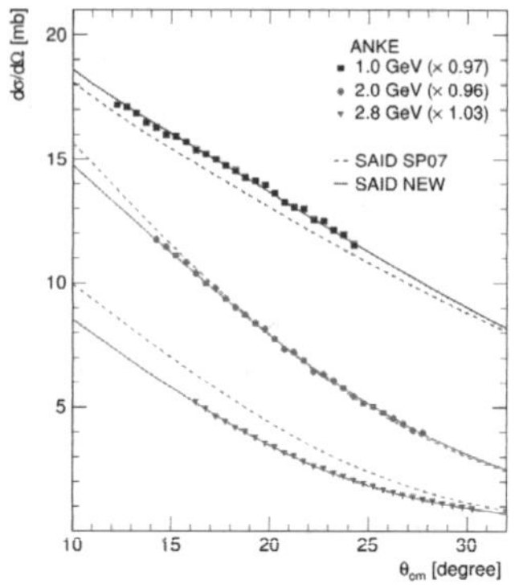

Figure 11. Scaled ANKE data at 1.0, 2.0, and 2.8 GeV compared to the SAID SP07 solution [2] and a modified (new) partial wave solution where the ANKE data have been taken into account. The 2.0 and 2.8 GeV data and curves have been reduced by factors of 0.5 and 0.25, respectively.

5. Analysing power in quasi-elastic proton-neuteron scattering at small angles

The analysing power in $\vec{p}n$ quasi elastic scattering has been measured at small angles at COSY-ANKE at 796 MeV and five other beam energies between 1.6 and 2.4 GeV using a polarised proton beam and unpolarised deuteron target. The asymmetries were obtained by detecting the fast proton in the ANKE forward detector and the slow recoil proton in a silicon tracking telescope.

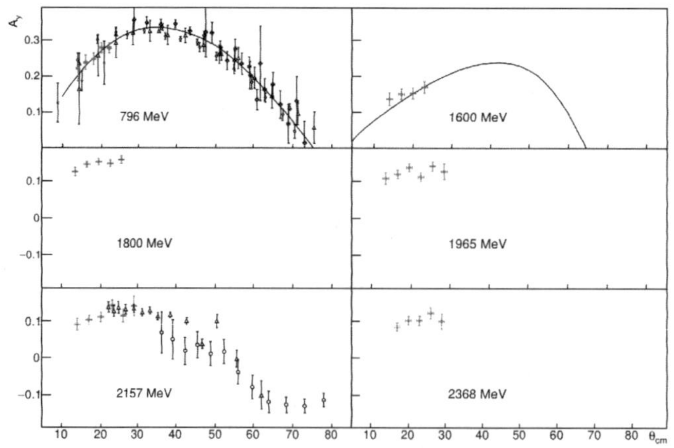

Figure 12. (Preliminary) ANKE measurements of the analysing power in pn quasi-elastic scattering (red filled circles). The results at 0.8 GeV are compared with the curves corresponding to the SAID SP07 (solid black line) [2] and other existing measurements [26, 27, 28, 29, 30, 31] (black open symbols). Only statistical errors are shown. Also shown are selected results from Argonne National Laboratory at 2.2 GeV energy [30, 31] (black open symbols).

The results of all the ANKE measurements are shown for the six energies in Fig. 12. The SAID SP07 solution [2], which is only valid up to 1.3 GeV kinetic energy, is shown by the solid black line at the lowest energy. The agreement between the new ANKE data and other existing data at 796 MeV and also with the SP07 SAID predictions at this energy is very good. Using our data, the SAID group was able to provide the preliminary modified solution for pn quasi-elastic scattering at 1.6 GeV, also shown in black in Fig. 12. The dominant systematic error is that arising from the determination of the beam polarisation in the EDDA polarimeter, which was estimated to be 3% [16].

[1] J. Bystricky et al., *Nucl. Phys.* A **444** (1985) 597.
[2] R.A. Arndt, I.I. Strakovsky, R.L. Workman, *Phys. Rev.* C **62** (2000) 034005;
R.A. Arndt, W.J. Briscoe, I.I. Strakovsky, R.L. Workman, *Phys. Rev.* C **76** (2007) 025209;
SAID data base, available from `http://gwdac.phys.gwu.edu`.
[3] D. Albers et al., *Phys. Rev. Lett.* **78** (1997) 1652;
M. Altmeier et al., *Phys. Rev. Lett.* **85** (2000) 1819;
F. Bauer et al., *Phys. Rev. Lett.* **90** (2003) 142301;
D. Albers et al.,*Eur. Phys. J.* A **22** (2004) 125;
M. Altmeier et al., *Eur. Phys. J.* A **23** (2005) 351.
[4] D. Chiladze et al., Absolute measurement of pp-elastic scattering at ANKE-COSY using the Schottky technique, COSY proposal/Beam Request #200.0, Spokesperson: D. Chiladze (2010).
Available from `www.fz-juelich.de/ikp/anke`.
[5] D. V. Bugg and C. Wilkin, *Nucl. Phys.* A **467** (1987) 575.
[6] C. Ellegaard et al., *Phys. Rev. Lett.* **59** (1987) 974.
[7] S. Kox et al., *Nucl. Phys.* A **556** (1993) 621.
[8] J. Carbonell, M. B. Barbaro, and C. Wilkin, *Nucl. Phys.* A **529** (1991) 653.
[9] D. Chiladze et al., *Phys. Lett.* B **637** (2006) 170.
[10] D. Chiladze et al., *Eur. Phys. J.* A **40** (2009) 23.
[11] D. Mchedlishvili et al., *Eur. Phys. J.* A **49** (2013) 49.
[12] B. Gou et al., *Phys. Lett.* B **741** (2015) 305.
[13] S. Dymov et al., *Phys. Lett.* B **744** (2015) 391.
[14] Z. Bagdasarian et al., *Phys. Lett.* B **739** (2014) 152.
[15] A. Lehrach et al., *AIP Conf. Proc.* **675** (2003) 153.
[16] E. Weise, PhD thesis, University of Bonn, 2000.
[17] M.W. McNaughton et al., *Phys. Rev.* C **23** (1981) 1128.
[18] F. Irom, G.J. Igo, J.B. McClelland, C.A. Whitten,*Phys. Rev.* C **25** (1982) 373.
[19] P.R. Bevington et al., *Phys. Rev. Lett.* **41** (1978) 384.
[20] C.E. Allgower et al., *Nucl. Phys.* A **637** (1998) 231.
[21] A.V. Dobrovolsky et al., *Nucl. Phys.* B **214** (1983) 1.
[22] A.V. Dobrovolsky et al., report LNPI-1454 (1988).
[23] I. Ambats et al., *Phys. Rev.* D **9** (1974) 1179.
[24] D. Chiladze, Proc. 8th Int. Conf. Nucl. Phys. at Storage Rings, PoS(STORI11) (2011) 039.
[25] H.J. Stein et al., *Phys. Rev.* ST Accel. Beams **11** (2008) 052801.
[26] J. Ball et al., *Nucl. Phys.* A **559** (1993) 489.
[27] G. Glass et al., *Phys. Rev.* C **41** (1990) 2732.
[28] G. Glass et al., *Phys. Rev.* C **47** (1993) 1369.
[29] M.L. Barlett et al.,*Phys. Rev.* C **27** (1983) 682.
[30] R. Diebold et al., *Phys. Rev. Lett.* **35** (1975) 632.
[31] Y. Makdisi et al., *Phys. Rev. Lett.* **45** (1980) 1529.

Journal of Physics: Conference Series **678** (2016) 012015 doi:10.1088/1742-6596/678/1/012015

Superconducting racetrack booster for the ion complex of MEIC

Yu Filatov[2,3], A M Kondratenko[1], M A Kondratenko[1], A Kovalenko[2], Ya S Derbenev[4], F Lin[4], V S Morozov[4] and Y Zhang[4]

[1]Science and Technique Laboratory "Zaryad", 630090, Novosibirsk, Russia
[2]Joint Institute for Nuclear Research, 141980, Dubna, Russia
[3]Moscow Institute of Physics and Technology, 141700, Dolgoprudny, Russia
[4]Jefferson Lab,12000 Jefferson Avenue, Newport News, VA 23606, USA

E-mail: kondratenkom@mail.ru

Abstract. The current design of the Medium-energy Electron-Ion Collider (MEIC) project at Jefferson lab features a single 8 GeV/c figure-8 booster based on super-ferric magnets. Reducing the circumference of the booster by switching to a racetrack design may improve its performance by limiting the space charge effect and lower its cost. We consider problems of preserving proton and deuteron polarizations in a superconducting racetrack booster. We show that using magnets based on hollow high-current NbTi composite superconducting cable similar to those designed at JINR for the Nuclotron guarantees preservation of the ion polarization in a racetrack booster up to 8 GeV/c. The booster operation cycle would be a few seconds that would improve the operating efficiency of the MEIC ion complex.

1. Introduction

A figure-8 booster in the ion complex of MEIC completely solves the problem of polarization preservation during acceleration of any ion species including protons and deuterons [1]. A racetrack booster has a smaller circumference, which may reduce its cost and relax the limitation related to space charge. For the same length of the booster straights, reduction of the field integral of the arc magnets equals about 70 T·m at the ion momentum of 8 GeV/c, i.e., when using magnet with a maximum field of 3 T, the total arc length is reduced by 23 m, which is about 12% of the total booster circumference. Let us consider the question of preservation of the proton and deuteron polarizations in a racetrack booster.

2. Spin resonances in a racetrack booster

As an example, let us choose a racetrack design most similar to the existing figure-8 booster design, i.e. keep the optical beam characteristics in the straight and in one FODO cell period of the booster arc. Superperiodicity of the booster lattice is N=2 and the radial and vertical betatron tunes equal $\nu_x = 5.95$ and $\nu_y = 4.84$, respectively.

In a racetrack booster, the stable polarization points along the vertical axis and the spin tune is proportional to the beam energy $\nu = \gamma G$ (G is the anomalous magnetic moment) that unavoidably leads to crossing of spin resonances during acceleration and, as a consequence, to resonant depolarization of the beam. Table 1 gives the numbers of linear resonances along with conditions for their occurrence in the booster.

Content from this work may be used under the terms of the Creative Commons Attribution 3.0 licence. Any further distribution of this work must maintain attribution to the author(s) and the title of the work, journal citation and DOI.
Published under licence by IOP Publishing Ltd

Journal of Physics: Conference Series **678** (2016) 012015 doi:10.1088/1742-6596/678/1/012015

Table 1. Number of linear resonances in a racetrack booster.

Particles	**Number of resonances**			
	integer $\nu = k$	intrinsic $\nu = kN \pm \nu_y$	coupling $\nu = kN \pm \nu_x$	non-super-periodic $\nu = k \pm \nu_y (k \neq mN)$
protons	16	16	32	16
deuterons	0	0	0	1

Our calculations show that proton resonance strengths lie within the ranges of $10^{-4} \div 10^{-3}$ for integer resonances, $10^{-4} \div 10^{-3}$ for intrinsic resonances, and $10^{-7} \div 10^{-5}$ for coupling and non-super-periodic resonances. The strength of a single deuteron resonance is $w \sim 10^{-5}$. The calculations assumed that normalized emittances of the betatron motion are 5π mm mrad.

3. Crossing of spin resonances

Polarization after crossing of a resonance with a strength w is determined by the rate of crossing ε'. A fast crossing $\varepsilon' \gg w^2$ keeps the spin aligned with the field. A slow resonance crossing $\varepsilon' \ll w^2$ flips the spin direction along the field. In an intermediate situation $\varepsilon' \sim w^2$, the spin orientation significantly deviates from the field direction. In the considered example, the crossing rate can be expressed through the field ramp rate dB/dt as:

$$\varepsilon' \approx 4.5 \times 10^{-7} \frac{dB\,[T]}{dt\,[s]} \quad \text{for protons,} \qquad \varepsilon' \approx 1.8 \times 10^{-8} \frac{dB\,[T]}{dt\,[s]} \quad \text{for deuterons.}$$

There are various techniques for crossing of spin resonances: resonance strength compensation, intentional enhancement of the spin resonance strength using specially introduced magnetic fields (a partial Siberian snake), betatron tune jump, spin tune jump [2, 3, 4, 5] . It is worth noting that the aforementioned techniques allow one only to reduce the depolarization for each resonance crossing. Thus, crossing of a large number of dangerous resonances may eventually lead to a significant polarization loss. References [6, 7] proposed and experimentally tested a transparent crossing technique, which, in principle, allows one to eliminate polarization loss during a crossing. The limiting factors for the transparent and fast resonance crossings are effects of the spin and betatron tune spreads in the beam. For slow resonance crossings, preservation of the polarization is a complex task that requires consideration of the synchrotron energy oscillations of the beam particles and higher-order resonances.

4. Acceleration of polarized protons in a racetrack booster

Polarized protons are successfully accelerated up to about 25 GeV in the AGS. The AGS uses warm magnets with a sufficiently high field ramp rate that provides fast crossing of essentially all resonances. To preserve the polarization, the AGS uses a weak partial snake for integer resonances and betatron tune jumps for intrinsic resonances. Polarization losses at the end of the acceleration cycle in the AGS also depend on the beam emittances and constitute 10-30%. A racetrack booster can be built with warm magnets using the AGS experience.

Application of superconducting magnets opens new opportunities when working with polarized beams. Due to a lower field ramp rate in superconducting magnets (dipoles, quadrupoles), one cannot apply the techniques used at the AGS. However, in a superconducting racetrack booster, it is possible to use a solenoidal snake, which completely eliminates resonant beam depolarization.

Journal of Physics: Conference Series **678** (2016) 012015 doi:10.1088/1742-6596/678/1/012015

To preserve proton polarization up to 8 GeV/c, the required maximum longitudinal field integral of the snake is 30 T·m. Placement of the snake solenoids in a straight of the booster is shown in figure 1. For the snake length of 2×4.6 m, the maximum solenoid field is 3.3 T.

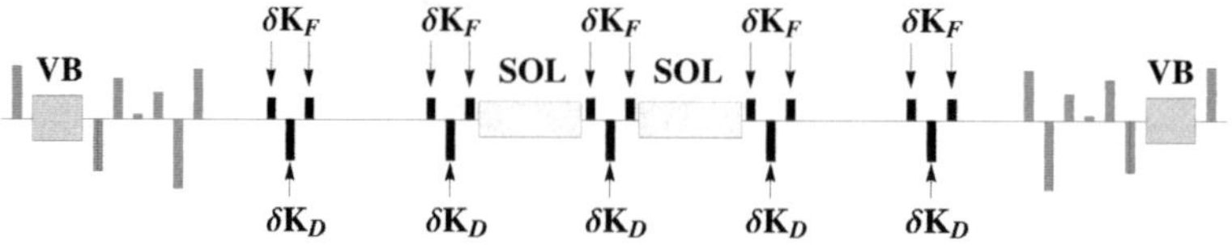

Figure 1. Placement of the snake in a booster's straight. SOL are solenoids, VB are arc dipoles.

The solenoids mainly introduce coupling of the betatron oscillations and shift the betatron tunes. For example, when matching the snake solenoids to the optics of the straight, it is sufficient to correct only the betatron tune shifts and leave the betatron oscillations coupled, similarly to what was proposed for the solenoidal snake in Nuclotron [8]. The tune shift can be compensated by adjusting the gradients of the two focusing and defocusing quadrupole families in the booster straight's triplets by small values δK_F and δK_D. Figures 2 and 3 show graphs of the β-functions for the cases of the booster without a snake and the booster with a snake with compensation of the betatron tune shifts introduced by the snake. A comparison of the graphs yields that the change in the β-functions in the booster with the snake is quite acceptable.

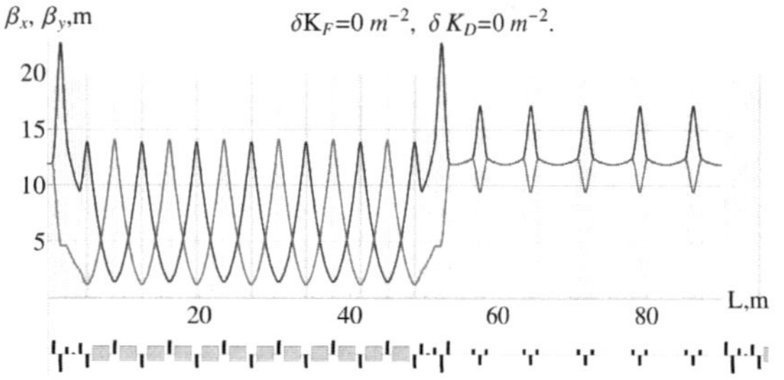

Figure 2. β-functions in one super-period of the booster without a snake.

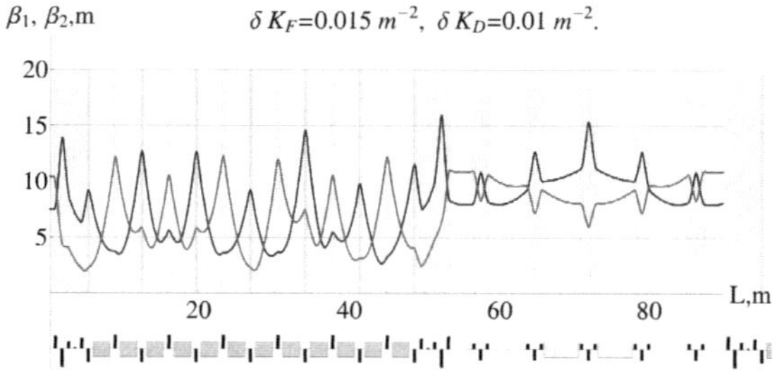

Figure 3. β-functions in one super-period of the booster with a snake.

5. Acceleration of polarized deuterons in a racetrack booster

As for protons, deuteron polarization can be preserved in the booster using a snake. However, use of a snake is not efficient for deuterons, since it requires a maximum longitudinal field integral of ~100 T·m and ~30 m of free space in a booster straight.

Due to small anomalous magnetic moment of the deuteron, its resonance grid is very sparse. In our example, there are no integer and intrinsic resonances at all. Also, the strength of the single non-super-periodic linear resonance related to lattice imperfections is quite small $w \sim 10^{-5}$. In this case, even when using superconducting magnets, it is possible to employ conventional acceleration of vertically polarized deuterons as for protons in the AGS.

Use of magnets with a field ramp rate of ~1 T/s guarantees fast crossing of the indicated resonance as well as of all higher-order spin resonances. Polarized deuterons were successfully accelerated in the Nuclotron, which uses superconducting magnets with a high field ramp rate [9].

6. Superconducting magnets for a racetrack booster

Superconducting pulsed solenoid and dipole with a field ramp rate of $dB/dt \approx 2 \div 4$ T/s can be manufactured using hollow high-current forced-helium-flow-cooled NbTi composite superconducting cable similar to that designed and manufactured at JINR for the Nuclotron [10]. The maximum magnetic field of the Nuclotron dipoles is limited to 2 T. Dipoles with fields of up to 3÷4.5 T and a field ramp rate of 1 T/s were proposed in references [11, 12]. A design work is presently being done on a solenoid for a proton snake for the Nuclotron. Use of Nuclotron-type magnets guarantees preservation of deuteron polarization in a racetrack booster.

When using conventional super-ferric magnets with a field ramp rate of ~1 T/min, preservation of polarization in the booster becomes a complex task, since the linear resonance is crossed at intermediate rate. In this case, one has to use fast quadrupoles to arrange a betatron tune jump. Higher-order resonances must be analyzed. One also has to carefully optimize the booster lattice and account for more subtle effects, such as those related to synchrotron energy oscillations, which lead to splitting of a resonance into a series of synchrotron side-band resonances. One should also take care of the orbital motion stability at the moments of betatron tune jumps.

Conclusion

In summary, one can draw the following conclusions:

- a racetrack booster built of superconducting magnets allows for acceleration of polarized protons and deuterons,
- acceleration of polarized protons requires a solenoidal Siberian snake,
- for acceleration of polarized deuterons, it is sufficient to use conventional spin resonance crossing techniques,
- fast-cycling Nuclotron-type magnets guarantee preservation of deuteron polarization in a racetrack booster up to 8 GeV/c and shorten the acceleration time to a few seconds.

Acknowledgments

Authored by Jefferson Science Associates, LLC under U.S. DOE Contracts No. DE-AC05-06OR23177 and DE-AC02-06CH11357. The U.S. Government retains a non-exclusive, paid-up, irrevocable, world-wide license to publish or reproduce this manuscript for U.S. Government purposes.

References

[1] Derbenev Ya S *et al* 2015 *Baseline Scheme for Polarization Preservation and Control in the MEIC Ion Complex* Proc. IPAC2015 TUPWI029 pp 2301-03

[2] Courant E D and Ruth R D 1980 *The Acceleration of Polarized Protons in Circular Accelerators* BNL 51270, Particle Accelerators and High Voltage Machines TIC-4500 ISA 80-5

[3] Ratner L G 1983*Polarized Beam at the AGS* AIP Conf. Proc. 95 pp 412-18

[4] Khiari F Z 1989 *Phys. Rev.* D **39** p 45-85

[5] Golubeva N I, Kondratenko A M and Filatov Yu N 1994 *A Jump in Spin Precession Frequency as a Method to Pass Spin Resonance* Proc. of the Int. Workshop Deuteron-93 p 374

[6] Kondratenko A M, Kondratenko M A and Filatov Yu N 2004 *Pepan Lett.* **1** 5 pp 266-69

[7] Morozov V S *et al* 2009 Experimental Test of a New Technique to Overcome Spin-Depolarizing Resonances *Phys. Rev. Lett.* **102** 244801

[8] Kondratenko M A *et al* 2013 *Orbital Parameters Of Proton Beam In Nuclotron With Solenoid Siberian Snake* Proc. XV Workshop on HESP DSPIN 2013 Dubna pp 371-74

[9] Kovalenko A D *et al* 2002 *Superconducting Fast-Cycling Dipole Magnets for the GSI Future Accelerator Facility* Proc. of EPAC2002 MOPLE022 pp 2406-08

[10] Smirnov A A *et al* 1984 A pulsed super-conducting dipole magnet for the Nuclotron *J. de Physique Colloques* **45 (C1)** pp C1-279-82.

[11] Agapov N N *et al* 2000 *Rapid Cycling Superconducting Booster Synchrotron* Proc. of 7th EPAC MOP3A17 pp 560-62

[12] Kovalenko A D *et al* 2010 Design of a twin aperture 4 T curved dipole based on a high current hollow superconducting cable for the NICA facility at JINR *J.Phys.Conf.Ser.* **234** 3 032033

Journal of Physics: Conference Series **678** (2016) 012016 doi:10.1088/1742-6596/678/1/012016

Lightlike solitons with spin

Alexander A. Chernitskii

[1] Department of Mathematics
St. Petersburg State Chemical Pharmaceutical Academy
Prof. Popov str. 14, St. Petersburg, 197022, Russia
[2] A. Friedmann Laboratory for Theoretical Physics
St. Petersburg, Russia

E-mail: AAChernitskii@mail.ru

Abstract. New exact solution class of Born – Infeld type nonlinear scalar field model is obtained. These solutions are three-dimensional solitons propagating with speed of light. The appropriate twisted solitons are considered. Energy and absolute value of momentum for such soliton are proportional to absolute value of its angular momentum or spin in some approximation. The model of ideal gas for the twisted lightlike solitons with lowest twist parameter is considered. Explicit conditions provide that the spin of each soliton equals Planck constant. It is shown that these solitons look like photons.

1. Introduction

A nonlinear space-time scalar field model considered here is known for a long time sufficiently. This model is related to well known Born – Infeld nonlinear electrodynamics[1]. It is sometimes called Born – Infeld type scalar field model.

This model is attractive because it has relatively simple and geometrically clear form. It can be considered as a relativistic generalization of the minimal surface or minimal thin film model in three-dimensional space. In this generalization we have an extremal four-dimensional hypersurface or film in five-dimensional space-time. Thus this model can be called the extremal space-time film one.

The model under consideration has much in common with Born – Infeld nonlinear electrodynamics. It can be shown that the methods for description of soliton interaction used in nonlinear electrodynamics [2, 3] applying to the scalar model give the similar results.

Here we present a short report of investigation for the lightlike soliton solutions of the extremal space-time film model. More details can be found in article [4].

2. Extremal space-time film

We consider the action in the following world volume form:

$$\mathcal{A} = \int\limits_{\overline{V}} \sqrt{|\mathfrak{M}|}\,(\mathrm{d}x)^4\,, \quad \text{where} \quad \mathfrak{M} \doteqdot \det(\mathfrak{M}_{\mu\nu})\,, \quad \mathfrak{M}_{\mu\nu} = \mathfrak{m}_{\mu\nu} + \chi^2\,\frac{\partial\Phi}{\partial x^\mu}\,\frac{\partial\Phi}{\partial x^\nu}\,, \tag{1}$$

$(\mathrm{d}x)^4 \doteqdot \mathrm{d}x^0\mathrm{d}x^1\mathrm{d}x^2\mathrm{d}x^3$, $\overline{V}$ is a space-time volume, $\mathfrak{m}_{\mu\nu}$ are components of metric tensor for flat four-dimensional space-time, Φ is scalar real field function, χ is a dimensional constant. The Greek indices take values $\{0, 1, 2, 3\}$.

Content from this work may be used under the terms of the Creative Commons Attribution 3.0 licence. Any further distribution of this work must maintain attribution to the author(s) and the title of the work, journal citation and DOI.
Published under licence by IOP Publishing Ltd

The tensor $\mathfrak{M}_{\mu\nu}$ can be called also the world tensor. The variational principle $\delta\mathcal{A} = 0$ with the world volume action (1) corresponds to extremal four-dimensional film $\Phi(\{x^\mu\})$ in five-dimensional space-time $\{\Phi, x^0, x^1, x^2, x^3\}$.

It is evident that the model under consideration keeps invariance for space-time rotation and scale transformation. Thus any solution of the model field equation give birth to the appropriate class of solutions with the following transform:

$$\Phi(\{x^\mu\}) \rightarrow a\,\Phi(\{L^\mu_{.\nu}\,x^\nu/a\})\,, \tag{2}$$

where $L^\mu_{.\nu}$ are components of space-time rotation matrix, a is scale parameter.

Here we use the Minkowski metric $\mathfrak{m}^{00} = 1$, $\mathfrak{m}^{11} = \mathfrak{m}^{22} = \mathfrak{m}^{33} = -1$. For the case of another metric signature $\{-,+,+,+\}$ in (1) we have the field model with some distinctions [4]. But the appropriate details are inessential here.

3. Lightlike soliton

We consider solutions in the form of wave propagating along x^3 axis of Cartesian coordinate system with the speed of light. Thus we can write

$$\Phi = \Phi\left(\theta, x^1, x^2\right)\,, \quad \theta = \omega\,x^0 - k_3\,x^3\,, \quad k_3 = \pm\omega\,, \quad \omega > 0\,. \tag{3}$$

The substitution this function (3) to the model field equation gives the following one:

$$\left(1 - \chi^2\left(\frac{\partial\Phi}{\partial x^2}\right)^2\right)\frac{\partial^2\Phi}{(\partial x^1)^2} + 2\,\chi^2\,\frac{\partial\Phi}{\partial x^1}\,\frac{\partial\Phi}{\partial x^2}\,\frac{\partial\Phi^2}{\partial x^1\partial x^2} + \left(1 - \chi^2\left(\frac{\partial\Phi}{\partial x^1}\right)^2\right)\frac{\partial^2\Phi}{(\partial x^2)^2} = 0\,. \tag{4}$$

We introduce new independent variables $\xi = x^1 + \imath\,x^2$, $\overset{*}{\xi} = x^1 - \imath\,x^2$ ($\imath^2 = -1$) and obtain the general solution of the transverse equation (4) in the form

$$\Phi = \Xi_1(\tilde{\xi}_1) + \Xi_2(\tilde{\xi}_2)\,, \tag{5a}$$

where the functions $\Xi_1(\tilde{\xi}_1)$ and $\Xi_2(\tilde{\xi}_2)$ are arbitrary with restriction by reality of the field function Φ. The connection between variables $\{\tilde{\xi}_1, \tilde{\xi}_2\}$ and $\{\xi, \overset{*}{\xi}\}$ is defined by relations

$$\begin{pmatrix} \mathrm{d}\xi \\ \mathrm{d}\overset{*}{\xi} \end{pmatrix} = \begin{pmatrix} 1 & \chi^2\left(\Xi_2'\right)^2 \\ \chi^2\left(\Xi_1'\right)^2 & 1 \end{pmatrix} \begin{pmatrix} \mathrm{d}\tilde{\xi}_1 \\ \mathrm{d}\tilde{\xi}_2 \end{pmatrix}\,. \tag{5b}$$

According to (2) we have the following solution class from solution (5):

$$\Phi(\xi, \overset{*}{\xi}) \rightarrow \sqrt{\eth\,\overset{*}{\eth}}\;\Phi\left(\xi/\eth, \overset{*}{\xi}/\overset{*}{\eth}\right)\,, \tag{6}$$

where $\eth = \overline{\rho}\,\mathrm{e}^{\imath\overline{\varphi}}$ is arbitrary complex constant with respect to coordinates $\{\xi, \overset{*}{\xi}\}$. Here $\overline{\rho}$ and $\overline{\varphi}$ are real constants with respect to coordinates $\{x^1, x^2\}$. But in general case these constants can be depend on the phase θ of soliton: $\overline{\rho} = \overline{\rho}(\theta)$, $\overline{\varphi} = \overline{\varphi}(\theta)$.

We take the arbitrary functions Ξ_1 and Ξ_2 in (5) which are appropriate to the solution

$$\Phi = \frac{\rho\!\!\!/^{\,m+1}}{\chi\,m}\left(\tilde{\xi}^{-m} + \overset{*}{\tilde{\xi}}{}^{-m}\right)\,, \tag{7a}$$

where m is natural number and constant $\rho\!\!\!/$ has a physical dimension of length.

We consider the twisted Gaussian lightlike soliton with the following phase functions

$$\overline{\varphi} = \pm\frac{\theta}{m}\,, \quad \overline{\rho} = \exp\left(-\frac{\theta^2}{2\,\overline{\theta}^2}\right)\,, \tag{7b}$$

where the signs '+' and '−' correspond to right and left twisted soliton accordingly, $\overline{\theta}$ is non-dimensional characteristic length of the soliton.

For slowly varying scale function $\overline{\rho}(\theta)$ ($\overline{\theta} \gg 1$) and sufficiently high frequency ω ($\overline{\rho}\omega \gg 1$) we have the following notable relations between energy, momentum, and angular momentum of the twisted lightlike soliton:

$$\mathbb{E} = \mathbb{P} = \tilde{\omega}\,\mathbb{J}\,, \quad \text{where} \quad \mathbb{P} \div |\mathbf{P}|\,, \quad \mathbb{J} \div |\mathbf{J}|\,, \quad \tilde{\omega} \div \frac{\omega}{m}\,, \tag{8}$$

$\tilde{\omega}$ is the angular velocity of the twisted soliton.

The reversibility condition for the variable transformation (5b) is violated on a singular line which is a singular surface in three-dimensional space. Thus we have a shell soliton with excluded inner area.

Some graphics relating to solution (7) with twist parameter $m = 1$ are shown on Figures 1 – 3. The twist of soliton is well seen on these figures. We see two-sheeted helical surface with excluded cavity on Fig. 2 for $m = 1$. In general case we have the helical surfaces with $2\,m$ sheets.

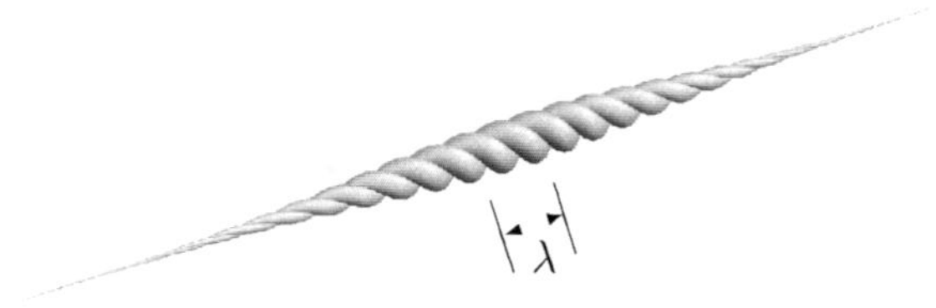

Figure 1. The shell of Gaussian twisted soliton for $m = 1$ and $\overline{\theta} = 10$.

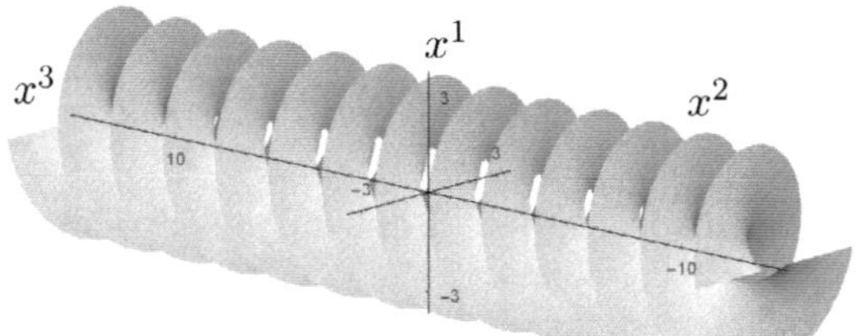

Figure 2. Zero level surfaces of the field function Φ for the Gaussian twisted soliton with $m = 1$ and $\overline{\theta} = 10$.

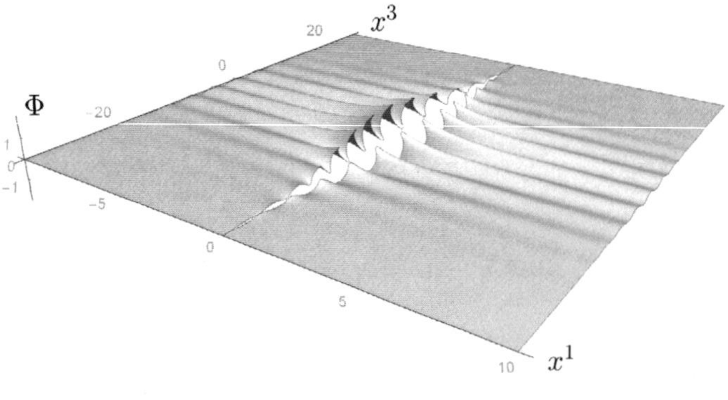

Figure 3. The field function Φ of Gaussian twisted soliton for $m = 1$ and $\overline{\theta} = 10$ on the plane $\{x^1, x^3\}$ for $x^2 = 0$.

4. Twisted solitons and photons

Let us consider ideal gas of the twisted lightlike solitons in a bounded three-dimensional volume V. Let us suppose also that absorptive and emissive capacities of the volume walls are provided by soliton-particles having the constant absolute value of angular momentum $\mathbb{J}_e = \hbar/2$, where $\hbar$ is Planck constant.

It can be shown [4] that in this case the absolute value of angular momentum or spin of the twisted lightlike solitons in the volume V must be equal to $\hbar$, because of the angular momentum conservation in absorption and emission events.

Because of notable connection between energy, momentum, and spin of the twisted lightlike solitons (8), for the case $m = 1$ we obtain the equilibrium distribution function by soliton frequencies which has the form of Planck formula for photons in high-frequency approximation.

A beam of the twisted lightlike solitons with $m = 1$ can provide the polarization property of light, because of an appropriate symmetry of the soliton solution. But the solitons with higher twist parameters $m \geqslant 2$ can not provide this polarization property [4].

Thus the lightlike solitons with twist parameter $m = 1$ can be considered as usual photons in some approximation. But the solitons of higher twist $m \geqslant 2$ have qualitative differences from the solitons of the lowest twist $m = 1$.

5. Conclusions

Thus the field model of extremal space-time film which is sometimes called Born – Infeld type scalar field model has considered in the presented investigation.

We have obtained new exact solution class for this model that is lightlike solitons. The soliton under consideration has a singularity which is a moving two-dimensional surface or shell. The general lightlike soliton can have a set of tubelike shells with the appropriate cavities which are excluded from the space.

We have considered the appropriate significant solution subclass that is twisted lightlike solitons. It is notable that energy and absolute value of momentum for the twisted lightlike soliton is proportional to absolute value of its angular momentum or spin in high-frequency approximation and for slowly varying scale function. We have the appropriate relations which are characteristic for photon in the case of lowest twist parameter $m = 1$ for explicit conditions.

Then we have investigated relations of the twisted lightlike solitons with $m = 1$ to photons. The model of ideal gas of the twisted lightlike solitons in a bounded volume has considered for this purpose. Planck formula for the soliton energy spectral density in the volume has obtained with explicit assumptions in some approximation.

An experimental check for a validity of the obtained soliton energy exact formula for real photon is proposed.

A beam of twisted lightlike solitons has considered. We have shown that this beam provides the effect of mechanical angular momentum transfer to absorbent by circularly polarized beam. This effect well known for photon beam. It has been found that the twisted lightlike soliton beam with $m = 1$ can provide polarization property of light as well as photon beam.

Thus we have a correspondence between photon and lightlike twisted soliton with the minimal value of twist parameter.

References

[1] Chernitskii A A 2005 *Encyclopedia of Nonlinear Science* ed Scott A (New York and London: Routledge) pp 67–69 (*Preprint* hep-th/0509087)

[2] Chernitskii A A 1999 Dyons and interactions in nonlinear (Born-Infeld) electrodynamics *J. High Energy Phys.* JHEP12(1999)010 (*Preprint* hep-th/9911093)

[3] Chernitskii A A 2012 *Nonlinear electrodynamics: singular solitons and their interactions* (Saint-Petersburg: ENGECON) in Russian

[4] Chernitskii A A 2015 Lightlike shell solitons of extremal space-time film *Preprint* 1506.09137

Journal of Physics: Conference Series **678** (2016) 012017 doi:10.1088/1742-6596/678/1/012017

Generalized DGLAP evolution

D Strózik-Kotlorz[1,2], S V Mikhailov[2] and O V Teryaev[2]

1 Opole University of Technology, Division of Physics, Prószkowska 76, 45-758 Opole, Poland
2 Bogoliubov Laboratory of Theoretical Physics, JINR, 141980 Dubna, Russia

E-mail: dorota@theor.jinr.ru

Abstract. We present progress in development of the truncated Mellin moments approach (TMMA). We show our recent results on the generalization of DGLAP evolution equations and discuss some their applications in spin physics.

1. Introduction

According to the factorization theorem, the cross sections for DIS reactions and some classes of hadron - hadron collisions can be expressed as convolution of two parts: a short-distance perturbative and a long-distance nonperturbative ones. The perturbative part, describing partonic cross sections at sufficiently high scale of the momentum transfer Q can be calculated within perturbative chromodynamics (pQCD). The non-perturbative part contains universal, process independent parton distribution functions $f(x)$ (PDF) and fragmentation functions $D_q^h(x)$ (FF), which can be measured experimentally. The evolution of these functions with the interaction scale Q^2 is again described with the use of the perturbative QCD methods. The standard DGLAP approach [1–3] enables one to calculate parton densities which characterize the internal nucleon structure at a given scale Q^2 when these densities are known for a certain input scale Q_0^2. We have shown that also the truncated Mellin moments of the PDFs, $\int_z^1 x^{n-1} f(x)dx$, satisfy the DGLAP evolution equations and can be an additional tool in the QCD analysis of structure functions. The major advantage of the TMMA is a possibility to adapt theoretical QCD analysis to the experimentally available region of the Bjorken-x variable. In this way, one can avoid the problem of dealing with the unphysical region $x \to 0$ corresponding to the infinite energy of interaction. A number of important issues in particle physics, e.g., solving of the 'nucleon spin puzzle', quark - hadron duality or higher twist contributions to the structure functions refers directly to moments. Note that TMM, contrary to standard moments, may be directly extracted from the accurate (JLab) data by appropriate binning (keeping Q^2 fixed). These issues initiate a large number of experimental projects and theoretical studies as well. Below we present the generalization of DGLAP evolution equations within TMMA and discuss some applications in spin physics.

2. Truncated Mellin moments approach

The main finding of the TMMA is that the generalized truncated (cut) moments (CMM) obtained by multiple integrations as well as multiple differentiations of the original parton distribution also satisfy the DGLAP equations with the simply transformed evolution kernel [4–6]. A similar generalized evolution equation, with the correspondingly modified coefficient functions, can also

Content from this work may be used under the terms of the Creative Commons Attribution 3.0 licence. Any further distribution of this work must maintain attribution to the author(s) and the title of the work, journal citation and DOI.
Published under licence by IOP Publishing Ltd

Journal of Physics: Conference Series **678** (2016) 012017 doi:10.1088/1742-6596/678/1/012017

Table 1. CMM (first column) and the corresponding evolution kernels (second column).

	Generalized CMM	DGLAP kernel $\mathcal{P}$
1.	$f(x)$	$P(y)$
2.	$x^n f(x)$	$P(y) \cdot y^n$
3.	$\int_z^1 dx\, x^{n-1} f(x)$	$P(y) \cdot y^n$
4.	$\int\limits_z^1 z_k^{n_k-1} dz_k \ldots \int\limits_{z_2}^1 z_1^{n_1-1} f(z_1)\, dz_1$	$P(y) \cdot y^{n_1+n_2+\ldots+n_k}$
5.	$f_\omega(z,n) = (\omega * f x^n)$	$P(y) \cdot y^n$
6.	$\left(-\frac{d}{dx}\right)^k [x^n f(x)]$	$P(y) \cdot y^{n-k}$

be obtained for structure functions. In Table 1, we summarize the generalized CMM together with the correspondingly transformed DGLAP evolution kernels.

3. Applications of TMMA

Below we present examples of applications of TMMA to analysis of the Bjorken sum rule [7] and fragmentation functions.

3.1. Generalized Bjorken sum rule

For any normalized function $\omega(x)$, $\int_0^1 \omega(t)\, dt = 1$, one can construct generalized CMM $f_\omega(z,n)$ as a Mellin convolution with the function f, which obeys the DGLAP evolution equation with the rescaled kernel [8]:

$$f_\omega(z,n) = (\omega * f x^n) \equiv \int_z^1 \omega\,(z/x)\;\; f(x)\, x^n\, dx/x\,, \tag{1}$$

$$\mathcal{P}(y) = P(y) \cdot y^n \tag{2}$$

The special case of f_ω is suitable for the generalized Bjorken sum rule (BSR):

$$G_\omega(x,Q^2) = \left(\omega * g_1^{NS}\right)(x). \tag{3}$$

G_ω has the same evolution kernel as g_1^{NS} and the generalized BSR is equal to the ordinary BSR:

$$\int_0^1 G_\omega(x,Q^2)\, dx = \int_0^1 g_1^{NS}(x,Q^2)\, dx = \text{BSR}. \tag{4}$$

The corresponding cut first moments of G_ω go to the BSR limit as the cut point x_0 goes to zero. This allows one to study behaviour of the generalized cut moments near $x_0 = 0$ and estimate the value of the BSR from the cut integrals $\int_{x_0}^1 G_\omega(x,Q^2)dx$ at $x_0 \neq 0$. The attempts for the case $\omega(t) = n\; t^{n-1}$ are shown in Fig. 1 and can be tested experimentally. We have also calculated contributions to the BSR itself and compared them to the experimental data. In Fig. 2 we compare TMMA predictions for the contributions to the BSR to recent COMPASS data. One

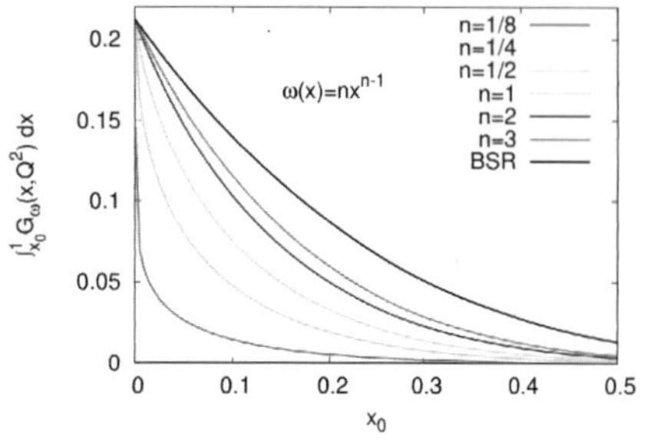

Figure 1. The cut first moments of the generalized CMM G_n (3), where $\omega = n\, t^{n-1}$, for different n versus the cut point x_0.

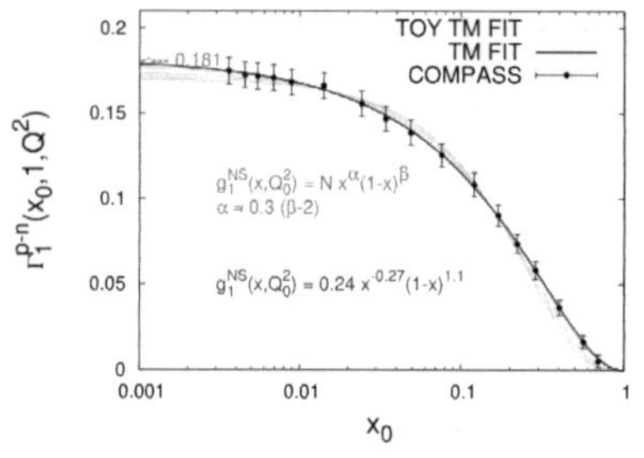

Figure 2. Contributions to the Bjorken sum rule obtained within TMMA. Comparison to COMPASS data.

Table 2. Truncated contributions to the Bjorken sum rule with use different input parameterizations. Comparison with experimental data.

N	INPUT	x -range	Q^2 [GeV^2]	Γ_1^{NS}	EXP Γ_1^{NS}
1.	$(1-x)^3$			0.161	HERMES
2.	$x^{-0.2}(1-x)^3$	0.021-0.9	5	0.149	$0.1479 \pm 0.0055 \pm 0.0142$
3.	$x^{-0.4}(1-x)^3$			0.131	
4.	$(1-x)^3$			0.177	COMPASS
5.	$x^{-0.2}(1-x)^3$	0.004-0.7	3	0.173	$0.175 \pm 0.009 \pm 0.015$
6.	$x^{-0.4}(1-x)^3$			0.163	
7.	$x^{\alpha}(1-x)^{\beta}$			0.168-	COMPASS
8.	$\alpha \approx 0.3(\beta-2)$	0.0025-0.7	3	0.170	0.170 ± 0.008

can see that simple input parametrization 7, where $\alpha = 0.3(\beta - 2)$ can satisfactorily reproduce the experimental data. This relation between α and β, together with the positivity constraint can provide knowledge on the small-x behaviour of the polarized structure function g_1^{NS}. From our analysis the favoured small-x behaviour of g_1^{NS} is x^{α}, where $\alpha = -0.2 \div -0.3$. Table 2 contains the truncated contributions to the Bjorken sum rule in the experimentally available x-region,

$$\Gamma_1^{p-n}(x_1, x_2, Q^2) = \int_{x_1}^{x_2} g_1^{NS}(x, Q^2)\, dx \tag{5}$$

obtained for different input parametrizations

$$g_1^{NS}(x, Q_0^2) = N\, x^{a_1}(1-x)^{a_2}. \tag{6}$$

Our predictions are compared with the HERMES [9] and COMPASS [10] data.

3.2. Fragmentation functions

Finally, it is worthy to mention that TMMA can be very useful in analysis of the hadron fragmentation functions as in the small-x region behaviour of FF is known very poorly. In this way, one can restrict the analysis to well determined models for $x \geq x_0$. FF also obey the corresponding DGLAP evolution and their CMM can provide new insight into the hadron structure.

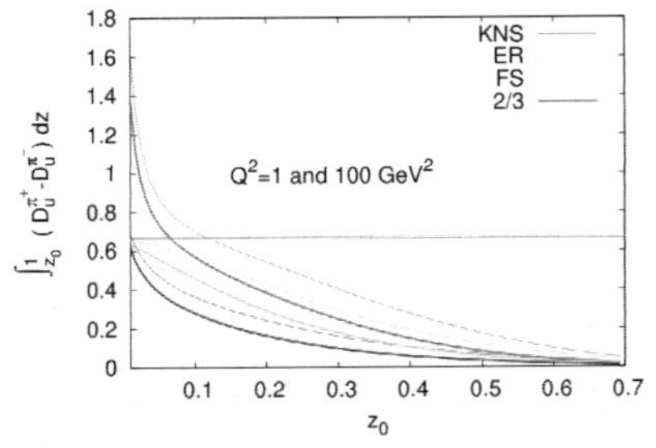

Figure 3. TMM of the pion FF, $Q^2 = 1, 100$ GeV^2 (thin, thick). Inputs: KNS [11], ER [12], FS [13]

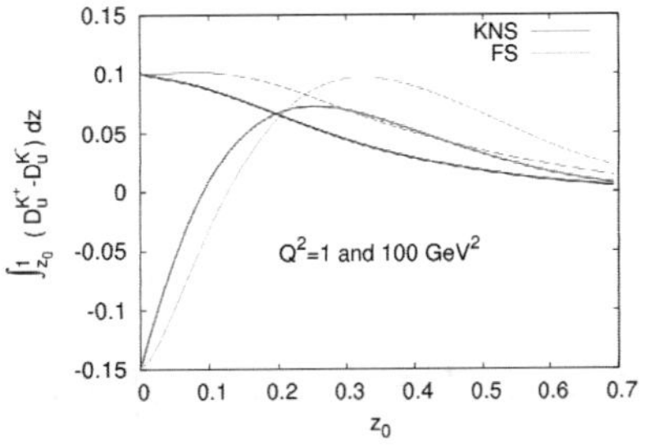

Figure 4. TMM of the kaon FF, $Q^2 = 1, 100$ GeV^2 (thin, thick). Inputs: KNS [11], FS [13]

In Figs. 3,4 we present evolution of the truncated moments of FF contributing to the quark charge conservation,

$$\sum_h Q_h \int_0^1 dz D_q^h(z, Q^2) = Q_q, \tag{7}$$

where D_q^h denotes a fragmentation function of the hadron h from a parton q and Q_h, Q_q are the charges of the hadron h and parton q. We use different parametrization of the FF at the initial scale [11–13]. The excess of the obtained moments for pions, providing the main contributions to sum rule, over the charge conservation values at $z_0 \sim 0.2$ may be considered as a support of inapplicability of independent fragmentation picture at this region. This fact is also exhibited in the large differences in the presented predictions in this region, depending on the input parametrization, for both pions and kaons. Therefore TMM approach can be a natural tool also in the study of the fragmentation functions, requiring further investigations.

Acknowledgments

We are grateful to A.V. Efremov to stimulating discussions. O.T. is indebted to S. Kumano for useful discussions. This work is supported by the Bogoliubov-Infeld Program, Grant No. 01-3-1113-2014/2018 and by the Russian Foundation for Fundamental Research, Grant No. 14-01-00647a.

References

[1] Gribov V N and Lipatov L N 1972 *Sov. J. Nucl. Phys.* **15** 438, *ibid.* 675
[2] Yu. L. Dokshitzer Yu L 1977 *Sov. Phys. JETP* **46** 641
[3] Altarelli G and Parisi G 1977 *Nucl. Phys.* B **126** 298
[4] Kotlorz D and Kotlorz A 2007 *Phys. Lett.* B **644** 284
[5] Kotlorz D and Kotlorz A 2011 *Acta Phys. Pol.* B **42** 1231
[6] Kotlorz D and Mikhailov S V 2014 *JHEP* **06** 065
[7] J. D. Bjorken J D 1966 *Phys. Rev.* **148** 1467, *ibid.* 1970 D **1** (1970) 1376
[8] Strózik-Kotlorz D, Mikhailov S V and Teryaev O V 2014 *Proc. of Science, Baldin ISHEPP XXII* **033**
[9] Airapetian A *et al.* [HERMES Collaboration] 2005 *Phys. Rev.* D **71** 012003, *ibid.* 2007 D **75** 012007
[10] Alexakhin V Yu *et al.* [COMPASS Collaboration] 2007 *Phys. Lett.* B **647** 8, Alekseev M *et al.* [COMPASS Collaboration] 2008, *Phys. Lett.* B **660** 458, Adolph C *et al.* [COMPASS Collaboration] 2015 [arXiv:1503.08935][hep-ex]
[11] Hirai M, Kumano S, Nagai T -H and Sudoh K 2007 *Phys. Rev.* D **75** 094009
[12] Efremov A V and Radyushkin A V 1980 *Rivista Del Nuovo Cimento* **3** 1, Sec. 1.6
[13] de Florian D, Sassot R and Stratmann M 2007 *Phys. Rev.* D **75** 114010

Journal of Physics: Conference Series **678** (2016) 012018 doi:10.1088/1742-6596/678/1/012018

Infrared modified QCD couplings and Bjorken sum rule

V L Khandramai[1], O V Teryaev[2], I R Gabdrakhmanov[2]

[1] International Center for Advanced Studies, Technical University, Gomel, 246746, Belarus
[2] Bogoliubov Laboratory of Theoretical Physics, JINR, Dubna, 141980, Russia

E-mail: v.khandramai@gmail.com

Abstract. We test the recently proposed "Massive" Perturbation Theory (MPT) for the description of the Γ_1^{p-n} data at low momentum transfers. The MPT constructed on the two grounds: the first is pQCD with only one parameter added, an effective "glueball mass" $m_\rho \lesssim M_{gl} \lesssim 1$ GeV; serving as an infrared "regulator"; the second stems out of the ghost-free Analytic Perturbation Theory comprising non-power perturbative expansion that makes it compatible with linear integral transformations. It is regular in the low-energy region and could serve as a practical means for the analysis of data below 1 GeV up to the IR-limit. We study the non-perturbative Bjorken sum rule higher twists correction by using the MPT, the integral representation for infinite sum of higher twists coefficients and the QCD-inspired model for the Q^2-dependence of the generalized Gerasimov-Drell-Hearn sum rule.

1. Introduction

The perturbative QCD (pQCD) is a firmly established part of the particle interaction theory. Starting with gauge-non-invariant quantization, it correlates several dozen of experiments at quite different scales from a few up to hundreds of GeV. At the same time, pQCD meets troubles in the low-energy domain, below a few GeV, at the scales marked by the QCD parameter Λ_{QCD}.

To avoid the unwanted singularity of the QCD running coupling in the low energy region, several modifications (for example, [1–3]) of the pQCD have been devised. Recently, one of them, the Analytic Perturbation Theory ([4] and a latter review paper [5]) (APT), has proved to be good [6] in describing the polarized $\Gamma_1^{p-n}(Q^2)$ moment of the Bjorken Sum Rule (BSR) down to a few hundred MeV.

To approach the global fitting of data, one needs a modified perturbation theory (MPT) with two essential properties: correspondence with common pQCD in ultra-violet limit (that is above a few GeV) and regularity and finiteness of the modified effective coupling $\alpha^{MPT}(Q^2)$ and matrix elements in the low-energy domain. As a primary launch pad for this construction, the above-mentioned APT seems good. It satisfies the first condition and, partially, the second one. To exempt the APT-like scheme from its last drawback – the singularity (infinite derivatives) in the infra-red limit, one has to disentangle it from the ultra-violet logs. To this goal the infra-red regulator has been introduced just by the shift of the Q^2 scale [7], $\ln\left(Q^2/\Lambda^2_{QCD}\right) \to \ln\left[(Q^2+M^2_{gl})/\Lambda^2_{QCD}\right]$, with the only fitting parameter added, an effective glueball mass, M_{gl}.

Content from this work may be used under the terms of the Creative Commons Attribution 3.0 licence. Any further distribution of this work must maintain attribution to the author(s) and the title of the work, journal citation and DOI.
Published under licence by IOP Publishing Ltd

2. Description of the methods

Let us briefly discuss methods of the QCD analysis of the Γ_1^{p-n} data base on the MPT.

Away from the large Q^2 limit, the BSR is given by a double series in powers of α_s and in powers of $1/Q^2$ and can be written as

$$\Gamma_1^{p-n}(Q^2) = \frac{|g_{A/V}|}{6}\left[1 - \Delta_{\rm Bj}(Q^2)\right] + \sum_{i=2}^{\infty} \frac{\mu_{2i}}{Q^{2i-2}}, \tag{1}$$

where $|g_{A/V}| = 1.2723$ [8] is the nucleon axial charge, μ_{2i} are the higher twist (HT) coefficients, $\Delta_{\rm Bj}(Q^2)$ is the perturbative correction, which at the four-loop (N^{3}LO) level in the massless case reads [9]

$$\Delta_{\rm Bj}(Q^2) = 0.318\,\alpha_s + 0.363\,\alpha_s^2 + 0.652\,\alpha_s^3 + 1.804\,\alpha_s^4\,. \tag{2}$$

In the framework of MPT perturbative α_s-series replaced by expansions over MPT functions $\mathcal{A}_k$:

$$\sum_k c_k\,\alpha_s^k \to \sum_k c_k\,\mathcal{A}_k, \tag{3}$$

where $\mathcal{A}_1(Q^2) \equiv \alpha^{MPT}(Q^2)$ is MPT running coupling, which is the two-loop massive renormalization group solution in the denominator representation (for details, see [10]) has a following form

$$\mathcal{A}_1(Q^2) \equiv \alpha^{MPT}(Q^2) = \frac{\alpha_0}{1 + \alpha_0\beta_0 L^* + \alpha_0\,\beta_1 \ln\left[1 + \alpha_0\beta_0 L^*\right]/\beta_0}, \quad L^* = \ln\left(\frac{Q^2 + M_{gl}^2}{\Lambda_{QCD}^2}\right).$$

The MPT preserves an essential APT feature, namely, the non-polynomiality of perturbative expansion over a set of higher functions $\mathcal{A}_k$ $(k > 1)$. These functions are connected by the differential recurrent relations (at NLO)

$$\beta_0\,\mathcal{A}_{k+1}(Q^2) = -\frac{Q^2 + M_{gl}^2}{k}\frac{d}{dQ^2}\mathcal{A}_k\,(Q^2) - \beta_1\,\mathcal{A}_{k+2}(Q^2). \tag{4}$$

In the following analysis, we use the natural condition

$$\Gamma_1^{p-n}(Q^2 = 0) = 0, \tag{5}$$

which is motivated by finiteness of the the spin-dependent cross-sections in real photon limit.

In the spirit of [11, 12], where Gerasimov–Drell–Hearn and Burkhardt–Cottingham sum rules considered for the purpose of a smooth continuation of $\Gamma_1^{p,n}(Q^2)$ to the non-perturbative region $0 \le Q^2 \lesssim \Lambda_{QCD}^2$, we obtain slope of Γ_1^{p-n} moment at the IR-limit $Q = 0$:

$$\frac{d}{dQ^2}\Gamma_1^{p-n}(Q^2 = 0) = \frac{-(\mu_p - 1)^2 + \mu_n^2}{8M^2}, \tag{6}$$

where $\mu_p = 2.79$ and $\mu_p = -1.91$ are proton and neutron magnetic moments [8], respectively, and $M = 0.938$ GeV is a nucleon mass.

We use the non-perturbative series summation procedure proposed in [13]. This procedure allows to represent a non-perturbative operator product expansion series introducing a single free parameter m_{ht}

$$\sum_{n=1}^{\infty} \mu_{2n+2}^{p-n}\left(\frac{m_{ht}^2}{Q^2}\right)^n \to \frac{\mu_4^{p-n}\,m_{ht}^2}{Q^2 + m_{ht}^2}. \tag{7}$$

Journal of Physics: Conference Series **678** (2016) 012018 doi:10.1088/1742-6596/678/1/012018

We use the conditions (5) and (6) to connect the free parameters M_{gl}, m_{ht} and μ_4^{p-n}. As a result we obtain an approach with only one free parameter – effective "glueball mass" M_{gl}

$$\Gamma_1^{p-n}(Q^2, M_{gl}^2) = \frac{|g_{A/V}|}{6}\Big[1 - 0.318\mathcal{A}_1(Q^2, M_{gl}^2) - 0.363\mathcal{A}_2(Q^2, M_{gl}^2) - \quad (8)$$

$$- 0.652\mathcal{A}_3(Q^2, M_{gl}^2) - 1.804\mathcal{A}_4(Q^2, M_{gl}^2) + ...\Big] + \frac{\mu_4^{p-n}\, m_{ht}^2}{Q^2 + m_{ht}^2}. \quad (9)$$

3. Results of fit and discussion

Using expression (1) fitted to the low Q^2 data [14–16], we can extract the HT coefficients μ_4^{p-n} and parameters of "effective glueball mass" M_{gl} and parameter m_{ht}^2 in the HT sum.

In the Table 1 we present our result for the coefficient μ_4 obtained in different PT orders.

Table 1. Central values of the extracted parameters.

Order	M_{gl}^2, GeV^2	m_{ht}^2, GeV^2	μ_4, GeV	$\chi^2_{d.o.f}$
LO	0.782	0.736	-0.170	6.24
NLO	0.954	0.724	-0.146	6.71
N^{2}LO	0.648	0.039	0.025	5.11
N^{3}LO	0.546	0.143	-0.055	0.73

One can see that the value $\mu_4^{MPT} = -0.055$ is compatible with previously extracted $\mu_4^{APT} = -0.050(2)$ [6].

Figure 1 shows the fits curves in various orders of MPT and for comparison four-loop APT curve from [6].

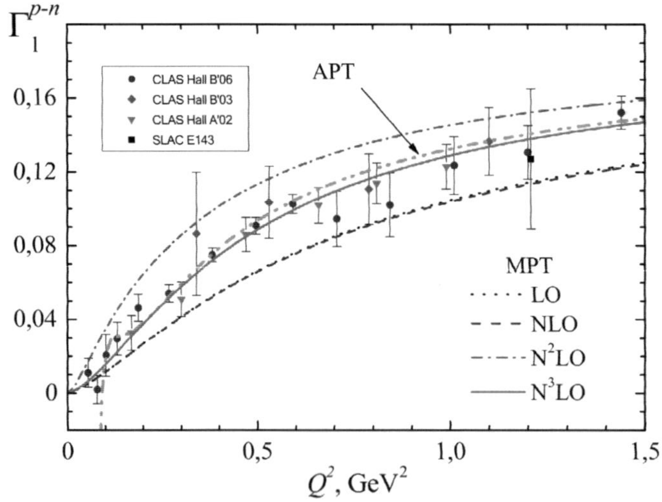

Figure 1. The fit of the Γ_1^{p-n} data in different orders of the MPT and the for-loop APT.

Journal of Physics: Conference Series **678** (2016) 012018 doi:10.1088/1742-6596/678/1/012018

In conclusion we stress, that MPT approach combined the IR-frozen and the non-polynomiality of perturbative expansions is a next step for understanding low-energy QCD. The MPT together with a duly modified HT sum allows one to fit the data on Γ_1^{p-n} down to the IR limit $Q^2 = 0$.

4. Acknowledgments

It is a pleasure for the authors to thank D. V. Shirkov and S. V. Mikhailov for interest in this work and helpful discussions. This research was supported by the JINR-BelRFBR grants F15RM-072, F14D-007 and by the RFBR Grant 14-01-00647.

References

[1] Shirkov D V, Mikhailov S.V. 1994 *Z. Phys.* C **63** 463
[2] Simonov Yu A 1995 *Phys. Atom. Nucl.* **58** 107
[3] Nesterenko A V, Papavassiliou J 2006 *J. Phys.* **32** 1025
[4] Solovtsov I L, Shirkov D V 1999 *Theor. Math. Phys.* **120** 1220
[5] Shirkov D V, Solovtsov I L 2007 *Theor. Math. Phys.* **150** 132
[6] Khandramai V L, Pasechnik R S, Shirkov D V, Solovtsova O P, Teryaev O V 2012 *Phys. Lett.* B **706** 340
[7] Shirkov D V 2013 *Phys. Part. Nucl. Lett.* **10** 186
[8] Olive K A *et al.* 2014 *Chin. Phys.* C **38** 090001
[9] Baikov P A, Chetyrkin K G, Kühn J H 2010 *Phys. Rev. Lett.* **104** 132004
[10] Shirkov D V Mass 1982 *Theor. Math. Phys.* **49** 1039
[11] Soffer J, Teryaev O V 1993 *Phys. Rev. Lett.* **70** 3373
[12] Soffer J, Teryaev O V 2004 *Phys. Rev.* **70** 116004
[13] Teryaev O V 2013 *Nucl. Phys. Proc. Suppl.* **245** 195
[14] Deur A *et al.* 2004 *Phys. Rev. Lett.* **93** 212001
Deur A *et al.* 2007 *Phys. Lett.* B **650** 244
Deur A *et al.* 2008 *Phys. Rev.* D **78** 032001
Deur A *et al.* 2008 *Phys. Lett.* B **665** 349
Deur A *et al.* 2014 *Phys. Rev.* D **90** 012009
[15] Alekseev M G *et al.* 2010 *Phys. Lett.* B **690** 466
[16] Abe K *et al.* 1997 *Phys. Rev. Lett.* **79** 26
Anthony P L *et al.* 2000 *Phys. Lett.* B **493** 19

Journal of Physics: Conference Series **678** (2016) 012019 doi:10.1088/1742-6596/678/1/012019

Spin observables in antihyperon-hyperon production at $\overline{\text{P}}$ANDA at FAIR

Michael Papenbrock
on behalf of the $\overline{\text{P}}$ANDA collaboration

Department of Physics and Astronomy, Uppsala University, Box 516, S-751 20 Uppsala, Sweden

E-mail: michael.papenbrock@physics.uu.se

Abstract. Hyperon production is an excellent probe of QCD in the confinement domain, and spin observables are a powerful tool in understanding the underlying physics. For the Ω hyperon, seven polarisation parameters can be extracted from the angular distributions of its decay products with the future $\overline{\text{P}}$ANDA experiment at FAIR. Simulation studies reveal great prospects for strange and single charmed hyperon channels with $\overline{\text{P}}$ANDA. Software tools supporting these investigations are currently under development.

Hyperon production in $\bar{p}p \to \overline{Y}Y$ reactions gives important insights into strangeness and charm production. In particular, the production of single and multi-strange as well as single-charmed hyperons from the light quarks in $\bar{p}p$ implies processes where the latter are converted into heavier quarks. For any given process the relevant degrees of freedom are given by its energy scale, which for strangeness production is determined by the mass of the strange quark, $m_s \approx 100\,\text{MeV}$. This is close to the QCD cut-off, $\Lambda_{QCD} \approx 200\,\text{MeV}$ [1], where the coupling constant α_s grows too large for perturbative QCD to be applicable. Therefore, the production of strange hyperons presents an excellent tool to study QCD in the region below the perturbative regime, *i.e.* the confinement domain, about which very little is known so far. On the other hand, the scale of charm production is governed by the mass of the charm quark, $m_c \approx 1300\,\text{MeV}$, being more than ten times larger than m_s. In this region, the strong coupling constant becomes $\alpha_s \approx 0.3$, which is close to the standard limit for perturbative QCD of 0.1. A perturbative treatment could therefore be used but with some caution. Comparing the production of strange and charmed hyperons could give important insights into the differences in the production mechanism at these two energy scales. The production of hyperons in $\bar{p}p \to \overline{Y}Y$ is often described by theoretical models based on a constituent quark-gluon picture [2]. In case of strange hyperons, models based on the exchange of one or more t-channel kaons exist as well [3]. There have also been efforts to combine both approaches [4]. In order to discriminate between different models, spin observables are very powerful. For hyperons these can be related to the spin of the individual quarks. The Λ hyperon is often modelled as consisting of a spin 0 ud di-quark and an s-quark carrying the hyperon spin. Since the Λ_c^+ hyperon can be described in a similar way, comparing the spin observables of Λ and Λ_c^+ could give deeper insight into the role of spin degrees of freedom in the production of s- and c-quarks. Strangeness production is often discussed in terms of the Okubo-Zweig-Iizuka (OZI) rule, stating that all processes with disconnected quark lines are suppressed [5]. For baryons, the OZI rule implies that single strangeness hyperon

Content from this work may be used under the terms of the Creative Commons Attribution 3.0 licence. Any further distribution of this work must maintain attribution to the author(s) and the title of the work, journal citation and DOI.
Published under licence by IOP Publishing Ltd

production is suppressed with respect to production of non-strange hadrons, and multistrange production suppressed with respect to single strangeness production [6]. Deviations from the OZI rule have been interpreted as *e.g.* gluonic intermediate states [7] or a polarised strangeness component of the nucleon [8]. In the same spirit, a model of the $\bar{p}p \to \bar{\Lambda}\Lambda$ process was suggested in Ref. [9], where a $s\bar{s}$ pair inside the initial proton or antiproton wave function dissociates and rearrange into the $\Lambda\bar{\Lambda}$ pair in the final state.

The density matrix for arbitrary spin j is given by

$$\rho = \frac{1}{2j+1}\mathscr{I} + \sum_{L=1}^{2j}\frac{2j}{2j+1}\sum_{M=-L}^{L} Q^L_M r^L_M, \tag{1}$$

in terms of the hermitian matrices Q^L_M and the polarisation parameters r^L_M [10]. Here, the first term represents the unpolarised differential cross section I, and the second the polarised part with the r^L_M parameters, L being the angular momentum and M its third component. For spin $\frac{1}{2}$ particles the Q^L_M are the Pauli matrices σ_i and the polarisation parameters correspond to the vector polarisations P_l, P_m and P_n. The indices are defined in the coordinate system shown in the left panel of Figure 1. For a single spin $\frac{1}{2}$ particle the spin density matrix is given by

$$\rho(1/2) = \frac{1}{2}(\mathscr{I} + \bar{P}\cdot\bar{\sigma}) = \frac{1}{2}\begin{bmatrix} 1+P_l & P_m + iP_n \\ P_m - iP_n & 1-P_l \end{bmatrix}. \tag{2}$$

In parity conserving processes, *e.g.* $\bar{p}p \to \bar{\Lambda}\Lambda$ with unpolarised beam and target, P_l and P_m equal zero due to symmetries of the spin density matrix, yielding

$$\rho(1/2) = \frac{1}{2}\begin{bmatrix} 1 & iP_n \\ -iP_n & 1 \end{bmatrix}. \tag{3}$$

The only non-zero polarisation component is the one perpendicular to the production plane, namely P_n.

Since the hyperons have a weak, parity violating decay, the decay products are preferentially emitted in the direction of the spin of the parent hadron. The angular distribution of the final state is given by

$$I(\theta,\phi) = \mathrm{Tr}(T\rho T^*), \tag{4}$$

in which T is the decay matrix consisting of parity conserving s-wave part T_s and a parity violating p-wave part T_p. Furthermore, we define the decay parameters α, β and γ as

$$\begin{aligned} \alpha &= 2\Re(T_s^* T_p) & (5)\\ \beta &= 2\Im(T_s^* T_p) & (6)\\ \gamma &= |T_s|^2 - |T_p|^2 & (7) \end{aligned}$$

with the relation

$$\alpha^2 + \beta^2 + \gamma^2 = |T_s|^2 + |T_p|^2 = 1. \tag{8}$$

For $\Lambda \to p\pi^-$, the polarisation is accessible by measuring the angular distribution of the proton, both of which are related by

$$I(\cos\theta_p) = \frac{1}{4\pi}(1 + \alpha_\Lambda P_n \cos\theta_p), \tag{9}$$

with $\alpha_\Lambda = 0.64$ [11] being the asymmetry parameter. Some hyperons decay into other weakly decaying hyperons, *e.g.* $\Xi^- \to \Lambda\pi^-$ (see right panel of Figure 1), in which case the additional

Journal of Physics: Conference Series **678** (2016) 012019 doi:10.1088/1742-6596/678/1/012019

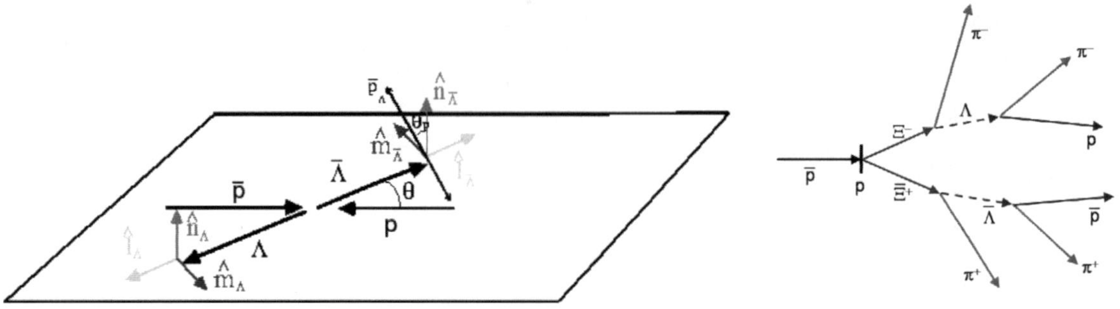

(a) (b)

Figure 1. (a) The reference system of $\bar{p}p \to \bar{Y}Y$. **(b)** The $\bar{p}p \to \bar{\Xi}\Xi, \Xi \to \Lambda\pi, \Lambda \to p\pi$ decay.

asymmetry parameters β and γ become accessible through the angular distribution of the protons [12] via

$$I(\theta_p, \phi_p) = \frac{1}{2\pi}\left[1 + \alpha_\Xi \alpha_\Lambda \cos\theta_p + \frac{\pi}{4}\alpha_\Lambda P \sin\theta_p \left(\beta_\Xi \sin\phi_p - \gamma_\Xi \cos\phi_p\right)\right]. \quad (10)$$

The spin structure of spin $\frac{3}{2}$ hyperons is more complicated. Fifteen Q^L_M matrices with corresponding r^L_M parameters become relevant. Eight polarisation parameters were found to be zero using symmetries imposed by the strong interaction [13]. The spin density matrix is given by

$$\rho(3/2) = \frac{1}{4}\begin{bmatrix} 1+\sqrt{3}r^2_0 & i\frac{3}{\sqrt{5}}r^1_{-1}-\sqrt{3}r^2_1 & \sqrt{3}r^2_2-i\sqrt{3}r^3_{-2} & -i\sqrt{6}r^3_{-3} \\ -i\frac{3}{\sqrt{5}}r^1_{-1}-\sqrt{3}r^2_1 & 1-\sqrt{3}r^2_0 & i2\sqrt{\frac{3}{5}}r^1_{-1}+i3\sqrt{\frac{2}{5}}r^3_{-1} & \sqrt{3}r^2_2+i\sqrt{3}r^3_{-2} \\ \sqrt{3}r^2_2+i\sqrt{3}r^3_{-2} & -i2\sqrt{\frac{3}{5}}r^1_{-1}-i3\sqrt{\frac{2}{5}}r^3_{-1} & 1-\sqrt{3}r^2_0 & i\frac{3}{\sqrt{5}}r^1_{-1}+\sqrt{3}r^2_1 \\ i\sqrt{6}r^3_{-3} & \sqrt{3}r^2_2-i\sqrt{3}r^3_{-2} & -i\frac{3}{\sqrt{5}}r^1_{-1}+\sqrt{3}r^2_1 & 1+\sqrt{3}r^2_0 \end{bmatrix}. \quad (11)$$

In the $\Omega^- \to \Lambda K^-$ decay the parameters r^2_0, r^2_1, and r^2_2 can be obtained from the angular distribution of the Λ [13], assuming that $\alpha_\Omega = 0$ according to previous measurements [11]. They have been derived as

$$r^2_0 = \frac{15}{2\sqrt{3}}\left(\frac{1}{3} - \langle\cos^2\theta_\Lambda\rangle\right) \quad (12)$$

$$r^2_2 = \frac{8}{3}\left(1 - \langle\cos^2\theta_\Lambda\rangle - 2\langle\sin^2\theta_\Lambda \sin^2\phi_\Lambda\rangle\right) \quad (13)$$

$$r^2_1 = 5\langle\cos\theta_\Lambda \sin\theta_\Lambda \cos\phi_\Lambda\rangle. \quad (14)$$

The combined angular distribution $I(\theta_\Lambda, \phi_\Lambda, \theta_p, \phi_p)$ of the Λ hyperons from Ω^- decay and the protons from the subsequent Λ decay grants access to the other four non-zero polarisation parameters r^1_{-1}, r^3_{-1}, r^3_{-2}, and r^3_{-3}:

$$r^1_{-1} = -\frac{20\sqrt{10}\langle(3\cos\theta_\Lambda - 1)\sin\phi_p\rangle}{3\pi\alpha_\Lambda\gamma_\Omega} \quad (15)$$

$$r^3_{-1} = \frac{2\sqrt{5}\langle(15\cos\theta_\Lambda - 1)\sin\phi_p\rangle}{\sqrt{3}\pi\alpha_\Lambda\gamma_\Omega} \quad (16)$$

$$r^3_{-2} = -\frac{1024\langle\sin\phi_\Lambda \cos\phi_p\rangle}{3\pi^2\alpha_\Lambda\gamma_\Omega} \quad (17)$$

$$r^3_{-3} = -\frac{1}{5\sqrt{6}}\left(\frac{640}{\pi\alpha_\Lambda\gamma_\Omega}\langle\sin\phi_\Lambda \cos\phi_\Lambda \sin\phi_p\rangle + 4\sqrt{15}r^3_{-1} + 3\sqrt{10}r^1_{-1}\right) \quad (18)$$

Journal of Physics: Conference Series **678** (2016) 012019 doi:10.1088/1742-6596/678/1/012019

With the $\overline{p}p \to \overline{Y}Y$ reaction CP violation can be studied in a clean way. Being a CP eigenstate, the initial state has the same symmetry as the final state. Therefore, no hadronic final state interactions can produce a false signal. Also, due to baryon number conservation there is no mixing between the antihyperon and the hyperon. If CP is conserved, then $\alpha = -\overline{\alpha}$ and the CP violation parameters

$$A = \frac{\Gamma\alpha + \overline{\Gamma}\overline{\alpha}}{\Gamma\alpha - \overline{\Gamma}\overline{\alpha}} \simeq \frac{\alpha + \overline{\alpha}}{\alpha - \overline{\alpha}} \tag{19}$$

$$B = \frac{\Gamma\beta + \overline{\Gamma}\overline{\beta}}{\Gamma\beta - \overline{\Gamma}\overline{\beta}} \simeq \frac{\beta + \overline{\beta}}{\beta - \overline{\beta}} \tag{20}$$

$$B' = \frac{\Gamma\beta + \overline{\Gamma}\overline{\beta}}{\Gamma\alpha - \overline{\Gamma}\overline{\alpha}} \simeq \frac{\beta + \overline{\beta}}{\alpha - \overline{\alpha}}, \tag{21}$$

where Γ is the partial decay width, should be zero. These parameters can be obtained for the Ξ since the polarisation of the decay products can be retrieved. None such measurement has been conducted before.

The major part of the existing data on $\overline{p}p \to \overline{Y}Y$ are collected with beam momenta below 2 GeV/c. In this region, the PS185 experiment provided a high-quality data set for Λ and Σ hyperons [14]. The spin structure of the $\overline{p}p \to \overline{\Lambda}\Lambda$ reaction was explored in detail, but neither quark-gluon models nor hadron models were in agreement with the data on the spin observables [15]. For multi-strange and charmed hyperons produced in $\overline{p}p$ annihilation the existing data are much more limited, with only a few bubble chamber events on $\overline{p}p \to \overline{\Xi}^+\Xi^-$ and no data on $\overline{p}p \to \overline{\Omega}^+\Omega^-$ nor $\overline{\Lambda}_c^-\Lambda_c^+$. New experimental data will serve as a guideline for the theoretical investigation which in turn could widen our understanding of the underlying processes in this energy regime.

The foreseen $\overline{\text{P}}$ANDA experiment at FAIR provides unique possibilities for the investigation of hyperon physics. The HESR storage ring can be operated with antiprotons in a momentum range from 1.5 to 15 GeV/c, interacting with an internal hydrogen target. It is foreseen to run in two luminosity modes, one at $1{\cdot}10^{31}$ cm^{-2} s^{-1} for high resolution measurements which will be available from day-1, and later on also a mode with an order of magnitude higher luminosity in order to investigate rare processes. The $\overline{\text{P}}$ANDA experiment will have nearly full angular coverage with precise tracking and vertex reconstruction, particle identification, calorimetry, and will benefit from the high luminosity of HESR [16]. Simulation studies with a simplified Monte Carlo framework show excellent prospects with high signal and low background rates and a good detection efficiency across the full phase space for all strange and single-charmed hyperon channels [13, 20]. Since these measurements are feasible from start of the experiment, the estimated signal rates in Table 1 are calculated using the day-1 luminosity mode of HESR and will increase significantly once the high luminosity mode becomes available. The only well known cross section is the one for the $\overline{\Lambda}\Lambda$ channel at 1.64 GeV/c, whereas the cross sections for the other strange channels are extrapolated from data at other energies. For the heavier hyperons, *i.e.* Ω and Λ_c, where no data exist from $\overline{p}p$ annihilation, the values are taken from theoretical predictions which can differ significantly [18, 19].

Simulation studies of spin observables for Λ, Σ, Ξ, Ω, and Λ_c hyperons have been performed as well. Since there exist no data on multi-strange or charmed hyperon channels, a parametrisation based on sine functions was introduced at the event generation stage. These fulfil the requirement of the definition of the production plane (see Figure 1) of being equal to zero at $\cos\theta_Y = \pm 1$. Figure 2 shows some examples of spin observables of Ξ and Ω, parametrised by $P_{y,\Xi} = \sin 2\theta_\Xi$, $C_{xx,\Xi} = \sin\theta_\Xi$ and $r^2_{0,\Omega} = \sin 2\theta_\Omega/\sqrt{3}$. It is clear that these can be accurately reconstructed with $\overline{\text{P}}$ANDA.

Table 1. Estimated production rates of various hyperon channels [13, 20]. The cross sections marked with a * have a very large uncertainty. Details are given in the text.

Beam mom. (GeV/c)	Reaction	σ (μb)	Eff (%)	Decay	Rate at $1{\cdot}10^{31}$ cm^{-2}s^{-1}
1.64	$\bar{p}p \to \bar{\Lambda}\Lambda$	64	11	$\Lambda \to p\pi^-$	29 s^{-1}
4	$\bar{p}p \to \bar{\Lambda}\Sigma^0$	≈40	≈30	$\Sigma^0 \to \Lambda\gamma$	50 s^{-1}
4	$\bar{p}p \to \bar{\Xi}^+\Xi^-$	≈2	≈20	$\Xi^- \to \Lambda\pi^-$	1.5 s^{-1}
12	$\bar{p}p \to \bar{\Omega}^+\Omega^-$	≈0.002*	≈30	$\Omega \to \Lambda K^-$	≈4 h^{-1}
12	$\bar{p}p \to \bar{\Lambda}_c^-\Lambda_c^+$	≈0.1*	≈35	$\Lambda_c \to \Lambda\pi^+$	≈2 d^{-1}

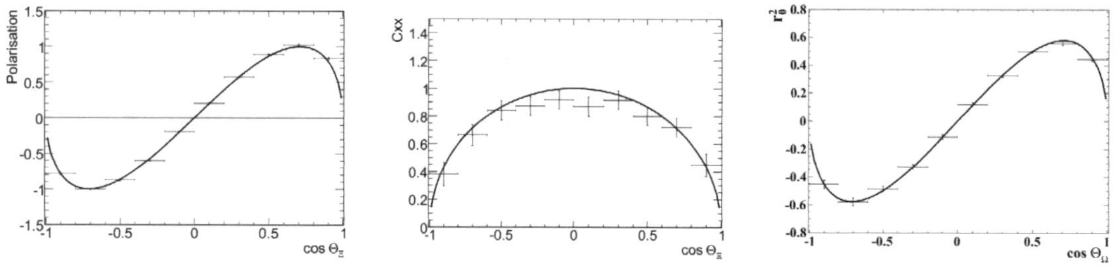

(a) (b) (c)

Figure 2. **(a)** The polarisation of the Ξ^- as a function of $\cos\theta_\Xi$. **(b)** The spin correlation C_{xx} in the x direction ($\hat{m}$ in Fig. 1) of the $\bar{\Xi}^+$ and the Ξ^-. **(c)** The polarisation parameter r_0^2 of the Ω as a function of $\cos\theta_\Omega$. The lines represent the input sine functions (see text) and the points with error bars the reconstructed values for 10000 simulated events.

The weak decay of the hyperons goes along with comparatively long lifetimes. Their decay vertices can easily be displaced by several tens of centimetres from the production point, setting high demands on precise track reconstruction. The $\overline{\text{P}}$ANDA experiment will include a variety of subdetectors designed for charged track reconstruction, including a Micro Vertex Detector (MVD), a Straw Tube Tracker (STT), a Gas Electron Multiplier (GEM) in the central spectrometer, and a Forward Tracking System (FTS) based on straw tubes in the forward spectrometer. The reconstruction can also be supported by a Scintillator Tile Hodoscope (SciTil). Multiple algorithms for pattern recognition and tracking are currently under development. A special focus is given here on the STT, which is a large barrel of 4636 straw tubes. Since its main purpose is to gather information about the transverse component of the flight path of charged tracks, most of these are aligned parallel to the beam axis, while a few layers of straws are skewed by $\pm 2.9°$, allowing to measure the longitudinal component with a worse resolution as well. Simulation studies show that in particular tracks originating from long living particles may completely miss one or more of the aforementioned subdetectors, creating only hits within the STT. In these cases, the skewed straws of the STT are the only viable source for a longitudinal momentum information of the track. When charged particle passes through a tube of the STT, this hit can only be measured as the minimum distance of its trajectory to the central wire. Therefore, each hit is represented as a circle, or isochrone. Since these isochrones are derived from the drift time of the electrons in the tube, knowledge of point in time when the particle passes through the tube is required. The start time of the incident collision of each event t_0 provides the necessary information. A tool to support a standalone track finder for the STT is under development; it matches the isochrones from the skewed straws with transverse information gained from the parallel straws to gather p_z information [21].

Journal of Physics: Conference Series **678** (2016) 012019 doi:10.1088/1742-6596/678/1/012019

Once $\overline{\text{P}}$ANDA is running with the high luminosity mode of HESR, *i.e.* $\mathcal{L} = 10^{32}\text{cm}^{-2}\text{s}^{-1}$, the mean time between events can be as small as 100 ns while the drift times in the STT can be up to 250 ns, usually causing hits from multiple events to overlap. Events are reconstructed on a software level, using time gaps between hits in one or more of the faster subdetectors, *e.g.* the MVD. This procedure also yields a first approximation of t_0, which in turn is required to determine the isochrones from the straw tube signals. With this first approximation, preliminary tracks will be propagated to detectors with excellent time resolution, *e.g.* the SciTil, in order to acquire a precise t_0.

Hyperon production is an excellent probe of the strong interaction in the confinement domain and spin observables provide a powerful tool in understanding the underlying physical processes. Hyperon decays can be studied to test CP violation to great precision which can provide evidence for physics beyond the standard model. The seven accessible polarisation parameters of the Ω baryon have been derived and simulation studies show that all strange and single charmed hyperon production channels have great prospects with the $\overline{\text{P}}$ANDA detector.

References

[1] Ellis R K, Stirling W J and Webber B R, *QCD and Collider Physics* (Cambridge Monographs on particle physics, Nuclear Physics and Cosmology, Cambridge, 1996)
[2] Kohno M and Weise W, Phys. Lett. B **179**, 15 (1986); Rubinstein H R and Snellman H, Phys. Lett. B **165**, 187 (1985); Furui S and Faessler A, Nucl. Phys. A **486**, 669 (1987); Burkardt M and Dillig M, Phys. Rev. C **37**, 1362 (1988); Alberg M A et al., Z. Phys. A **331**, 207 (1988)
[3] Tabakin F and Eisenstein R A, Phys. Rev. C **31**, 1857 (1985); Kohno M and Weise W, Phys. Lett. B **179**, 15 (1986); La France P et al., Phys. Lett. B **214**, 317 (1988); Timmermanns R G E et al., Phys. Rev. D **45**, 2288 (1992); Haidenbauer J et al., Phys. Rev. C **46**, 2516 (1992)
[4] Ortega P G et al., Phys. Lett. B **696**, 352 (2011)
[5] Okubo S, Phys. Lett. **5**, 165 (1963); Zweig G, CERN report TH-401; Iizuka J, Prog. Theor. Phys. Suppl **37**, 21 (1966)
[6] Kunihiro T and Hatsuda T, Phys. Lett. B **240**, 209 (1990)
[7] Lindenbaum S J, Nuovo Cimento A **65**, 222 (1981)
[8] Ellis J R et al., Phys. Lett. B **353**, 319 (1995); Ellis J R et al., Nucl. Phys. A **673**, 256 (2000)
[9] Alberg M A et al., Phys. Lett. B **356**, 113 (1995)
[10] Doncel M G et al., Nucl. Phys B **38**, 477 (1972); Doncel M G et al., Phys. Rev. D **7**, 815 (1973)
[11] Yao W M et al., J. Phys. G **33**, 1 (2006)
[12] Koch W ed. Nikolic M, *Analysis of scattering and decay* (Gordon and Breach, New York-London-Paris, 1968)
[13] Thomé E, *Multi-Strange and Charmed Antihyperon-Hyperon Physics for PANDA* (Ph. D. Thesis, Uppsala Univeryity, 2012)
[14] Johansson T, AIP Conf. Proc. 8th Int. Conf. on Low Energy Antiproton Physics, 95 (2003)
[15] Klempt E et al., Phys. Rep. **368**, 119 (2002)
[16] The PANDA collaboration, Technical Performance Report, (2005)
[17] The PANDA collaboration, Physics Performance Report, 43 (2009)
[18] Kaidalov A B and Volkovitsky P E, Z. Phys. C **63**, 51 (1994)
[19] Titov A I and Kämpfer B 2008, Phys. Rev. C **78**, 025201 (2008); Gornitschnig A T et al., Eur. Phys. J A **42**, 43 (2009); He J et al., Phys. Rev. D **84**, 114010 (2011); Khodjamirian A et al., Eur. Phys. J A **48**, 31 (2012)
[20] Grape S, *Studies on PWO Crystals and Simulations of the* $\bar{p}p \rightarrow \bar{\Lambda}\Lambda, \bar{\Lambda}\Sigma^0$ *Reactions for the PANDA experiment* (Ph. D. Thesis, Uppsala University, 2009)
[21] Ikegami Andersson W K, *private communication*

Journal of Physics: Conference Series **678** (2016) 012020 doi:10.1088/1742-6596/678/1/012020

Double-Polarized pd Scattering and Test of Time-reversal Invariance

Yuriy Uzikov

Joint Institute for Nuclear Researches, Laboratory of Nuclear Problems, Dubna, Moscow reg. 141980 Russian Federation

E-mail: uzikov@jinr.ru

Abstract. The integrated proton-deuteron cross section $\widetilde{\sigma}$ for the case of transversal polarization p_y^p of the proton and tensor polarization P_{xz} of the deuteron provides a null test signal for time-reversal invariance violating but P-parity conserving effects. The corresponding experiment is planned at COSY to measure the observable $\widetilde{\sigma}$. Here we consider in general case the status of the null-test observable in *pd* scattering, calculate $\widetilde{\sigma}$ within the Glauber theory of the double-polarized *pd* scattering at 100 - 1000 MeV and study the contribution of the deuteron S- and D-waves for several type of the T-odd NN interactions.

1. Introduction

Discrete symmetries play an important role in physics of fundamental interactions. In particular, the CP-violation (or T-invarinace violation under CPT symmetry) is responsible for the baryon asymmetry of the Universe [1]. However, within the standard model CP-violation effects observed in physics of kaons and B-mesons are far not sufficient to explain this asymmetry [2]. Therefore, other sources of CP violation have to be found. Time-invariance-violating (T-odd) P-parity conserving (P-even) (TVPC) interactions do not arise on the fundamental level within the standard model. This type of interaction can be generated by weak radiative corrections to the T-odd P-odd interaction discovered in physics of kaons and B-mesons. However in such a case its intensity is too small to be observed in experiments at present [3]. Thus, observation of TVPC effects would be considered as indication to physics beyond the standard model.

The total polarized cross section $\widetilde{\sigma}$ of the proton-deuteron scattering with vector polarization of the proton p_y^p and tensor polarization of the deuteron P_{xz} constitutes a null-test observable for TVPC effects [4]. The dedicated experiment is planned at COSY [5] at proton beam energy 135 MeV. The first analysis of the TVPC null-test signal [6] was done within the nonmesonic deuteron breakup channel $pd \to ppn$ estimated in the single scattering approximation. Recently we used the spin-dependent formalism [7] of the Glauber theory to calculate the cross section $\widetilde{\sigma}$ [8] and also "null-combinations" of some differential spin observables of the *pd* elastic scattering which deviate from zero only in case of presence of the TVPC effects [9]. The formalism includes full spin dependence of elementary pN-amplitudes and S- and D-components of the deuteron wave function. This formalism allows one to explain existing data on the non-polarized differential cross section and spin observables of the elastic *pd* scattering [10] at energies of the experiment planned at COSY [5]. We calculated energy dependence of $\widetilde{\sigma}$ for different type of TVPC interactions in the S-wave approximation for the deuteron [8]. Influence of the Coulomb

Content from this work may be used under the terms of the Creative Commons Attribution 3.0 licence. Any further distribution of this work must maintain attribution to the author(s) and the title of the work, journal citation and DOI.
Published under licence by IOP Publishing Ltd

Journal of Physics: Conference Series **678** (2016) 012020 doi:10.1088/1742-6596/678/1/012020

interaction on the cross section $\widetilde{\sigma}$ was found to be negligible [8]. Here we consider in general case the status of the null-test observable in *pd* scattering and study for several type of the TVPC NN interactions the contribution of the D-wave of the deuteron.

2. Null-test signal of TVPC interactions

Time-reversal symmetry conserving and P-parity conserving (TCPC or T-even P-even) interactions lead to the following transition amplitude of the elastic *pd* scattering at zero degree [11]

$$e'^*_\beta M(0)^{TCPC}_{\alpha\beta} e_\alpha = g_1[\mathbf{e}\,\mathbf{e}'^* - (\mathbf{me})(\mathbf{me}'^*)] + g_2(\mathbf{me})(\mathbf{me}'^*) + \\ ig_3\{\boldsymbol{\sigma}[\mathbf{e}\times\mathbf{e}'^*] - (\boldsymbol{\sigma}\mathbf{m})(\mathbf{m}\cdot[\mathbf{e}\times\mathbf{e}'^*])\} + ig_4(\boldsymbol{\sigma}\mathbf{m})(\mathbf{m}\cdot[\mathbf{e}\times\mathbf{e}'^*]), \quad (1)$$

where $\mathbf{e}$ ($\mathbf{e}'$) is the polarization vector of the initial (final) deuteron, $\mathbf{m}$ is the unit vector along the beam momentum, $\boldsymbol{\sigma}$ is the Pauli matrix, g_i ($i = 1, \ldots, 4$) are complex amplitudes. To the right-hand side of Eq.(1) one can add the TVPC (T-odd P-even) term in a very general form

$$e'^*_\beta M(0)^{TVPC}_{\alpha\beta} e_\alpha = \widetilde{g}\{(\boldsymbol{\sigma}\cdot[\mathbf{m}\times\mathbf{e}])(\mathbf{m}\cdot\mathbf{e}'^*) + (\boldsymbol{\sigma}\cdot[\mathbf{m}\times\mathbf{e}'^*])(\mathbf{m}\cdot\mathbf{e})\}, \quad (2)$$

where $\widetilde{g}$ is the TVPC transition amplitude.

Using the generalized optical theorem, the total cross section of the *pd* scattering can be written in the form [10]

$$\sigma_{tot} = \sigma_0 + \sigma_1\mathbf{p}^p\cdot\mathbf{p}^d + \sigma_2(\mathbf{p}^p\cdot\mathbf{m})(\mathbf{p}^d\cdot\mathbf{m}) + \sigma_3 P_{zz} + \widetilde{\sigma}p^p_y P^d_{xz}, \quad (3)$$

where $\mathbf{p}^p$ ($\mathbf{p}^d$) is the vector polarization of the initial proton (deuteron) and P_{zz} and P_{xz} are the tensor polarizations of the deuteron. The OZ axis is directed along the proton beam momentum $\mathbf{m}$, OY$\uparrow\uparrow$ $\mathbf{p}^p$, OX $\uparrow\uparrow$ $[\mathbf{p}^p\times\mathbf{m}]$. In Eq. (3) the terms σ_i with $i = 0, 1, 2, 3$ are non-zero only for T-even P-even interactions corresponding to Eq. (1) and the last term $\widetilde{\sigma}$ constitutes a null-test signal of T-invariance violation with P-parity conservation. The generalized optical theorem gives $\widetilde{\sigma} = -4\sqrt{\pi}Im\frac{2}{3}\widetilde{g}$ [8].

Hadronic amplitudes of pN scattering are taken as [7]

$$M_N(\mathbf{p},\mathbf{q};\boldsymbol{\sigma},\boldsymbol{\sigma}_N) = A_N + C_N\boldsymbol{\sigma}\hat{\mathbf{n}} + C'_N\boldsymbol{\sigma}_N\hat{\mathbf{n}} + \sum_{l=n,q,k} B^l_N(\mathbf{q})(\boldsymbol{\sigma}\hat{\mathbf{l}})(\boldsymbol{\sigma}_N\hat{\mathbf{l}}), \quad (4)$$

where $\hat{\mathbf{q}}$, $\hat{\mathbf{k}}$ and $\hat{\mathbf{n}}$ are defined as unit vectors along the vectors $\mathbf{q} = \mathbf{p} - \mathbf{p}'$, $\mathbf{k} = \mathbf{p} + \mathbf{p}'$ and $\mathbf{n} = [\mathbf{k}\times\mathbf{q}]$, respectively; $\mathbf{p}$ ($\mathbf{p}'$) is the initial (final) proton momentum; $\boldsymbol{\sigma}_N$ is the Pauli matrix acting on the spin state of the nucleon N. We consider the following terms of the TVPC NN interaction which were under discussion in Ref. [6]:

$$t_{pN} = h_N[(\boldsymbol{\sigma}\cdot\mathbf{k})(\boldsymbol{\sigma}_N\cdot\mathbf{q}) + (\boldsymbol{\sigma}_N\cdot\mathbf{k})(\boldsymbol{\sigma}\cdot\mathbf{q}) - \frac{2}{3}(\boldsymbol{\sigma}_N\cdot\boldsymbol{\sigma})(\mathbf{k}\cdot\mathbf{q})]/m_p^2 + \\ +g_N[\boldsymbol{\sigma}\times\boldsymbol{\sigma}_N]\cdot[\mathbf{q}\times\mathbf{k}]/m_p^2 + g'_N(\boldsymbol{\sigma}-\boldsymbol{\sigma}_N)\cdot i\,[\mathbf{q}\times\mathbf{k}][\boldsymbol{\tau}\times\boldsymbol{\tau}_N]_z/m_p^2. \quad (5)$$

Here $\boldsymbol{\sigma}$ ($\boldsymbol{\sigma}_N$) is the Pauli matrix acting on the spin state of the proton (nucleon $N = p, n$), $\boldsymbol{\tau}$ ($\boldsymbol{\tau}_N$) is the corresponding matrix acting on the isospin state; m_p is the proton mass. In the framework of the phenomenological meson exchange interaction the term g' corresponds to ρ-meson exchange, and h-term provides the axial meson h_1 exchange.

The crucial point is the following. The strong (T-even P-even) background is excluded from the null-test observable $\widetilde{\sigma}$. This means that when calculating (or measuring at ideal conditions) the observable $\widetilde{\sigma}$ one does not deal with a sum of the strong NN amplitude and the weak TVPC NN amplitude. Only products of these amplitudes may appear in $\widetilde{\sigma}$ (see below Eq. (6)). Furthermore, uncertainties in our knowledge of the strong amplitudes of the NN-scattering could not affect the final result for $\widetilde{\sigma}$ drastically.

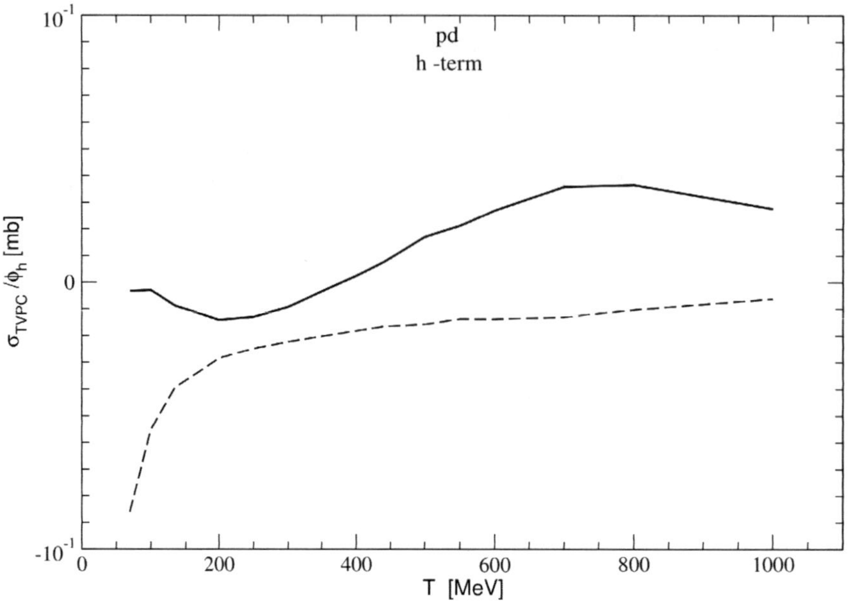

Figure 1. The calculated energy dependence of the TVPC cross section $\widetilde{\sigma}/\phi_h$ for the h-term in units of the unknown h_1NN constant ϕ_h for the separate contribution of the deuteron S-wave (dashed line), and total result with the S- and D-wave included (full). The energy dependence for the g-term has the same behaviour.

2.1. g' term

As was shown in our papers [8] the g'-term give zero contribution to the null-test observable $\widetilde{\sigma}$ within the Glauber theory of the *pd* elastic scattering. The reason for this is the specific spin-isospin structure of the g' term given in Eq. (5) which allows only the charge-exchange double pN-scattering mechanism for the elastic pd scattering amplitude.

2.2. h- and g-terms and S-D interference

For the single scattering mechanism the amplitude $\widetilde{g}$ vanishes within the Glauber theory. This follows from Eq. (5) where all terms are zeros at zero transferred momentum $\mathbf{q} = 0$. For double scattering mechanism considered for the h- and g-terms in Eq. (5) we derived in Ref. [8] a formula for $\widetilde{g}$ in the S-wave approximation. Taking into account the D-wave we find for the TVPC amplitude the following result

$$\widetilde{g} = \frac{i}{4\pi m_p} \int_0^\infty dqq^2[S_0^{(0)}(q) - 2\sqrt{2}S_2^{(1)}(q)][C_n'(q)(g_p - h_p) + C_p'(q)(g_n - h_n)], \tag{6}$$

where $S_0^{(0)} = \int_0^\infty dru^2(r)j_0(qr)$ and $S_2^{(1)}(q) = 2\int_0^\infty dru(r)w(r)j_2(qr)$ are the elastic form factors of the deuteron including the S-wave $u(r)$ and D-wave $w(r)$ of the deuteron in notations of Ref. [8].

2.3. Numerical results

For the C_N' amplitude of the pN scattering we use the SAID date base [12]. The deuteron wave function is taken in the CD Bonn model. The results of calculation of the observable $\widetilde{\sigma}$ for the h-term in units of the unknown constant $\phi_h = \bar{G}_h/G_h$, where $\bar{G}_h$ (G_h) is the TVPC (strong)

Journal of Physics: Conference Series **678** (2016) 012020 doi:10.1088/1742-6596/678/1/012020

coupling constant of the h_1 meson and nucleon hNN, are shown in Fig.1 versus the proton beam energy T_p. One can see that the D-wave of the deuteron taken into account changes considerably the result for the observbale $\widetilde{\sigma}$ as compared to the S-wave result obtained previously in [8, 13]. The destructive S-D interference diminishes considerably the null-test signal $\widetilde{\sigma}$ at energies of the planned COSY experiment [5] ~100 MeV as compared to the pure S-wave contribution and provides an enhancement at 700-800 MeV.

3. Summary

In the first theoretical analysis of the null-test observable $\widetilde{\sigma}$ performed in Ref. [6] only the single scattering mechanism was considered for the straightforward calculation of the inelastic and elastic pd-scattering. We use the optical theorem for calculation of the $\widetilde{\sigma}$, that requires to get only the forward elastic pd scattering amplitude. On this way we find within the Glauber theory that the single scattering mechanism gives zero contribution to the null-test cross section $\widetilde{\sigma}$. The double scattering mechanism for the h and g terms of the TVPC NN interaction leads to Eq. (6) for the TVPC amplitude of the pd scattering and shows that the only hadronic pN amplitude C'_N from Eq. (4) modulates the TVPC observable $\widetilde{\sigma}$. Furthermore, we have shown for the case of the h- and g-type of interaction that the deuteron D-wave gives a valuable contribution to the null-test signal due to interference between the deuteron S- and D- waves. Of course, there are many other TVPC terms in the NN interaction [14] which we are going to consider later on.

The g'-term caused by the $\rho-$ meson exchange in the TVPC NN-interaction does not contribute to $\widetilde{\sigma}$ within the Glauber theory with T-even P-even NN-interaction in the deuteron [8]. However, the g'-term can give contribution to the null-test signal $\widetilde{\sigma}$ if this interaction will be included into the deuteron bound state [8]. One-pion exchange is excluded from TVPC NN-interaction [15], however, two-pion exchange probably does contribution [16]. Furthermore, TVPC NN forces can contribute to the electromagnetic p-d interaction due to the toroidal quadrupole form factor of the deuteron [17]. Another T-odd P-even effect very similar to the $\widetilde{\sigma}$ observable in pd-scattering can be observed in inclusive interaction of unpolarized leptons with polarized nucleons [18].

[1] Sakharov A D 1967 *Pis'ma v ZHETF* **5** 32
[2] Riotto A, Troden M 1999 *Annu. Rev. Nucl. Part. Sci* **49** 35
[3] Khriplovich I B 1991 *Nucl. Phys. B* **352** 385
[4] Conzett H E 1993 *Phys. Rev. C* **48** 423
[5] COSY Proposal N 215, Spokespersons: Eversheim P D, Lorentz B and Valdau Yu [http://apps.fz-juelich.de/pax/paxwiki/images/8/8c/215-TRI_Prop_sum.pdf]
[6] Beyer M 1993 *Nucl. Phys. A* **560** 895
[7] Platonova M N, Kukulin V I 2010 *Phys. Rev. C* **81** 014004
[8] Uzikov Yu N, Temerbayev A A 2015 *Phys. Rev.* C **92** 014002
[9] Temerbayev A A, Yu. N. Uzikov Yu N 2015 *Izv. Ross. Akad. Nauk Ser. Fiz.* **79** N4 535 [*Bull. Russ. Acad. Sci. Phys.* **79** N4 493]
[10] Temerbayev A A, Uzikov Yu N 2015 *Yad. Fiz.* **78** 38
[11] Rekalo M P, Piskunov N M, Sitnik I M 1998 *Few-Body Syst.* **23** 187
[12] Arndt R A, Briscoe W J, Strakovsky I I, Workman R L 2007 *Phys. Rev.* C **76** 025209; http://gwdac.phys.gwu.edu
[13] Uzikov Yu N, Temerbayev A A 2014 *Proc. 22nd Int. Baldin Sem. on High Energy Physics Problems: Relativistic Nuclear Physics and Quantum Chromodynamics, 15-20 September, Dubna, Russia*, PoS (Baldin ISHEPP XXII) 094
[14] Herczeg P 1966 *Nucl. Phys.* **75** 655
[15] Simonius M 1997 *Phys. Rev. Lett.* **78** 4161
[16] De Vries J, Meissner Ulf-G, Epelbaum E, Keiser N 2013 *Eur. Phys. J.* A **49** 149
[17] Mereghetti E, de Vries J, Timmermans R G E, van Kolck U 2013 *Phys. Rev.* C **88** 034001
[18] Teryaev O V 2003 *Chech. J. Phys.* **53** 47A

Journal of Physics: Conference Series **678** (2016) 012021 doi:10.1088/1742-6596/678/1/012021

Di-hadron asymmetry and interplay between transversity induced asymmetries in hadron leptoproduction at COMPASS

Giulio Sbrizzai on behalf of the COMPASS Collaboration
Trieste University, Physcics Departement, Via Valerio 2 34127, Trieste, IT
E-mail: giulio.sbrizzai@ts.infn.it

Abstract. New results on the transverse spin azimuthal asymmetries in semi-inclusive DIS reactions extracted by the COMPASS Collaboration from the data collected with a transversely polarised proton target are presented. A noticeable similarity between the Collins asymmetry and the dihadron asymmetry, already been observed and reported, triggered a more deep investigation on the angular correlations and the relevant kinematical variables. The resulting phenomenological analysis of the transversity induced asymmetries, presented in this talk, allows to establish quantitative relationships, providing strong indication that the underlying fragmentation mechanisms are all driven by a common physical process.

1. Introduction

The analysis of the data collected by the COMPASS Collaboration during the years 2007 and 2010 by scattering 160 GeV longitudinally polarised positive muons on a transversely polarised proton target (NH^3) has provided several interesting measurements on the transverse spin structure of the proton. In particular the year 2010 was fully dedicated to these measurements and COMPASS published precise results on the Collins asymmetries [1] obtained for positive and negative hadrons in SIDIS reactions $\mu p \rightarrow \mu' hX$ and on the di-hadron asymmetry [2] obtained using all the possible pairs formed by one positive and one negative hadron in $\mu p \rightarrow \mu' h^+h^-X$.

The similarity between the Collins and di-hadron asymmetry shown in figure 1 inspired a new analysis in which the comparison between these asymmetries has been studied on a common data sample. In particular the angular correlation between the positive and negative hadrons forming the pairs has been investigated in the reaction $\mu p \rightarrow \mu' h^+h^-X$ and the dependence of both the Collins and the di-hadron asymmetry on the difference between the azimuthal angles of the two hadrons has been measured. The corresponding formalism has been developed in parallel [3] and the results [4] are very interesting and essentially show that the same process gives origin to these asymmetries.

2. Transversity induced single spin asymmetries

2.1. *Collins-like asymmetries*

A common data sample is used for the analysis of the interplay between one-hadron and di-hadron asymmetries, namely the "two-hadrons sample" which contains DIS events, selected by the cuts $Q^2 > 1$ $(\mathrm{GeV}/c)^2$, $0.1 < y < 0.9$ and $W > 5$ GeV/c^2, in which at least two oppositely

Content from this work may be used under the terms of the Creative Commons Attribution 3.0 licence. Any further distribution of this work must maintain attribution to the author(s) and the title of the work, journal citation and DOI.
Published under licence by IOP Publishing Ltd

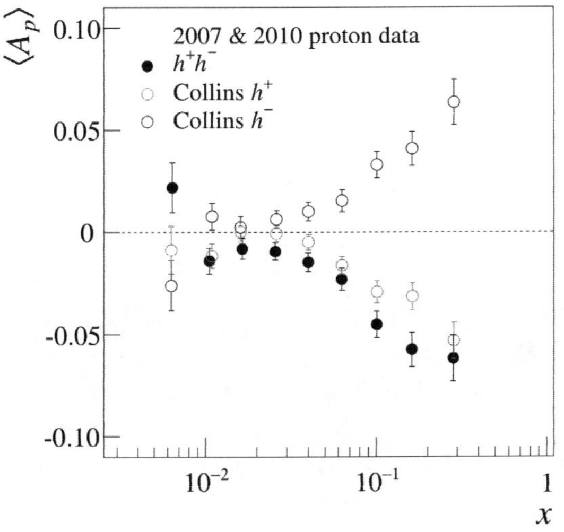

Figure 1. Collins asymmetries from [5, 1] for positive (red circles) and negative (black circles) hadrons as function of x compared with the di-hadron asymmetry from [6, 2] (black points). The measurements are obtained combining the 2007 and 2010 COMPASS data on transversely polarised proton target (NH^3). The di-hadron and the positive hadrons results show a similar x dependence and the di-hadron signal is slightly larger. The figure was first produced for the paper in [2].

charged hadrons having $z > 0.1$ and $p_T^h > 0.1$ GeV/c are detected. The variable Q^2 is the virtuality of the virtual photon excheanged in the reaction, x is the Bjorken scaling variable and y is the energy fraction lost by the incoming muon. The invariant mass of the hadron system after the interaction is W, z si the fraction of the available energy carried by the hadron and p_T^h its transverse momentum with respect to the virtual photon direction. The azimuthal angles ϕ_h of each hadron are the angle of the vectors p_T^h with respect to the scattering plane. The Collins angle of the positive (negative) hadron are then defined as $\phi_{h+(-)} + \phi_S - \pi$ where ϕ_S is the azimuthal angle of the target polarisation. All quantities are calculated in the Gamma Nucleon System where the z axis is defined by the virtual photon and the $x - z$ plane is the lepton scattering plane.

The amplitudes of the modulations in the Collins angles calculated in the "two-hadrons sample" give the Collins-like (CL) asymmetries $A_{1CL}^{\sin(\phi_{h^+}+\phi_S-\pi)}$ and $A_{2CL}^{\sin(\phi_{h^-}+\phi_S-\pi)}$ for positive and negative hadrons respectively. We have named them CL to distinguidh them from the "standard" Collins asymmetries, which are defined for the hadrons in the "one-hadron sample", namely when at least one hadron is detected. They are very similar to the "standard" Collins asymmetries as it is shown in figure 2 where their dependence on x is compared.

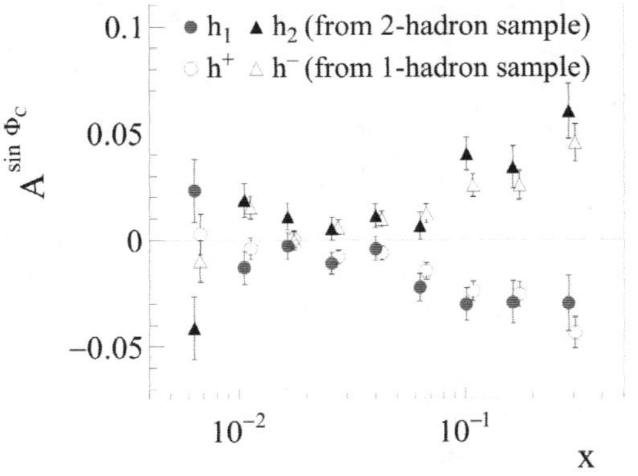

Figure 2. Collins-like asymmetries as function of x obtained from the "two hadrons sample" for positive hadrons (red points) and negative hadrons (black full triangles) compared with the Collins asymmetries for positive hadrons (red circles) and negative hadrons (balck open triangles). The asymmetries extracted from the "two hadrons sample" are essentially the same as the standard ones.

Starting from the general expression for the cross-section of the process $\mu p \rightarrow \mu' h^+ h^- X$ [4], integrated over x, Q^2, z_{h^+}, z_{h^-}, $p_T^{h^+}$ and $p_T^{h^-}$

Journal of Physics: Conference Series **678** (2016) 012021 doi:10.1088/1742-6596/678/1/012021

$$\frac{d\sigma}{d\phi_{h^+}d\phi_{h^-}d\phi_S} = \sigma_U^{h^+h^-} + S_T \cdot [\sigma_{1C}^{h^+h^-} \cdot \sin(\phi_{h^+} + \phi_S - \pi) + \sigma_{2C}^{h^+h^-} \cdot \sin(\phi_{h^-} + \phi_S - \pi)], \quad (1)$$

the CL asymmetries for positive and negative hadrons can be written as $A_{1CL}^{\sin(\phi_{h^+}+\phi_S-\pi)} = \frac{\sigma_{1C}^{h^+h^-}+\sigma_{2C}^{h^+h^-}\cos(\Delta\phi)}{\sigma_U^{h^+h^-}}$ and $A_{2CL}^{\sin(\phi_{h^-}+\phi_S-\pi)} = \frac{\sigma_{2C}^{h^+h^-}+\sigma_{1C}^{h^+h^-}\cos(\Delta\phi)}{\sigma_U^{h^+h^-}}$ where $\Delta\phi = \phi_1 - \phi_2$ and the labels 1 and 2 refer to the positive and negative hadrons respectively.

The asymmetries $A_{1CL}^{\sin(\phi_{h^+}+\phi_S-\pi)}$ and $A_{2CL}^{\sin(\phi_{h^-}+\phi_S-\pi)}$ measured as function of $\Delta\phi$ are shown in figure 3. They are both larger in magnitude when $\Delta\phi$ gets near to $\pm\pi$, namely when the two hadrons have opposite directions in the transverse plane, as expected in the $^3P_0$ recursive string fragmentation model [7, 8]. There is also a clear mirror effect between the two asymmetries which have the same magnitude and opposite sign. This implies $\sigma_{1CL}^{h^+h^-} = -\sigma_{2C}^{h^+h^-}$ and the expression of the asymmetries simplifies into $A_{1CL}^{\sin(\phi_{h^+}+\phi_S-\pi)} = \frac{\sigma_{1CL}^{h^+h^-}(1-\cos(\Delta\phi))}{\sigma_U^{h^+h^-}}$ and $A_{1CL}^{\sin(\phi_{h^+}+\phi_S-\pi)} = \frac{-\sigma_{2C}^{h^+h^-}(1-\cos(\Delta\phi))}{\sigma_U^{h^+h^-}}$. The red dashed and the dot-dashed black lines show the curves obtained from the fits using the function $\pm c\cdot(1-\cos(\Delta\phi))$ where c is a constant parameter. Its values given by the fits are 0.014 ± 0.003 and 0.016 ± 0.003.

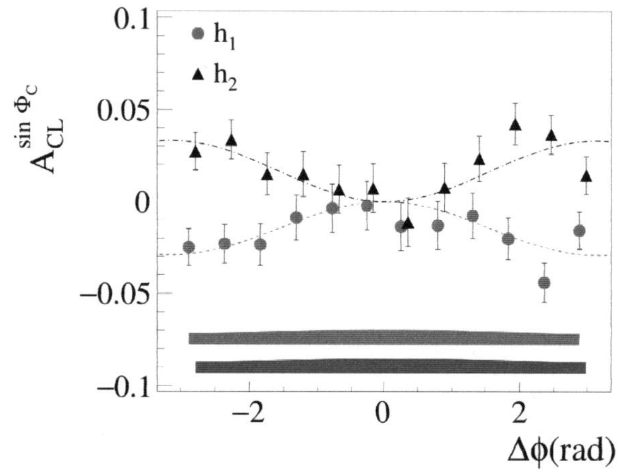

Figure 3. Collins-like asymmetries as function of $\Delta\phi$ the difference of the azimuthal angles of the two hadrons. The red points give the $A_{1CL}^{\sin(\phi_{h^+}+\phi_S-\pi)}$ asymmetry measured for the positive hadrons and the black triangles give the $A_{2CL}^{\sin(\phi_{h^-}+\phi_S-\pi)}$ asymmetry for the negative hadrons. The signal is clearly bigger for values of $\Delta\phi$ which are near $\pm\pi$ for both. There is also a mirror effect between the two asymmetries. The bands give the size of the systematic error.

2.2. Interplay between one-hadron and di-hadron asymmetries

A new definition of the azimuthal angle of the pair has been introduced to determine the relation between di-hadron and one-hadron asymmetries, namely $\phi_{2h} = [\phi_{h^+} + \phi_{h^-} + \pi \cdot sgn(\Delta\phi)]/2$, where sgn is the signum function. It is a very simple definition and allows, together with $\sigma_{1C}^{h^+h^-} = -\sigma_{2C}^{h^+h^-}$, to rewrite the differential cross section of equation 1 as follows:

$$\frac{d\sigma}{d\phi_{2h}d\Delta\phi d\phi_S} = \sigma_U^{h^+h^-} + S_T \cdot \sigma_{1C}^{h^+h^-} \cdot \sqrt{2(1-\cos(\Delta\phi))} \cdot \sin(\phi_{2h} + \phi_S - \pi), \quad (2)$$

which exhibits a $\sin(\phi_{2h} + \phi_S - \pi)$ modulation with the amplitude $A_{2h,CL}^{\sin(\phi_{2h}+\phi_S-\pi)} = \frac{\sigma_{1C}^{h^+h^-}\cdot\sqrt{2(1-\cos(\Delta\phi))}}{\sigma_U^{h^+h^-}}$.

In [2, 4] it is shown that the $A_{2h,CL}^{\sin(\phi_{2h}+\phi_S-\pi)}$ asymmetry and the one measured in [6, 2] using ϕ_R as the azimuthal angle of the vector $R = \frac{z_{h^-}P_T^{h^+} - z_{h^+}P_T^{h^-}}{z_{h^-}+z_{h^+}}$ [7] are essentially the same. To establish a relation between the CL and the di-hadron asymmetries it is then enough to unveil the relationship between the CL and the ϕ_{2h} asymmetries.

The figure 4 shows the $A_{2h,CL}^{\sin(\phi_{2h}+\phi_S-\pi)}$ asymmetry with the curves of figure 3 superimposed togehter with the one obtained from the fit using a $-c \cdot \sqrt{2 \cdot (1-\cos(\Delta\phi))}$ function where $c = 0.017 \pm 0.002$ (continuos line). It is clear that the behaviour expected from the calculations is well in agreement with the di-hadron data. Also, the values of the constants c appearing in the fits of the $\Delta\phi$ dependence of the CL asymmetries are in agreement with the value of the constant c fitted on the di-hadron asymmetry. This implies that the Collins-like (and hence the Collins) mechanism and the di-hadron mechanism are all generated by the same physics process.

The factor by which the magnitude of the di-hadron asymmetry measured as function of x is larger than the one of the Collins asymmetries, figure 1, can be calculated by integrating the curves over $\Delta\phi$ and it is $4/\pi$.

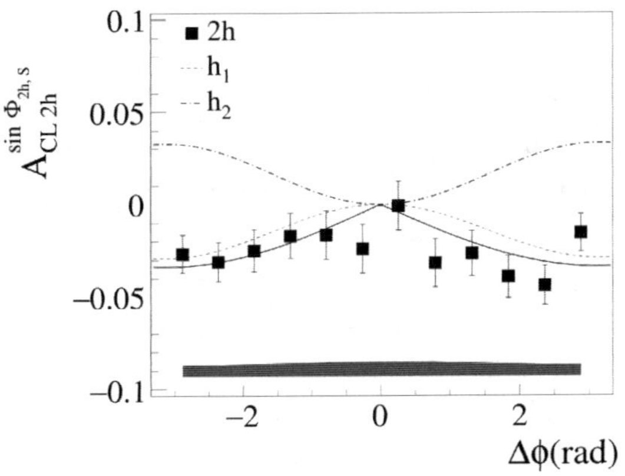

Figure 4. The di-hadron asymmetry in the ϕ_{2h} angle $A_{2h,CL}^{\sin(\phi_{2h}+\phi_S-\pi)}$ as function of $\Delta\phi$ (black full squares). The red dashed line is the curve $-c\cdot(1-\cos(\Delta\phi))$ fitted on the Collins-like asymmetries for positive hadrons $A_{1CL}^{\sin(\phi_{h^+}+\phi_S-\pi)}$, the black dot-dashed line is the curve $c\cdot(1-\cos(\Delta\phi))$ fitted on the Collins-like asymmetries for negative hadrons $A_{2CL}^{\sin(\phi_{h^-}+\phi_S-\pi)}$ and the grey continous line is the curve $-c \cdot \sqrt{2 \cdot (1-\cos(\Delta\phi))}$ fitted on the di-hadron asymmetry $A_{2h,CL}^{\sin(\phi_{2h}+\phi_S-\pi)}$. The values obtained for the parameter c from the three fits are: 0.014 ± 0.003, 0.016 ± 0.003 and 0.017 ± 0.002 respectively. The band gives the systematic error for $A_{2h,CL}^{\sin(\phi_{2h}+\phi_S-\pi)}$.

References

[1] Adolph C *et al.* (COMPASS Collaboration) 2012 *Phys. Lett.* B **717** 376
[2] Adolph C *et al.* (COMPASS Collaboration) 2014 *Phys. Lett.* B **736** 124
[3] Kotzinian A 2015 *Physics Review* D **91** 054001
[4] Adolph C *et al.* (COMPASS Collaboration) 2015 hep-ex/1507.07593
[5] Alekseev M G *et al.* (COMPASS Collaboration) 2010 *Phys. Lett.* B **692** 240
[6] Adolph C *et al.* (COMPASS Collaboration) 2012 *Phys. Lett.* B **713** 10
[7] Artru X and Collins J C 1996 *Z. Phys.* C **69** 277
[8] Artru X, Czyzewski J and Yabuki H 1997 *Z. Phys.* C **73** 527

Journal of Physics: Conference Series **678** (2016) 012022 doi:10.1088/1742-6596/678/1/012022

Orbital parameters of proton and deuteron beams in the NICA collider with solenoid Siberian snakes

A D Kovalenko[1], A V Butenko[1], V D Kekelidze[1], V A Mikhaylov[1], M A Kondratenko[2], A M Kondratenko[2] and Yu N Filatov[1,3]

[1]Joint Institute for Nuclear Research, 141980, Dubna, Russia
[2]Science and Technique Laboratory "Zaryad", 630090, Novosibirsk, Russia
[3]Moscow Institute of Physics and Technology, 141700, Dolgoprudny, Russia

E-mail: `kovalen@dubna.ru`

Abstract. Two solenoid Siberian snakes are required to obtain ion polarization in the "spin transparency" mode of the NICA collider. The field integrals of the solenoid snakes for protons and deuterons at maximum momentum of 13.5 GeV/c are equal to 2×50 T·m and 2×160 T·m respectively. The snakes introduce strong betatron oscillation coupling. The calculations of orbital parameters of proton and deuteron beams in NICA collider with solenoid snakes are presented.

1. The NICA collider in the spin transparency mode

The scheme of proton and deuteron polarization control at the NICA collider in the spin transparency mode has been presented in 2012 [1]. The scheme of polarization control is based on the use of two types of solenoids, namely, strong-field solenoids and weak-field solenoids (see figure 1). The strong-field solenoids are used for Siberian snakes, which create the spin transparency mode in the NICA collider. Two solenoidal snakes are inserted in the opposite straight sections of the racetrack-shaped collider. The snakes are divided symmetrically onto two parts by MPD and SPD setups. Such snake design allows one to control the beam polarization in vertical plane of SPD and MPD detectors while the polarization leis in the plane of the collider in the NICA arcs.

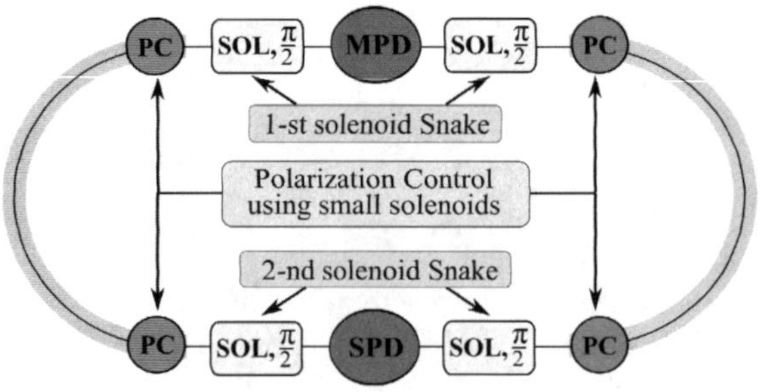

Figure 1. Polarization control scheme in the NICA spin transparency mode.

Content from this work may be used under the terms of the Creative Commons Attribution 3.0 licence. Any further distribution of this work must maintain attribution to the author(s) and the title of the work, journal citation and DOI.
Published under licence by IOP Publishing Ltd

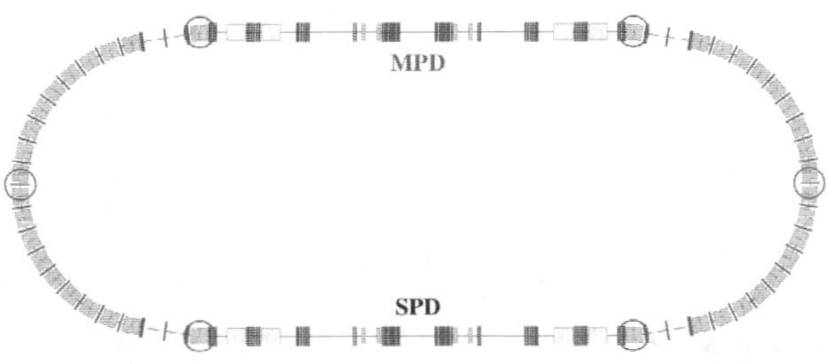

Figure 2. Placement of the snake solenoids in the NICA collider.

The spin transparency mode means that the influence of the fields generated by the two solenoidal snakes and the collider's arcs does not change the spin direction from turn to turn, i.e. the magnetic system is transparent for the spin. The NICA collider with two solenoidal snakes becomes similar to the figure-8 collider project at JLAB [2]

The weak-field solenoids with small integral of the longitudinal field are used directly for the control of the beam polarization by means of polarization control insertions (PC insertions). Thus, there is unique opportunity for efficient spin manipulation of any particle species (p, d, ^3He,...) in any orbit place without affecting of the collider orbital characteristics [3]. For example, it is possible to arrange both vertical and longitudinal directions of the beam polarization in MPD and SPD detectors. One can provide any direction in the collider arcs, which is necessary for the matching of the polarizations during the beam injection from Nuclotron. One can provide the spin reversals in the detectors during an experiment (spin flipping systems) [4]. It allows to carry out high quality experiments with polarized proton and deuteron beams.

2. Orbit matching of the snake solenoids to the NICA collider lattice

The control solenoids with weak fields do not influence on the collider orbital characteristics. On the contrary, the snake solenoids introduce a strong coupling of the betatron oscillations. The snake solenoid fields are proportional to the momentum and the collider optics remains unchanged during the beam acceleration. At maximum collider momentum of 13.5 GeV/c, the solenoid integral field of the one snake is 50 T·m for protons and is 160 T·m for deuterons. So, one has to provide the matching of the snake solenoids to the collider lattice for proton and deuteron beams.

2.1. Placement of the snake solenoids in the NICA collider

The schematic of placement of the solenoids in the NICA collider lattice is presented in figure 2. Here snake solenoids are marked yellow bars and PC insertions marked in circles. The one snake consist of four solenoids of 5 m length each. At maximum momentum the solenoid field is 2.5 T for protons and is 8 T for deuterons.

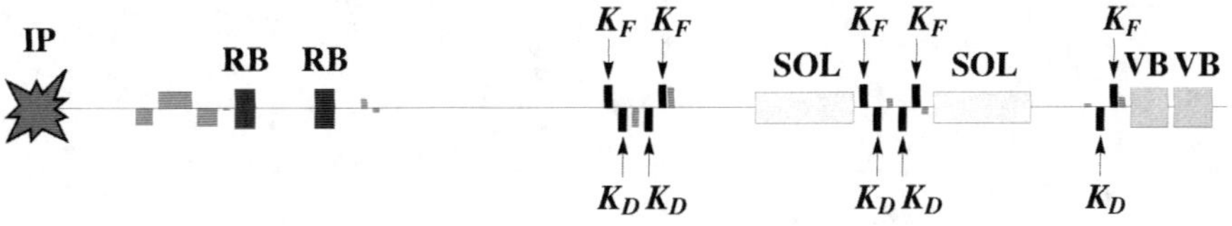

Figure 3. Placement of the snake solenoids in the NICA half straight.

The placement of the snake solenoids in the collider half straight is shown in figure 3 in more details. Here SOL are solenoids, VB are arc's vertical-field bending magnets and RB are radial-field bending magnets, which provide beams collisions at interaction point IP.

Two families of focusing K_F and defocusing K_D quadrupoles are used for matching the snake solenoids to the collider optics. Additional quadrupoles for compensation of the coupling of the betatron oscillations are not used in our matching scheme [5].

2.2. Optical functions in the NICA collider

We present an example, when matching the snake solenoids to the NICA collider lattice occurs at the same values of the quadrupole gradients K_F and K_D for three cases, namely, for NICA without snakes, NICA with proton snakes and NICA with deuteron snakes. The β-functions and dispersion functions for these cases are shown in figures 4-6. Also these figures show unit circles with eigenvalues $\lambda_{x,y} = \exp(i\nu_{x,y})$ of the transfer matrix, where $\nu_{x,y}$ are radial and vertical betatron tunes.

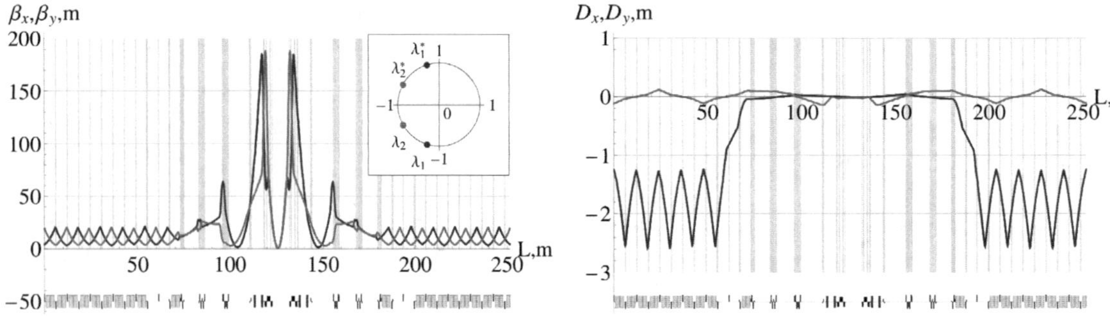

Figure 4. β-functions and dispersion functions in NICA without snakes.

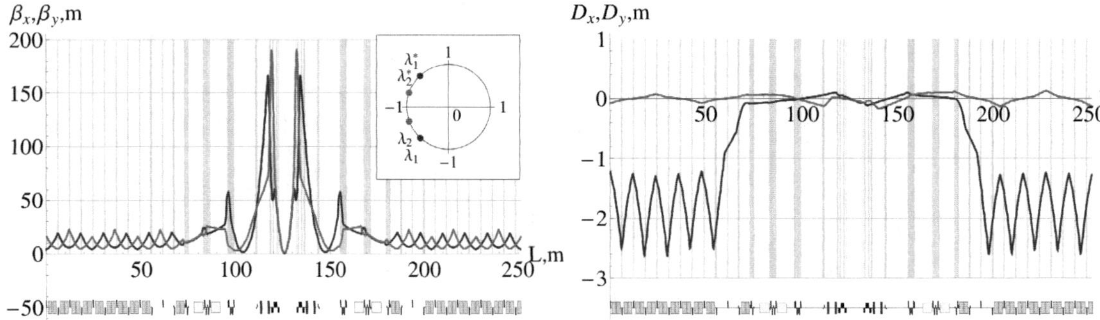

Figure 5. β-functions and dispersion functions in NICA with proton snakes.

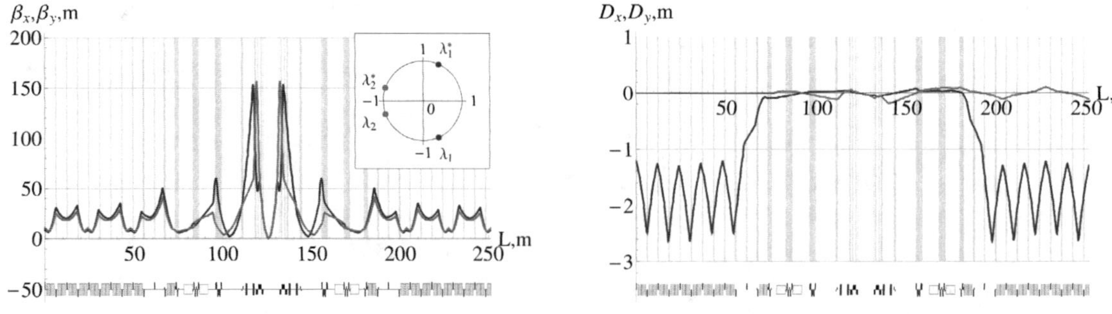

Figure 6. β-functions and dispersion functions in NICA with deuteron snakes.

The value of gradients in unit of the magnetic rigidity is 0.597 m^{-2} for focusing quads K_F and is 0.586 m^{-2} for defocusing quads K_D.

2.3. Optical parameters at the interaction point

The values of β-functions and dispersion functions at the interaction points of MPD and SPD detectors are presented in table 1.

Table 1. Optical parameters at the interaction points.

Mode	β_x^*, cm	β_y^*, cm	D_x^*, cm	D_y^*, cm
without snakes	36	34	-1	-2
proton snakes	33	36	-1	-2
deuteron snakes	54	52	-1	-1

Proton snakes practically do not change beam's orbital parameters at the interaction points. Deuteron snakes also practically do not change the dispersion functions in the given examples, but increase values of β-functions about of 20 cm. Such β-functions increasing can be compensated by optimization of lattice of the collider straights.

Conclusion

In summary, one can draw the following conclusions:

- the possibility of matching the snake solenoids to the NICA collider lattice without compensation of betatron oscillations coupling is demonstrated for proton and deuteron beams,
- proton snakes practically do not change orbital parameters of the NICA collider,
- influence of deuteron snakes on collider orbital parameters can be compensated at interaction points by optimization of lattice of the collider straights,
- the spin transparency mode in the NICA collider is feasible and can provide high quality experiments with polarized proton and deuteron beams.

Acknowledgments

We would like to thank all the participants of the seminar at Veksler and Baldin Laboratory of High Energy Physics (JINR, Dubna) for the useful discussions of some questions, considered in this paper.

References

[1] Kovalenko A D, Filatov Yu N, Kondratenko A M, Kondratenko M A and Mikhaylov V A 2014 Polarized deuterons and protons at NICA@JINR *PEPAN Lett.* **45** 1 pp 321-22
[2] Abeyratne S *et al* 2012 arXiv:1209.0757 [physics.acc-ph]
[3] Filatov Yu N, Kovalenko A D, Kondratenko A M, Kondratenko M A and Mikhaylov V A 2013 *Proton And Deuteron Polarization Control In NICA Collider Using Small Solenoids* Proc.XV Workshop on HESP DSPIN2013 Dubna pp 351-54
[4] Derbenev Ya S, Kondratenko A M, Kondratenko M A, Morozov V S and Filatov Yu N 2011 *Spin-Flipping Systems For Storage Rings* Proc. XIV Workshop on HESP DSPIN-11 Dubna pp 377-84
[5] Kondratenko M A, Kondratenko A M, Kovalenko A D, Mikhaylov V A and Filatov Yu N 2013 *Orbital Parameters Of Proton Beam In Nuclotron With Solenoid Siberian Snake* Proc.XV Workshop on HESP DSPIN2013 Dubna pp 371-74

Numerical calculation of ion polarization in the NICA collider

A D Kovalenko[1], A V Butenko[1], V D Kekelidze[1], V A Mikhaylov[1], M A Kondratenko[2], A M Kondratenko[2] and Yu N Filatov[1,3]

[1]Joint Institute for Nuclear Research, 141980, Dubna, Russia
[2]Science and Technique Laboratory "Zaryad", 630090, Novosibirsk, Russia
[3]Moscow Institute of Physics and Technology, 141700, Dolgoprudny, Russia

E-mail: kovalen@dubna.ru

Abstract. The NICA Collider with two solenoid Siberian snakes is "transparent" to the spin. The collider transparent to the spin provides a unique capability to control any polarization direction of protons and deuterons using additional weak solenoids without affecting orbital parameters of the beam. The spin tune induced by the control solenoids must significantly exceed the strength of the zero-integer spin resonance, which contains a coherent part associated with errors in the collider's magnetic structure and an incoherent part associated with the beam emittances. We present calculations of the coherent part of the resonance strength in the NICA collider for proton and deuteron beams.

1. Introduction

The NICA collider is setup in the "spin transparency" mode using two solenoidal snakes inserted in its opposite straights (see figure 1) [1]. The spin transparency mode means that the influence of the fields generated by the two solenoidal snakes and the collider's arcs does not change the spin direction from turn to turn, i.e. the magnetic system is transparent for the spin. The NICA collider with two solenoidal snakes becomes similar to the figure-8 collider project at JLAB [2].

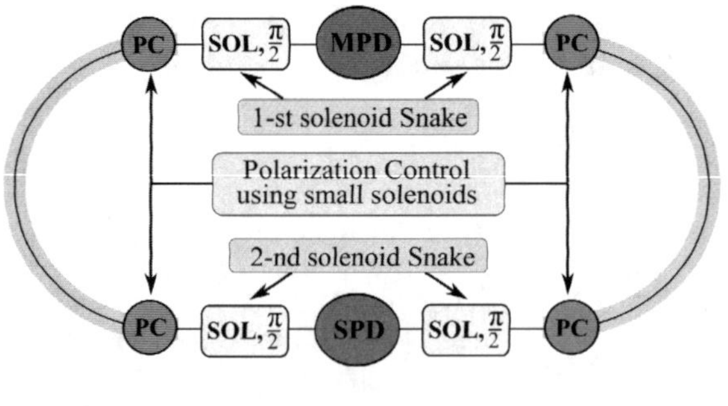

Figure 1. Polarization control scheme in the NICA spin transparency mode.

The control of the beam polarization is provided by means of polarization control insertions (PC insertions) based on the weak-field solenoids [3, 4].

Content from this work may be used under the terms of the Creative Commons Attribution 3.0 licence. Any further distribution of this work must maintain attribution to the author(s) and the title of the work, journal citation and DOI.
Published under licence by IOP Publishing Ltd

- to provide polarization control of any ion species (p, d, ^3He, ...),
- to provide any direction of polarization in the vertical plane of SPD and MPD detectors,
- to solve the problems of spin matching at injection in the NICA collider and polarization measurement as well,
- to eliminate resonant beam depolarization during acceleration,
- to realize spin-flipping system,
- to control polarization at SPD and MPD detectors without any change of beam orbital characteristics.

The required weak field integrals are limited by the strength of the zero-integer spin resonance ω. For stability of the beam polarization, the spin tune ν induced by the PC solenoids must significantly exceed the strength of the zero-integer spin resonance: $\nu \gg \omega$.

Let us calculate the spin resonance strength to estimate required weak field integrals of PC solenoids for proton and deuteron beams.

2. Strength of the zero-integer spin resonance in the NICA collider

The resonance strength is the value of the average spin field, which is determined by deviation of the trajectory from the ideal design orbit. The resonance strength consists of two parts: a coherent part arising due to additional dipole and longitudinal fields on a trajectory deviating from the design orbit and an incoherent part associated with betatron and synchrotron oscillations of ions (beam emittances). The coherent part of the resonance strength is determined by the linear approximation in orbit deviations and lies in the orbit plane. The incoherent part of the resonance strength is determined by the second approximation in orbit deviations and is significantly less than the coherent part.

The coherent part of zero-integer spin resonance induced by perturbing radial field h_x can be calculated using periodical response function $F(z)$ (z is a particle coordinate along the orbit):

$$\omega_{coherent} = \gamma G \langle h_x(z) F(z) \rangle .$$

The response function is only determined by the design collider lattice. Perturbing radial fields arise, for example, due to dipole roll errors, vertical quadrupole misalignments, etc.

Figure 2 shows graphs of the response functions in the NICA collider at momentum of 13.5 GeV/c for proton and deuteron beams, respectively. The lattices of the NICA collider with proton and deuteron solenoidal snakes are presented in paper [5].

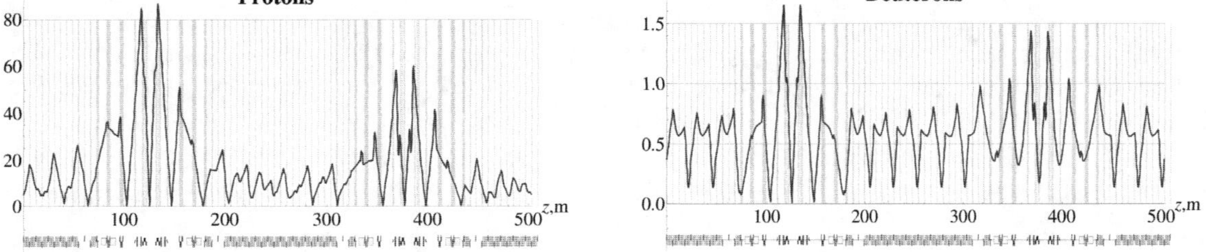

Figure 2. Response functions for protons and deuterons at 13.5 GeV/c in the NICA collider.

Let us calculate the coherent part of the spin resonance strength using a statistical model of quadrupole misalignments in vertical direction. Figure 3 shows calculated dependence of the resonance strength $\omega_{coherent}$ on the proton and deuteron momenta, respectively, when the orbit

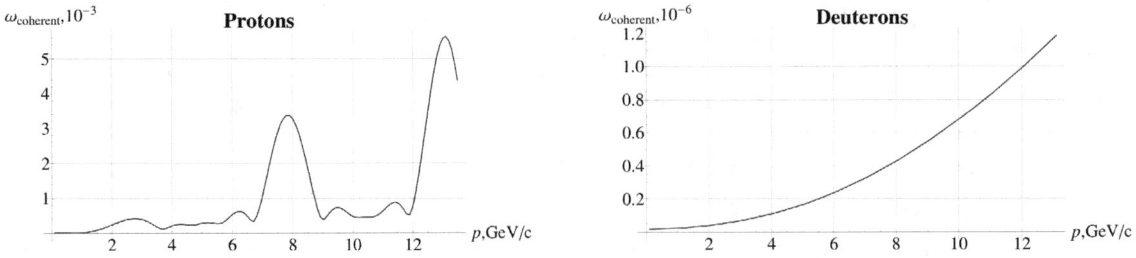

Figure 3. Coherent parts of the zero-integer spin resonance strength for protons and deuterons.

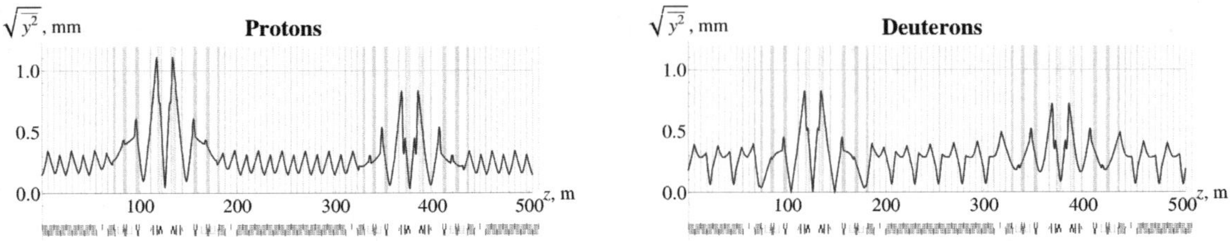

Figure 4. Vertical beam excursions with random misalignment of all quads in the NICA collider.

excursion in the arcs does not exceed ± 0.5 mm for protons and deuterons. Graphs of the rms closed orbit distortions in the collider under the same parameters are shown in figure 4.

For protons in the presented examples, the coherent part of the resonance strength does not exceed $\sim 10^{-3}$ almost in all momentum range, except for two ranges near of 8 and 13 GeV/c with peak values of $3.5 \cdot 10^{-3}$ and $6 \cdot 10^{-3}$. It is due to that the response function depends on the collider lattice and the spin rotation angle in each arc's dipole. One can suppress the peak values down to $10^{-4} \div 10^{-3}$ by means of collider lattice optimization. Besides, it is useful to minimize the response function in the experimental straights to reduce their contribution to the resonance strength, which is about 85% in the considered example (see figure 2). Thus, one can use PC solenoids with integral field of 1 T$\times$0.6 m for the protons polarization control. Such PC solenoids can provide the spin tune value of 10^{-2} in the all collider energy range.

Let us pay attention to the control of deuteron beam polarization. The coherent part of the resonance strength for deuterons does not exceed $\sim 10^{-6}$ in the all collider momentum range and is approximately three orders of magnitude lower than for protons. Thus, the spin tune value of 10^{-4} is surely sufficient for the deuteron polarization control. The field of the one PC solenoid is only 0.03 T at length of 0.6 m. The ramp field time of such a solenoid can be a few milliseconds. There is unique opportunity to arrange fast spin reversals in the detector for all bunches simultaneously during the experiment. The fast spin-flipping system would cardinally solve the problem of experimental systematic errors related to the uncertainty of a relative bunch-by-bunch luminosity.

3. The NICA collider for high precision experiments with polarized proton and deuteron beams

There is additional option for improvement of a polarized beam quality by the compensation of the coherent part of the zero-integer spin resonance strength. It is sufficient to use the existing collider control PC insertions (see figure 5). The first PC insertions located in the straight containing the interaction point directly control the polarization. The second PC insertions with constant solenoid fields are located in the other straight and are used to compensate the coherent part of resonance strength.

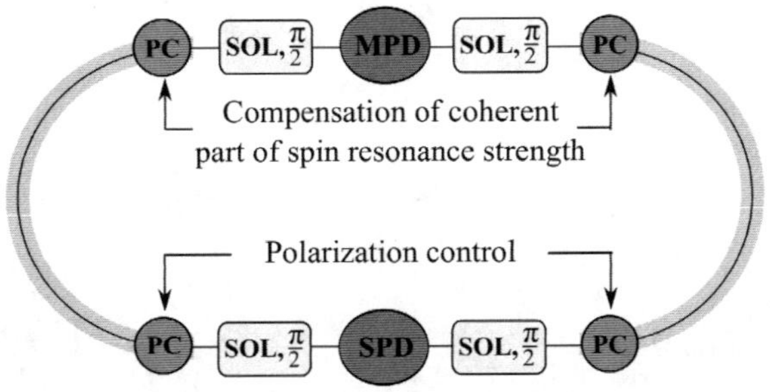

Figure 5. The scheme of compensation of coherent part of the zero-integer spin resonance strength.

The technique for compensation of integer resonance harmonics is well known and has been successfully utilized, for example, at the AGS [6].

Conclusion

In summary, one can draw the following conclusions:

- for stability of ion polarization at the NICA collider in spin transparency mode, the spin tune induced by the PC solenoids must significantly exceed the strength of the zero-integer spin resonance,
- PC solenoids with 0.03 T field are sufficient for stabilization and control of deuteron polarization in NICA collider,
- such PC solenoids fields allow one to realize spin-flipping system with reversal time of a few seconds for protons and a few miliseconds for deuterons,
- the NICA collider in spin transparency mode allows one to carry out ultra-high precision experiments with polarized beams.

Acknowledgments

We would like to thank all the participants of the seminar at Veksler and Baldin Laboratory of High Energy Physics (JINR, Dubna) for the useful discussions on some questions and results considered in the paper.

References

[1] Kovalenko A D, Filatov Yu N, Kondratenko A M, Kondratenko M A and Mikhaylov V A 2014 Polarized deuterons and protons at NICA@JINR *PEPAN Lett.* **45** 1 pp 321-22

[2] Abeyratne S et al 2012 arXiv:1209.0757 [physics.acc-ph]

[3] Filatov Yu N, Kovalenko A D, Kondratenko A M, Kondratenko M A and Mikhaylov V A 2013 *Proton And Deuteron Polarization Control In NICA Collider Using Small Solenoids* Proc. XV Workshop on HESP DSPIN 2013 Dubna pp 351-54

[4] Kovalenko A D, Butenko A V, Kekelidze V D, Mikhaylov V A, Kondratenko A M, Kondratenko M A and Filatov 2015 *Ion Polarization Control in the MPD and SPD Detectors of the NICA Collider* Proc. IPAC2015 TUPTY017

[5] Kondratenko M A, Kovalenko A D, Butenko A V, Kekelidze V D, Mikhaylov V A, Kondratenko A M and Filatov 2015 Orbital parameters of proton and deuteron beams in NICA collider with solenoid Siberian snakes *IOP Conf. Ser. In these Proc.* XVI Workshop on HESP DSPIN-2015 Dubna

[6] Khiari F Z *et al* 1989 Acceleration of polarized protons to 22GeV and the measurement of spin-spin effects in $p\uparrow + p\uparrow \rightarrow p + p$ *Phys. Rev.* D **39** pp 45-48

Journal of Physics: Conference Series **678** (2016) 012024 doi:10.1088/1742-6596/678/1/012024

Momentum transfer dependence of GPDs

O.V. Selyugin

BLTPh, Joint Institute for Nuclear Research, Dubna, Russia

E-mail: selugin@theor.jinr.ru

Abstract. Based on the factorization representation of the General Parton Distributions (GPDs) the momentum transfer dependence was determined by the analysis of the different representations of parton distribution functions (PDFs) and all possible experimental data of the electromagnetic form factors of the proton and neutron. The obtained t-dependence of the GPDs is checked by analysis of the different hadronic reactions (including exclusive and elastic hadron scattering) in a wide energy region with minimum free fitting parameters.

1. Introduction

The parton picture of the hadron structure is represented, in most part, by the parton distribution functions (PDFs). They are determined in the deep inelastic processes. The next step in the development of the picture of the hadron structure was made by introducing the non-forward structure functions - general parton distributions - GPDs [1, 2, 3] with the spin-independent $H(x,\xi,t)$ and the spin-dependent $E(x,\xi,t)$ parts. Some of the advantages of GPDs were presented by the sum rules [2]

$$F_1^q(t) = \int_0^1 dx\ \mathcal{H}^q(x,\xi=0,t), \quad F_2^q(t) = \int_0^1 dx\ \mathcal{E}^q(x,\xi=0,t). \tag{1}$$

Using the different momenta of GPDs as a function of x^{n-1} we can obtain the different form factors.

$$F_{n-1}^q(t) = \int_0^1 dx\ x^{n-1}\mathcal{F}^q(x,\xi=0,t) \tag{2}$$

$n=0$, (zero moment) give the Compton form factors $R_V(t), R_A(t), R_T(t)$
$n=1$, (first moment) give the electromagnetic form factors $F_1(t), F_2(t)$
$n=2$, (second moment) give the gravimagnetic form factors $A_1(t), B_2(t)$

Good knowledge of the momentum transfer dependence of the GPDs is required. Now we cannot obtain the t-dependence of GPDs from the first principles, but it can be obtained from the phenomenological description by *GPDs* of the nucleon electromagnetic form factors. Many different forms of the t-dependence of GPDs were proposed. In the quark diquark model [4, 5] the form of GPDs consists of three parts - PDFs, function distribution and Regge-like. In other works (see e.g. [6]), the description of the t-dependence of GPDs was developed in a more

Content from this work may be used under the terms of the Creative Commons Attribution 3.0 licence. Any further distribution of this work must maintain attribution to the author(s) and the title of the work, journal citation and DOI.
Published under licence by IOP Publishing Ltd

complicated picture using the polynomial forms with respect to x. In [7], it was shown that at large $x \to 1$ and momentum transfer the behavior of GPDs requires a larger power of $(1-x)^n$ in the t-dependent exponent:

$$\mathcal{H}^q(x,t) \ \sim exp[a \ (1-x)^n \ t] \ q(x). \tag{3}$$

with $n \geq 2$. It was noted that $n = 2$ naturally leads to the Drell-Yan-West duality between parton distributions at large x and form factors.

2. New momentum transfer dependence of GPDs

Let us modify the original Gaussian ansatz and choose the t-dependence of GPDs in a simple form

$$\mathcal{H}^q(x,t) \ = q(x) \ exp[a_+ \ (1-x)^2/x^m \ t]. \tag{4}$$

The value of the parameter $m = 0.4$ is fixed by the low t experimental data while the free parameters $a_\pm$ (a_+ - for $\mathcal{H}$ and a_- - for $\mathcal{E}$) were chosen to reproduce the experimental data in the whole t region. The isotopic invariance can be used to relate the proton and neutron GPDs. Hence, we do not change any parameter and keep the same t-dependence of GPDs as in the case of proton.

In all calculations we restrict ourselves to the contributions of only valence u and d quarks. In our first work [8] the function $q(x)$ is based on the MRST02 global fit [9].

Further development of the model requires a careful analysis of the momentum transfer form of GPDs and properly chosen PDFs form. In [10], the analysis of more than 24 different PDFs was made. The complex analysis of the corresponding description of the electromagnetic form factors of the proton and neutron by the different PDF sets (24 cases) was carried out . These PDFs include the leading order (LO), next leading order (NLO) and next-next leading order (NNLO) determination of the parton distribution functions. They used the different forms of the x dependence of PDFs. We slightly complicated the form of GPDs in comparison with eq.(4), but it is the simplest one as compared to other works (for example [11]).

$$\mathcal{H}^u(x,t) \ = q(x)^u_{nf} \ e^{2a_H \ \frac{(1-x)^{2+\epsilon_u}}{(x_0+x)^m} \ t}; \quad \mathcal{H}_\lceil{}^d(x,t) \ = q(x)^d_{nf} \ e^{2a_H(1+\epsilon_0)(\frac{(1-x)^{1+\epsilon_d}}{(x_0+x)^m}) \ t}. \tag{5}$$

$$\mathcal{E}^u(x,t) \ = q(x)^u_{fl} \ e^{2a_E \ \frac{(1-x)^{2+\epsilon_u}}{(x_0+x)^m} \ t}; \quad \mathcal{E}_\lceil{}^d(x,t) \ = q(x)^d_{fl} \ e^{2a_E(1+\epsilon_0)(\frac{(1-x)^{1+\epsilon_d}}{(x_0+x)^m}) \ t}. \tag{6}$$

where $q(x)^{u,d}_{fl} = q(x)^{u,d}_{nf}(1.-x)^{z_1,z_2}$

To obtain the form factors, we have to integrate over x in the whole range $0-1$. Hence, the form of the x-dependence of PDF affects the form and size of the form factor. But the PDF sets are determined from the inelastic processes only in some region of x, which is only approximated to $x = 0$ and $x = 1$. Some PDFs have the polynomial form of x with different power. Some other have the exponential dependence of x. As a result, the behavior of PDFs, when $x \to 0$ or $x \to 1$, can impact the form of the calculated form factors. The analysis was carried out with different forms of the t dependence of GPDs. The minimum number of free parameters was six and maximum was ten. The obtained electromagnetic form factors and the ratio of the electromagnetic form factors for the proton $\mu_p G^p_E/G^p_M$ and for the neutron $\mu_n G^n_E/G^n_M$ were described quantitatively well. Our calculation reproduces the data obtained by the polarization method.

On the basis of our GPDs with ABM12 [12] PDFs we calculated the hadron form factors by the numerical integration

$$F_1(t) = \int_0^1 dx[\frac{2}{3}q_u(x)e^{2\alpha_H t(1.-x)^{2+\epsilon_u}/(x_0+x)^m} - \frac{1}{3}q_d(x)e^{2\alpha_H t(1.-x)^{1+\epsilon_d}/((x_0+x)^m)}] \tag{7}$$

and then by fitting these integral results by the standard dipole form with some additional parameters for $F_1(t)$

$$F_1(t) = \frac{4m_p - \mu t}{4m_p - t}\frac{1}{(1+q/a_1+q^2/a_2^2+q^3/a_3^3)^2} \tag{8}$$

The matter form factor

$$A(t) = \int_0^1 x\ dx[q_u(x)e^{2\alpha_H t(1.-x)^{2+\epsilon_u}/(x_0+x)^m} + q_d(x)e^{2\alpha_H t(1.-x)^{1+\epsilon_d}/((x_0+x)^m)}] \tag{9}$$

is fitted by the simple dipole form $A(t) = \Lambda^4/(\Lambda^2 - t)^2$.

Our description is valid up to a large momentum transfer with the following parameters: $a_1 = 16.7$ GeV, $a_2^2 = 0.78$ GeV^2, $a_3^3 = 12.5$ GeV^3 and $\Lambda^2 = 1.6$ GeV^2. These form factors will be used in our model of the proton-proton and proton-antiproton elastic scattering.

3. The Compton cross sections

Our calculations are based on the works [13, 11]. The differential cross section for that reaction can be written as

$$\frac{d\sigma}{dt} = \frac{\pi\alpha_{em}^2}{s^2}\frac{(s-u)^2}{-us}[R_V^2(t) - \frac{t}{4m^2}R_T^2(t) + \frac{t^2}{(s-u)^2}R_A^2(t)], \tag{10}$$

where $R_V((t)$, $R_T(t)$, $R_A(t)$ are the form factors given by the $1/x$ moments of the corresponding GPDs $H^q(x,t)$, $E^q(x,t)$, $\tilde{H}^q(x,t)$. The last is related with the axial form factors. As noted in [11], this factorization, which bears some similarity to the handbag factorization of DVCS, is formulated in a symmetric frame where the skewness $\xi = 0$. For $H^q(x,t)$, $E^q(x,t)$ we used the PDFs obtained from the works [14] with the parameters obtained in our fitting procedure of the description of the proton and neutron electromagnetic form factors in [10].

$$R_V(t) = \int_0^1 \frac{dx}{x}\left\{\frac{4}{9}u(x)\ exp[2\alpha_1\frac{(1-x)^{p_1}}{(x_0+x)^{p_[2}}t] + \frac{1}{9}d(x)\ exp[2\alpha_1\frac{(1-x)^{p_1(k_d)}}{(x_0+x)^{p_2}}t]\right\}, \tag{11}$$

$$R_T(t) == \int_0^1 \frac{dx}{x}\left\{\frac{4}{9}u^e(x)\ exp[2\alpha_1\frac{(1-x)^{p_1}}{(x_0+x)^{p_[2}}t] + \frac{1}{9}d^e(x)\ exp[2\alpha_1\frac{(1-x)^{p_1(k_d)}}{(x_0+x)^{p_2}}t]\right\}, \tag{12}$$

$$R_A(t) = \int_0^1 \frac{dx}{x}\left\{\frac{4}{9}\Delta u^e(x)\ exp[2\alpha_1\frac{(1-x)^{p_1}}{(x_0+x)^{p_[2}}t] + \frac{1}{9}\Delta d^e(x)\ exp[2\alpha_1\frac{(1-x)^{p_1(k_d)}}{(x_0+x)^{p_2}}t]\right\}, \tag{13}$$

For $\tilde{H}^q(x,t)$ we take Δq^e in the form

$$\Delta q^e = N_i x_1^a(1 + a_2\sqrt{x} + a_3 x), \tag{14}$$

with the parameters determined in [15]. The calculations of R_i on the whole, correspond to the calculations [11], but the integrals with our t dependence of GPDs do not divergence at momentum transfer $-t > 2$ GeV^2. In [11], R_i is presented beginning with $-t = 4$ GeV^2.

The calculations of the differential cross sections of RCS are shown in Fig.15 at three energies $s = 9.8, 10.92$ and 20 GeV^2. The calculations have a sufficiently good coincidence with the existing experimental data and, in some part, coincide with calculations [11].

Journal of Physics: Conference Series **678** (2016) 012024 doi:10.1088/1742-6596/678/1/012024

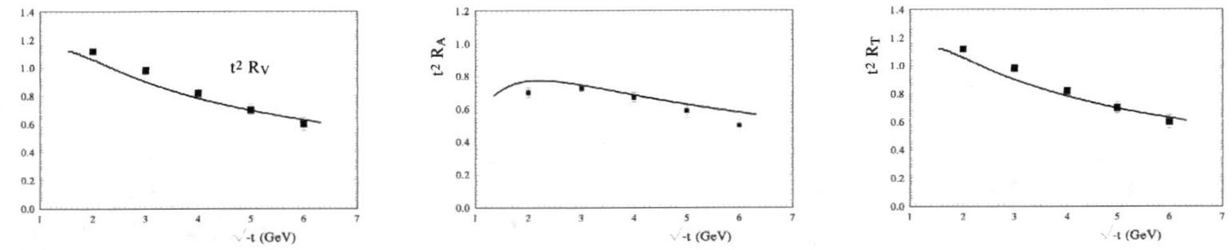

Figure 1. Compton form factors (squares - show the model calculations [11]
a) [left] $t^2 R_V(t)$, b) [middle] $t^2 R_T(t)$, c) [right] $t^2 R_A(t)$

.

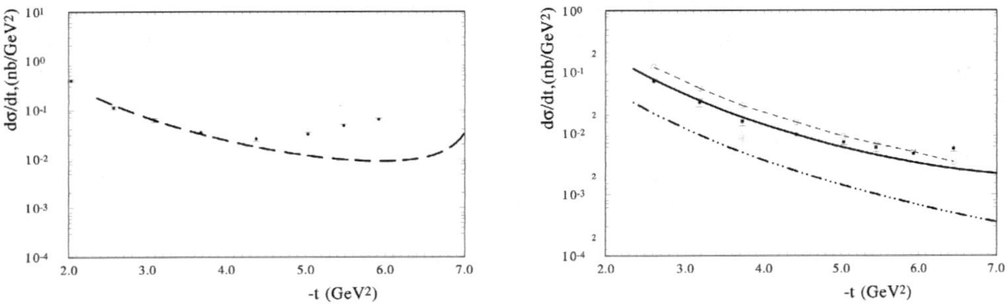

Figure 2. Differential Compton cross sections $\gamma p \to \gamma p$; the curves are our calculations at a) [left] $s = 8.9$ GeV^2, b) [right] $s = 10.92$ GeV^2 (dashed line with circles - the calculations [11]), and $s = 20$ GeV^2 the data points are for $s = 8.9$ GeV^2 (circles); $s = 10.92$ GeV^2 (squares) [23].

4. Elastic nucleon scattering (HEGS model)

Let us use the obtained momentum transfer dependence of GPDs in calculations of the electromagnetis and gravimagnetic form factors of the nucleons as the first and second moments of GPDs. The obtained form factors are related to the charge and matter distributions. The new High Energy General Structure (HEGS) model of the elastic proton-proton and proton-antiproton scattering was proposed [16, 17] with the two form factors determined early with the fixed parameters.

The model is very simple from the viewpoint of the number of parameters and functions. There are no any artificial functions or any cuts which bound the separate parts of the amplitude by some region of momentum transfer or energy. One of the most remarkable properties is that the real part of the hadron scattering amplitude is determined only by complex energy $\hat{s}$ that satisfies the crossing-symmetries.

The differential cross sections of nucleon-nucleon elastic scattering can be written as the sum of different helicity amplitudes:

$$\frac{d\sigma}{dt} = \frac{2\pi}{s^2}(|\Phi_1|^2 + |\Phi_2|^2 + |\Phi_3|^2 + |\Phi_4|^2 + 4|\Phi_5|^2). \tag{15}$$

The total helicity amplitudes can be written as $\Phi_i(s,t) = F_i^h(s,t) + F_i^{\mathrm{em}}(s,t)e^{\varphi(s,t)}$, where $F_i^h(s,t)$ comes from the strong interactions, $F_i^{\mathrm{em}}(s,t)$ from the electromagnetic interactions

Journal of Physics: Conference Series **678** (2016) 012024 doi:10.1088/1742-6596/678/1/012024

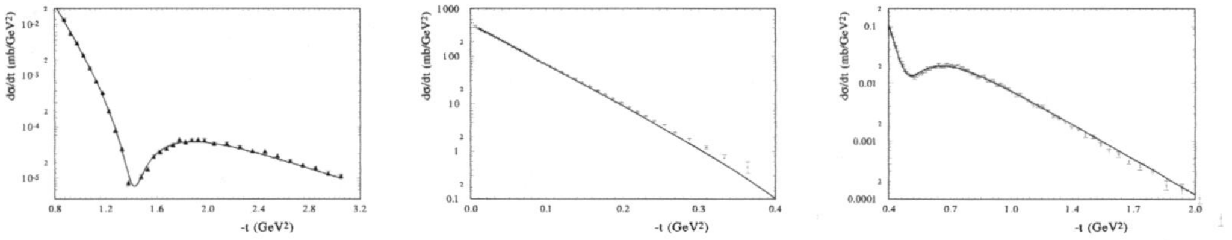

Figure 3. Differential cross sections of the elastic pp scattering a) [left] at $\sqrt{s} = 30.5$ GeV b)[middle] at $\sqrt{s} = 7$ TeV (small t) and c) [right] at $\sqrt{s} = 7$ TeV (large t)

and $\varphi(s,t)$ is the interference phase factor between the electromagnetic and strong interactions [18, 19, 20].

The (HEGS) model [17] gives a quantitative description of the elastic nucleon scattering at high energy with only 5 fitting high energy parameters. A successful description of the existing experimental data by the model shows that the elastic scattering is determined by the generalized structure of the hadron. The model leads to a coincidence of the model calculations with the preliminary data at 8 TeV. We found that the standard eikonal approximation [21] works perfectly well from $\sqrt{s} = 9$ GeV up to 8 TeV.

In the model the Born term of the elastic hadron amplitude is determined

$$F_h^{Born}(s,t) \ = h_1 \ F_1^2(t) \ F_a(s,t) \ (1 + r_1/\hat{s}^{0.5}) \ + h_2 \ A^2(t) \ F_b(s,t) \ (1 + r_2/\hat{s}^{0.5}), \tag{16}$$

where $F_a(s,t)$ and $F_b(s,t)$ have the standard Regge form

$$F_a(s,t) \ = \hat{s}^{\epsilon_1} \ e^{B(s) \ t}; \quad F_b(s,t) \ = \hat{s}^{\epsilon_1} \ e^{B(s)/4 \ t}. \tag{17}$$

The final elastic hadron scattering amplitude is obtained after unitarization of the Born term. So, first, we have to calculate the eikonal phase and then, using the standard eikonal representation, obtained the final hadron elastic scattering amplitude.

Our complete fit of 3416 experimental data in the energy range $9.8 \leq \sqrt{s} \leq 8000$ GeV and the region of the momentum transfer $0.000375 \leq \ -t \ \leq 14.75$ GeV^2 gave $\sum_{i=1}^{N} \chi_i^2/N = 1.28$ with the parameters $h_1 = 3.67$; $h_2 = 1.39$; $h_{odd} = 0.76$; $k_0 = 0.16$; $r_0^2 = 3.82$; and the low energy parameters $h_{sf} = 0.05$; $r_1 = 53.7$; $r_2 = 4.45$.

Obviously, for such a huge energy region we have a very small number of free parameters. Noteworthy also is a good description of the CNI region of momentum transfer in a very wide energy region, approximately three orders, with the same slope of the scattering amplitude. As example, the differential cross sections of the proton-proton elastic scattering in the region of the diffraction minimum and different energies are presented in Fig. 3 at $\sqrt{s} = 30.5$ GeV and $\sqrt{s} = 7.$ TeV. In most part, the form and energy dependence of the diffraction region is determined by the real part of the scattering amplitude. Note that in the model only the Born term is determined. The complicated diffractive structure is obtained only after the unitarization procedure. The extended variant of the model [10] shows the contribution of the "maximal" odderon with specific kinematic properties and does not show a visible contribution of the hard pomeron, as in [22].

Only our model takes into account the whole region of the momentum transfer ($3.75 \ 10^{-4} \geq |t| \geq 15$ GeV^2, which includes the high precision experimental data in the Coulomb-hadron interference region.

Journal of Physics: Conference Series **678** (2016) 012024 doi:10.1088/1742-6596/678/1/012024

5. Conclusion

The analysis of PDFs of many Collaborations with different forms of the x-dependence show that additional parameters of the function $f(x)$ in the exponential t-dependence part of GPDs lead to the identical results in the final fitting procedure. Hence, such additional parameters equalize the different form of PDFs to obtain the same results for the electromagnetic form factors of the nucleons.

Our examination of the obtained momentum transfer dependence of GPDs was checked in the case of the real Compton scattering. The calculated Compton form factors $R_V(t)$, $R_T(t)$, $R_A(t)$ gave a somewhat larger slope than such form factors calculated in [11] (see Fig.1). Our form factors reproduce the differential cross sections of the real Compton scattering at large momentum transfer reasonably well (see Fig.2).

GPDs let us calculate not only electromagnetic form factor (which reflects the charge distribution) but also the second moment of the GPDs give the gravimagnetic form factors (which reflects the matter distribution of the nucleons). The new HEGS model of the elastic pp and $\bar{p}p$ scattering was built taking into account both these form factors. It gave the possibility to essentially decrease the number of the fitting parameters. As a result, the model with a minimum of the fitting parameters describes quantitatively the maximum number of the experimental data in a wide region of the energy (9.8 GeV $\leq \sqrt{s} \leq$ 8 TeV) and the momentum transfer ($0.00037 \leq |t| \leq 15$ GeV^2). Such a unique description of the differential cross sections on the basis of the GPDs gave large support of the determining momentum transfer dependence of the GPDs.

Acknowledgments *The authors would like to thank J.-R. Cudell and O.V. Teryaev for fruitful discussion of some questions considered in the paper.*

[1] D. Muller *et al.*, Fortsch. Phys. **42**, (1994) 101;
[2] X.D. Ji, Phys. Lett. **78** , (1997) 610; Phys. Rev D **55** (1997) 7114;
[3] Radyushkin, A.V., Phys. Rev. D **56**, 5524 (1997).
[4] G.R. Goldstein, J.O. Hernandez, S. Liuti, Phys.Rev. **D84** 034007 (2011).
[5] J.O. Gonsales-Hernandes *et al.*, arXiv:1206.1876 v3.
[6] M.Diehl *et al.*, Eur.Phys. J. C **39** (2005) 1.
[7] F. Yuan, Phys. Rev. D, **69**, 051501(R) (2004) .
[8] O. Selyugin, O. Teryaev, Phys. Rev. **D 79** 033003 (2009);
[9] A.D. Martin *et al.*, Phys. Lett. B **531** (2002) 216.
[10] O.V. Selyugin, Phys. Rev. **D 89** 093007 (2014) .
[11] M. Diehl and P. Kroll, Eur.Phys.J. **C73** 2397 (2013).
[12] S. Alekhin, J. Blu"mlein, and S. Moch, Phys.Rev. D86 , 054009 (2012).
[13] Radyushkin, A.V., Phys. Rev. D **58**, 114008 (1998).
[14] H. Khanpour *et al.*, arXiv:1205.5194
[15] D. de Florian, R. Sassot, M. Strtmann, W. Vogelsang, Phys. Rev. **D 80** 034030 (2009) .
[16] O.V. Selyugin, Eur.Phys.J. **C72**, 2073 (2012).
[17] O.V. Selyugin, Phys. Rev. **D 91** 113003 (2015) .
[18] O.V. Selyugin, Mod. Phys. Lett. A**9** (1994) 1207.
[19] O.V. Selyugin, Mod. Phys. Lett. A**14**, 223 (1999).
[20] O. V. Selyugin, Phys. Rev. D **60** (1999) 074028
[21] J.-R. Cudell, E. Predazzi, O. V. Selyugin, Phys.Rev. **D 79** , 034033 (2009).
[22] O. V. Selyugin, Nucl.Phys. A **903** 54 (2013).
[23] A. Danagoulian, et. al. (Jefferson Lab Hall A Collaboration), Phys.Rev.Lett., **98** 152001 (2007).

Journal of Physics: Conference Series **678** (2016) 012025 doi:10.1088/1742-6596/678/1/012025

The gluon masses

S.A. Larin

Institute for Nuclear Research, 60-th October Anniversary prospect, 7-A, Moscow 117312, Russia

E-mail: larin@inr.ac.ru

Abstract. It is shown that the fundamental Lagrangian of Quantum Chromodynamis should be modified by the adding gluon masses to produce the mass-gap in accordance with the Källen-Lehmann spectral representation. On mass-shell renormalizability and unitarity of the resulting theory is demonstrated.

The discovery [1] of asymptotic freedom in Quantum Chromodynamics (QCD) has lead to the establishment of QCD as the theory of strong interactions. The gauge bosons of the theory, the gluons, are considered to be massless to have gauge invariance and correspondingly renormalizability. The introduction of the non-zero Lagrangian gluon masses into the fundamental QCD Lagrangian was considered to be forbidden because of the violating either renormalizability or untarity of the corresponding theory. Giving masses to gauge bosons via the Higgs mechanism [2] is also not allowed since colored Higgs particles are rejected by experiments.

Quite recently it was found that the non-abelian Yang-Mills theory [3] with masses of the Proca type is in fact on mass-shell renormalizable [4]. Unitarity of the theory of the Proca type is obvious since it contains only physical degrees of freedom (no ghosts).

In the present paper it is shown that it is impossible to obtain the necessary mass-gap in QCD with zero Lagrangian gluon masses. The fundamental QCD Lagrangian should be modified by the adding non-zero Lagrangian gluon masses of the Proca type to produce the mass-gap in accordance with the Källen-Lehmann spectral representation [5]. On mass-shell renormalizability of the resulting theory is discussed.

The Lagrangian of QCD is

$$L_{QCD} = -\frac{1}{4}F^a_{\mu\nu}F^a_{\mu\nu} + i\overline{\psi}_f\gamma_\mu D_\mu\psi_f - m_f\overline{\psi}_f\psi_f \tag{1}$$

$$-\frac{1}{\xi}(\partial^\mu A^a_\mu)^2 + \partial^\mu\overline{c}^a(\partial_\mu c^a - gf^{abc}c^b A^c_\mu) + counterterms,$$

where $F^a_{\mu\nu} = \partial_\mu A^a_\nu - \partial_\nu A^a_\mu + gf^{abc}A^b_\mu A^c_\nu$ is the gluon field strength tensor, $D_\mu = \partial_\mu - igA^a_\mu T^a$ is the covariant derivative. The quark fields ψ_f transform as the fundamental representation of the colour group $SU(3)$, $f = u, d, s, c, b, t$ is the flavour index. The gluons A^a_μ transform as the adjoint representation of this group. c^a are the ghost fields, ξ is the gauge parameter of the usually chosen general covariant gauge, f^{abc} are the structure constants of the group, T^a are the generators of the fundamental representation. $g = g(\mu)$ is the renormalized strong coupling

Content from this work may be used under the terms of the Creative Commons Attribution 3.0 licence. Any further distribution of this work must maintain attribution to the author(s) and the title of the work, journal citation and DOI.
Published under licence by IOP Publishing Ltd

Journal of Physics: Conference Series **678** (2016) 012025 doi:10.1088/1742-6596/678/1/012025

constant, $g^2/(16\pi^2) \equiv a_s$, $m_f = m_f(\mu)$ is the Lagrangian (renormalized) mass of a quark with a flavor f, and μ is the renormalization point. The summations over repeated indexes are assumed.

Let us consider the vacuum polarization function $\Pi(q^2)$

$$(-q^2 g_{\mu\nu} + q_\mu q_\nu)\Pi(q^2) = i \int dx e^{iqx} \langle 0|\, T\, j_\mu(x) j_\nu(0)\, |0\rangle. \tag{2}$$

where $j_\mu = \sum_f q_f \overline{\psi}_f \gamma_\mu \psi_f$ is the electromagnetic quark current and $q_f = 2/3, -1/3, ...$ is the electromagnetic charge of the quark with a flavor f.

According to general principles of local quantum field theory the function $\Pi(q^2)$ satisfies the Källen-Lehmann [5] spectral representation

$$\Pi(q^2) = \frac{1}{12\pi^2} \int_{4m_\pi^2}^{\infty} ds\, \frac{R(s)}{s - q^2 - i0}, \tag{3}$$

where the ratio $R(s) = \sigma_{tot}(e^+e^- \to hadrons)/\sigma(e^+e^- \to \mu^+\mu^-)$ is the normalized total cross-section of electron-positron annihilation into hadrons, m_π is a pion mass.

The Källen-Lehmann representation determines the analytic properties of $\Pi(q^2)$ which should be an analytic function in the complex q^2-plane with the cut starting from the first physical threshold, i.e. as it is dictated by experiments from the two-pion threshold $q^2 = 4m_\pi^2$. In particular, one gets for the discontinuity of $\Pi(q^2)$ on the cut

$$\Delta\Pi(q^2) \equiv \Pi(q^2 + i0) - \Pi(q^2 - i0) = \begin{cases} i\, R(q^2)/(6\pi) & at \;\; s > 4m_\pi^2 \\ 0 & at \;\; s < 4m_\pi^2. \end{cases} \tag{4}$$

Perturbative QCD produces the following expression for the discontinuity

$$\Delta\Pi(q^2)_{pQCD} = \theta(q^2)\, \rho_{gluon}(q^2) + \theta(q^2 - 4M_u^2)\, \rho_{quark}(q^2). \tag{5}$$

The gluon spectral density $\rho_{gluon}(q^2)$ contributes for $q^2 > 0$ as it is indicated by the theta-function $\theta(q^2)$. This is the known zero threshold. It arises from those absorptive parts of Feynman diagrams of $\Pi(q^2)$ which are produced by purely gluonic cuts of the diagrams (i.e. by the Cutcosky cuts which cross only gluon propagators of diagrams). As it is well known such diagrams appear for the first time at the four-loop level in the order a_s^3 (corresponding cuts cross 3 gluon propagators).

The quark spectral density $\rho_{quark}(q^2)$ arises from the quark cuts of the diagrams (i.e. from the cuts which cross two or more quark propagators of the diagrams). It contributes for $q^2 > 4M_u^2$ where M_u is the perturbative pole mass of the lightest u-quark, defined as the pole of the quark propagator within perturbation theory. A perturbative quark pole mass $M_f = m_f(\mu) + O(a_s)$ naturally appears after summation of perturbative loop corrections to a quark propagator. It is a renormalization group invariant quantity, i.e. independent on the renormalization point μ and on the choice of the subtraction scheme. In this sense it behaves as a physical object and that is why it is natural to use this definition of a quark mass to parametrize the theory.

We will not discuss here the important by themselves questions of convergence or divergence of corresponding perturbative QCD series at low or at high energies. Here we will just accept the constructive approach that our conventional perturbation theory is adequate to the exact solution of the theory, i.e. it correctly reproduces the perturbative expansion of the exact solution. (We assume that the exact solution is in principle obtainable if we know enough mathematics.)

Hence one gets within QCD that $\Delta\Pi(q^2)$ is non-zero in the energy interval $0 < q^2 < 4m_\pi^2$ since the perturbative contribution $\Delta\Pi(q^2)_{pQCD}$ is non-zero in this interval. And we would like to stress here that one should get in QCD an exact zero below the two-pion threshold as it is dictated by experiments. There are of course also non-perturbative contributions, i.e. contributions of the type of e^{-1/a_s} which are invisible in the perturbative expansion at $a_s = 0_+$

$$e^{-1/a_s} = 0 \cdot a_s + 0 \cdot a_s^2 + ...$$

But here we note that non-perturbative contributions can not exactly cancel the perturbative contribution in the continuous interval $0 < q^2 < 4m_\pi^2$ because of the different analytucal dependence on on the coupling constanta_s. To get that $\Delta\Pi(q^2) = 0$ at $0 < q^2 < 4m_\pi^2$ in agreement with experiments one should move perturbative gluon and quark thresholds above $q^2 = 4m_\pi^2$. That is why we should introduce the non-zero Lagrangian gluon masses.

One could object that peturbative contributions in the exact solution for $\Delta\Pi(q^2)$ could be moved above the two-pion threshould with the help of the suitably arranged θ-function. E.g. the gluon perturbative contribution could be moved above the two-pion threshould in the following way

$$\theta\left(q^2 - M^2 f(a_s)\right)\ \rho_{gluon}(q^2), \tag{6}$$

where $f(a_s)$ is some function which is zero at $a_s = 0$ and M^2 is a dimensional parameter such that $M^2 f(a_s)$ is a renormalization group invariant quantity which produces the necessary mass-gap below the two-pion threshold.

But perturbative expansion of the contribution of eq.(6) would contain terms with $\delta(q^2)$ and its derivatives arising from the differentiating the θ-function and we do not see such terms in real perturbative expansion. Thus the contributions of the type of eq.(6) are excluded in the exact solution of the theory.

The naive objection here is that nobody trusts perturbation theory below the two-pion threshold, i.e. that the corresponding perturbative series is heavily divergent in this energy region. But for us only the principal existence of the pertubative series with finite coefficients below the two-pion threshold is of importance here independently on the question of its divergence.

Thus one obtains the following restrictions on the (perturbative pole) masses of gluons and quarks

$$(3M_{gl})^2 > 4m_\pi^2, \tag{7}$$
$$4M_u^2 > 4m_\pi^2.$$

Although the restriction on M_u seems to be quite strong for the lightest u-quark it is not excluded from the first principles.

To construct QCD with massive gluons we will follow the approach of [4]. Presently this is the only known way to get (on mass-shell) renormalizable theory of massive gluons without color scalars (color scalars are rejected by experiments). Within this approach one starts from a renormalizable theory with scalar fields using the Englert-Brout-Higgs mechanism of spontaneous symmetry breaking [2] and after transition to the unitary gauge removes remaining massive scalar fields. Thus we add to the massless QCD Lagrangian (1) the scalar part to begin with the following general Lagrangian

$$L_{QCD+scalars} = -\frac{1}{4}F^a_{\mu\nu}F^a_{\mu\nu} + i\overline{\psi}_f\gamma_\mu D_\mu\psi_f - m_f\overline{\psi}_f\psi_f + \tag{8}$$

$$(D_\mu\Phi)^+ D_\mu\Phi + (D_\mu\Sigma)^+ D_\mu\Sigma - \lambda_1\left(\Phi^+\Phi - v_1^2\right)^2 - \lambda_2\left(\Sigma^+\Sigma - v_2^2\right)^2$$
$$-\lambda_3\left(\Phi^+\Phi + \Sigma^+\Sigma - v_1^2 - v_2^2\right)^2 - \lambda_4\left(\Phi^+\Sigma\right)\left(\Sigma^+\Phi\right)$$
$$+L_{gf} + L_{gc} + counterterms,$$

Journal of Physics: Conference Series **678** (2016) 012025 doi:10.1088/1742-6596/678/1/012025

where we introduced two triplets $\Phi(x)$ and $\Sigma(x)$ of complex scalar fields in the fundamental representation of the $SU(3)$ color group to get all gluon massive. L_{gf} is the gauge fixing part of the Lagrangian in some chosen gauge and L_{gc} is the corresponding gauge compensating part with the Faddeev-Popov ghost fields.

We can choose the following shifts of scalar fields by the quantities v_1 and v_2 to generate masses of all eight gluons

$$\Phi(x) = \begin{pmatrix} \phi_1(x) + i\phi_2(x) + v_1 \\ \phi_3(x) + i\phi_4(x) \\ \phi_5(x) + i\phi_6(x) \end{pmatrix}, \quad \Sigma(x) = \begin{pmatrix} \sigma_1(x) + i\sigma_2(x) \\ \sigma_3(x) + i\sigma_4(x) + v_2 \\ \sigma_5(x) + i\sigma_6(x) \end{pmatrix}. \tag{9}$$

Choosing for simplicity $v_1 = v_2 \equiv v$ one obtains the following massive terms for gluons in the Lagrangian

$$L_M = m_{gl}^2 \Big[(A^1)^2 + (A^2)^2 + (A^3)^2 + \frac{1}{2}(A^4)^2 + \tag{10}$$
$$\frac{1}{2}(A^5)^2 + \frac{1}{2}(A^6)^2 + \frac{1}{2}(A^7)^2 + \frac{1}{3}(A^8)^2 \Big],$$

where $m_{gl}^2 \equiv g^2 v^2$ is the gluon mass parameter of the theory.

After the chosen shifts the following four combinations of scalar fields

$$\phi_1 + \frac{\lambda_3}{\lambda_1 + \lambda_3}\sigma_3, \quad \sigma_3, \quad \sigma_1 + \phi_3, \quad \sigma_2 - \phi_4 \tag{11}$$

become massive Higgs particles.

The following eight combinations

$$\sigma_1 - \phi_3, \quad \phi_4 + \sigma_2, \quad \phi_2 - \sigma_4, \quad \phi_2 + \sigma_4, \quad \phi_5, \quad \phi_6, \quad \sigma_5, \quad \sigma_6 \tag{12}$$

become massless Goldstone ghosts.

Now one can make transition to the unitary gauge. All ghost fields as usual disappear from the Lagrangian. Following the approach of [4] one can remove in the unitary gauge all Higgs fields from the Lagrangian preserving on mass-shell renormalizability of the theory.

To give the derivation of this statement let us consuder as an example the simplified case (the generalization to the above case will be straightforward). Let us consider the known model given by the initial $SU(2)$-invariant Lagrangian of interaction of vector bosons and scalar fields possessing the spontaneously broken symmetry

$$L = -\frac{1}{4}F_{\mu\nu}^a F_{\mu\nu}^a + (D_\mu\Phi)^+ D_\mu\Phi - \lambda\left(\Phi^+\Phi - v^2\right)^2 \tag{13}$$

with the doublet of scalar fields $\Phi(x)$ in the fundamental representation of the group.

Here $D_\mu\Phi = \left(\partial_\mu - ig\frac{\tau}{2}^a W_\mu^a\right)\Phi$ is the covariant derivative, τ^a are the Pauli matrices, $\lambda > 0$, $v^2 > 0$. (This model can be considered as the Standard Model of electroweak interactions without $U(1)$-interaction and fermions. The derivation given below can be applied also to the complete Standard Model, the γ_5- matrix being treated within dimensional regularization according to the technique of [6].)

To get the complete Lagrangian one makes the standard shift of the scalar field fixes the gauge and adds ultraviolet counterterms. Let us consider two gauges: the widely used R_ξ-gauge [7], [8] with an arbitrary parameter ξ and the unitary gauge.

The theory in the R_ξ-gauge describes three physical massive vector bosons with the mass $m = gv/\sqrt{2}$, and the physical Higgs field χ with the mass $M = 2\lambda v$. Here are also Goldstone

Journal of Physics: Conference Series **678** (2016) 012025 doi:10.1088/1742-6596/678/1/012025

ghosts ϕ^a and Faddeev-Popov ghosts c^a with masses ξm^2. The structure of the counterterms (consistent with gauge invariance and Slavnov-Taylor identities [9, 10] to ensure unitarity) is well known, see e.g. [11].

The propagators in the unitary gauge defined by the gauge condition $\phi^a = 0$ are obtained from those of the R_ξ-gauge by taking the limit $\xi \to \infty$. The theory in the unitary gauge is renormalizable only on mass-shell, i.e. Green functions are divergent at $\epsilon \to 0$ (ϵ being the parameter of dimensional regularization) but the S-matrix elements are finite.

To consider renormalization for our purpose it is convenient to use the Bogoliubov-Parasiuk-Hepp subtraction scheme [12]. As it is well known in this scheme a counterterm of e.g. a primitively divergent Feynman diagram is the truncated Taylor expansion of the diagram itself at some fixed values of external momenta. Hence counterterms of mass dependent diagrams are also mass dependent. Needless to say that subtractions should respect Slavnov-Taylor identities.

Let us consider S-matrix elements in the R_ξ-gauge without external Higgs bosons (i.e. with external W-bosons only in this simplified model). We will analyze the dependence of diagrams on the Higgs mass M by using for convinience the expansion in large M (after renormalization but before the removing regularization). The algorithm for the large mass expansion of Feynman diagrams is given e.g. in [13] (where it is quite reasonably checked in calculations of the 4-loop diagrams for the Z-boson decay into hadrons). It can be rigorously derived with the technique of [14].

We separate all diagrams into physical ones which are not nullified in the limit $\xi \to \infty$ and unphysical ones which are nullified. In this limit the propagator of the W-boson reduces to the known unitary form

$$\lim_{\xi\to\infty} < T(W^a_\mu W^b_\nu) >= -i\delta^{ab} \lim_{\xi\to\infty} \left(\frac{g_{\mu\nu} - k_\mu k_\nu/m^2}{k^2 - m^2} + \frac{k_\mu k_\nu/m^2}{k^2 - \xi m^2} \right) = -i\delta^{ab} \frac{g_{\mu\nu} - k_\mu k_\nu/m^2}{k^2 - m^2}$$

The propagators of the Goldstone bosons ϕ^a and ghosts c^a vanish in this limit and correspondingly all diagrams which contain these propagators are also nullified.

Thus in our notations the physical diagrams are the diagrams which do not contain Goldstone bosons propagators or ghosts propagators and the unphysical diagrams are the diagrams which contain such propagators.

Within the large-M expansion the physical diagrams with χ-propagators contain either terms with integer negative powers of M^2: $\frac{1}{M^{2n}}$, $n = 1, 2, 3, ...$ or terms with non-integer powers of M^2 (non-integer powers contain ϵ): $\frac{1}{M^{2(k+l\epsilon)}}$, $k - integer,\ l - positive\ integer$. This is because each vertex with the large factor M^2 has three or four attached χ-propagators due to the structure of the Higgs boson self-coupling.

In contrast, unphysical diagrams can have polynomial in M terms due to the four-ϕ vertex with the large factor M^2. But they are ξ-dependent (they are nullified in the limit $\xi \to \infty$) and this polynomial terms cancel in S-matrix elements.

In the renormalizable R_ξ-gauge one can present ultraviolet renormalization in a standard form of the Bogoliubov-Parasiuk R-operation for individual diagrams. This ensures that after renormalization the M-dependent terms are finite at $\epsilon \to 0$ separately from M-independent terms. Thus if one removes all M-dependent terms one is left with a finite expression.

On the Lagrangian level it means in the unitary gauge that one removes from L_U all terms containing the field χ and also all M-dependent terms in the counterterms. This should be done in the unitary gauge because in the R_ξ-gauge some diagrams containing propagators of Higgs particle can give contributions not depending on the Higgs mass M. In contrast in the unitary gauge all diagrams containing Higgs propagators give only contributions depending on M so there is one to one correspondence between M-dependent diagrams and terms in the Lagrangian containing the Higgs field χ

The resulting theory is on mass-shell finite. This is the massive Yang-Mills theory of the Proca type

$$L_{YM} = -\frac{1}{4}(\partial_\mu W^a_\nu - \partial_\nu W^a_\mu + \frac{z_1}{z_2} g f^{abc} W^b_\mu W^c_\nu)^2 + m^2 W^a_\mu W^a_\mu + counterterms.$$

Thus the Higgs mechanism can be considered as an efficient mathematical tool to observe on mass-shell renormalizability of the massive Yang-Mills theory of the Proca type which is far from to be obvious directly.

Let us now return to our Lagrangian (8) with two scalar triplets which after spontaneous symmetry breaking has four Higgs particles (11). Following the above approach we can remove in the unitary gauge all four Higgs fields from the Lagrangian preserving on mass-shell renormalizability of the theory. The Lagrangian of the resulting QCD with massive gluons is

$$L_{massive\ QCD} = L_M - \frac{1}{4} F^a_{\mu\nu} F^a_{\mu\nu} + i\overline{\psi}_f \gamma_\mu D_\mu \psi_f - m_f \overline{\psi}_f \psi_f + counterterms, \qquad (14)$$

where L_M is given in eq.(10).

Let us note that on mass-shell renormalizability does not mean that one should consider quarks and gluons as free external particles. It means that in the $SU(3) \times SU(2) \times U(1)$ theory the S-matrix elements with the physical external particles will be finite.

One can calculate the one-loop β-function in this theory to obtain for a massless renormalization scheme (i.e. a scheme where renormalization group functions do not depend on masses, e.g. the MS-scheme) the following result

$$\beta(a_s) = \mu^2 \frac{\partial a_s}{\partial \mu^2} = \sum_{i \geq 0} \beta_i a_s^{i+2}, \qquad \beta_0 = -\frac{7}{2} C_A + \frac{4}{3} T_F n_f,$$

here $C_A = 3$ is the Casimir operator of the adjoint representation of the $SU(3)$ color group, $T_F = 1/2$ is the trace normalization of the fundamental representation, n_f is the number of active quark flavors. Thus asymptotic freedom remains valid in the considered theory with massive gluons.

The author is grateful to collaborators of the Theory division of INR for helpful discussions and to the Organizing Committee of the Workshop for kind hospitality.

References

[1] D.J. Gross, F. Wilczek, Phys. Rev. Lett. 30 (1973) 1343;
H.D. Politzer, Phys. Rev. Lett. 30 (1973) 1346.
G. 't Hooft, report at the Marseille Conference on Yang-Mills Fields, 1972.
[2] P.W. Higgs, Phys. Lett. 12 (1964) 132.
F.Englert and R. Brout, Phys. Rev. Lett. 13 (1964) 321.
[3] C.N. Yang and R.L. Mills, Phys. Rev. 96 (1954) 191.
[4] S.A. Larin, e-Print: hep-ph/0503198; Phys.Part.Nucl. 44 (2013) 386; PoS BaldinISHEPP XXII (2015) 018.
[5] G. Källen, Helv. Phys. Acta 25 (1952) 417; H. Lehmann, Nuovo Cim. 11 (1954) 342.
[6] S.A. Larin, Phys. Lett. B 303 (1993) 113; e-Print: hep-ph/9302240.
[7] G. 't Hooft, Nucl. Phys. B 35 (1971) 167.
[8] K. Fujikawa, B.W. Lee and A.I. Sanda, Phys. Rev. D6 (1972) 2923.
[9] A.A. Slavnov, Theor. Math. Phys. 10 (1972) 99.
[10] J.C. Taylor, Nucl. Phys. B 33 (1971) 436.
[11] L.D. Faddeev and A.A. Slavnov, Gauge fields. Introduction to quantum theory, Front. Phys. 83 (1990) 1.
[12] N.N. Bogoliubov and O.S. Parasiuk, Acta Math. 97 (1957) 227.
K. Hepp, Comm. Math. Phys. 2 (1966) 301.
[13] S.A. Larin, T. van Ritbergen and J.A.M. Vermaseren, Nucl. Phys. B 438 (1995) 278.
[14] S.A. Larin, Phys. Lett. B 469 (1999) 220.

XVI Workshop on High Energy Spin Physics (D-SPIN2015) IOP Publishing
Journal of Physics: Conference Series **678** (2016) 012026 doi:10.1088/1742-6596/678/1/012026

The Q_{weak} Experiment: First Direct Measurement of the Weak Charge of the Proton

J.F. Dowd for the Q_{weak} Collaboration
College of William and Mary, Williamsburg, VA 23185 USA
E-mail: jdowd@jlab.org

Abstract. The recently completed Q_{weak} experiment at Jefferson Laboratory made the first direct determination of the proton's weak charge, Q_W^p, via a measurement of the parity-violating asymmetry in elastic electron-proton scattering at low four-momentum transfer. The Standard Model (SM) makes a precise prediction of $Q_W^p(SM) = 0.0710 \pm 0.0007$. A deviation from this prediction could be an indicator of new physics. A longitudinally polarized electron beam was scattered off a liquid hydrogen target and detected in eight azimuthally symmetric fused silica detectors. The small asymmetry, $A_{ep} = -279 \pm 35$ (stat) ± 31 (syst) ppb, was measured by observing the difference in rates seen in the detectors when the helicity of the electron beam was rapidly reversed. The measured asymmetry is the most precise and smallest asymmetry ever measured in an $\vec{e}p$ scattering experiment. Combining this asymmetry with previous parity-violating electron scattering (PVES) data, we obtained a value of $Q_W^p(PVES) = 0.064 \pm 0.012$, which agrees well with the SM value. The results of the experiment's commissioning run, which constitutes about 4% of the total data set, are reported here. Analysis of the remainder of the data set is ongoing and will significantly reduce the statistical and systematic uncertainties; several aspects of this analysis will be highlighted.

1. Introduction

The Standard Model (SM) is an undeniably effective theoretical framework of particle physics. Due to the existence of dark matter and the inability to explain the large observed matter/anti-matter asymmetry in the universe, the SM is thought to be an effective low-energy theory of more fundamental higher-energy physics. Exploration of possible physics beyond the SM include direct searches at the energy frontier and indirect searches at the precision frontier. The weak charge of the proton is suppressed and well-predicted in the SM, which increases its sensitivity to possible new physics and makes it an excellent candidate for a precision measurement [1].

The weak charge of the proton emerges from the SM Lagrangian through the axial electron, vector quark weak couplings, $C_{1i} = 2g_A^e g_V^i$. At tree-level, Q_w^p is given by

$$Q_w^p = -2(2C_{1u} + C_{1d}) = 1 - 4\sin^2\theta_W \tag{1}$$

where θ_W is the Weinberg angle, or electroweak mixing angle. The proton's weak charge is the neutral-weak analog to the proton's electric charge.

The observable quantity measured in this experiment is the parity-violating (PV) asymmetry, A_{ep}, formed from the difference over the sum of elastic $\vec{e}p$ scattering cross sections with positive and negative helicity, given by

$$A_{ep} = \frac{\sigma_+ - \sigma_-}{\sigma_+ + \sigma_-}. \tag{2}$$

Content from this work may be used under the terms of the Creative Commons Attribution 3.0 licence. Any further distribution of this work must maintain attribution to the author(s) and the title of the work, journal citation and DOI.
Published under licence by IOP Publishing Ltd

At tree-level and in terms of electromagnetic, and vector and axial-vector neutral-weak form factors, the PV asymmetry is

$$A_{ep} = \left[\frac{-G_F Q^2}{4\pi\alpha\sqrt{2}}\right]\left[\frac{\varepsilon G_E^\gamma G_E^Z + \tau G_M^\gamma G_M^Z - (1-4\sin^2\theta_W)\varepsilon' G_M^\gamma G_A^Z}{\varepsilon (G_E^\gamma)^2 + \tau (G_M^\gamma)^2}\right] \quad (3)$$

where

$$\varepsilon = \frac{1}{1+2(1+\tau)\tan^2\frac{\theta}{2}}, \qquad \varepsilon' = \sqrt{\tau(1+\tau)(1-\varepsilon^2)} \quad (4)$$

are kinematic quantities. G_F is the Fermi constant, θ_W is the weak mixing angle, θ is the laboratory electron scattering angle, $-Q^2$ is the four-momentum transfered squared, M is the proton mass, and $\tau = Q^2/4M^2$. At low Q^2 and in the forward angle limit, $\theta \to 0$, the asymmetry simplifies and can be recast into the reduced asymmetry

$$A_{ep}/A_0 = Q_w^p + Q^2 B(Q^2, \theta), \qquad A_0 = \left[\frac{-G_F Q^2}{4\pi\alpha\sqrt{2}}\right]. \quad (5)$$

The second term in the reduced asymmetry, $Q^2 B(Q^2,\theta)$, contains all of the hadronic structure and is suppressed at low Q^2. It was determined from previous PVES experiments at higher Q^2. The first term, Q_w^p, is then the intercept when the asymmetry is plotted versus Q^2.

Looking beyond tree level, with electroweak radiative corrections applied, the SM prediction of Q_w^p at $Q^2 = 0$ is

$$Q_w^p(SM) = [\rho_N C + \Delta_e][1 - 4\sin^2\hat{\theta}_W(0) + \Delta_e'] + \Box_{WW} + \Box_{ZZ} + \Box_{\gamma Z}(0). \quad (6)$$

Here, ρ_{NC} renormalizes the ratio of neutral to charged-current interactions at low energies [1]. Δ_e and Δ_e' are corrections to the Zee and γee couplings at the electron vertex. $\Box_{WW}$ and $\Box_{ZZ}$ are exchanges of two of the same weak boson that are easily calculable with perturbative quantum chromodynamics (QCD). The final correction, $\Box_{\gamma Z}$, is the dominant energy-dependent correction to equation 5 at the experiment's kinematics. This energy-dependent contribution must be subtracted from the Q_{weak} measurement to be compared with the SM prediction. Refinement of the $\Box_{\gamma Z}$ calculation is ongoing and has been evaluated by several groups (see Table 1). The most recent calculation [7], used to correct the data introduced here, uses parton distribution functions in combination with recent $\vec{e}d$ scattering data [9] to make the most precise calculation of $\Box_{\gamma Z}$ to date.

Presented here is the analysis of the commissioning run (approximately 4% of our total dataset) of the Q_{weak} experiment, which measured the parity-violating asymmetry in $\vec{e}p$ scattering to high precision. Combined with results from previous parity-violating electron scattering experiments (PVES), these data were used as a direct measurement of Q_w^p. The combination of C_{1u} and C_{1d} contained in Q_w^p is nearly orthogonal to that found in atomic parity violation (APV) experiments on ^{133}Cs [2], where $Q_w(^{133}\text{Cs}) = -2(188\text{C}_{1\text{u}} + 211\text{C}_{1\text{d}})$.

Table 1. $\Box_{\gamma Z}^V(E, Q^2)$ contribution to Q_w^p at Q_{weak} kinematics

Gorchtein & Horowitz [3]	0.0026 ± 0.0026
Sibirtsev, Blunden, Melnitchouk, & Thomas [4]	$0.0047^{+0.0011}_{-0.0004}$
Rislow & Carlson [5]	0.0057 ± 0.0009
Gorchtein, Horowitz, & Ramsey-Musolf [6]	0.0054 ± 0.0020
Hall, Blunden, Melnichouk, Thomas, & Young [7]	0.00557 ± 0.00036

Journal of Physics: Conference Series **678** (2016) 012026 doi:10.1088/1742-6596/678/1/012026

The complementarity of these two measurements provides a clean extraction of the individual light quark weak charges and, therefore, also allows a determination of the weak charge of the neutron, $Q_w^n = -2(C_{1u} + 2C_{1d})$.

2. Experimental Overview

The custom Q_{weak} apparatus [10] (Figure 1) was installed in Hall C of Jefferson Lab, a continuous electron beam accelerator facility [11]. The main components were a 34.4 cm long LH_2 target, a triple collimator system, toroidal magnetic spectrometer, and eight fused silica Cerenkov detectors arrayed around the beam axis. Retractable wire drift chambers, before and after the magnet, were used at low beam current to track individual scattered electrons and characterize the Q^2 of the experiment.

At the target center, the 89 $\pm$ 1.8% longitudinally polarized 145 μA electron beam had an average energy of 1.155 $\pm$ 0.003 GeV. The helicity of the electrons was rapidly reversed at 960 Hz in a pseudo-random quartet pattern, either (+ - - +) or (- + + -). The rapid reversal limited noise due to target density fluctuations, and the quartet reversal pattern reduced noise due to slow linear drifts. Precision beam polarimetry for the commissioning period was provided by an existing Møller polarimeter. This measurement was invasive and was limited to low beam currents (about 1 μA). For the final result, the Møller polarimeter measurement will be crosschecked with a new Compton polarimeter built specifically for the Q_{weak} experiment. The Compton polarimetery measurement is non-invasive and can be performed continuously at beam currents up to 180 μA.

The 34.4 cm long, conical, aluminum target cell was designed using computational fluid dynamics to minimize density fluctuations from the deposition of heat from the high energy and high current beam. The electron beam was incident on the high-power (3 kW) LH_2 target and was uniformly rastered over an area of 3.5 x 3.5 mm^2 to improve heat dissipation at the thin aluminum target windows, yielding a very high luminosity, of order 10^{39} cm^{-2}s^{-1}. The target density fluctuations contributed about 40 ppm to the measured asymmetry width per quartet.

The acceptance of the experiment was defined by the triple-collimator system, with an effective elastic scattering angle of 7.9 $\pm$ 3°. Upon scattering from the target, forward angle electrons pass through the collimators, where the toroidal magnet focuses elastically scattered electrons onto the detectors, while simultaneously diverting inelastics to larger radii. In the azimuthal direction, the experiment's acceptance is about 50% of 2π.

Eight radiation-hard quartz Cerenkov detectors were arrayed symmetrically about the beam axis. Each detector was comprised of two 100 x 16 x 1.25 cm bars glued together to make a 2 m long bar, two photo-multiplier tubes (PMTs), and a pre-radiator. The 2 cm thick lead

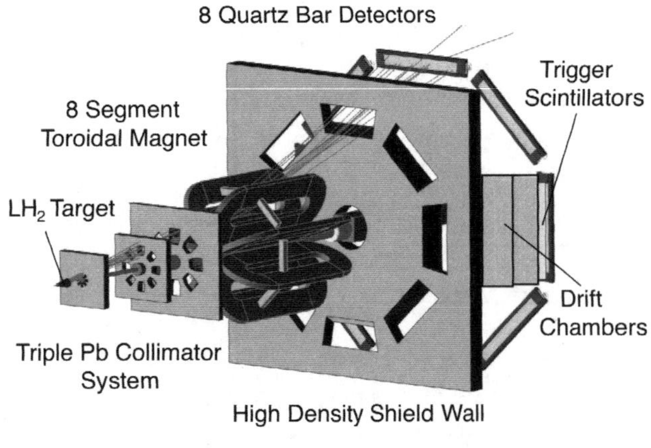

Figure 1. Schematic view of the Q_{weak} experiment showing the target, collimators, toroidal magnet, and detectors. Elastically scattered electrons (red lines) are focused at the detectors, while inelastically scattered electrons (not shown) are bent away to higher radii.

Journal of Physics: Conference Series **678** (2016) 012026 doi:10.1088/1742-6596/678/1/012026

pre-radiators were installed to suppress soft backgrounds and amplify the electron signal. The average light yield of an elastically scattered electron event was about 90 photo-electrons. The octagonal symmetry arrangement was chosen to suppress helicity-correlated beam motion effects when extracting the asymmetry.

In addition, there exists a smaller Q_{weak} data set with a beam current of 3.35 GeV, a Q^2 of 0.09 (GeV)^2, and a missing mass $W \approx 2.2$ GeV. The chosen kinematics allows access to non-resonant inelastic $\vec{e}p$ scattering, where the asymmetry depends on the $F_{1,3}^{\gamma Z}$ structure functions used in $\Box_{\gamma Z}$ calculations. The large missing mass leads to a significant background of pions that enter the acceptance of the experiment and hit the Cerenkov detectors. A 10 cm thick lead wall was placed in front of one Cerenkov detector to attenuate the scattered electrons and isolate the pions. Combining this with a pulse height analysis of all eight detectors, the pion background will be separated from the inelastic signal of interest. The polarized beam had a large transverse component which dilutes the parity-violating inelastic asymmetry. This contribution has been measured and characterized with dedicated transversely polarized beam ($\approx$100%). Analysis of this data will provide experimental validation of the theoretical models used to predict the $\Box_{\gamma Z}$ contribution to the weak charge of the proton and will be presented in an upcoming thesis.

3. Analysis

The charge normalized detector light yield was integrated over each stable helicity state, $Y_{\pm}$, in each quartet to form a raw asymmetry,

$$A_{raw} = \frac{Y_+ - Y_-}{Y_+ + Y_-}. \tag{7}$$

The raw asymmetry was corrected for sources of false asymmetries to extract the measured asymmetry,

$$\begin{aligned} A_{msr} &= A_{raw} + A_T + A_L - \sum_{i=1}^{5} \left(\frac{\partial A}{\partial \chi_i}\right) \Delta\chi_i \\ &= A_{raw} + A_T + A_L + A_{reg}. \end{aligned} \tag{8}$$

A_T is due to the small residual transverse polarization in the nominally longitudinally polarized beam. A_T is highly suppressed by the azimuthal symmetry of the detector system and is explicitly measured with dedicated transversely polarized beam. A_L accounts for possible non-linearities in the PMT response. In the final term, the $\Delta\chi_i$ are helicity-correlated differences in beam trajectory or energy in a quartet. The slopes $\frac{\partial A}{\partial \chi_i}$ were determined using linear regression of natural beam motion. For the data presented here, $A_T = 0 \pm 4$ ppb, $A_L = 0 \pm 3$ ppb, and $A_{reg} = -35 \pm 11$ ppb. After these corrections are applied the resulting measured asymmetry is $A_{meas} = -204 \pm 31$ ppb (stat) ± 13 ppb (syst).

The PV elastic $\vec{e}p$ asymmetry, A_{ep}, was then extracted from A_{meas} by,

$$A_{ep} = R_{Tot} \frac{A_{meas}/P - \sum_{i=0}^{4} f_i A_i}{1 - \sum f_i}. \tag{9}$$

The overall factor R_{Tot} accounts for the combined effects of radiative corrections, the non-uniformities in light and Q^2 distribution across the detectors, and kinematics normalization. P is the longitudinal polarization of the beam. A_i and f_i are the asymmetries and corresponding dilution factors (fraction of total signal due to background i) for each background. The measured asymmetry was diluted by backgrounds that arose from electrons that scattered off the aluminum target cell windows, the beamline, soft neutral background, and inelastically scattered electrons.

Journal of Physics: Conference Series **678** (2016) 012026 doi:10.1088/1742-6596/678/1/012026

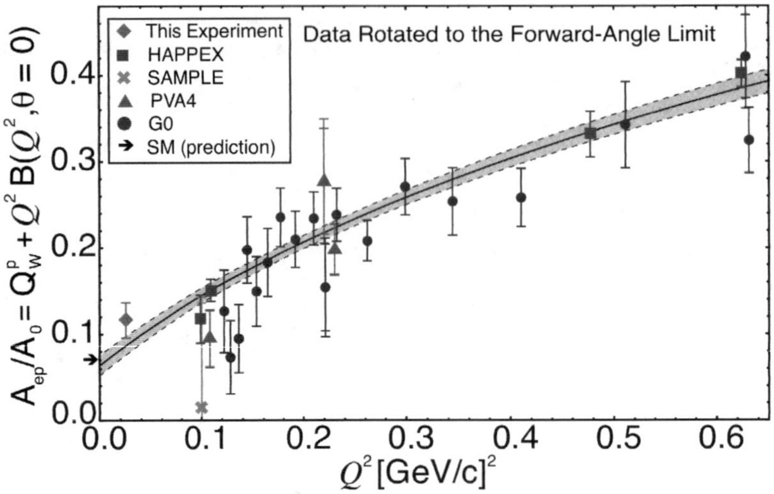

Figure 2. Global fit (black line) of world PVES data including the Q_{weak} commissioning data (red diamond). The outer band (dashed line) represents the uncertainty of the fit. This fit incorporated all PVES measurements including proton, helium, and deuterium data up to $Q^2 = 0.63(\text{GeV})^2$. All data were rotated to the forward angle limit, with the outer error bars indicating the additional uncertainty from the rotation. Q^p_w is the intercept of the fit and the SM prediction is indicated by the black arrow.

Because the detector signals are integrated, we cannot exclude background signals from the signal of interest. Therefore, each background was accounted for, with explicit measurements of their asymmetries and dilution factors and subtracted from the measured asymmetry. By far, the largest background to the measured asymmetry was from the aluminum target cell windows, due to the neutron content of the aluminum, where the measured asymmetry was 1.76 ppm and the dilution factor was 3.2%.

4. Results

Reported here are the results of the Q_{weak} experiment's commissioning run, which represents 4% of the total dataset. The fully corrected asymmetry was $A_{ep} = -279 \pm 35$ (stat) ± 31 (syst) ppb. Following the procedure in references [12] and [13], the asymmetry reported here was combined with existing world PVES data on hydrogen, deuterium, and ^{4}He targets [14–25] in a global fit (Figure 2) of the reduced asymmetry in equation 5. All PVES data up to $Q^2 = 0.63(\text{GeV})^2$ were used. The free parameters of the fit include the light quark weak charges C_{1u} and C_{1d}, the strange charge radius ρ_s, the strange magnetic moment μ_s, and the isovector axial form factor $G_A^{Z(T=1)}$. The value of the isoscalar axial form factor $G_A^{Z(T=0)}$ was constrained by theoretical calculations [26]. The Kelly parametrization [27] was used for the electromagnetic form factors. In addition, all data were corrected for the energy dependence that arises from $\Box_{\gamma Z}$ contributions. The intercept of the fit ($Q^2 = 0$) is $Q^p_w(PVES) = 0.064 \pm 0.012$ in excellent agreement with the SM prediction, $Q^p_w(SM) = 0.0710 \pm 0.0007$. This represents the first direct measurement of the weak charge of the proton.

This measurement of the weak charge of the proton was then used to constrain the weak isoscalar and isovector couplings (see figure 3). Combined with the APV experiments mentioned in section 1, the light quark weak charges were determined to be $C_{1u} = -0.1835 \pm 0.0054$ and $C_{1d} = +0.3355 \pm 0.0050$. These results are in good agreement with the SM predictions, $C_{1u}(SM) = -0.188$ and $C_{1d}(SM) = +0.341$ [28]. The light quark weak charges are in turn used

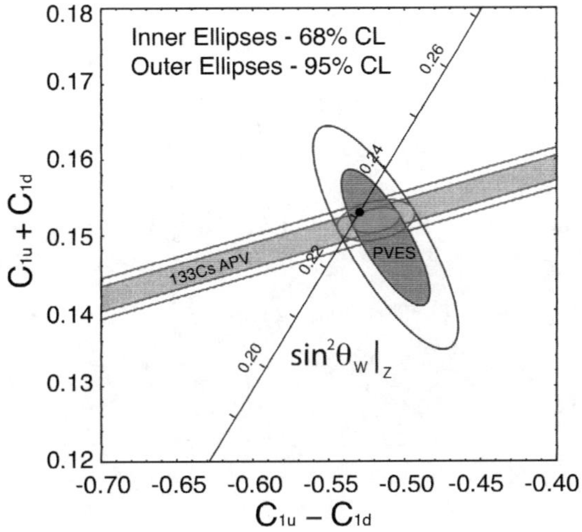

Figure 3. The constraints on the isoscalar $(C_{1u} + C_{1d})$ and isovector $(C_{1u} - C_{1d})$ weak currents from PVES and APV experiments are shown here. The almost horizontal (green) APV band is nearly orthogonal to PVES (blue) ellipse, which allows for a relatively clean extraction of C_{1u} and C_{1d}. The red ellipse shows the result of combining the PVES and APV data. The SM prediction [28] as a function of $\sin^2\theta_W$ in the $\overline{MS}$ scheme is plotted (diagonal black line) with the SM best fit value indicated by the (black) point.

to extract the neutron's weak charge, $Q_w^n(PVES + APV) = 2(C_{1u} + 2C_{1d}) = -0.975 \pm 0.010$, which is also in good agreement with the SM prediction, $Q_w^p(SM) = -0.9890 \pm 0.0007$. This represents the first determination of the weak charge of the neutron.

The commissioning data reported here represents 4% of the total data acquired during the running of the Q_{weak} experiment. The final result when published will benefit from an asymmetry anticipated to have significantly reduced systematic uncertainties and a five times smaller statistical uncertainty.This high-precision asymmetry measurement will provide a test of both the standard model and parity-violating new physics beyond the standard model at the TeV scale [1, 12].

References

[1] J. Erler, A. Kurylov and M. J. Ramsey-Musolf, Phys. Rev. D **68**, 016006 (2003) [hep-ph/0302149].
[2] C. S. Wood, S. C. Bennett, D. Cho, B. P. Masterson, J. L. Roberts, C. E. Tanner and C. E. Wieman, Science **275**, 1759 (1997).
[3] M. Gorchtein and C. J. Horowitz, Phys. Rev. Lett. **102**, 091806 (2009) [arXiv:0811.0614 [hep-ph]].
[4] A. Sibirtsev, P. G. Blunden, W. Melnitchouk and A. W. Thomas, Phys. Rev. D **82**, 013011 (2010) [arXiv:1002.0740 [hep-ph]].
[5] B. C. Rislow and C. E. Carlson, Phys. Rev. D **83**, 113007 (2011) [arXiv:1011.2397 [hep-ph]].
[6] M. Gorchtein, C. J. Horowitz and M. J. Ramsey-Musolf, Phys. Rev. C **84**, 015502 (2011) [arXiv:1102.3910 [nucl-th]].
[7] N. L. Hall, P. G. Blunden, W. Melnitchouk, A. W. Thomas and R. D. Young, Phys. Rev. D **88**, no. 1, 013011 (2013) [arXiv:1304.7877 [nucl-th]].
[8] N. L. Hall, P. G. Blunden, W. Melnitchouk, A. W. Thomas and R. D. Young, arXiv:1504.03973 [nucl-th].
[9] D. Wang *et al.* [Jefferson Lab Hall A Collaboration], Phys. Rev. Lett. **111**, no. 8, 082501 (2013) [arXiv:1304.7741 [nucl-ex]].
[10] T. Allison *et al.* [Qweak Collaboration], Nucl. Instrum. Meth. A **781**, 105 (2015) [arXiv:1409.7100 [physics.ins-det]].
[11] C. W. Leemann, D. R. Douglas and G. A. Krafft, Ann. Rev. Nucl. Part. Sci. **51**, 413 (2001).
[12] R. D. Young, R. D. Carlini, A. W. Thomas and J. Roche, Phys. Rev. Lett. **99**, 122003 (2007) [arXiv:0704.2618 [hep-ph]].
[13] R. D. Young, J. Roche, R. D. Carlini and A. W. Thomas, Phys. Rev. Lett. **97**, 102002 (2006) [nucl-ex/0604010].
[14] D. T. Spayde *et al.* [SAMPLE Collaboration], Phys. Lett. B **583**, 79 (2004) [nucl-ex/0312016].
[15] T. M. Ito *et al.* [SAMPLE Collaboration], Phys. Rev. Lett. **92**, 102003 (2004) [nucl-ex/0310001].
[16] K. A. Aniol *et al.* [HAPPEX Collaboration], Phys. Rev. Lett. **82**, 1096 (1999) [nucl-ex/9810012].
[17] K. A. Aniol *et al.* [HAPPEX Collaboration], Phys. Rev. Lett. **96**, 022003 (2006) [nucl-ex/0506010].
[18] K. A. Aniol *et al.* [HAPPEX Collaboration], Phys. Lett. B **635**, 275 (2006) [nucl-ex/0506011].

[19] A. Acha *et al.* [HAPPEX Collaboration], Phys. Rev. Lett. **98**, 032301 (2007) [nucl-ex/0609002].
[20] Z. Ahmed *et al.* [HAPPEX Collaboration], Phys. Rev. Lett. **108**, 102001 (2012) [arXiv:1107.0913 [nucl-ex]].
[21] D. S. Armstrong *et al.* [G0 Collaboration], Phys. Rev. Lett. **95**, 092001 (2005) [nucl-ex/0506021].
[22] D. Androic *et al.* [G0 Collaboration], Phys. Rev. Lett. **104**, 012001 (2010) [arXiv:0909.5107 [nucl-ex]].
[23] F. E. Maas *et al.* [A4 Collaboration], Phys. Rev. Lett. **93**, 022002 (2004) [nucl-ex/0401019].
[24] F. E. Maas *et al.*, Phys. Rev. Lett. **94**, 152001 (2005) [nucl-ex/0412030].
[25] S. Baunack *et al.*, Phys. Rev. Lett. **102**, 151803 (2009) [arXiv:0903.2733 [nucl-ex]].
[26] S. L. Zhu, S. J. Puglia, B. R. Holstein and M. J. Ramsey-Musolf, Phys. Rev. D **62**, 033008 (2000) [hep-ph/0002252].
[27] J. J. Kelly, Phys. Rev. C **70**, 068202 (2004).
[28] J. Beringer *et al.* [Particle Data Group Collaboration], Phys. Rev. D **86**, 010001 (2012).

Journal of Physics: Conference Series **678** (2016) 012027 doi:10.1088/1742-6596/678/1/012027

The physics program of PAX at COSY

Yury Valdau[1], Alexander Aksentyev[2,3], Dieter Eversheim[1], and Bernd Lorentz[2] for the PAX collaboration

[1] Helmholtz-Institut für Strahlen- und Kernphysik, Universität Bonn, Nussallee 14-16, D-53115 Bonn, Germany

[2] Institute für Kernphysik, Forschungszentrum Jülich, Jülich, D-52425 Jülich, Germany

[3] National Research Nuclear University - Moscow Engineering Physics Institute, Dept. of Electrophysical Facilities, Kashirskoe shosse, 31, 115409 Moscow, Russia

E-mail: y.valdau@fz-juelich.de

Abstract. The construction of the PAX installation was inspired by the idea to make a beam of polarized antiprotons available for the experiments at the HESR FAIR. A spin filtering experiment with transversally polarized protons was realized using the new PAX installation at COSY. The results of this measurement are in perfect agreement with the FILTEX experiment. Hence, filtering is a viable method to produce a stored beam of polarized antiprotons. Another experiment which can be pursued using the PAX installation is the test of Time Reversal Invariance at COSY (TRIC). The goal of the TRIC experiment is to improve the present upper limit on violation of the T-odd P-even interaction by an order of magnitude using a genuine null observable available in a double polarized *pd* scattering. The status of the PAX spin filtering experiments as well as present understanding of the possible systematic uncertainties in TRIC are presented in this contribution.

PAX spin filtering experiments

High intensity polarized antiproton beams are not yet available for the experiments, although the potential of studies with polarized antiproton beams for physics is enormous. A detailed physics program with polarized beams of protons and antiprotons can be found in Ref. [1]. The Polarized Antiproton eXperiments (PAX) collaboration has taken over this challenge, and formulated the way to produce high intensity beams of polarized antiprotons. On the first stage of the PAX program all the methods and equipment should be tested for use with proton beams. As a second step, experiments with antiproton beams should be performed to demonstrate that it is possible to produce the first ever high intensity beam of polarized antiprotons. Finally, the HESR installation should be extended to realize the ambitious physics program with double polarized proton-antiproton experiments proposed by the PAX collaboration [1].

There are two ways to polarize an initially unpolarized beam of spin 1/2 particles stored in the accelerator: particles with undesired directions of spin either should change the orientations of their spins (spin flip) or should be eliminated from the beam (spin filtering). The spin flip method would allow to preserve beam intensity and hence would be preferable over spin filtering. However, no reliable method for spin flipping has been proposed up to now. In Ref. [2] the possibility of producing beams of polarized antiprotons using the interaction between antiproton beams and polarized beams of positrons was suggested. To test the feasibility of this method the PAX collaboration performed a measurement of the spin flip cross section using an

Content from this work may be used under the terms of the Creative Commons Attribution 3.0 licence. Any further distribution of this work must maintain attribution to the author(s) and the title of the work, journal citation and DOI.
Published under licence by IOP Publishing Ltd

Journal of Physics: Conference Series **678** (2016) 012027 doi:10.1088/1742-6596/678/1/012027

initially polarized proton beam and an unpolarized electron beam. The observed *ep* spin flip depolarization cross section measured by PAX [4] is many orders of magnitude lower than the prediction from [2]. Hence, it is not feasible to polarize beams of antiprotons by using polarized positrons as was suggested in Ref. [2].

The spin filtering method is based on the spin-dependence of the strong interaction. The total cross section for the scattering of unpolarized protons on a polarized target contains a term which differs depending on whether the spins of the beam and the target are codirectional or not. Hence, in case a stored beam of unpolarized protons continuously interacts with a polarized proton target, after a certain time an initially unpolarized beam becomes polarized. Due to the lack of experimental data on antiproton-proton scattering, it is hard to predict the level of the effect for antiprotons; but there are very solid calculations for proton-proton scattering available in the literature [5]. The FILTEX experiment at the TSR Heidelberg has demonstrated the feasibility of the spin filtering method for protons at 23 MeV [3]. The only facility where spin filtering experiments with antiprotons can be realized presently is the Antiproton Decelerator (AD) storage ring at CERN. The CERN SPSC have requested the PAX collaboration to perform a complete construction and commissioning of the PAX installation and realize a spin filtering experiment with protons stored in the storage ring at a different energy before the preparation for the experiments at the CERN AD will be started [6].

The PAX installation has been constructed for the COoler-SYnchrotron COSY-Jülich at the Forschungszentrum Jülich [7]. In parallel, a complete PAX section for the AD storage ring, fully compatible with the PAX detector system from COSY, has been prepared. The PAX installation at COSY consist of: low-β section, Atomic Beam Source (ABS), Breit-Rabi target Polarimeter (BRP), holding field system, and target chamber, which accommodates a storage cell and a beam polarimeter. The PAX installation with a closed storage cell was used for the spin filtering experiment at COSY [8] in 2011, except PAX silicone detector which is presently being set up and commissioned. The experiment was performed at COSY using an unpolarized proton beam at 49.3 MeV. An initially unpolarized proton beam was injected, accelerated and stored in the storage ring. After its acceleration to 49.3 MeV a polarized hydrogen gas from the PAX ABS was injected into the storage cell located in the target chamber for the duration of filtering of 12000 and 16000 s. During a complete spin filtering cycle the PAX holding field system serves as a guiding field for the hydrogen gas in the target storage cell, while the BRP is used to continuously monitor gas polarization using a fraction of the gas from the beam-target interaction region. At the end of a spin-filtering period a polarized hydrogen flux at PAX is switched off and beam polarization is measured using unpolarized cluster-jet target and Silicone Tracking Telescopes (STT's) placed at ANKE [9]. The resulting total spin filtering cross section measured at 49.3 MeV is in good agreement with the theoretical model which consistently describes both FILTEX and PAX results [8]. The physics program of PAX at COSY can be further extended with a longitudinal spin filtering experiment in case a Siberian-snake becomes available.

After the successful completion of the transversal spin-filtering experiment at COSY [8] it is possible to conclude that spin-filtering is an established way to polarize protons, which can potentially be applicable to antiprotons. Due to the lack of experimental data on antiproton-proton scattering, it is very hard to make solid predictions for the spin-filtering cross section for the antiprotons [5]. Only experiments with antiproton beams will allow one to estimate the parameters of the antiproton polarizing ring reliably, which should be built to provide polarized antiproton beams for the experiments at HESR, as was suggested by PAX [1].

Recently, the CERN SPSC have informed spokespersons of PAX that collaboration plans to modify the AD ring are now not compatible with the current physics program pursued at the ring. Hence, the PAX activities will be postponed. In this situation the PAX collaboration have started to extend their experimental program at COSY. One of the important pieces of

Journal of Physics: Conference Series **678** (2016) 012027 doi:10.1088/1742-6596/678/1/012027

this program is an experiment to test Time Reversal Invariance at COSY (TRIC) which can be realized using PAX at COSY.

The test of Time Reversal Invariance at COSY (TRIC)

The Big Bang theory assumes that right after the Big Bang matter and antimatter were produced in equal amounts. However, there is very little antimatter observed in the visible part of the Universe. One of the explanations of the Baryon Asymmetry of the Universe presupposes strong CP and C violations [11]. All the CP and T violating effects discovered up to now are relatively weak and can not explain the predominance of matter in the universe. Hence, it is necessary to search for the new sources of the CP or T violations in baryon systems. The discovery of such an effect would be a direct indication for the physics beyond the Standard Model.

The TRIC experiment aims to improve the present upper limit on the strength of the T-odd P-even interaction by an order of magnitude [10]. It is planned to perform a measurement of a genuine null observable, $A_{y,xz}$, available in a double-polarized proton-deuteron scattering. In this experiment a polarized proton beam of COSY will continuously interact with a tensor polarized deuterium target located at PAX. The measurement will be done using a transmission technique utilizing an optical theorem for the total cross section measurement in a double polarized pd scattering. The spin-dependent part of the total cross section will be extracted by comparing the difference in the beam current slopes for the cycles with different beam-target spin configurations. The PAX installation at COSY is ideally suited for the realization of the TRIC experiment because of the availability of the high density polarized target, holding field system, and beam and target polarimeters. It is possible to identify the following advantages in the way the TRIC experiment is constructed:

- In comparison to the present upper limit, obtained from the experiment on Ho^{165} [12] target, the TRIC experiment will be free from any model dependent corrections associated with the target nuclei because of the use of the simplest tensor polarized nuclei (deuteron).
- The limit on the T-odd P-even interaction obtained from the experiment with a proton beam will be complementary to the present upper limit obtained using a beam of polarized neutrons.
- The TRIC experiment will utilize an optical theorem measuring the difference in the total cross sections. Hence, this experiment is free from any corrections due to the final state interactions significant for other T-symmetry tests.
- The possibility to independently flip the polarization of the beam and the target in TRIC will allow to keep systematic errors under control and hence provide a reliable upper limit.

1. Possible systematic uncertainties in TRIC

As with any precision measurement, the TRIC experiment requires significant and continuous efforts toward understanding the possible systematic uncertainties present in it. In this contribution our present idea of the possible systematic uncertainties in the experiment is summarized [13].

1.1. General formulas

In the most general form, the total cross section for pd scattering σ can be written as

$$\sigma = \sigma_0 \left(1 + A_{y,xz} P_y^b P_{xz}^t + \sum_l A_{y,l} P_y^b P_l^s + \sum_{mn \neq xz} A_{y,mn} P_y^b P_{mn}^s \right), \quad (1)$$

where σ_0 is the unpolarized total cross section, $A_{y,xz}$ the genuine T-odd P-even null observable, $A_{y,l}$ and $A_{y,mn}$ other spin correlation coefficients, P_y^b transversal beam polarization, P_{xz}^t desired

Journal of Physics: Conference Series **678** (2016) 012027 doi:10.1088/1742-6596/678/1/012027

tensor polarization of the deuterium target, P_l^s and P_{mn}^s vector and tensor polarizations of rest gas in the ring. The TRIC experiment is planned as a transmission experiment, which means that the $A_{y,xz}$ from Eq. (1) will be extracted from the beam current measurement performed during subsequent cycles with different configurations of beam and target polarizations.

The beam current losses in the accelerator can be associated with different factors; in the most general form, beam current in the storage ring at any moment of time can be written in the form:

$$I(t) = I_0(1 - \langle\sigma,\rho\rangle|_{ring})^{\nu t} = I_0 \exp\left(\nu t \times \ln(1 - \langle\sigma,\rho\rangle|_{ring})\right), \tag{2}$$

where: I_0 is the initial beam current, σ total cross section, ρ effective target thickness integrated over the complete ring, ν beam revolution frequency, and t the moment of measurement. Performing a Taylor series expansion of Eq. (2) and restricting our consideration only to the first part of series one can end up with a simple formula for the beam current losses:

$$I(t) \approx I_0 \exp\left(-\nu t \langle\sigma,\rho\rangle|_{ring}\right) = I_0 \exp\left(-t/\tau\right). \tag{3}$$

On the basis of the expected conditions of the TRIC experiment, it can be demonstrated that the difference between equations (2) and (2) is as small as 10^{-18} [13]. Beam losses due to the interaction with an internal target are continuous, if the beam current has no time structure, and discrete, if the beam is bunched. If a bunched beam is used, both the beam current and beam losses have some time structure, hence it only makes sense to discuss an averaged beam current. Here and later in the discussion we are only going to operate with the averaged beam current $I(t)$ which can be defined for both types of the beams. Using equations (1) and (3), a formula for the $A_{y,xz}^{+-}$ measured using the difference in $1/\tau$ of the cycles with different polarization states (+ and −) can be written:

$$A_{y,xz}^{+-} = \frac{A_{y,xz}}{\nu(\langle\sigma,\rho\rangle|_{ring}^{+} - \langle\sigma,\rho\rangle|_{ring}^{-})} \times \left[1/\tau^{+} - 1/\tau^{-}\right] \tag{4}$$

where $A_{y,xz}^{+-}$ is the genuine null observable. Equation (4) can be rewritten in a form which directly contains the averaging time for each state t_s:

$$A_{y,xz}^{+-} = \frac{A_{y,xz}}{\nu(\langle\sigma,\rho\rangle|_{ring}^{+} - \langle\sigma,\rho\rangle|_{ring}^{-})} \times \left[\frac{\langle\sigma,\rho\rangle|_{period}^{+}}{t_s^{+}} - \frac{\langle\sigma,\rho\rangle|_{period}^{-}}{t_s^{-}}\right] \tag{5}$$

1.2. Possible systematic errors in TRIC

The coefficient $A_{y,xz}$ is a null observable, meaning that it should be equal to zero in case T-symmetry is conserved, and any deviation of its value from zero would be either a discovery of a T-symmetry violation or a systematic error. Our goal is to keep all the possible systematic biases associated with any kind of effects in the experiment suppressed and under control to the accuracy better than the desired precision for the observable of interest. From the simple analysis of Eq. (5) the following classification of biases not associated with particular measurement apparatus can be used: bias associated with cross section, bias associated with beam polarization, bias associated with target polarization, bias associated with gas density, and bias associated with time measurements.

It is important to stress that the vacuum system of the PAX installation is fully integrated into the COSY vacuum system, hence there are no clear and solid borders of the target region, and essentially all the rest gas in the vacuum system of COSY will give some contribution to beam loss. For the same reason, target polarization must be modeled not only in the region of the nominal holding field but also in the region where it is inhomogeneous [13]. Since TRIC is a transmission experiment it is sensitive to the total of beam loss, only a fraction of which will be happening in the beam-target interaction region, where there are polarized deuterium gas and guiding field present. Due to the fact that $A_{y,xz}$ is a unique null observable in a double polarized *pd* scattering, all the other contributions to the total beam loss in COSY will be the same for the two beam-target polarization configurations and hence canceled out in Eq. (5).

Using equations from Sec. 1.1 the following equation for the $A_{y,xz}$ observable measured in the experiment can be written:

$$A_{y,xz} = A^0_{y,xz}(1 + B_{time})[1 + B_{beam} + B_{target}] + \Delta_{y,y} \tag{6}$$

here: $A^0_{y,xz}$ is the genuine value of the observable of interest, B_{time}-bias associated with time measurements, B_{beam}-bias associated with beam polarization, B_{target}-bias associated with target polarization, $\Delta_{y,y}$-bias associated with a background observable $A_{y,y}$.

It is clear from Eq. (6) that the most crucial task is to suppress the $\Delta_{y,y}$ bias to the level of at least 10^{-6}. All the other biases are multiplied by the value of $A_{y,xz}$ and hence it is sufficient to keep them below 10^{-1}. It is important to stress here that in case only the beam polarization is flipped, to simulate two spin configurations for the $A_{y,xz}$ measurements, the holding field profile in the target region remains the same and hence $\Delta_{y,y}$ should be equal to zero. Hence, the possibility to flip beam and target polarizations independently allows to control $\Delta_{y,y}$ in the experiment.

1.2.1. Bias associated with the $A_{y,y}$ observable

From Eq. (6) it is clear that the precision to which $A_{y,xz}$ can be measured is in direct correspondence with the level to which the $\Delta_{y,y}$ bias can be suppressed in the experiment. There is no measurement for the $A_{y,y}$ under the zero degree available in the literature. Hence, it is planned to utilize a dedicated beam time, during which a complete installation for the TRIC experiment will be commissioned and used for the $A_{y,y}$ measurements. The bias $\Delta_{y,y}$ will be reduced in the TRIC experiment using the following experimental techniques: vector polarization in tensor polarized gas from ABS will be suppressed, the holding field in the storage cell region will be provided in the horizontal plane only, and adjusted to reduce the effect from $A_{y,y}$ in the experiment. Much more detail to the way $\Delta_{y,y}$ will be estimated can be provided, but this not a subject of this contribution.

1.2.2. Bias associated with time measurements

The precision of time measurements will have influence on all the biases in Eq. (6). However, the expected accuracy of about 100 μs, which will be provided using a high stability generator and scaler system, is sufficient to reduce B_{time} to 10^{-4} [13].

1.2.3. Bias associated with beam

The $A_{y,xz}$ depends on the polarization of the beam, hence all the beam losses which do not depend on the orientation of the proton spin are canceled out. For the TRIC experiment only the averaged value of beam polarization during measurements is crucial, and this is why the experiment is not sensitive to the particular orbit of the polarized particles in the storage ring (Stern-Gerlach experiment). It is essential for the experiment that beam polarization life time should be kept much longer than the averaging time for one beam-target configuration. According to the current experience at COSY, polarization life-time at COSY lasts for as long as 10^5 s [14], but several measurements of this value should be realized during the experiment to keep this effect under control. The effect of the beam's synchrotron oscillation, together with the finite size of the distribution of its momentum, can lead to a bias B_{beam}, which, according to our estimates, is as small as 10^{-4} [13]. The beam spin flip efficiency, important for the control over $\Delta_{y,y}$, is as high as 99 % [8], and hence should not lead to any sizable bias.

1.2.4. Bias associated with the target
The total cross section of the *pd* scattering is relatively small, hence beam loss due to the beam-target interaction is relatively slow. Due to the fact that the PAX polarized target is located inside the accelerator, a polarized proton beam sees a complete vacuum system of the accelerator as a continuous gas target. The COSY rest gas outside of the PAX target region will be unpolarized and hence will have no influence on the $A_{y,xz}$. The rest gas in the accelerator is under the influence of a constant vertical guiding field from the COSY dipoles, which does not change direction. Therefore, any effect associated with rest gas in the $A_{y,xz}$ measurement will be the same for the cycles with different beam-target spin configurations, and hence cancel out. Since the PAX target chamber is part of the COSY vacuum system it is not possible to set a clear border between the polarized target region and the rest of the vacuum system. Hence, by the (polarized) target in this experiment we mean a region in which a horizontal holding field is provided. To reduce any kind of uncertainties associated with the holding field, it is crucial to provide a guiding field in the horizontal plane in the storage cell region with as little as possible of the vertical component, which will be symmetrically reflected in case target polarization is flipped. A detailed modeling of the effects associated with the guiding field is complicated and dependent on the real field configuration which is presently unknown. Simple estimates for the holding field configuration and target gas density distribution show $B_{target} < 10^{-3}$ [13].

Conclusions

After the successful realization of the spin-filtering experiment at COSY the PAX collaboration is ready to start filtering experiments with antiprotons. Unfortunately, the CERN SPSC conclude that modifications to the CERN AD, suggested by the PAX collaboration, are not compatible with the present physics program at the AD. Hence, the PAX spin filtering experiments with antiprotons are put on hold. The test of Time Reversal Invariance at COSY (TRIC) can be realized using the present PAX installation. This experiment aims to improve the present upper limit on the T-odd P-even interaction using a genuine null observable $A_{y,xz}$ available in a double-polarized *pd* scattering. The status of the spin filtering experiments, as well as our present understanding of the possible systematic uncertainties in the TRIC experiment, are presented in this contribution.

This work was funded in part by an ERC Advanced Grant POLPBAR (№ 246980).

References

[1] P. Lenisa and F. Rathmann for the PAX collaboration, technical proposal for FAIR, Antiproton-Proton Scattering Experiments with Polarization, January 2006.
[2] H. Arenhöven, Eur. Phys. J. A **34**, (2007) 303; Th. Walcher et al., Eur. Phys. J. A 34 (2007) 447.
[3] F. Rathmann *et al.*, Phys. Rev. Lett. **71**, (1993) 1379.
[4] D. Oellers *et al.*, Phys. Lett. B **674**, (2009) 269; D. Oellers *et al.*, Nucl. Instr. Methods A **759**, (2014) 6.
[5] V. F. Dmitriev, A. I. Milstein, S. G. Salnikov, Phys. Lett. B **690**, (2010) 427.
[6] Communication between CERN SPSC and PAX collaboration; Letter of intent available via: P. Lenisa and F. Rathmann for the PAX collaboration, arXiv:nucl-ex/0512021
[7] R. Maier, Nucl. Instr. Methods A **390**, (1997) 1.
[8] W. Augustyniak *et al.*, Phys. Lett. B **718**, (2012) 64.
[9] R. Schleichert *et al.*, IEEE Trans. Nucl. Sci. **50**, (2003) 268.
[10] D. Eversheim, B. Lorentz and Yu. Valdau for the PAX collaboration, Cosy Proposal #215 (2012).
[11] A. D. Sakharov, Pisma Zh. Eksp. Teor. Fiz. **5** (1967) 32 [JETP Letters **5**, (1967) 24].
[12] P. R. Huffman *et al.*, Phys. Rev. C **55**, (1997) 2684.
[13] A. Aksentyev, Diploma thesis, National Research Nuclear University - Moscow Engineering Physics Institute (2015), in russian.
[14] C. Weidemann *et al.*, Phys. Rev. ST Accel. Beams **18**, (2015) 020101.

Journal of Physics: Conference Series **678** (2016) 012028 doi:10.1088/1742-6596/678/1/012028

Polarimetry with inclusive charged pions at U-70 accelerator of IHEP

V L Rykov[1†], V V Abramov[2], A A Bogdanov[1], V V Mochalov[1,2], S B Nurushev[1,2], P A Semenov[1,2], M N Strikhanov[1] and A N Vasiliev[1,2]

[1] National Research Nuclear University (Moscow Engineering Physics Institute), Moscow, 115409, Russia

[2] Institute for High Energy Physics National Research Centre Kurchatov Institute, Protvino, 142284, Russia

[†] Corresponding author, E-mail: rykov@bnl.gov

Abstract. A feasibility of the proton beam polarimetry based on exploiting the observed large single-spin asymmetry A_N in inclusive charged pion production is evaluated for the SPASCHARM experiment at IHEP, Protvino, Russia. It is shown that, for a few hours of beam exposition time, the beam polarization could be measured at the statistical errors of ~3-5%. The inclusive pion polarimetry at SPASCHARM does not require construction of a special polarimeter. The polarimetry will be carried out entirely by the main SPASCHARM spectrometer simultaneously with physics data taking.

1. Introduction

The new experimental program SPASCHARM [1] for the 70 GeV proton synchrotron U-70 is currently under development at the Institute for High Energy Physics (IHEP), Protvino, Russia. The main physics motivation for the SPASCHARM experiment is the systematic study of spin phenomena for a wide range of inclusive and exclusive reactions in collisions of high-energy polarized hadrons. The experiment will be carried out in two stages. At the first stage, the measurements of single-spin asymmetries will be performed, using unpolarized meson and proton beams interacting with the transversely polarized protons of a "frozen" target. At the second stage, polarized proton and antiproton beams in the energy range of ~10-45 GeV will be available along with the polarized target at the new beam channel 24A. The SPASCHARM physics program, experimental setup and characteristics of polarized beams have been presented in details elsewhere [1–3].

The SPASCHARM polarized proton and antiproton beams will be formed by the method similar to the one already used in two experiments: E581/704 at Fermilab [4] and FODS at IHEP [5]. The parity-violating decays of Λ-hyperons, $\Lambda \to p\pi^-$ and $\overline{\Lambda} \to \overline{p}\pi^+$, serve as a source of a tertiary (anti)proton beam. A transversely polarized beam is produced by selecting a portion of these Λ-decays. Such a selection is based on strong correlations of decay (anti)proton's transverse polarization and its emission angle in Λ-decay. The selection is done either by tagging each beam trajectory with tracking detectors [4], or using a collimation [5].

The beam tagging method is very robust and reliable because it fully relies on the well known characteristics of Λ-decays and the spin-$\frac{1}{2}$ evolution in magnetic field. The beam absolute

Content from this work may be used under the terms of the Creative Commons Attribution 3.0 licence. Any further distribution of this work must maintain attribution to the author(s) and the title of the work, journal citation and DOI.
Published under licence by IOP Publishing Ltd

polarization is calculated at a quite descent accuracy from a beam-line computer simulations. The authors of the E704 analysis [6] estimated their beam polarization $P = 0.45 \pm 0.03$, with $\Delta P/P \simeq \pm 6.7\%$. The respective values at FODS experiment were [7, 8]: $P = 0.39 \pm 0.02$, $\Delta P/P \simeq \pm 5.1\%$. Still, the complexity of a beam channel geometry, of a magnetic field configuration, a potential presence of errors in sophisticated on-line tagging algorithms, uncertainties in Λ-spectra, etc. make it highly desirable the verification and confirmation of tagging results with independent beam polarization measurements as well as continuous polarization monitoring for various instabilities. One option for the proposed SPASCHARM beam polarimeter is based on using the known analyzing power A_N in elastic pp-scattering. It is presented in details in Refs. [3, 9]. Here we discuss the feasibility of a complementary approach based on exploiting the experimentally observed large analyzing power A_N in inclusive charged pion production.

2. Experimental data on analyzing power A_N in inclusive pion production

For the last decades, the measurements of spin asymmetries in inclusive processes at high energies have been in the focus of a number of experiments across a wide energy range, $\sqrt{s}$ from ~5 to 200 GeV. To the surprise of early pQCD predictions, the asymmetry A_N in pion production in forward direction at $x_F \gtrsim 0.4 - 0.6$ and moderate $p_T \sim 0.5 - 1$ GeV/c had been found large, for charged pions at the level of up to ~40%, and its large value persists in the entire explored energy range. This is illustrated by the compilation [10] of some experimental data shown in figure 1.

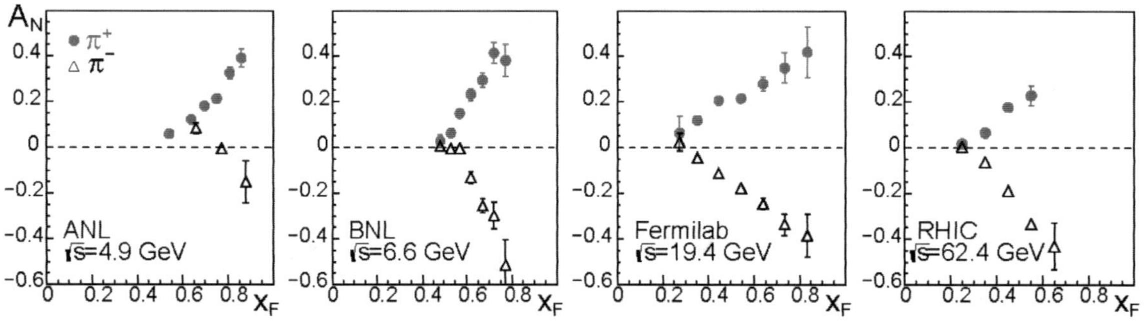

Figure 1. The asymmetry A_N in forward $\pi^{\pm}$ production as measured in polarized proton-proton collisions across a range of center-of-mass energies. From left to right, the data are from Refs. [11], [12], [13] and [14].

The similar A_N behavior has been revealed also in $p^{\uparrow}p \to \pi^0 X$ processes[1]. But A_N's absolute values for inclusive $\pi^{\pm}$ appeared to be more than twice as large as for π^0. Moreover, more data and of the better accuracy are currently available on charged pions in the energy range of interest for SPASCHARM. It was also found that the asymmetry A_N in $p^{\uparrow}A \to \pi^{\pm}X$ processes with a carbon target stays remarkably close to the one in $p^{\uparrow}p \to \pi^{\pm}X$ reaction [12], and at least up to copper, it is virtually independent of a target material [7].

3. Experimental setup for polarimetry and measurement time estimates

The measurements of single-spin asymmetries, including A_N in inclusive production of light mesons, is important part of the SPASCHARM program [1]. Therefore the main SPASCHARM

[1] See, for example, the relevant sections in RHIC Spin Collaboration write-ups [15] and references therein.

Journal of Physics: Conference Series **678** (2016) 012028 doi:10.1088/1742-6596/678/1/012028

spectrometer will perfectly do the job for inclusive pion polarimetry, and no any additional detectors are required. The detailed descriptions of the SPASCHARM setup can be found elsewhere [1]. It is the wide aperture spectrometer for charged and neutral particles with good identification, covering 2π in azimuth, p_T up to 2.5 GeV/c, $0 < x_F < 1$ for the beam momenta in the range of ~10-45 GeV/c. Its key properties are briefly listed below:

Tracking: Wide aperture spectrometer magnet; 5 thin-wall drift tube stations; $\Delta p/p \simeq 0.4\%$ at 10 GeV/c; Acceptance: 2π in azimuth and up to 250 mrad vertically and 350 mrad horizontally in polar angle θ.

Particle ID: 2 multi-channel Cherenkov counters ($\pi^{\pm}$ ID 3-23 GeV; $K^{\pm}$ ID 11-23 GeV; proton ID > 11 GeV).

Electromagnetic calorimeter: Fine-segmented Lead-Scintillator 'shashlyk'-type, cell size 55×55 mm^2, total area ~2×3 m^2, energy resolution $\sigma_E/E \simeq 1.3\% \oplus 2.8\%/\sqrt{E}$ (E in Gev).

Hadron calorimeter: Compensated Lead-Scintillator sandwich, ~7 interaction lengths, $\sigma_E/E \simeq 60\%/\sqrt{E}$, ($E$ in Gev).

Table 1. Event rate and time estimates for the beam polarization measurements, using the reaction $p^{\uparrow}p \rightarrow \pi^+X$ with a 20 cm long liquid hydrogen target. See text for more details. Only statistical errors for A_N from Ref. [12] are shown. The scale error of 14.5% is not included.

Parameter name for $p^{\uparrow}p \rightarrow \pi^+X$	$0.55 < x_F < 0.6$ $<p_T>= 0.7$ GeV/c	$0.6 < x_F < 0.65$ $<p_T>= 0.7$ GeV/c	$0.65 < x_F < 0.7$ $<p_T>= 0.8$ GeV/c
Analyzing power A_N, %	15.7 ± 1.5	23.7 ± 2.0	29.1 ± 2.9
Required N_{events}	$\sim 5.5 \times 10^4$	$\sim 2.5 \times 10^4$	$\sim 1.6 \times 10^4$
Cross section σ, μb	~15	~7	~4
N_{events}/spill	~12.7	~6.0	~3.4
Measurement time	~10.8 hours	~10.4 hours	~11.8 hours

Table 2. Same as table 1 but for the reaction $p^{\uparrow}C \rightarrow \pi^+X$ with a 4 cm thick graphite target. Only statistical errors for A_N from Ref. [12] are shown. The scale error of 25% is not included.

Parameter name for $p^{\uparrow}C \rightarrow \pi^+X$	$0.55 < x_F < 0.6$ $<p_T>= 0.7$ GeV/c	$0.6 < x_F < 0.65$ $<p_T>= 0.7$ GeV/c	$0.65 < x_F < 0.7$ $<p_T>= 0.8$ GeV/c
Analyzing power A_N, %	12.5 ± 1.1	22.8 ± 1.5	30.2 ± 2.4
Required N_{events}	$\sim 8.7 \times 10^4$	$\sim 2.6 \times 10^4$	$\sim 1.5 \times 10^4$
Cross section σ, μb	~110	~50	~30
N_{events}/spill	~48	~22	~13
Measurement time	~4.5 hours	~3 hours	~3 hours

The estimates for event rates and beam exposition times which are necessary for measuring the polarization of a 22 GeV/c proton beam at the statistical errors $\Delta P \simeq 4\%$ ($\Delta P/P \simeq 10\%$ for an expected $P \simeq 40\%$) are presented in tables 1 and 2. In these estimates, an intensity of a polarized proton beam has been assumed at 10^6 protons/spill with the spill repetition frequency of 0.11 Hz. The measured A_N values for inclusive π^+ production in $p^{\uparrow}p$ (table 1) and $p^{\uparrow}C$

Journal of Physics: Conference Series **678** (2016) 012028 doi:10.1088/1742-6596/678/1/012028

(table 2) collisions are from Ref. [12]. The cross sections σ have been estimated, using the measurements [12] normalized with the data from Ref. [16]. The results are presented for three kinematic bins, and the statistical error of 4% is required to be achieved for each bin of the size $\Delta x_F \times \Delta p_T$=0.05×0.1 GeV/c. These errors will be further reduced by the simultaneous spin asymmetry measurements with negative pions although the production cross sections for inclusive π^--mesons in *pp* and *pA* collisions are noticeably lower than for π^+ [12].

4. Conclusion

A feasibility of the proton beam polarimetry based on exploiting the observed large single-spin asymmetry A_N in inclusive charged pion production is evaluated for the SPASCHARM experiment at U-70 accelerator of IHEP. It is shown that, for a few hours of beam exposition time, the beam polarization could be measured at the statistical errors of ~3-5%. The measurements could be performed as with cryogenic hydrogen targets as well as with much more easier to handle solid nuclear, for example, graphite target.

The accuracy of the currently available experimental A_N data in inclusive production of charged pions is good enough for utilizing them for verification and confirmation of the beam-tagging operation and for continuous monitoring the time stability of beam polarization. The A_N accuracy for inclusive pions will undoubtedly be improved by the SPASCHARM experiment on the course of its physics program realization which is, however, beyond the scope of this paper.

It is important to underline that the inclusive charged pion polarimetry at SPASCHARM does not require construction of a special polarimeter. The polarimetry will be carried out entirely by the main SPASCHARM spectrometer simultaneously with physics data taking.

Acknowledgments

The authors are thankful to I. A. Savin and Y. Makdisi for the useful communications. This work has been supported in part by the RFBR Grant No.15-02-1669.

References

[1] Mochalov V V 2013 *Phys. Part. Nucl.* **44** No 6 930; Mochalov V V et al 2014 Systematic Study of Spin Effects at the SPASCHARM Experiment at 70-GeV Accelerator in Protvino *To be published in Int. J. Mod. Phys.: Conf. Series* (Proc. of 21st Int. Symp. on Spin Phys. "Spin2014", Beijing, China, 2014)
[2] Nurushev S B et al 2015 Polarized antiproton beam at U-70 accelerator of IHEP *To be published in J. Phys.: Conf. Series* (these proceedings)
[3] Semenov P A et al 2014 Polarimeters for the SPASCHARM experiment *To be published in Int. J. Mod. Phys.: Conf. Series* (Proc. of "Spin2014")
[4] Grosnick D P et al 1990 *Nucl. Inst. Meth.* A **290** 269
[5] Abramov V V et al 1997 *Nucl. Phys.* B **492** 3
[6] Bravar A et al 1996 *Phys. Rev. Lett.* **77** 2626
[7] Abramov V V et al 2007 *Yad. Phys.* **70** No 9 1561 [*Phys. Atom. Nucl.* **70** No 9 1515]
[8] Abramov V V et al 2007 *Yad. Phys.* **70** No 10 1799 [*Phys. Atom. Nucl.* **70** No 10 1752]
[9] Bogdanov A A et al 2015 Elastic scattering polarimeter for a polarized antiproton beam at U-70 accelerator of IHEP *To be published in J. Phys.: Conf. Series* (these proceedings)
[10] Aidala C A et al 2013 *Rev. Mod. Phys.* **85** 655
[11] Klem R D et al 1976 *Phys. Rev. Lett.* **36** 929
[12] Allgower C E et al 2002 *Phys. Rev.* D **65** 092008
[13] Adams D L et al 1991 *Phys. Lett.* B **264** 462
[14] Arsene I et al 2008 *Phys. Rev.Lett.* **101** 042001
[15] Aschenauer E-C et al 2013 *Preprint arXiv:nucl-ex/1304.0079*; Aschenauer E-C et al 2015 *Preprint arXiv:nucl-ex/1501.01220*
[16] Allaby J V et al 1970 *Preprint CERN 70-12*; Dekker D et al. 1965 *Phys. Rev.* **137** B962

Journal of Physics: Conference Series **678** (2016) 012029 doi:10.1088/1742-6596/678/1/012029

Renormdynamics, Discrete Dynamics and Spin

Nugzar Makhaldiani
Joint Institute for Nuclear Research
Dubna, Moscow Region, Russia
E-mail: mnv@jinr.ru

Abstract. In the Standard Model of Particle Physics (SM), minimal supersymmetric extension of the SM (MSSM), standard pion-nucleon field theory and other models is shown how to define the values of coupling constants and masses. Discrete dynamics, spins, constituent supersymmetric model for W and Higgs bosons considered, supersymmetric mechanism of confinement in QCD described.

We say that we find **New Physics** (NP) when either we find a phenomenon which is forbidden by SM in principal - this is the qualitative level of NP - or we find a significant deviation between precision calculations in SM of an observable quantity and a corresponding experimental value. **Renordynamics** (RD) unifies different renormgroups in one society. Quantum field theory (QFT) and Fractal calculus (FC) provide Universal language of fundamental physics (see e.g. [1]). In QFT the existence of a given theory means that we can control its behavior at some scales (short or large distances) by renormalization theory [2]. If the theory exists, we want to solve it, which means to determine what happens on other (large or short) scales. This is the problem (and content) of Renormdynamics. The result of the Renormdynamics, the solution of its discrete or continual motion equations, is the effective QFT on a given scale (different from the initial one). We will call RDF functions $g_n = f_n(t)$ which are solutions of the RD motion equations

$$\dot{g}_n = \beta_n(g), 1 \leq n \leq N. \quad (1)$$

In the simplest case of one coupling constant the function $g = f(t)$ is constant, $g = g_c$ when $\beta(g_c) = 0$, or is invertible (monotone). Indeed,

$$\dot{g} = f'(t) = f'(f^{-1}(g)) = \beta(g). \quad (2)$$

Each monotone interval ends by UV and IR fixed points and describes corresponding phase of the system. Note that the simplest case of the classical dynamics, the Hamiltonian system with one degree of freedom, is already two dimensional, so we have no analog of one charge renormdynamics. The regular Hamiltonian systems of the classical mechanics are defined on the even dimensional phase space, so there is no analog of the three dimensional renormdynamics for the coupling constants of the SM. The fixed points of renormdynamics belong to a set of zeros of the polynomial system of equations $\beta_n(g) = 0, 1 \leq n \leq N$, in the perturbative renormdynamics. Describing qualitative and numerical properties of the set, finding the number and multiplicity, evaluation and finding the numerical values of the zeros is the task of contemporary algebraic

Content from this work may be used under the terms of the Creative Commons Attribution 3.0 licence. Any further distribution of this work must maintain attribution to the author(s) and the title of the work, journal citation and DOI.
Published under licence by IOP Publishing Ltd

Journal of Physics: Conference Series **678** (2016) 012029 doi:10.1088/1742-6596/678/1/012029

and computational geometry. QCD is the theory of the strong interactions with, as only inputs, one mass parameter for each quark species and the value of the QCD coupling constant at some energy or momentum scale in some renormalization scheme. The quantitative values and qualitative content of the given field theory depend on the scale. In QCD e.g. the effective action has the following form:

$$S(\mu) = \frac{1}{g^2(\mu)} \int d^D x \mathcal{L}(A(\mu)), \tag{3}$$

variation with respect to the change of scale gives

$$\delta S = -2\frac{\beta(g)}{g}\delta\mu S + \frac{1}{g^2}\int d^D x \frac{\delta \mathcal{L}}{\delta A}\delta A \tag{4}$$

and the following two statements are equivalent:

$$\delta S = 0,\ \beta(g) = 0 \Leftrightarrow \delta S = 0,\ \frac{\delta \mathcal{L}}{\delta A} = 0. \tag{5}$$

So, from renorminvariance of the effective action follows that at the conformal symmetric points, the motion equations for fields are satisfied. Generalization for the several coupling constants and other models is obvious. In string theory, the connection between conformal invariance of the effective theory on the parametric world sheet and the motion equations of the fields on the embedding space is well known [3]. A more recent topic in this direction is AdS/CFT Duality [4]. In this approach for QCD coupling constant the following expression was obtained [5]

$$\alpha_{AdS}(Q^2) = \alpha(0)e^{-Q^2/4k^2}. \tag{6}$$

A corresponding β-function is

$$\beta(\alpha_{AdS}) = \frac{d\alpha_{AdS}}{\ln Q^2} = -\frac{Q^2}{4k^2}\alpha_{AdS}(Q^2) = \alpha_{AdS}(Q^2)\ln\frac{\alpha_{AdS}(Q^2)}{\alpha(0)} \leq 0 \tag{7}$$

So, this renormdynamics of QCD interpolates between the IR fixed point $\alpha(0)$, which we take as $\alpha(0) = 2$, and the UV fixed point $\alpha(\infty) = 0$. Based on real experiments and computer simulations, quantum nonabelian gauge theory in four dimensions is believed to have a mass gap. This is one of the most fundamental facts that makes the Universe the way it is.

For the QCD running coupling considered in [6]

$$\alpha(q^2) = \frac{4\pi}{9\ln(\frac{q^2+m_g^2}{\Lambda^2})}, \tag{8}$$

where $m_g = 0.88GeV,\ \Lambda = 0.28GeV$, the $\beta-$function of renormdynamics is

$$\beta(\alpha) = -\frac{\alpha^2}{k}(1 - c\exp(-\frac{k}{\alpha})),$$
$$k = \frac{4\pi}{9} = 1.40,\ c = \frac{m_g^2}{\Lambda^2} = (3.143)^2 = 9.88, \tag{9}$$

for a nontrivial (IR) fixed point we have

$$\alpha_{IR} = \frac{k}{\ln c} = 0.61 \tag{10}$$

For $\alpha(m) = 2$, at valence quark scale m we predict the gluon (or valence quark) mass as

$$m_g = \Lambda e^{\frac{k}{2\alpha(m)}} = 1.42\Lambda = m_N/3, \;\; \Lambda = 220MeV. \tag{11}$$

It is nice to have a nonperturbative $\beta-$function like (9), but it is more important to see which kind of nonperturbative corrections we need to have a phenomenological coupling constant dynamics. It was noted [7] that in valence quark parametrization $\alpha_s^{-1}(m) = 2$, at a valence quark scale m.

In the 1870's G.J. Stoney [8], the physicist who coined the term "electron" and measured the value of elementary charge e, introduced as universal units of Nature for L, T, M :

$$\begin{aligned} l_S &= \frac{e}{c^2}\sqrt{G}, \\ t_S &= \frac{e}{c^3}\sqrt{G} \\ m_S &= \frac{e}{\sqrt{G}}, \end{aligned} \tag{12}$$

M. Planck introduced [9] as universal units of Nature for L, T, M:

$$\begin{aligned} m_P &= \sqrt{\frac{hc}{G}} = \frac{m_S}{\sqrt{\alpha}}, \\ l_P &= \frac{h}{cm_P} = \frac{l_S}{\sqrt{\alpha}} = 11.7 l_S, \\ t_P &= \frac{l_P}{c} = \frac{t_S}{\sqrt{\alpha}}, \end{aligned} \tag{13}$$

Stoney's fundamental constants are more fundamental just because they are less than Planck's constants :) Due to the value of $\alpha^{-1} = 137$, we can consider relativity theory and quantum mechanics as deformations of the classical mechanics when deformation parameter $c = 137$ (in units $e = 1, \hbar = 1$) and $\hbar = 137$ (in units $e = 1, c = 1$), correspondingly. These deformations have an analytic sense of p-adic convergent series. The number 137 has a very interesting geometric sense. The Babylonians used a base 60 number system which is still used for measuring time - 60 seconds in a minute, 60 minutes in an hour - and for measuring angle - 360 degrees in a full turn. The base 60 number system has its origin in the ration of the Sumerian mina (m) and Akkadian shekel (s), $m/s \simeq 60 = 3 \cdot 4 \cdot 5$. We also can consider base 137 system for fundamental theories. For the nuclear physics strong coupling phenomena description we may take as a base $p = 13$. For the hadronic physics, valence scale QCD, and graphen QED_{2+1} strong coupling phenomena description we may take as a base $p = 2$. For the weak coupling physics SM m_Z scale and MSSM unification scale phenomena description we may take as a base $p = 29$. There are different opinions about the number of fundamental constants [10]. According to Okun, there are three fundamental dimensionful constants in Nature: Planck's constant, $\hbar$; the velocity of light, c; and Newton's constant, G. According to Veneziano, there are only two: the string length l_s and c. According to Duff, there are not fundamental constants at all. In mathematics we have two kind of structures, discrete and continuous one. If a physical quantity has discrete values, it might have no dimension. If the values are continuous - the quantity might have a dimension, a unit of measure. These structures may depend on scale, e.g. on macroscopic scale condensed state of matter (and time) is well described as continuous medium, so we use dimensional units of length (and time). On the scale of atoms, the matter has a discrete structure, so we may count lattice sites and may not use a unit of length. If at small (e.g. at Plank) scale space (and/or time) is discrete, then we do not need a unit of length (time) for measuring, there is a fundamental length and we can just count.

Journal of Physics: Conference Series **678** (2016) 012029 doi:10.1088/1742-6596/678/1/012029

Let us consider the following **discrete dynamics**:

$$S_{n+1} + S_{n-1} = \Phi(S_n), \tag{14}$$

which is obviously a (discrete) time (n) invertible in this implicit form. In the explicit form

$$S_{n+1} = F(S_n, S_{n-1}) = \Phi(S_n) - S_{n-1} \tag{15}$$

it is not obviously time invertible. If we take two step time lattice-make simplest discrete RD step and from one component-scalar S(n) construct two component-spinor $\Psi(n)$, we obtain explicit time invertible dynamics

$$\Psi_{n+1} = \Omega(\Psi_n),\ \Psi_{n+1} = \begin{pmatrix} S_{n+2} \\ S_{n+1} \end{pmatrix},\ \Psi_n = \begin{pmatrix} S_n \\ S_{n-1} \end{pmatrix} \tag{16}$$

This **dynamical mechanism of origin spin** which connects time inversion symmetry and spin were invented when were constructed the theory of quanputers [11]. This mechanism indicates that with time inversion symmetry we can have only composed scalar fields. With the discovery of the Higgs particle with mass 125 GeV, a nice number $m_W/m_H \simeq 2/3$ appear, which, at least for me, indicates for composed nature of W and H, with a same mass of about 40 GeV two and three valence constituents correspondingly. The fermion constituents ψ_n^a of W and scalar constituents φ_n^a of H compose scalar super multiplet (φ_n^a, ψ_n^a) with a flavor index n and color index a. Another notation is (h,sh)-(He,She). SQCD with valence $u, d - v$ quarks and their superpartner scalar $h-$quarks all of them with valence mass about 40 GeV may produce the observable W and H particles. With exact SUSY we have cofinement by dimentional counting: superspace dimension is zero on the hadronic scale, hadrons are pointlike, color is confined inside hadrons. For SM QCD this picture indicates that at the hadronic scale we have effective SQCD, which contains scalar quarks. The 40 GeV constituents may be good candidates in dark matter particles. Coupling constant unification at $\alpha_u^{-1} = 29.0$ and scale $10^{16} Gev$ in MSSM [12] has a relict on the SM scale: $\alpha_2^{-1}(m_Z) \simeq 29$, $\alpha_2^{-1}(m) = 29.0$ at $m = 41 GeV$. If we extrapolate the SM value of $\alpha^{-1}(m_Z)$ to electron masse scale, we find $\alpha^{-1}(m_e) = 137.0$

Note that in condensed state physics we have an analogous phenomenon, a localization of a particle in the random environment which can be described as a superspace dynamics of the pointlike particle on the scale larger than the size of the localized particle. For QCD this picture corresponds to the localization of the (valence) quarks in the random (gluon) vacuum inside hadrons. A mathematically rigorous description can be done within SQCD model.

References

[1] Makhaldiani N 2011 *Fractal Calculus (H) and some Applications Physics of Particles and Nuclei Letters* **8** 325

[2] Collins J C 1984 *Renormalization* (London: Cambrige University Press)

[3] Green M B, Schwarz J H and Witten E 1987 *Super String Theory* (London: Cambrige University Press)

[4] Maldacena J M 1998 *Adv. Theor. Math. Phys.* **2** 231

[5] Brodsky Stanley J, de Tèramond Guy F, Deur Alexandre 2010 *Phys. Rev. D* **81** 096010

[6] Diakonov D 2003 *Prog. Par. Nucl. Phys.* **51** 173

[7] Voloshin M B and Ter-Martyrosian K A 1984 *Gauge Theory of Elementary Particles* (Moscow: Atomizdat)

[8] Stoney G J 1881 *The philosophical magazine and journal of science* **11** 381

[9] Planck M 1900 *Ann. d. Phys.* **1** 69

[10] Duff M, Okun L and Veneziano G 2002 *Trialogue on the number of fundamental constants J. High Energy Phys.* **0203** 023 arXiv:physics/0110060

[11] Makhaldiani N 2011 *Regular method of construction of the reversible dynamical systems and their linear extensions - Quanputers, Atomic Nuclei* **74** 1040

[12] Makhaldiani N 2014 *Renormdynamics, Unified Field Theories and Universal Distributions of the Multipartical Production Processes, Physics of Particles and Nuclei Letters* **11** 1019

Journal of Physics: Conference Series **678** (2016) 012030 doi:10.1088/1742-6596/678/1/012030

Unbiased polarized PDFs upgraded with new inclusive DIS data

Emanuele R. Nocera

Rudolf Peierls Centre for Theoretical Physics, University of Oxford, 1 Keble Road, OX1 3NP Oxford, United Kingdom

E-mail: emanuele.nocera@physics.ox.ac.uk

Abstract. I present a determination of longitudinally-polarized parton distribution functions of the proton from inclusive deep-inelastic scattering data: NNPDFpol1.0+. This determination, based on the NNPDF methodology, upgrades a previous analysis, NNPDFpol1.0, in two respects: first, it includes all new data sets which have recently become available from the COMPASS experiment at CERN and from the E93-009, EG1-DVCS and E06-014 experiments at JLAB; second, it uses the state-of-the-art unpolarized parton set NNPDF3.0 as a baseline for the reconstruction of fitted observables and for the determination of positivity constraints. I discuss the impact of both these new inputs on the uncertainty of parton distribution functions.

In the last years, the NNPDF collaboration put a great deal of effort into the determination of minimally biased longitudinally-polarized Parton Distribution Functions (PDFs) of the proton and their associated uncertainties. Two sets of longitudinally-polarized PDFs based on the NNPDF methodology have been released so far: NNPDFpol1.0 [1] and NNPDFpol1.1 [2]. The former was obtained only from inclusive Deep-Inelastic Scattering (DIS) data with longitudinally-polarized beams and targets; the latter was obtained, in addition to the DIS data included in NNPDFpol1.0, also from open-charm production data by the COMPASS experiment at CERN and from $W^{\pm}$ and high-p_T jet production data by the STAR and PHENIX experiments at RHIC. Respectively, RHIC data sets have been proven to provide a first hint of flavor symmetry breaking for polarized sea quarks, and a first evidence of a sizable, positive gluon polarization in the proton [2].

Despite very promising, available RHIC data are not so many, and they cover a rather small kinematic range so far. Hence, the bulk of the experimental information on the longitudinally-polarized PDFs is still provided by inclusive DIS data: the leading observable, reconstructed from spin asymmetries, *i.e.*, differences between cross sections with opposite target polarizations, is the polarized structure function of the nucleon $g_1(x, Q^2)$. Following factorization, this reads, up to power-suppressed corrections, as

$$g_1(x,Q^2) = \frac{\sum_{q=1}^{n_f} e_q^2}{2n_f}\left[\mathcal{C}_{\rm NS}\otimes\Delta q_{\rm NS} + \mathcal{C}_{\rm S}\otimes\Delta\Sigma + 2n_f\mathcal{C}_g\otimes\Delta g\right] + \frac{h^{\rm TMC}}{Q^2} + \frac{h}{Q^2} + \mathcal{O}\left(\frac{1}{Q^4}\right)\,, \quad (1)$$

where: n_f is the number of active flavors; e_q is the fractional electric charge for the $q^{\rm th}$ quark flavor; $\otimes$ denotes the usual convolution product; $\Delta q_{\rm NS} \equiv \sum_q^{n_f}\left(n_f e_q^2/\sum_{q=1}^{n_f} e\_1\right)(\Delta q + \Delta\bar{q})$, $\Delta\Sigma = \sum_{q=1}^{n_f}(\Delta q + \Delta\bar{q})$ and Δg are the nonsinglet, singlet and gluon longitudinally-polarized PDFs; $\mathcal{C}_{\rm NS}$, $\mathcal{C}_{\rm S}$ and $\mathcal{C}_g$ are the corresponding leading-twist coefficient functions; and $h^{\rm TMC}$ and

Content from this work may be used under the terms of the Creative Commons Attribution 3.0 licence. Any further distribution of this work must maintain attribution to the author(s) and the title of the work, journal citation and DOI.
Published under licence by IOP Publishing Ltd

Figure 1. Kinematic coverage of data.

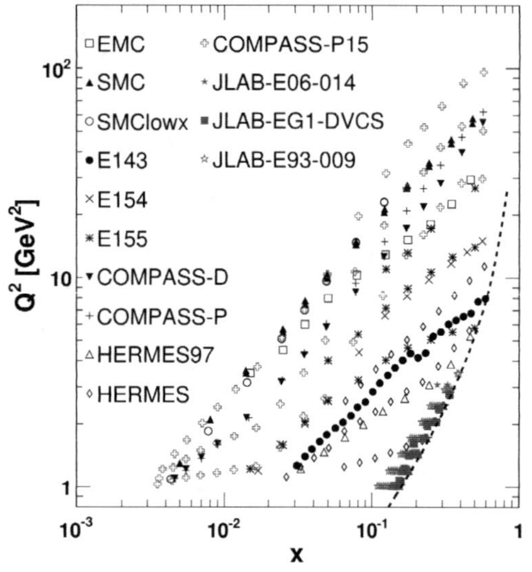

Table 1. Values of $\chi^2/N_{\rm dat}$ for each data set.

Experiment	$N_{\rm dat}$	$\chi^2/N_{\rm dat}$ 1.0	1.1	1.0+
EMC	10	0.44	0.43	0.43
SMC	24	0.93	0.90	0.92
SMClowx	16	0.97	0.97	0.94
E143	50	0.64	0.67	0.63
E154	11	0.40	0.45	0.34
E155	40	0.89	0.85	0.98
COMPASS-D	15	0.65	0.70	0.57
COMPASS-P	15	1.31	1.38	0.93
HERMES97	8	0.34	0.34	0.23
HERMES	56	0.79	0.82	0.69
COMPASS-P15	51	0.68*	0.69*	0.65
JLAB-E93-009	148	1.26*	1.23*	0.94
JLAB-EG1-DVCS	18	0.45*	0.59*	0.29
JLAB-E06-014	2	2.81*	3.20*	1.33
* Experiment not included		0.77	0.78	0.74

h denote respectively (kinematic) target-mass corrections (TMCs) and (dynamic) higher-twist corrections. In Eq. (1), the dependence of PDFs, coefficient functions, TMCs and higher-twist terms on both the scaling variable x and the energy Q^2 has been omitted for brevity.

In this write-up, I present `NNPDFpol1.0+`, an update of the analysis in Ref. [1], in which:

- new data sets for the polarized structure function of the proton g_1^p from COMPASS (COMPASS-P15 [3]), for the ratio of polarized to unpolarized structure functions of the proton and deuteron $g_1^{p,d}/F_1^{p,d}$ from CLAS (JLAB-E93-009 [4, 5] and JLAB-EG1-DVCS [6]), and for the virtual photoabsorption asymmetry of the neutron A_1^n from HALL-A (JLAB-E06-014 [7]) are fitted;
- the unpolarized PDF set used as a baseline both for the reconstruction of the structure function $g_1(x, Q^2)$ from experimental asymmetries and for the determination of positivity constraints (see respectively Secs. 2.1 and 4.4 in Ref. [1]) is updated from `NNPDF2.1` [8] to `NNPDF3.0` [9]; the unpolarized structure functions, if needed, are obtained with `APFEL` [10].

I will concentrate only on DIS data here; the `NNPDFpol1.0+` parton set could then be reweighted with proton-proton collision data from RHIC along the lines of the analysis of Ref. [2]. However, this will require an additional study which is beyond the scope of this write-up.

Except for the improvements listed above, the analysis presented here proceeds exactly as in Ref. [1]. The kinematic coverage of experimental data is shown in figure 1 (new data sets are listed in the right column), together with the kinematic cut $W^2 = m^2 + Q^2(1-x)/x \geq 6.25$ GeV^2, with m the nucleon mass. This was chosen so that the dynamic higher twist h in Eq. (1) become compatible with zero once fitted to experimental data, and can then be neglected; TMCs instead are included exactly, see Sec. 3.2 in Ref. [1] for details. The structure function $g_1(x, Q^2)$ is reconstructed from the measured observables according to the available experimental information: referring to Sec. 2.1 of Ref. [1], the COMPASS-P15 data set is treated as the EMC data set, JLAB-E93-009 and JLAB-EG1-DVCS as E155, and JLAB-E06-014 as E143.

The quality of the `NNPDFpol1.0+` analysis is assesed by the values of the χ^2 per data point, $\chi^2/N_{\rm dat}$, which are reported in table 1 for each data set, together with the number of included data points $N_{\rm dat}$. In table 1, I also show the values of $\chi^2/N_{\rm dat}$ obtained from the `NNPDFpol1.0` [1] and `NNPDFpol1.1` [2] parton sets. Inspection of table 1 allows for the following remarks.

Journal of Physics: Conference Series **678** (2016) 012030 doi:10.1088/1742-6596/678/1/012030

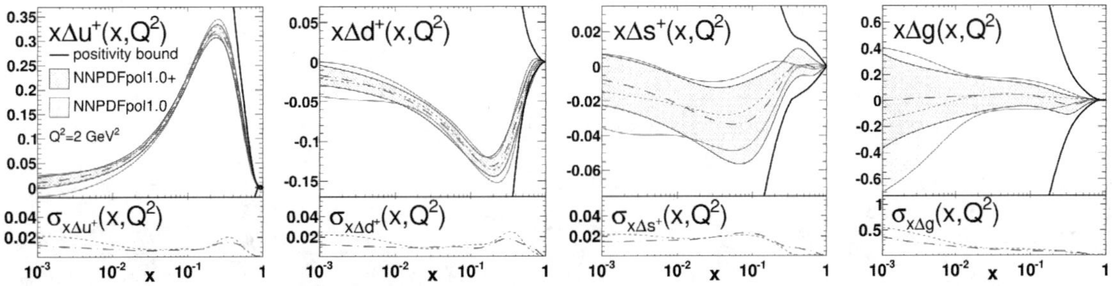

Figure 2. Comparison of $x\Delta u^+$, $x\Delta d^+$, $x\Delta s^+$, $x\Delta g$, and their uncertainties, from `NNPDFpol1.0` and `NNPDFpol1.0+` at $Q^2 = 2$ GeV^2. The positivity bound from `NNDPF3.0` is also shown.

- The quality of the `NNPDFpol1.0+` PDF determination, as measured by its total χ^2 per data point ($\chi^2_{\rm tot}/N_{\rm dat} = 0.74$), is good, and comparable to that achieved in previous determinations, both `NNPDFpol1.0` ($\chi^2_{\rm tot}/N_{\rm dat} = 0.77$) and `NNPDFpol1.1` ($\chi^2_{\rm tot}/N_{\rm dat} = 0.78$).
- In comparison to `NNPDFpol1.0` and `NNPDFpol1.1`, the value of $\chi^2/N_{\rm dat}$ for new experiments included in `NNPDFpol1.0+` improves substantially for all JLAB data sets, while the difference is less significant for the COMPASS data set. This suggests that JLAB data are likely to have a sizable impact on PDFs, while the effects of COMPASS data are expected to be moderate (see also the discussion in the sequel).
- The value of $\chi^2/N_{\rm dat}$ for single experiments are often well below the optimal value $\chi^2/N_{\rm dat} \sim 1$. This is a consequence of the lack of experimental information on correlations among systematics, which cannot be accounted for properly, and ostensibly lead to an overestimation of experimental uncertainties.

In figure 2, I show the total polarized quark distributions $\Delta q^+ = \Delta q + \Delta\bar{q}$ ($q = u, d, s$) and the polarized gluon distribution Δg at $Q^2 = 2$ GeV^2, obtained from both `NNPDFpol1.0+` and `NNPDFpol1.0` PDF sets; their absolute uncertainties, $\sigma_{\Delta q}$, and the positivity bound obtained from the `NNPDF3.0` PDF set are also shown. Inspection of figure 2 allows for the following remarks.

- In the small-x region ($x \lesssim 10^{-2}$), the uncertainty of all quark and gluon distributions is reduced by about a factor two. This is due to a better accuracy of the proton and neutron unpolarized structure functions $F_2^{p,n}$ and $F_L^{p,n}$, whose uncertainty is propagated to the polarized structure function $g_1^{p,n}$ when the latter is reconstructed from experimental asymmetries. The improvement between `NNPDF2.1` (used in the original `NNPDFpol1.0` analysis) and `NNPDF3.0` (used in this analysis) is displayed in figure 3, and is due to the significant amount of new LHC data included in `NNDPF3.0`. The impact of COMPASS-P15, the only new data set which covers the small-x region, is instead rather limited: I performed a fit including COMPASS-P15 data, but using the same unpolarized structure functions as in `NNPDFpol1.0`, and I found that PDF uncertainties are almost identical to those determined in `NNPDFpol1.0`.
- In the intermediate-to-large-x region ($10^{-2} \lesssim x \lesssim 0.6$), the uncertainty of Δu^+ and Δd^+ is about two thirds of the uncertainty obtained from `NNPDFpol1.0`. This is a genuine effect of new JLAB data sets, which attain slightly larger x values and are more accurate than those included in `NNPDFpol1.0`. The effects of updating the unpolarized structure functions from `NNPDF2.1` to `NNPDF3.0` are less prominent than in the small-x region: I checked explicitly that results are almost unchanged in this region if a fit including new data but old unpolarized structure functions from `NNPDF2.1` is performed. Note that the full potential of JLAB data, which includes the possibility to discriminate among different

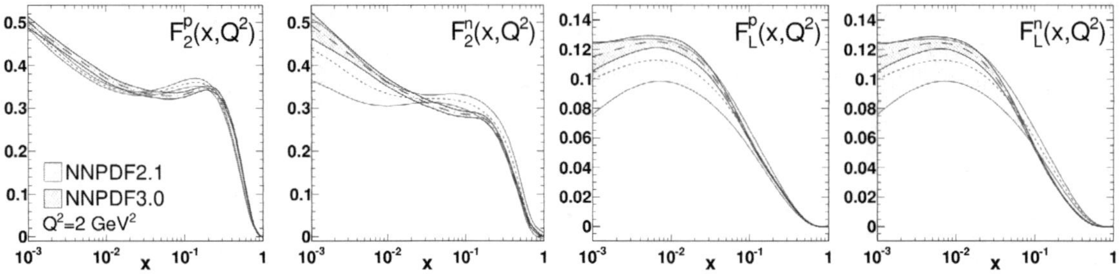

Figure 3. The proton and neutron unpolarized structure functions $F_2^{p,n}$ and $F_L^{p,n}$ from NNPDF2.1 and NNPDF3.0 at $Q^2 = 2$ GeV^2, as obtained with APFEL [10].

non-perturbative models of nucleon stucture [11], could be exploited only by lowering the kinematic cut on W^2. However, this will require a systematic inclusion of dynamic higher-twist corrections: I checked that, if h-terms in Eq. (1) are neglected but $W^2 \geq 4$ GeV^2 is taken, in principle the PDF uncertainty on Δu^+ and Δd^+ can be reduced even more; however, the value of the χ^2 deteriorates significantly and PDFs become no longer reliable.

- In the large-x region ($x \gtrsim 0.6$), no experimental data are presently available to determine longitudinally-polarized PDFs. However, their behavior is explicitly bounded by their unpolarized counterparts, as it follows from positivity constraints: at leading-order, $|\Delta f(x,Q^2)| \leq f(x,Q^2)$, $f = u^+, d^+, s^+, g$. No significant differences are found in the polarized PDFs by updating the baseline unpolarized PDF set from NNPDF2.1 to NNPDF3.0, except for Δs^+. For this distribution, the uncertainty is about ten times larger in NNPDFpol1.0+ than in NNPDFpol1.0. Note that $W+c$ data sets, sensitive to the unpolarized strange distribution s^+, have been included in NNPDF3.0, which lead to a larger s^+ than that found in NNPDF2.1. For this reason, the positivity bound on Δs^+ becomes less stringent, and, in absence of any experimental information, this allows the uncertainty on Δs^+ to grow significantly. A similar effect, though less prominent, can be noticed also for Δg.

In conclusion, longitudinally-polarized PDFs obtained in this analysis are well compatible with those obtained in the previous NNPDF determination based on DIS data, NNPDFpol1.0, but have slightly smaller uncertainties. I have explicitly shown that there is a sizable interplay between unpolarized and polarized PDFs, and that JLAB data are effective in unveiling the large-x behavior of PDFs; however, a determination of higher-twist corrections to g_1, Eq. (1), will be mandatory to exploit their full potential. In the future, an extensive study will be dedicated to a global determination of longitudinally-polarized PDFs, into which the new DIS data discussed here, the proton-proton collision data included in NNPDFpol1.1 and possibly other new data will be incorporated together.

References

[1] Ball R D *et al.* (The NNPDF Collaboration) 2013 *Nucl. Phys.* **B874** 36–84 (*Preprint* 1303.7236)
[2] Nocera E R *et al.* (The NNPDF Collaboration) 2014 *Nucl. Phys.* **B887** 276–308 (*Preprint* 1406.5539)
[3] Adolph C *et al.* (COMPASS) 2015 (*Preprint* 1503.08935)
[4] Dharmawardane K V *et al.* (CLAS) 2006 *Phys. Lett.* **B641** 11–17 (*Preprint* nucl-ex/0605028)
[5] Guler N *et al.* (CLAS) 2015 (*Preprint* 1505.07877)
[6] Prok Y *et al.* (CLAS) 2014 *Phys. Rev.* **C90** 025212 (*Preprint* 1404.6231)
[7] Parno D S *et al.* (Jefferson Lab Hall A) 2015 *Phys. Lett.* **B744** 309–314 (*Preprint* 1406.1207)
[8] Ball R D *et al.* (The NNPDF Collaboration) 2011 *Nucl. Phys.* **B849** 296–363 (*Preprint* 1101.1300)
[9] Ball R D *et al.* (The NNPDF Collaboration) 2015 *JHEP* **04** 040 (*Preprint* 1410.8849)
[10] Bertone V, Carrazza S and Rojo J 2014 *Comput. Phys. Commun.* **185** 1647–1668 (*Preprint* 1310.1394)
[11] Nocera E R 2015 *Phys. Lett.* **B742** 117–125 (*Preprint* 1410.7290)

Journal of Physics: Conference Series **678** (2016) 012031 doi:10.1088/1742-6596/678/1/012031

Study of polarization effects at Nuclotron

V P Ladygin[1†], M Janek[2], P K Kurilkin[1], N B Ladygina[1], P N Batyuk[1], S P Mertz[1]

[1] Joint Institute for Nuclear Research, Dubna, Russia
[2] Physics Department, University of Žilina, Žilina, Slovakia

E-mail: †vladygin@jinr.ru

Abstract. The major goal of the upgraded Nuclotron facility is to obtain the information on the equation-of-state for dense nuclear matter playing a key role in the under standing of the collapse supernovae and neutron stars stability. These studies can be performed either in heavy ion collisions or via the short-range few nucleon correlations. The obtained experimental results and future program with the use of polarized deuteron beam and the internal target station are discussed. The polarization studies for the NN, NA and dA reactions with the extracted deuteron beam at the BM@N setup are proposed. The further extension of the polarization program at BM@N is related with the study of the in-medium modification of the polarization for the strange and multi-strange baryons and the spin alignment for vector mesons decaying in hadronic modes.

1. Introduction

The Nuclotron at JINR will provide beams of heavy ions with energies up to 6 A·GeV for isospin symmetric nuclei, and 4.65 A·GeV for Au nuclei. In central heavy-ion collisions at these energies, nuclear densities of about 4 times nuclear matter density can be reached. These conditions are well suited to investigate the equation-of-state (EOS) of dense nuclear matter which plays a central role for the dynamics of core collapse supernovae and for the stability of neutron stars. The observation of multi-strange hyperons or of light hypernuclei at Nuclotron energies would represent a breakthrough in our understanding of strange matter, and would pave the road for the experimental exploration of the 3-rd dimension of the nuclear chart [1]. These studies are complimentary to the CBM experimental program at SIS100 [2].

Another tool to investigate the EOS at large densities is the study of the short range correlations (SRC) of nucleons in nuclei which is the subject of intensive theoretical and experimental works during last years. Since SRC have densities comparable to the density in the center of a nucleon which is about $\rho \sim 5\rho_0$ ($\rho_0 \approx 0.17$ fm^{-3}), they can be considered as the drops of cold dense nuclear matter [3]. These studies explore a new part of the phase diagram and very essential to understand the evolution of neutron stars.

The results obtained at BNL [4], SLAC [5] and JLAB [6, 7] clearly demonstrate that more than 90% all nucleons with momenta $k \geq 300$ MeV/c belong to 2N SRC; the probability for a given proton with momenta $300 \leq k \leq 600$ MeV/c to belong to pn correlation is ∼18 times larger than for pp correlations; the probability for a nucleon to have momentum ≥ 300 MeV/c in medium nuclei is ∼25%; 3N SRC are present in nuclei with a significant probability [8]. However, still many open questions persist and further investigations are required both from the

Content from this work may be used under the terms of the Creative Commons Attribution 3.0 licence. Any further distribution of this work must maintain attribution to the author(s) and the title of the work, journal citation and DOI.
Published under licence by IOP Publishing Ltd

experimental and theoretical sides. For instance, the experimental data on the spin structure of 2N (I=1) and 3N SRC are almost absent.

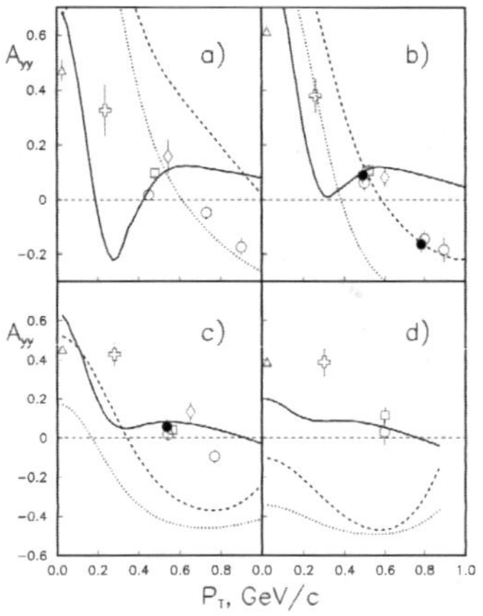

Figure 1. The dependence of the tensor analyzing power A_{yy} [9, 10, 11, 12, 13] as a function of the proton transverse momentum p_T at four different fixed values of $x_F \sim 0.61$, 0.67, 0.72 and 0.78, are shown in the a), b), c) and d) panels, respectively.

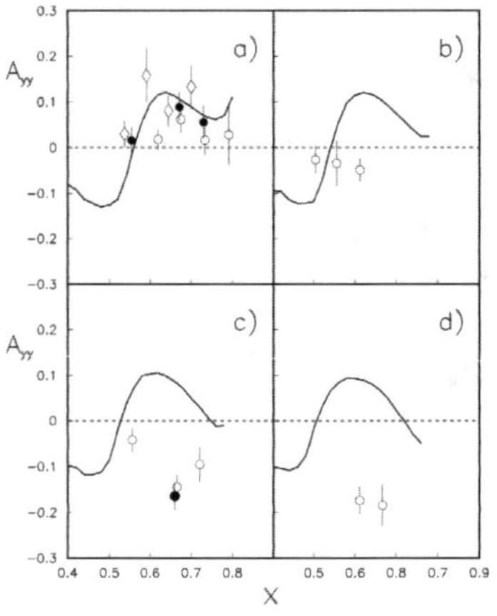

Figure 2. A_{yy} data[9, 10, 11, 12, 13] plotted as a function of longitudinal momentum fraction x_F obtained at fixed p_T values of ~550 MeV/c, ~700 MeV/c, ~800 MeV/c and ~900 MeV/c are presented in the a), b), c) and d) panels, respectively.

The main tools to study SRCs at hadronic facilities can be deuteron structure investigations at large internal momenta allowing to explore 2N SRC with $I = 0$; ^{3}He structure to understand the role of 2N SRC with $I = 1$ and 3N SRC; nuclei breakup $A(p,pp)X$, $A(p,pn)X$, $A(p,ppp)X$ etc. with the detection of few nucleons in the final state. The greate importance is the study of the spin effects in these reactions because the data on the SRCs spin structure are scarce. Nuclotron and NICA will allow to investigate the spin effects for multi-nucleon correlations in a wide energy range.

The spin structure of the np SRCs has been investigated at JINR via the measurements of the tensor analyzing power A_{yy} in deuteron inclusive breakup at different energies in the wide regions of the x_F and transverse proton momentum p_T [9, 10, 11, 12, 13]. The data on the tensor analyzing power A_{yy} obtained in the $A(d,p)X$ reaction at different values of x_F ~0.61, ~0.67, ~0.72 and ~0.78 and plotted as a function of the proton transverse momentum p_T are shown in panels a), b), c) and d) in Fig.1, respectively. The figure is taken from ref.[13]. It is seen that the A_{yy} data for different x_F are strongly dependent of the transverse momentum of the protons, p_T . Values of A_{yy} are positive at small p_T and monotonously decrease while transverse momentum increasing for all x_F values. On the other hand, A_{yy} values change the sign at p_T ~600 MeV/c independently on x_F and demonstrate kind of negative asymptotic at large p_T. The dashed, dash-dotted and solid curves are the results of the calculations using standard [14, 15] and covariant [16] deuteron wave functions (DWFs), respectively. The A_{yy} data obtained at different values of transverse momenta p_T as a function of x_F are plotted in Fig.2. The data shown in panels a), b), c) and d) correspond to the averaged values of p_T ~550 MeV/c, ~700 MeV/c, ~800 MeV/c and ~900 MeV/c, respectively. The figure is also taken from ref.[13]. The solid curves are the results of the calculations by using covariant DWF [16]. One can see that the A_{yy} data for different values of p_T demonstrate a weak dependence on x_F.

The data obtained at p_T ~550 ~MeV/c are in a good agreement with the calculations by using covariant DWF [16]. At higher p_T A_{yy} data have negative values, while the theory predicts a positive sign in the range of measurement Therefore, the A_{yy} data clearly demonstrate the dependence on two internal variables, x_F and p_T (or their combinations). However, the use of the deuteron structure function that depends on two variables [16] does not allow to describe the data.

The fundamental degrees of freedom in the frame of QCD are the quarks and gluons. These degrees (effective ones as $\Delta\Delta$, N^*N, NN or $6q$ and $9q$ components) begin to play a role at the internucleonic distances comparable with the size of the nucleon. At high energies s and large transverse momenta p_T the constituent counting rules (CCR) [17, 18] are working. These rules predict the dependence of the cross section of the binary reactions at the fixed scattering angle in the cms as a power-law of s. The analysis of the experimental data on the cross sections of the $dp \to pd$ and $dd \to^3 Hen$ reactions [19] has shown that the regime corresponding to CCR can occur already at $T_d \sim 500$ MeV. Therefore, the fundamental degrees of freedom can manifest in the deuteron induced reactions at Nuclotron energies.

2. Recent results and further plans for internal target experiments

The main goal of the Deuteron Spin Structure (DSS) experimental program is to obtain the information on the spin - dependent parts of two-nucleon ($2N$) and three-nucleon ($3N$) forces from two processes: *dp*- elastic scattering in a wide energy range and *dp*- nonmesonic breakup with two protons detection at energies 300 – 500 MeV [20, 21, 22]. The motivation of this program is based on theoretical analysis of the experimental results obtained at low and intermediate energies for the deuteron induced reactions (see recent reviews [23, 24] and references therein).

Such experimental program at Nuclotron was started by the measurements of the vector A_y and tensor A_{yy} and A_{xx} analyzing powers in *dp*- elastic scattering at T_d of 880 MeV [25] and 2000 MeV [26].

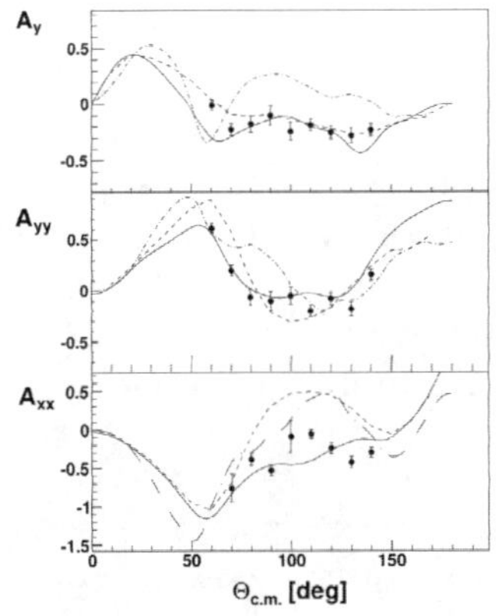

Figure 3. The angular dependence of the analyzing powers A_y, A_{yy} and A_{xx} of *dp*- elastic scattering at $T_d = 880$ MeV[25].

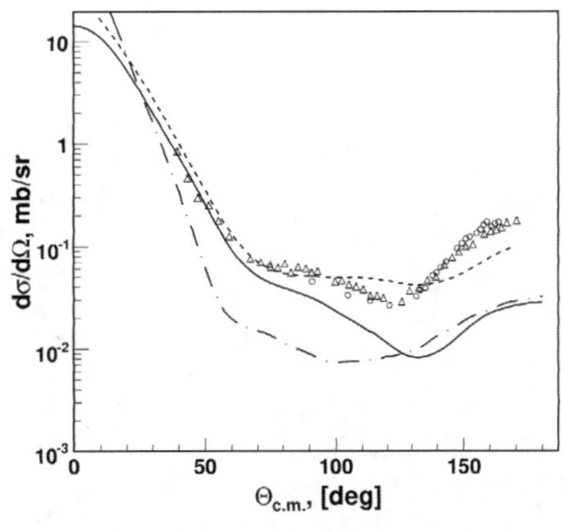

Figure 4. The angular dependence of the differential cross section of *dp*- elastic scattering at $T_d = 940$ MeV [27] and 850 MeV [28], respectively.

The results on the angular dependencies of the analyzing powers A_y, A_{yy} and A_{xx} of *dp*- elastic scattering obtained at Nuclotron at 880 MeV [25] are presented in Fig.3. The differential cross section obtained at 940 MeV [27] and 850 MeV [28] in the earlier experiments are shown in Fig.4 by the open circles and triangles, respectively. The solid, dashed and dot-dashed lines are the results of the non-relativistic Faddeev calculations [23], relativistic multiple scattering model [29, 30] and optical potential approach [31]. The Faddeev calculations without invoking $3N$ forces reproduce the behaviour of the analyzing powers, however, they fail to describe the cross section data at the scattering angles larger than 70° in the cms. The calculations performed within relativistic multiple scattering model [29, 30] describes the data on the vector analyzing power A_y and cross section. However, there are some problems in the description of the tensor analyzing powers at large angles in the cms. The optical potential approach fails to reproduce both cross section and analyzing powers. The observed deficiencies in the description of the differential cross section and deuteron analyzing powers at T_d ~880 MeV obtained at quite large transverse momenta require the consideration of the additional mechanisms, for instance, $3N$ forces. Since present $3N$ forces models cannot improve the agreement with the data obtained even at lower energies, new models of $3N$ forces (including their short-range part) should be considered. For instance, the next step in the relativistic multiple scattering model [29, 30] development could be taking into account the explicit Δ- isobar excitation.

As the first stage of the Deuteron Spin Structure (DSS) experimental program [20, 21, 22] the beam energy scan of *dp*- elastic scattering cross section at the deuteron energies 400 – 2000 MeV and measurements of *dp*- non-mesonic breakup at 300, 400 and 500 MeV in different kinematic configurations have been performed at Nuclotron. These measurements were performed using internal target station at Nuclotron [32] with new control and data acquisition system [33]. The continuation of the DSS experimental program at internal target is related with new polarized ion source developed at LHEP-JINR [34].

3. Spin studies with extracted polarized deuteron beam

Spin physics with extracted polarized deuteron beam from new polarized deuteron source [34] can be performed with the start version of the BM@N setup [1]. Since the multiplicity of the secondary particles is small (2 or 3 tracks) the forward and outer trackers only can be used. The BM@N experimental setup will be installed at the renovated 6V beamline [35] in the fixed-target hall of the Nuclotron. The 6V beamline contains the quadrupole lenses doublet, two dipole magnets allowing to correct the beam position in the vertical and horizontal planes, and SP41 dipole magnet with enlarged aperture for the momentum measurements [36].

The transportation line VP1 does not require modification in the sense of the installation or removing of magnetic elements. But spin program realization requires to install the polarimeters (with CH_2-C targets) at F3 and F5 focuses for polarimetry of the deuteron (or proton) beam. Also the place at F5 (or F3) is needed to install liquid hydrogen (deuterium) target, polarized $^3\vec{\mathrm{He}}$ target [37] and set of nuclear targets. It will be necessary also to have some place for the stop time-of-flight detector.

The spin physics can be divided on 3 part due to position of the target at VP1 transportation beam line. The physics with the target position at F5(or F3) focus is related with the measurements of the polarization observables in the reactions with the emission of the detected particle at forward angles. The solid angle is defined by the lenses doublet 7k100–8k100 of the VP1 transportation beam line. The separation of the primary deuteron and secondary beams is provided by the bending magnet 3SP40. The physics program can include the measurements of the tensor analyzing power T_{20} (and, possibly, vector polarization transfer coefficient κ_0) in the inclusive deuteron breakup, $A(d,p(0^\circ))X$, [9, 38, 39] at the highest available energy at Nuclotron; of the tensor analyzing power T_{20} in the inclusive pion production, $A(d,\pi^-(0^\circ))X$, [40] also at the highest available energy at Nuclotron; of the tensor analyzing power T_{20} (and, possibly, vector polarization transfer coefficient κ_0) in the inelastic deuteron scattering, $A(d,d')X$, in the vicinity of the baryonic resonances excitation [41]; of the tensor analyzing power T_{20} (and, possibly, vector polarization transfer coefficient κ_0 in $p(d,p)d$ [42, 43] and $d(d,p)t$ [44] reactions; of the tensor analyzing power T_{20} and spin correlation $C_{y,y}$ in the ^{3}He$(d,p)^4$He reaction in the kinetic energy range between 1.0 and 1.75 GeV [45, 46]; of the analyzing power T_{20} in the $A(d,^3He)X$ [47] reactions.

The measurements of the ^{3}He$(d,p)^4$He reaction is mostly challenging throughout the above mentioned experiments due to low density of the polarized ^{3}He target [37]. The goal of the ^{3}He$(d,p)^4$He reaction study at Nuclotron is to understand the reasons of the long staying puzzle, namely, the still unexplained strange structure in the behaviour of the tensor analyzing power T_{20} in *dp*- backward elastic scattering [42, 43] at the internal momentum k ~0.3–0.5 GeV/c in the vicinity of the D- wave dominance. The experiments performed at RIKEN at the energies below 270 MeV have shown that the polarization correlation coefficient, $C_{//} = 1 - \frac{1}{2\sqrt{2}}T_{20} + \frac{3}{2}C_{y,y}$, for the ^{3}He$(d,p)^4$He reaction may be a unique probe to the D-state admixture in deuteron [45]. The usefulness of this observable to investigate the D-state admixture is attributed to the strong spin-selectivity in neutron capture process by ^{3}He nucleus, i.e., spins of transferred neutron and ^{3}He must be anti-parallel to each other in order to form ^{4}He in the final state. In the one-nucleon exchange (ONE), the expression for $C_{//}$ is proportional to the D-state fraction in deuteron as

$$C_{//} = \frac{9}{4}\frac{w^2}{u^2 + w^2}, \tag{1}$$

where u and w are the S- and D-state wave functions of deuteron in momentum space. This is a marked contrast to T_{20} and κ_0 for *dp* backward elastic scattering which include S- and D-state interference term (uw-term) together with a w^2-term. It is thus expected that $C_{//}$ may be a candidate to provide an information on the deuteron structure complementary to those from T_{20} and κ_0 obtained in *dp*- backward elastic scattering [42].

The main goal of the proposed at Nuclotron experiment is to obtain the data on $C_{//}$ in the energy region of 1.0–1.75 GeV, where the contribution from the deuteron D-state is expected to reach a maximum in one-nucleon exchange approximation, to obtain new information on the strange structure observed in the behaviour of T_{20} in the *dp*- backward elastic scattering and to realize experiment on the full determination of the matrix element of the ^{3}He$(d,p)^4$He reaction in the model independent way. These data will help us also to understand the short-range spin structure of deuteron and effects of non-nucleonic degrees of freedom. For these purposes polarized deuteron beam from new PIS [34] and spin-exchange-type polarized ^{3}He target developed at CNS of Tokyo University [37] and modified for the experiment at Nuclotron can be used.

The simulation has been performed for the initial deuteron kinetic energy T_d= 1.5 GeV and the 30 cm ^{3}He target cell using the GEANT4 toolkit. The initial deuteron beam interacts with 30 cm ^{3}He target placed in one of the focuses of the V1 beam transportation line, for instance, in F5 focus. The secondary protons and deuterons, as well as primary deuterons pass along V1 transportation beam line with magnetic elements tuned for the optimal momentum of the secondary protons (~75% from the initial deuteron beam momentum). The identification of the secondary particles (protons and deuterons) in the experiment at extracted beam will be done by the measurements of their time-of-flight (TOF) over a baseline of about 20–30 m. The trigger counter with 1 mm of the quartz radiator and XP2020Q PMTs can be used as the start for TOF system. The other option for the TOF start is the 4 mm scintillation counter viewed from both sides by XP2020 PMT. This counter can be placed in front of the lenses doublet of the 6V beam line [35]. Both start counter options provide similar amount of the additional material. Two cases were considered: ^{3}He target only and ^{3}He target + 2 mm of quartz radiator of TOF start counter. Additionally only about 1 m of air was taken into account. Therefore, it was assumed that the pipes of the VP1 and 6V lines will be in vacuum. Only analyzing magnet, GEM tracker and one half of the mRPC-1 wall will be used for the proposed measurements.

The particle momentum will be analyzed using 12 stations GEM tracker placed in the magnetic field The 1-st and 12-st stations are positioned at the distance of 30 cm and 360 cm from the beginning of the magnetic pole, respectively. The simulation has been performed for magnetic field B_y value of ~0.9 T. The magnetic field integral $\int Bdl$ is equal to ~3 T·m at 1900 A with non-linearity of about 5% [36].

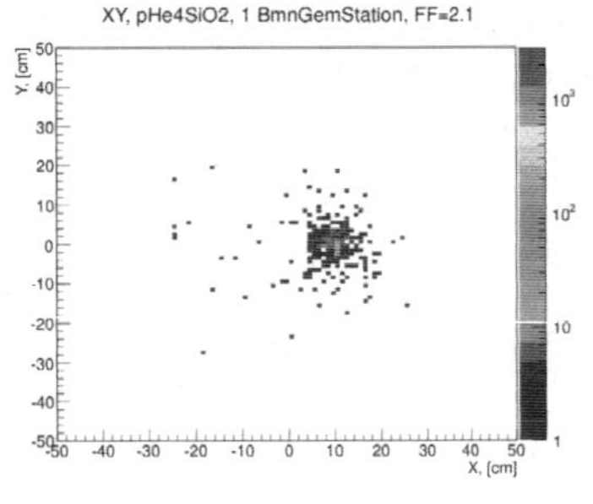

Figure 5. The XY secondary beam profile at the 1-st GEM station for the magnetic field B_y of ~0.9 T.

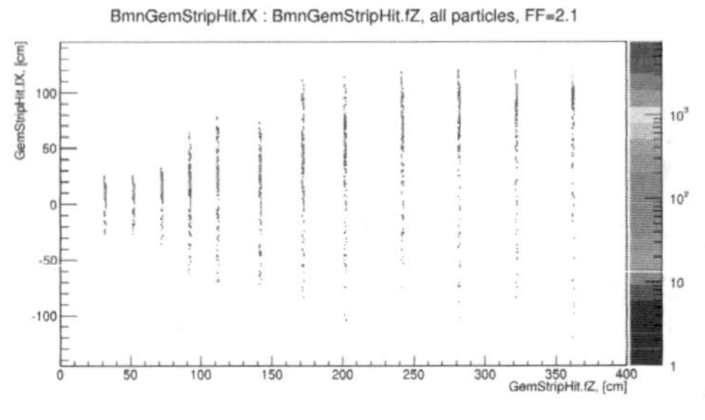

Figure 6. The XZ profile of the secondary beam hits in the GEM tracker stations for the magnetic field B_y of ~0.9 T.

The proton beam spot on the 1-st GEM tracker station is presented in Fig.5. One can see that the X- position of the beam is ~10 cm from the central axis of the spectrometer. The XZ profile of the secondary beam hits in the GEM tracker stations is shown in Fig.6. The deflection angle is about ~320 mr. The momentum resolution obtained for the current version of the tracking software is about several per cents. The removing of the additional material as well as the increasing of the space resolution of the tracking detectors (silicon tracker system) will significantly increase the momentum resolution. The time difference between mRPC and start detectors for protons and deuterons for the base line of ~24 m are separated by ~20 ns. This allows to have timing resolution of the TOF system at the level of 300–400 ps. Therefore, since the timing resolution of the mRPC wall is about 100 ps, the start counter can have the resolution of about 200–300 ps, what can be easily achieved by using scintillation counter with PMT readout.

The polarization observables in the $p(d,p)d$ [42, 43] and $d(d,p)t$ [44, 48, 49] reactions in the collinear geometry can be measured with the same experimental setup using liquid hydrogen or deuterium target or CH_2 and CD_2 solid targets with carbon background subtraction.

The solid nuclear target can be placed inside the pole of the 3SP40 magnet. It can be varied from the beginning to the middle of the pole of the 3SP40 magnet. The target position in the middle of the switched off magnet corresponds to ~100 mr. This configuration can be used for the measurements of the tensor A_{yy} and vector A_y analyzing powers (and, possibly, vector polarization transfer coefficient C_y^y) in inclusive deuteron breakup, $A(d,p)X$, at large transverse proton momenta [9, 10, 11, 12, 13] at the highest available energy at Nuclotron; of the tensor A_{yy} and vector A_y analyzing powers [50, 51, 52] (and, possibly, vector polarization transfer coefficient C_y^y [53]) in the inelastic deuteron scattering, $A(d,d')X$, in the vicinity of the baryonic resonances excitation; of the tensor A_{yy} and vector A_y analyzing powers in the inclusive pion production, $A(d,\pi^-)X$, [54] also at the highest available energy at Nuclotron. The measurements of the analyzing powers in inclusive deuteron breakup and the inelastic deuteron scattering in the vicinity of the baryonic resonances excitation can be performed simultaneously. All these experiments require the additional TOF detector placed between F5 and F6 focuses.

The third part of the measurements is the spin physics with the target position at F6 focus. The position of the F6 focus can be change by the magnetic optics of the 6V beam line. The target position for this part of the experiments can be varied depending on the readiness of the inner tracker. For the low multiplicity events (2 or 3 tracks) one can use the same forward tracker as for the discussed above experiments. The separation of the primary deuteron and secondary beams is provided by the modernized analyzing magnet SP41 [36]. The physiscs is related with the baryonic resonances spin properties studies at the energies between 2 and 6 GeV of the deuteron kinetic energy and includes the measurements of the tensor A_{yy} and vector A_y analyzing powers in quasi-elastic and inelastic $A(d,pp)X$ reaction; of the tensor A_{yy} and vector A_y analyzing powers in the inelastic deuteron scattering $A(d,d')X$ [53] and $A(d,d')\pi^{\pm}X$ reactions; investigation of the vector analyzing power in neutron induced reactions (with the proton spectator detection) like $np \to pn$, $np \to pp\pi^-$, $np \to np\pi^+\pi^-$ [55, 56] etc. For the these experiments the full size RPC wall is required. In parallel, the measurement of the dp- elastic scattering [25, 26], $dd \to {}^3\mathrm{He}n$ [57, 58, 59] and $dA \to {}^3\mathrm{He}X$ [47] processes can be performed.

The spin studies with BM@N requires also the advanced deuteron beam polarimetry at Nuclotron discussed in [60, 61].

4. Polarization effects in heavy ion collisions at Nuclotron

The change of the polarization properties of the secondary particles in the nucleus-nucleus collisions compared to the nucleon-nucleon collisions can be investigated at BM@N. A number of polarization observables have been proposed as a possible signature of phase transition, namely, decreasing of the Λ^0 transverse polarization in central collisions non-zero $\bar{\Lambda}^0$ longitudinal polarization, non-zero J/Ψ polarization at low p_T , anisotropy in dielectron production from vector mesons decay, global hyperon polarization and spin-alignment of vector mesons in non-central events, vorticity and hydrodynamical helicity in noncentral heavy-ion collisions etc.

5. Conclusions

New data on the analyzing powers A_y, A_{yy} and A_{xx} in dp- elastic scattering at various energies up to 2000 MeV and well as for the dp- nonmesonic breakup at the energies between 300 and 500 MeV for different kinematic configurations can be measured at ITS at the Nuclotron using new PIS [34].

First stage of the BM@N setup (without or with reduced version of the inner tracker) is well suited for the physics with polarized deuterons confirmed by the feasibility studies for the measurements of the polarization observables in the ${}^3\mathrm{He}(d,p){}^4\mathrm{He}$ reaction at 1.5 GeV.

Measurements of the polarization effects in the heavy ion collisions can significantly enrich the physics at BM@N.

Acknowledgments

The work has been supported in part by the RFBR under grant $N^0$13-02-00101a and JINR-Slovak Republic scientific cooperation program in 2015.

References

[1] Ladygin V *et al.* 2012 *Proc. of the XXI-st International Baldin Seminar on High Energy Physics Problems: Relativistic Nuclear Physics and Quantum Chromodynamics (Dubna)* PoS(Baldin-ISHEPP-XXI) p.038.
[2] Friman B, Hohne C, Kroll J, Leupold S, Randrup J, Rapp R and Senger P 2011, *Lect.Notes Phys.* **814** 1.
[3] Frankfurt L, Sargsian M and Strikman M 2008 *Int.J.Mod.Phys. A* **23** 2991.
[4] Piasetzky E, Sargsian M, Frankfurt L, Strikman M and Watson J W 2006 *Phys.Rev.Lett.* **97** 162504.
[5] Frankfurt L L, Strikman M I, Day D B and Sargsian M M 1993 *Phys.Rev. C* **48** 2451.
[6] Egiyan K Sh *et al.* 2003 *Phys.Rev. C* **68** 014313.
[7] Egiyan K Sh *et al.* 2006 *Phys.Rev.Lett.* **96** 082501.
[8] Frankfurt L, Sargsian M and Strikman M 2008 *AIP Conf.Proc.* **1056** p.322.
[9] Azhgirey L S *et al.* 1996 *Phys.Lett. B* **387** 37.
[10] Afanasev S V *et al.* 1998 *Phys.Lett. B* **434** 21; Azhgirey L S *et al.* 1999 *Phys.Atom.Nucl.* **62** 1673.

[11] Ladygin V P *et al.* 2002 *Few Body Syst.* **32** 127; Azhgirey L S *et al.* 2003 *Phys.Atom.Nucl.* **66** 690.
[12] Azhgirey L S *et al.* 2004 *Phys.Lett. B* **595** 151. Azhgirey L S *et al.* 2005 *Phys.Atom.Nucl.* **68** 991.
[13] Ladygin V P *et al.* 2005 *Phys.Lett. B* **629** 60; Azhgirey L S *et al.* 2008 *Phys.Atom.Nucl.* **71** 264.
[14] Machleidt R 2001 *Phys.Rev. C* **63** 024001.
[15] Lacombe M, Loiseau B, Vinh Mau R, Cote J, Pires P and de Tourreil R 1981 *Phys.Lett. B* **B101** 139.
[16] Karmanov V A and Smirnov A V 1992 *Nucl.Phys. A* **546** 691; 1994 *Nucl.Phys. A* **575** 520; Carbonell J and Karmanov V A 1995 *Nucl.Phys. A* **581** 625; 1995 *Nucl.Phys. A* **589** 713; Carbonell J, Desplanques B, Karmanov V A and Mathiot J F 1998 *Phys.Rep.* **300** 125.
[17] Brodsky S J and Farrar G R 1973 *Phys.Rev.Lett.* **31** 1153; Brodsky S J and Farrar G R 1975 *Phys.Rev.D* **11** (1975) 1309; Lepage G P. and Brodsky S J 1980 *Phys.Rev.D* **22** 2157.
[18] Matveev V A, Muradyan R M and Tavkhelidze A N 1973 *Lett. Nuovo Cim.* **7** 719.
[19] Uzikov Yu N 2005 *JETP. Lett.* **81** 303.
[20] Ladygin V P *et al.* 2011 *J.Phys.Conf.Ser.* **295** 012131.
[21] Ladygin V P *et al.* 2014 *Phys.Part.Nucl.* **45** 327.
[22] Ladygin V P *et al.* 2014 *Few Body Syst.* **55** 709.
[23] Glöckle W, Witala H, Hüber D, Kamada H and Golak J 1996 *Phys.Rept.* **274** 107.
[24] Kalantar-Nayestanaki N, Epelbaum E, Messchendorp J G and Nogga A 2012 *Rept.Prog.Phys.* **75** 016301.
[25] Kurilkin P K *et al.* 2012 *Phys.Lett. B* **715** 61.
[26] Kurilkin P K *et al.* 2011 *Phys.Part.Nucl.Lett.* **8** 1081.
[27] Alder J C *et al.* 1972 *Phys.Rev. C* **6** 2010.
[28] Booth N E *et al.* 1971 *Phys.Rev. D* **D4** 1261.
[29] Ladygina N B 2008 *Phys.Atom.Nucl.* **71** 2039.
[30] Ladygina N B 2009 *Eur.Phys.J. A* **42** 91.
[31] Shikhalev M A 2009 *Phys.Atom.Nucl.* **72** 588.
[32] Malakhov A I *et al.* 2000 *Nucl.Instrum.Meth. in Phys.Res. A* **440** 320.
[33] Isupov A Yu, Krasnov V A, Ladygin V P, Piyadin S M and Reznikov S G 2013 *Nucl.Instrum.Meth. in Phys.Res. A* **698** 127.
[34] Fimushkin V V *et al.* 2008 *Eur.Phys.J. ST* **162** 275.
[35] Piyadin S M *et al.* 2012 *Phys. Part. Nucl. Lett.* **9** 589.
[36] Akishin P G *et al* 2015 *Phys.Part.Nucl.Lett.* **12** 305.
[37] Uesaka T *et al.* 1998 *Nucl.Instr. and Meth. in Phys.Res. A* **402** 212.
[38] Punjabi V *et al.* 1989 *Phys.Rev. C* **39** 608.
[39] Kuhn B *et al.* 1994 *Phys.Lett. B* **334** 298.
[40] Afanasev S *et al.* 1998 *Phys.Lett. B* **445** 14.
[41] Azhgirey L S *et al.* 1995 *Phys.Lett. B* **361** 21.
[42] Punjabi V *et al.* 1995 *Phys.Lett. B* **350** 178.
[43] Azhgirey L S *et al.* 1997 *Phys.Lett. B* **391** 22.
[44] Ladygin V P *et al.* 2004 *Phys.Lett. B* **598** 47.
[45] Uesaka T *et al.* 1999 *Phys.Lett. B* **467** 199.
[46] Uesaka T *et al.* 2002 *Phys. Lett. B* **533** 1.
[47] Ladygin V P and Ladygina N B 1995 *Phys.Atom.Nucl.* **58** 1283.
[48] Ladygina N B 2012 *Few Body Syst.* **53** 253.
[49] Ladygina N B 2014 *Phys.Part.Nucl.* **45** 187.
[50] Ladygin V P *et al.* 2000 *Eur.Phys.J. A* **8** 409. Azhgirey L S *et al.* 2001 *Phys.Atom.Nucl.* **64** 1961.
[51] Azhgirey L S *et al.* 2005 *Phys.Atom.Nucl.* **68** (2005) 991.
[52] Ladygin V P *et al.* 2006 *Phys.Atom.Nucl.* **69** 852.
[53] Ladygin V P and Ladygina N B 2002 *Phys.Atom.Nucl.* **65** 182.
[54] Azhgirey L S *et al.* 2011 *Phys.Atom.Nucl.* **74** 1392.
[55] Agakishiev G. *et al.* 2015 *Phys.Lett. B* **750** 184.
[56] Jerusalimov A *et al.* 2015 *Eur.Phys.J. A* **51** 83.
[57] Janek M *et al.* 2007 Eur.Phys.J. A **33** 39.
[58] Kurilkin A K *et al.* 2009 *Int.J.Mod.Phys. A* **24** 526.
[59] Kurilkin A K *et al.* 2013 *Phys.Rev. C* **87** 051001(R).
[60] Kurilkin P K *et al.* 2011 *Nucl.Instr.Meth. in Phys.Res. A* **642** 45.
[61] Kurilkin P K and Ladygin V P 2014 *Phys.Part.Nucl.* **45** 265.

Journal of Physics: Conference Series **678** (2016) 012032 doi:10.1088/1742-6596/678/1/012032

Simulation of $^3He(d,p)^4He$ reaction and recent results of the experimental investigation of dp breakup reaction

M Janek[1], P K Kurilkin[2], G Tarjanyiova[1], V P Ladygin[2], Yu V Gurchin[2], A Yu Isupov[2], J-T Karachuk[3], A N Khrenov[2], A K Kurilkin[2], A N Livanov[2], G Martinska[4], S M Piyadin[2], S G Reznikov[2], S P Merts[2], P N Batyuk[2], A A Terekhin[2] and I E Vnukov[5]

[1]Physics Department, University of Zilina, Univerzitna 1, 01001 Zilina, Slovakia
[2]Veksler and Baldin Laboratory of High Energies, Joint Institute for Nuclear Research, Joliot-Curie 6, 141980 Dubna, Moscow region, Russia
[3]Advanced Research Institute for Electrical Engineering, Splaiul Unirii 313, Bucharest, Romania
[4]Institute of Physics, University of P. J. Shafarik, Park Angelinum 9, 04001 Kosice, Slovakia
[5]Belgorod State University, Pobedy 85, 308015 Belgorod, Russia

E-mail: janek@fyzika.uniza.sk

Abstract. The feasibility study for the measurements of the polarization observables in the $^3He(d,p)^4He$ reaction for DSS project using a part of the BM@N setup have been performed. Deuteron beam with energy of 1.5 GeV, magnet, 12 stations GEM tracker and TOF wall were used in simulations in order to obtain momentum resolution and to prove the separation of the secondary protons and deuterons. Summarized results of the *dp* breakup reactions with detection of two protons in coincidence for some kinematic configurations at energies from 300 to 500 MeV obtained in Nuclotron are presented.

1. Short introduction

Short range correlations (SRC) of nucleons in nuclei is the subject of intensive theoretical and experimental works during last decade. They can be considered as the drops of cold dense nuclear matter [1] with density five times larger than density in the centre of nucleon. One of the tools to study spin structure of SRC is the measurements of the cross section, tensor analyzing power T_{20} and spin correlation parameter $C_{y,y}$ in the $d^3He \to p(0°)^4He$ reaction [2] at the energies between 1000 and 1750 MeV using polarized ^3He target [3] and extracted polarized deuteron beam from new polarized ion source [4] at Nuclotron.

2. The $d^3He \to p^4He$ reaction at extracted beam at Nuclotron.

One of the goals of the $d^3He \to p^4He$ reaction study at Nuclotron is to understand the reason of the behaviour of the tensor analyzing power T_{20} in *dp*- backward elastic scattering [5]. Other goals are related to the accumulation of the data on polarisation observables, full determination of the matrix element of the reaction in the model independent way and the study of the short-range spin structure of the deuteron and non-nucleonic degrees of freedom. The experiments

Content from this work may be used under the terms of the Creative Commons Attribution 3.0 licence. Any further distribution of this work must maintain attribution to the author(s) and the title of the work, journal citation and DOI.
Published under licence by IOP Publishing Ltd

Journal of Physics: Conference Series **678** (2016) 012032 doi:10.1088/1742-6596/678/1/012032

performed at RIKEN at the energies below 270 MeV have shown that the polarization correlation coefficient, $C_{//} = 1 - 1/(2\sqrt{2})T_{20} + 3/2C_{y,y}$, for the $d^3He \to p^4He$ reaction can be a unique probe to the D-state admixture in deuteron [2]. In the one nucleon exchange approximation, the expression for $C_{//}$ is proportional to the D-state fraction in deuteron as $C_{//} = 9/4{*}w^2/(4u^2{+}w^2)$ where u and w are the S- and D-state wave functions of deuteron in momentum space.

Experiment is planned to be carried out with extracted deuteron beam of Nuclotron at the BM@N area at energy range of 1.0 - 1.75 GeV. The possibility of the $d^3He \to p^4He$ reaction study is investigated using GEANT4 and ROOT simulation package. There are serious requests on beam size and emittance in the target crossover point. In Fig.1 beam size in vertical and horizontal direction at crossover points F3, F4, F5 and F6 is shown. Deuteron beam hits the ^3He

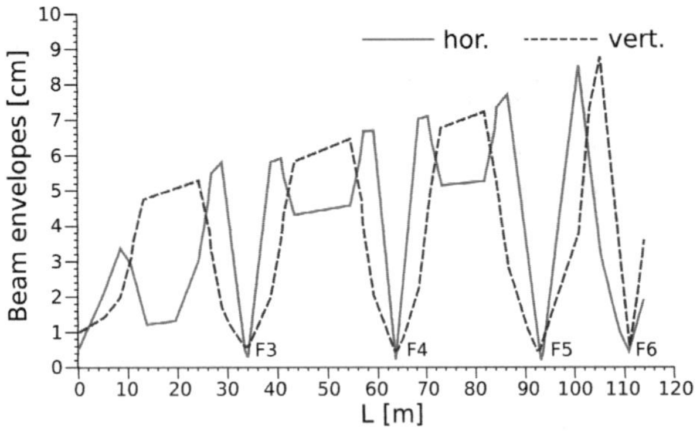

Figure 1. The size of the beam in horizontal and vertical directions along the transportation beam line versus the distance from the target. A calculated option of the beam envelopes for the horizontal and vertical beam emittance 10π mmmr. F3, F4, F5 and F6 are the focuses of the beam line.

target which is placed in the one of the focuses of the beam transportation line of Nuclotron. The length of polarized ^3He target is 30 cm. Scattered protons and deuterons, as well as primary deuterons pass along transportation beam line with magnetic elements tuned for the optimal momentum of the secondary protons $\sim$ 75% of the initial deuteron beam momentum. The simulation has been performed for the initial deuteron kinetic energy T_d = 1.5 GeV and 30 cm of the ^3He target cell using the GEANT4 toolkit. The software model reflects only substantial parts of BM@N setup. In the first variant of simulation ^3He target has only been used while in the second one the quartz radiator of 2 mm thick was added as a TOF start counter. The hadronic interaction of the secondary particles with the target, air and start counter was simulated using GEANT4 with using QBBC physics list with the Binary Cascade model. Quartz radiator spreads the size and the number of particles in the secondary beam due to additional multiply scattering on it in comparison with the first variant, however, the number of these events is quite small for both cases. The further simulation has been performed for the setup with quartz radiator. The momentum of the secondary protons has been taken within 7% around the momentum of proton coming from the binary $d^3He \to p(0°)^4He$ reaction. The momentum reconstruction of particles is based on the information derived from 12 GEM stations placed in the inhomogeneous magnetic field defined by a field map. The simulation has been performed at different field map scale factors: half of the magnetic field value and at maximal one. At this stage the BM@N setup with 12 stations of GEM tracker situated in the magnet can provide the momentum resolution of $\sim 1.5 - 2\%$. The time difference between mRPC and start detector will be used to separate protons from the background deuterons. At base line of $\sim$24 m the signal protons and background deuterons are separated by $\sim$20 ns. Additional improvement of the momentum resolution can be done by removing of the additional material along the beam line as well as by the increasing of the space resolution of the tracking detectors.

Thus, the BM@N setup can be used as for the studies of the short-range spin structure of the deuteron via deuteron fragmentation processes, $dA \to pX$ and $dA \to \pi^- X$, with the momentum resolution provided by the 12 stations GEM tracker [6].

3. Deuteron breakup reaction investigated on Carbon and CH_2 targets.

The experimental data accumulation for a large region of the phase space is requested, because in various region different effects, such as relativistic effects, non nucleonic degrees of freedom, three nucleon forces and also coulomb effects can be investigated. Deuteron breakup on nuclei have rich phase space therefore is interesting to investigate final states. Motivation to study three nucleon forces can be, e.g., observed discrepancy between experimental data and theoretical calculations based on nucleon-nucleon and three nucleon potentials in case of the dp breakup measured at 19 MeV [7] and 130 MeV [8].

The *dp* breakup reaction is investigated at intermediate energies from 300 - 500 MeV of deuteron energy with eight $\Delta E - E$ scintillation detectors [9]. Special setup of the detectors has been chosen in the experiment performed at 2014. Four detectors have been placed at certain position in the vicinity of the target to follow inverse kinematics of the reaction in which relativistic effects can play a role [10]. Polyethylene and Carbon targets enclosed in a spherical hull made from stainless steel are used to obtain *dp* reaction via subtraction of Carbon spectra from Polyethylene. Mean axis of the detector pass through the target and with incoming deuterons create the angle θ. Angle between detectors in plane perpendicular to the direction of the deuteron beam is denoted by ϕ_{12} and it is fixed to 180°. Solid angle of each of detector is 4.6°. Angles ϕ and θ of the first and second detector specify a detector configuration.

Calibration coefficients (function) have been obtained by calibration procedure which used amplitude and time difference information from PMT's and known values of proton energies of *pp*-quasi elastic reaction. Cross check of obtained calibration coefficients has been done in *dp* quasi elastic reaction configuration where except deuterons also protons from *dp* breakup exist. In Fig. 2 distribution of excitation energy for the case of dCH_2 and dC reactions when deuterons from *dp* quasi elastic reaction reach the detector at mean angle of 27°. Second arm is at angle of 43°. Azimuthal difference between arms is 180° Clear peak in dCH_2 spectra around zero value is caused by *dp* elastic events. The result confirms also obtained missing mass spectra if deuteron and proton is assumed to reach detector arms.

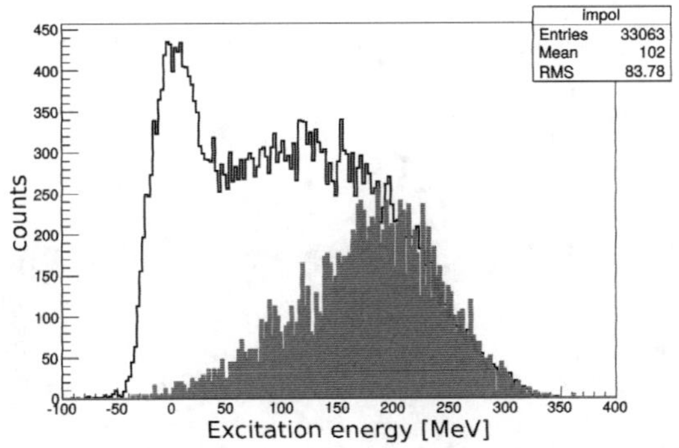

Figure 2. Distribution of excitation energy for the case of dCH_2 (non shaded spectra) and dC (shaded spectra) reactions when deuterons from *dp* quasi elastic reaction reach the detector at mean angle of 27°. Second arm is at angle of 43°. Azimuthal difference between arms is 180°.

The kinematical variable S corresponds to the arc-length along the kinematical variable S, with zero point chosen at the minimal value of deposited energy in the second arm. The number of breakup events in an interval $S - \Delta S/2$, and $S + \Delta S/2$ was obtained by projecting the events on a line perpendicular to the S-curve. The statistical error associated with the background subtraction is roughly given by $\Delta N_{dp} = \sqrt{N_{CH_2} + \Lambda^2 N_C}$, where the Λ is normalization coefficient of Carbon target. Preliminary result for the $\Delta S = 10$ MeV and for the arm's polar angles of $\theta_1 = 25°$, $\theta_2 = 34°$ and azimuthal angle between them $\phi_{12} = 135°$ at 300 MeV is presented in Fig. 3.

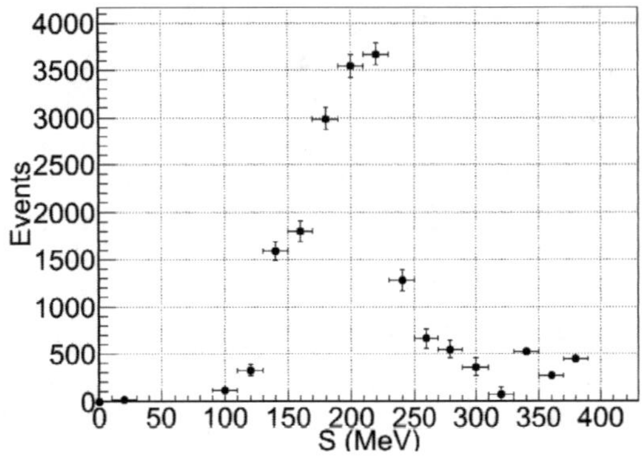

Figure 3. Preliminary result of S variable (MeV) is obtained after background subtraction at 300 MeV for the arm's polar angles of $\theta_1 = 25°$, $\theta_2 = 34°$ and azimuthal angle between them $\phi_{12} = 135°$.

4. Conclusions

The feasibility study of the measurements of the polarization observables in the $d^3He \rightarrow p^4He$ reaction at 1500 MeV by means of Geant4 simulation with using a part of the BM@N setup and start counter of the TOF system ~20 m upstream the analyzing magnet has been performed. Perfect separation of the secondary protons and deuterons is demonstrated. Partial improvement of the deuteron beam emittance can be done by removing unnecessary equipment along the transportation line. Mentioned experiment can be performed as a part of the commissioning of the BM@N setup in the complete configuration. Momentum resolution at current stage is $\sim 2\%$ due to software tracking system recognition. Feasibility study shows that BM@N setup can be used for the studies of the short-range spin structure of the deuteron via deuteron fragmentation processes with the momentum resolution provided by the 12 stations GEM tracker.

The Geant4 simulation has been performed to give insight in to dp breakup reaction via $dCH_2 \rightarrow ppX$ and $dC \rightarrow ppX$ reaction at 300, 400 and 500 MeV mainly for detector configurations used in experiment. Obtained energy and missing mass spectra have been compared with experimental one. Excitation energy for the case of deuteron proton quasi elastic scattering at detector configuration which satisfy dp quasi elastic reaction is obtained as a cross check for obtained calibration coefficients (function). Preliminary result of S variable is obtained after background subtraction at 300 MeV for the arm's polar angles of $\theta_1 = 25°$, $\theta_2 = 34°$ and azimuthal angle between them $\phi_{12} = 135°$ is presented.

Acknowledgments

The work has been supported in part by the VEGA 1/0556/16 and RFBR grant 13-02-00101a.

References

[1] L. Frankfurt, M. Sargsian, M. Strikman 2008 *Int.J.Mod.Phys.* **A23** 2991
[2] T. Uesaka et al. 1999 *Phys.Lett.* **B467** 199
[3] T. Uesaka et al. 1998 *Nucl.Instr. and Meth. in Phys.Res.* **A402** 212
[4] V.V. Fimushkin et al. 2008 *Eur.Phys.J. ST* **162** 275
[5] V. Punjabi et al. 1995 *Phys.Lett.* **B350** 178
[6] V.P. Ladygin et al. 2014 *PoS BaldinISHEPPXXII* 098
[7] J. Ley et al. 2006 *Phys. Rev. C* **73** 064001
[8] S. Kistryn et al. 2003 *Phys. Rev. C* **68** 054004
[9] S. M. Piyadin et al. 2011 *Physics of Particles and Nuclei Letters* **2** V.8 107
[10] H. Witala, J. Golak, R. Skibinski, W. Glockle, W. N. Polyzou and H. Kamada 2011 *Few-Body Syst.* **49** 61
[11] M. Janek, B. Trpisova, S. M. Piyadin and V. P. Ladygin 2014 *Physics of Particles and Nuclei Letters* **4** V.11 552559

Journal of Physics: Conference Series **678** (2016) 012033 doi:10.1088/1742-6596/678/1/012033

The impact of vector mesons polarization on meson-nucleon interaction.

Sergey Gevorkyan
LHEP, Joint Institute for Nuclear Research
E-mail: gevs@jinr.ru

Abstract. Production of unstable particles off nuclei allows to determine the total cross section of the unstable particle interaction with nucleons. The interaction of vector mesons $V(\rho,\omega,\phi)$ with nucleons are defined by a set of amplitudes corresponding to the transverse polarization of the vector meson (helicity $\lambda=\pm 1$) meson or longitudinal one ($\lambda=0$) . Whereas the total cross section for the interaction of the transversely polarized vector meson with nucleon $\sigma_T=\sigma(V_TN)$ can be extract from the coherent photoproduction, measurements of vector meson photoproduction in the incoherent region provide a unique opportunity to obtain the not-yet-measured total cross section for longitudinally polarized vector meson interaction with nucleon $\sigma_L=\sigma(V_LN)$.
As a good tool for such investigation we proposed the photoproduction of ω mesons on the set of nuclei in the incoherent region. Such experiment can be done using the feasibility of the new experiment GlueX at Jefferson Lab, designed to study the photoproduction of mesons in a large beam energy range up to 12 GeV.

1. Introduction

In the late 60's and early 70's many experiments on vector mesons $V(\rho,\omega,\phi)$ photoproduction off nuclei have been done [1] at SLAC, DESY, Cornell etc., allowed to check the predictions of vector dominance model and quark model, which for instance predict $\sigma(\rho N)=\sigma(\omega N)=\frac{\sigma(\pi^+N)+\sigma(\pi^-N)}{2}$. At that time the challenge of impact of vector meson polarization on its interaction with nucleons has not been considered as the naive quark model predicts that $\sigma_T(VN)=\sigma_L(VN)$. Moreover in the coherent photoproduction which is huge and has clear and unique theoretical predictions only tranverse vector mesons can be produced. As to the incoherent region to extract $\sigma_L(VN)$ one has to pick out the process, where the bulk of longitudinally vector mesons produced, which is not the case for ρ,φ photoproduction where s-channel helicity conservation takes place at moderate momenta transfer.
The first indications that interaction of vector meson with nucleon depends on its polarization come from ρ and φ electroproduction off protons [1]. The ratio of the production cross section can be represented as $R=\frac{\sigma(\gamma_Lp\to V_Lp)}{\sigma(\gamma_Tp\to V_Tp)}=\xi^2\frac{Q^2}{m_\rho^2}$, where the parameter ξ^2 corresponds to the ratio of longitudinal to transverse ρ^0 nucleon total cross sections $\xi^2=\frac{\sigma_L(\rho p)}{\sigma_T(\rho p)}$. As one can see from Fig.1a this ratio is $\xi\approx 0.5$, whereas for φ meson: $\xi^2=0.33\pm 0.08$

Content from this work may be used under the terms of the Creative Commons Attribution 3.0 licence. Any further distribution of this work must maintain attribution to the author(s) and the title of the work, journal citation and DOI.
Published under licence by IOP Publishing Ltd

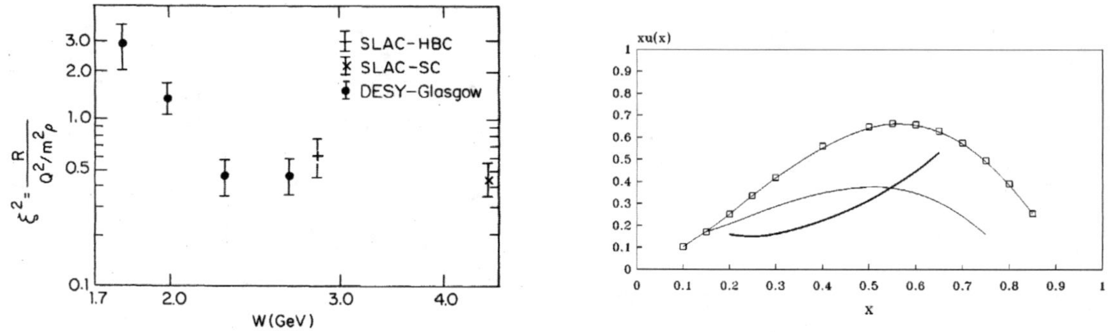

Figure 1. a) The ratio of longitudinal to transverse cross sections as a function of total cms energy; b) The valence quarks distribution in the ρ meson as a fraction of the light-cone momentum: □- the longitudinally polarized meson, the transversely polarized meson (thick line) , the pion (thin line).

At present it is known that valence quarks distribution in the transversely and longitudinally polarized vector mesons are significantly different [2] (see Fig.1b). The same distinction takes place [3] for the distribution of constituent quarks in vector mesons as depicted on the Fig.2. These differencies lead to impact of vector mesons polarization on their interaction with matter [4].

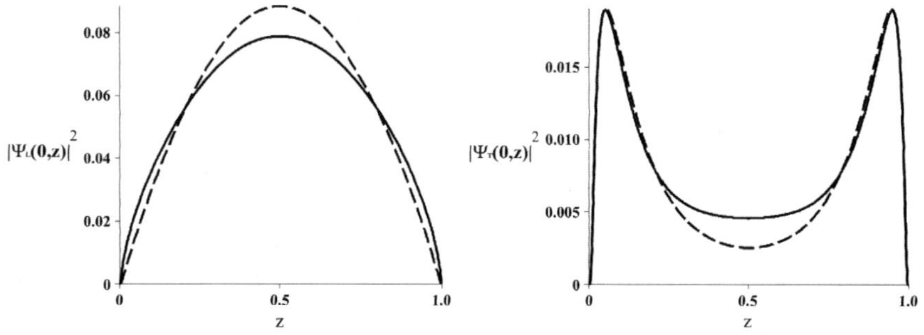

Figure 2. The longitudinal (left) and transverse (right) ρ meson light-cone wavefunctions squared at r=0 as a function of light-cone momentum x.

The dependence of the vector particle interaction on its polarization has been known for many years in the case when the constituents of the particle are in the D-wave state. A good example of such effect is the deuteron interaction with matter [5]. The D-wave component in the deuteron wave function leads to different absorption for transversely and longitudinally polarized deuterons. Recently this effect was experimentally measured [6]. There are also predictions that the interaction of mesons with nonzero orbital momentum with nucleons is strongly correlated with the meson polarization [7, 8]. On the other hand for ground-state S-wave vector mesons as ρ, ω, φ the D-wave component in their wave functions can emerge as a result of the Lorentz transformation [9].
As a good tool to obtain the information on the σ_L we propose the measurement of the ω mesons incoherent photoproduction on nuclei [10]. The deal is that in *omega* photoproduction

Journal of Physics: Conference Series **678** (2016) 012033 doi:10.1088/1742-6596/678/1/012033

at photon energies $4GeV \leq E_\gamma \leq 10GeV$ due to pion exchange appreciable amount of longitudinally polarized $\omega's$ can be produced. While in coherent photoproduction only transversely polarized mesons can be produced (the pion exchange leads to different signs of the photoproduction amplitudes on protons and neutrons which lead to reduction of pion exchange), the photoproduction in the incoherent region allows one to extract the σ_L from their absorption in the matter.
The photoproduction of ω mesons with different polarizations on nuclei in the reaction

$$\gamma + A \to \omega + A' \tag{1}$$

where A' denotes the excitation of the target nucleus or its break-up products. Generalizing the Glauber multiple scattering theory for potentially different absorptions of transversely and longitudinally polarized mesons [10] the incoherent cross section reads:

$$\begin{aligned} \frac{d\sigma_A(q)}{dt} &= \frac{d\sigma_0(q)}{dt}(\rho_{00}N(0,\sigma_L) + (1-\rho_{00})N(0,\sigma_T)) \\ N(0,\sigma) &= \int \frac{1-\exp(-\sigma\int\rho(b,z)dz)}{\sigma}d^2b, \end{aligned} \tag{2}$$

where $d\sigma_0(q)/dt$ is the differential cross section of the ω meson photoproduction on nucleon, $\sigma_{T,L}$ is the total ω-nucleon cross section for longitudinally and transversely polarized mesons, and ρ_{00} is the ω meson spin density matrix element defining the fraction of longitudinally polarized ω mesons. If $\sigma_T = \sigma_L$ the nuclear transparency has the well known form $A_{\text{eff}} = \frac{d\sigma_A}{dt}/\frac{d\sigma_0(q)}{dt} = N(0,\sigma)$. Spin density matrix elements for photoproduction on nuclei ρ^A_{00} and nucleons ρ_{00} are related as:

$$\rho^A_{00} = \frac{N(0,\sigma_L)}{\rho_{00}N(0,\sigma_L) + (1-\rho_{00})N(0,\sigma_T)}\rho_{00} \tag{3}$$

In the case of $\sigma_T = \sigma_L$ the spin density matrix elements on nuclei and nucleon coincides $\rho^A_{00} = \rho_{00}$. For $\sigma_T \neq \sigma_L$ ρ^A_{00} should depend on the nuclear mass number A.
On Fig.3 the A-dependence of nuclear transparency A_{eff} and spin density matrix element ρ^A_{00} at the value of $\sigma_T = 26mb$ and two values of σ_L are shown. The calculations have been done in the framework of more elaborated version of multiple scattering theory than considered above which leads to the energy dependence of nuclear transparency and ω meson spin density matrix elements on nuclei (for details see [10]).

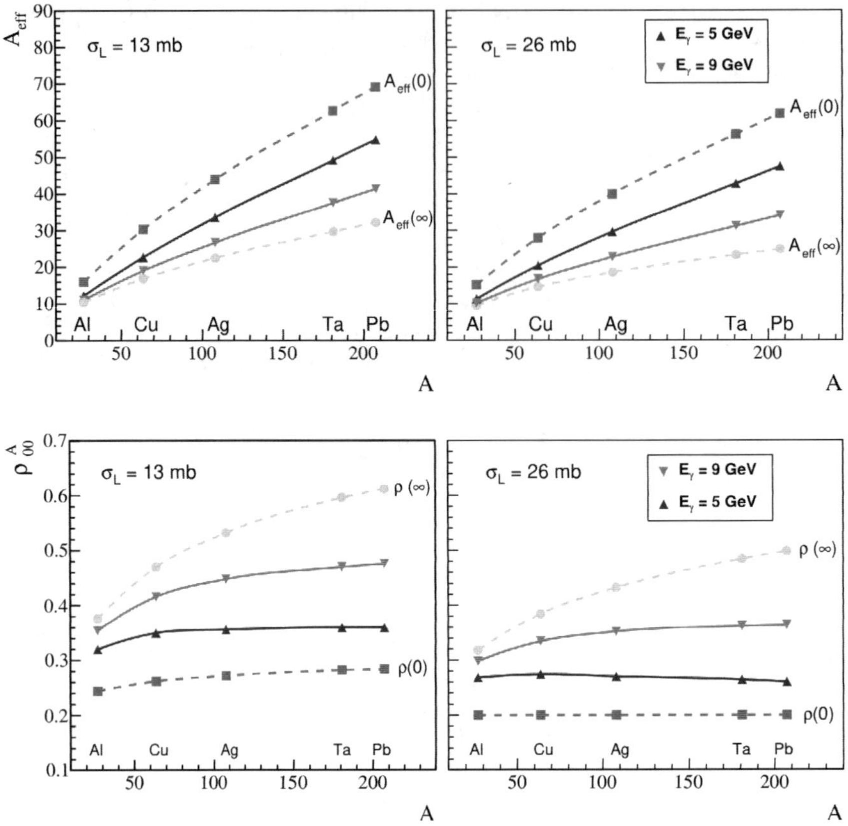

Figure 3. The A-dependence of the nuclear transparency and spin density matrix element.

Thus the measurement of the photoproduction cross section and omega meson spin density matrix elements in the incoherent region ($|t| \geq 0.1$ GeV) will allow to determine the total cross section of longitudinally polarized vector mesons with nucleons $\sigma_L(\omega N)$ and thus shed light on the exciting challenge of impact of vector meson polarization on strong interactions.
Let us note that neither the absorption of longitudinally polarized mesons, nor the spin density matrix elements on nuclei in the photoproduction have been measured so far.

Acknowledgments
I am grateful to my coauthors E. Chudakov and A. Somov and appreciates the hospitality of Jefferson Lab where this work has been done.

[1] Bauer T , Spital R ,Yennie D and Pipkin F 1978 *Rev.of Mod.Phys.* **50** 261
[2] Ioffe B and Oganesian A 2001 *Phys. Rev.* D **63** 096006
[3] Forshaw J and Sandapen R 2010 *JHEP* **11** 037
[4] Gevorkyan S in preparation
[5] Franco V. and Glauber R 1969 *Phys. Rev. Lett.***22** 370
[6] Azhgirey L et al. 2010 *Part. Nucl. Letters* **7** 49
[7] Gerland L et al. 1998 *Phys. Rev. Lett.***81** 762
[8] Huefner J Kopeliovich B and Tarasov A 2000 *Phys. Rev.* D **62** 094022
[9] Jaus W 1991 *Phys. Rev.* D **44** 2851
[10] Chudakov E Gevorkyan S and Somov A 2015 arXiv:hep-ph 1508.00422v2;
A Letter of Intend to Jefferson Lab *PAC-43*

Journal of Physics: Conference Series **678** (2016) 012034 doi:10.1088/1742-6596/678/1/012034

Elastic scattering polarimeter for a polarized antiproton beam at U-70 accelerator of IHEP

A A Bogdanov[1†], M A Chetvertkov[2], V A Chetvertkova[2], V I Garkusha[3], A P Meschanin[3], V V Mochalov[1,3], M B Nurusheva[2], S B Nurushev[1,3], A V Ridiger[2], V L Rykov[1], P A Semenov[1,3], M N Strikhanov[1], A N Vasiliev[1,3] and V N Zapolsky[3]

[1] National Research Nuclear University (Moscow Engineering Physics Institute), Moscow, 115409, Russia
[2] Federal state budget organization State center Interphyzika, Ministry of education and science, Moscow, 115409, Russia
[3] Institute for High Energy Physics National Research Centre Kurchatov Institute, Protvino, 142284, Russia
[†] Corresponding author, E-mail: bogdanov411@mail.ru

Abstract. The absolute polarimeter based on the elastic $\bar{p}p$-scattering in the diffraction kinematic regions with the total momentum transfer squared coverage of $0.1 < -t < 0.3$ (GeV/c)^2 is proposed for the polarized antiproton beam at the U-70 proton synchrotron of IHEP. It is shown that it would take ~200-400 hours for measuring the beam polarization at the statistical errors of $\Delta P_B/P_B \simeq$10–15%. These time estimates include also the time which is necessary for the measurements of an analyzing power A_N, using a polarized target. Besides the measurements of beam polarizations, the proposed polarimeter provides an opportunity for carrying out the experimental studies of the small momentum transfers physics which would be a valuable enrichment of the SPASCHARM experiment capabilities and its physics program.

1. Introduction

The polarized proton and antiproton beam-channel 24A for the SPASCHARM experiment [1] is to be built at the 70 GeV accelerator of IHEP, Protvino, Russia. The method of forming these polarized beams and their characteristics have been described in details elsewhere [1–3]. The beam protons and antiprotons originate from the parity-violating decays of Λ-hyperons. The transversely polarized portions of the total beam are selected, using the particle's trajectory tagging or collimation. An independent beam polarimetry is an important part of the beam-channel setup which serves the purposes of verification and confirmation of operation of the beam forming system. For the case of protons, it has been discussed in a number of publications and presentations[1].

The polarimetry for an antiproton beam has two additional complications. The intensity of antiprotons will be much lower compared to the proton beam. Moreover, the currently available data on spin analyzing power for antiprotons are very scarce and of a noticeably worse experimental accuracy than for protons. For the momentum transfers of interest here, we are aware of only two publications on measurements of analyzing power A_N in elastic $\bar{p}p$-scattering:

[1] See, for example, Refs. [3–6].

Content from this work may be used under the terms of the Creative Commons Attribution 3.0 licence. Any further distribution of this work must maintain attribution to the author(s) and the title of the work, journal citation and DOI.
Published under licence by IOP Publishing Ltd

Ref. [7] is devoted to the measurements at low momenta of 0.1–0.2 GeV/c, and Ref. [8] describes the experiment at the large beam momentum of 40 GeV/c. The similar A_N measurements have been carried out at 45 GeV/c in the experiment [9], but only for proton-proton elastic scattering. The bottom line from the written above is that, an antiproton polarimeter at SPASCHARM must be capable to measure first an analyzing power for the process of antiproton scattering, using a polarized target, in order to use then the results for measuring an antiproton beam polarization.

A number of physics processes have been considered for the polarimetry at SPASCHARM experiment. Among them are elastic pp- and $\bar{p}p$-scatterings in Coulomb-Nuclear Interference [3, 10] and diffraction [3] kinematic regions, polarimetry with inclusive neutral [4] and charged [6] pions as well as with inclusive photons [5]. In this report, we discuss the antiproton polarimetry based on $\bar{p}p$ elastic scattering in diffraction kinematic region.

2. Elastic scattering polarimeter in the diffraction kinematic region

The layout of the proposed SPASCHARM diffraction elastic scattering polarimeter is shown in figure 1. It consists of the beam counters and hodoscopes upstream the target, of two forward arms for detection of forward-scattered antiprotons as well as two recoil arms for detection of recoil protons. The detector acceptance for elastic pp- and $\bar{p}p$-events in momentum transfer squared is $0.1 < -t < 0.3$ (GeV/c)^2 for the beam momenta from ~10 to 45 GeV/c. The acceptance in azimuth is limited by the combined acceptance $\Delta\phi$ of two recoil arms which is equal to $2 \times 0.05 \times 2\pi = 0.1 \times 2\pi$ rad. The trigger for pp and $\bar{p}p$ elastic scattering would be a coincidence of signals from a forward and recoil arm.

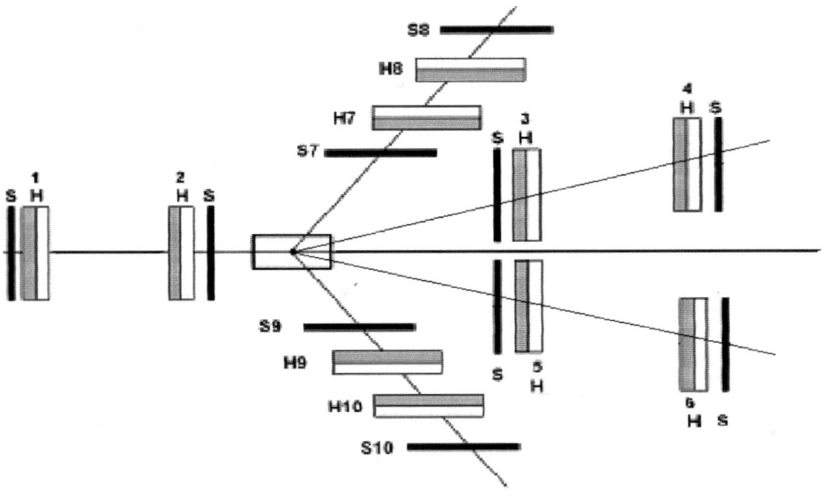

Figure 1. The layout of the SPASCHARM diffraction elastic scattering polarimeter. $S_1 - S_{10}$ are the scintillator counters for measuring total particle fluxes; H_1 and H_2 are the beam hodoscopes; $H_3 - H_6$ are the hodoscopes of the forward arms, and $H_7 - H_{10}$ are the hodoscopes of the recoil arms.

In order to make estimates of necessary beam exposition times for antiproton polarimetry, the knowledge of the cross section σ and analyzing power A_N is required. Unfortunately, as it has been already mentioned above, the reliable experimental data for $\bar{p}p$-scattering are not currently available either on σ, or A_N for the energies and in the kinematic region of interest.

Journal of Physics: Conference Series **678** (2016) 012034 doi:10.1088/1742-6596/678/1/012034

However, according to Ref. [11], the $\bar{p}p$ differential cross section is close to the one in pp-scattering. Using the results of measurements [11, 12], the elastic pp cross section integrated over the interval $0.1 < -t < 0.3$ $(\mathrm{GeV/c})^2$ could be estimated as $\sigma \approx 2.1$ mb. For the purpose of the considerations below, we will also assume the analyzing power in $\bar{p}p$ to be equal to the one in pp collisions. For pp, as it has been measured in Refs. [9, 13], the averaged over the t-interval above $A_N \approx 2.3\%$. We also assume here the intensity of the both, transversely polarized[2] and unpolarized antiproton beams $I \simeq 6 \times 10^4$ antiprotons/cycle of U-70 accelerator with the cycle repetition frequency of 0.11 Hz.

3. Measurement of analyzing power A_N in $\bar{p}p$-collisions with polarized target

The measurements of analyzing power A_N will be carried out with an unpolarized antiproton beam striking a 'frozen' propan-diol ($C_3H_8O_2$) target. The clean samples of elastic $\bar{p}p$ events will be separated from $\bar{p}C$ and $\bar{p}O$ events by reconstruction of the collision kinematics for the forward and recoil charged particles. The parameters of a polarized target are taken as follows: the length $l = 20$ cm; the density of hydrogen nuclei in the target $\rho = 0.12$ g/cm^3 (that corresponds to the proton density $n_p = 0.12 \times 6 \cdot 10^{23} = 7.2 \cdot 10^{22}$ protons/cm^3); the proton polarization in the target $P_T \simeq 70\%$.

We require the relative statistical error for the measurements of analyzing power A_N to be: $\Delta A_N/A_N = \Delta\epsilon/\epsilon = 0.1$ where ϵ is the raw asymmetry as it is measured in the polarimeter:

$$\epsilon = A_N \times P_T = 0.023 \times 0.7 \simeq 0.0161 \tag{1}$$

Thus, the required statistical error in raw asymmetry is:

$$\Delta\epsilon = 0.1 \times 0.0161 \simeq 0.00161 = 1/\sqrt{N_{req}} \tag{2}$$

where $N_{req} \simeq 4 \times 10^5$ is the required number of events to accumulate in order to achieve the relative error $\Delta\epsilon/\epsilon = \Delta A_N/A_N = 10\%$. An expected number of events per U-70 cycle in the acceptance of $\Delta\phi/2\pi = 2 \times 0.05 = 0.1$ and $0.1 < -t < 0.3$ $(\mathrm{GeV/c})^2$ could be estimated as:

$$N_{cycle} \simeq I \times \sigma \times n_p \times l \times \frac{\Delta\phi}{2\pi} \simeq 6 \cdot 10^4 \times 2.1 \cdot 10^{-27} \times 7.2 \cdot 10^{22} \times 20 \times 0.1 \simeq 18 \text{ events/cycle} \tag{3}$$

The required number of events can be accumulated for $N_{req}/N_{cycle} = 4 \cdot 10^5/18 = 2.2 \cdot 10^4$ cycles. At the cycle repetition frequency of 0.11 Hz, this corresponds to $\sim 2 \cdot 10^5$ seconds ≈ 54 hours.

4. Time estimates for measuring an antiproton beam polarization, using the known analyzing power A_N

For these estimates, we assume the transverse beam polarization $P_B \simeq 40\%$ [2]. The cross section, the analyzing power A_N and the statistical error requirement $\Delta P_B/P_B = \Delta\epsilon/\epsilon = 0.1$ are assumed to be same as in Sec. 3.

The beam polarization measurements can be carried out as with a propan-diol target[3] as well as with a liquid hydrogen target.

For the case of a propan-diol target, the only difference to the estimates of Sec. 3 is the beam polarization of 40% which is lower than the assumed above target polarization of 70%. Due to this difference, the raw asymmetry ϵ will be lower by a factor of 70/40=1.75. As a result, for the antiproton beam polarization measurements with the requirements and at the assumptions above, it would take $54 \times 1.75^2 \simeq 165$ hours of data taking time.

[2] One sign of polarization.

[3] Unpolarized though.

In the case of a liquid-hydrogen target of the same length 20 cm, the only difference from the propan-diol is the lower hydrogen density: 0.071 g/cm^3. With this modification, the event rate would be lower by a factor of 0.12/0.071=1.7 which would lead to prolongation of the required measurement time to 165×1.7=280 hours.

5. Conclusion

For the purpose of verification and confirmation of operation of the polarized antiproton beam forming system in the channel 24A at U-70 of IHEP, we propose to build the absolute polarimeter based on elastic $\bar{p}p$-scattering in the diffraction kinematic regions of $0.1 < -t < 0.3$ (GeV/c)^2. It is shown that the measurements of analyzing power A_N, using the polarized target, at the relative statistical errors of $\Delta A_N/A_N \simeq$10% could be completed for ∼50–60 hours. Then, it would take additional ∼150–300 hours for measuring an antiproton beam polarization at the same relative statistical errors of $\Delta P_B/P_B \simeq$10%.

It worth mentioning that the discussed here beam polarimeter can also be used for the experimental studies of the small momentum transfers physics which would be a valuable enrichment of the SPASCHARM experiment capabilities and its physics program.

Acknowledgments

This work has been supported in part by the RFBR Grant No.15-02-1669.

References

[1] Mochalov V V 2013 *Phys. Part. Nucl.* **44** No 6 930; Mochalov V V et al 2014 Systematic Study of Spin Effects at the SPASCHARM Experiment at 70-GeV Accelerator in Protvino *To be published in Int. J. Mod. Phys.: Conf. Series* (Proc. of 21st Int. Symp. on Spin Phys. "Spin2014", Beijing, China, 2014)
[2] Nurushev S B et al 2015 Polarized antiproton beam at U-70 accelerator of IHEP *To be published in J. Phys.: Conf. Series* (these proceedings)
[3] Semenov P A et al 2014 Polarimeters for the SPASCHARM experiment *To be published in Int. J. Mod. Phys.: Conf. Series* (Proc. of "Spin2014")
[4] Bogdanov A A et al 1998 The inclusive neutral pion polarimeter for high energy accelerator *Preprint IHEP 98-54*
[5] Bogdanov A A et al 2000 The local inclusive photon polarimeter for RHIC *Preprint IHEP 2000-42*
[6] Rykov V L et al 2015 Polarimetry with inclusive charged pions at U-70 accelerator of IHEP *To be published in J. Phys.: Conf. Series* (these proceedings)
[7] Klemp E, Bradamante F, Martin A and Richard J-M 2012 *Phys.Rep.* **368** 119
[8] Bruneton C et al 1975 *Phys. Lett.* B **57** 389
[9] Gaidot A et al 1976 *Phys. Lett.* B **61** 103
[10] Grosnick D P et al 1990 *Nucl. Inst. Meth.* A **290** 269
[11] Antipov A et al 1976 Elastic scattering of $\pi^\pm$, $K^\pm$ mesons and protons on protons at momenta 29-65 GeV/c *Preprint IHEP 76-95*
[12] Bruneton C 1977 *Nucl. Phys.* B **124** 391
[13] Corcoran M et al 1980 *Phys. Rev.* D **22** 2624; Erratum: 1981 *Phys. Rev.* D **24** 3010

XVI Workshop on High Energy Spin Physics (D-SPIN2015) IOP Publishing
Journal of Physics: Conference Series **678** (2016) 012035 doi:10.1088/1742-6596/678/1/012035

Role of pion pole in hard exlusive meson leptoproduction

Sergey Goloskokov

Bogoliubov Laboratory of Theoretical Physics, Joint Institute for Nuclear Research, Dubna 141980, Moscow region, Russia

E-mail: goloskkv@theor.jinr.ru

Abstract. We consider the pion pole contribution and transversity effects determined by the H_T and $\bar{E}_T$ Generalized Parton Distributions (GPDs) which are essential in hard pseudoscalar and vector meson leptoproduction. We investigate spin effects in the ω and ρ^0 reactions. It is shown that the pion pole contribution is very important in the ω production. Such effects in the ρ^0 channel are much smaller. Our results on spin asymmetries and spin density matrix elements in these reactions were found to be in good agreement with HERMES data.

1. Introduction

In this report, we discuss pion pole effects [1] and transversity twist-3 contributions from the H_T, $\bar{E}_T$ GPDs [2] which are important in explanation of spin effects in the light meson production at moderate Q^2. Our model of hard electroproduction of light mesons [3] is based on the handbag factorization at high photon virtuality Q^2 of the process amplitude into the hard meson electroproduction off partons, and GPDs [4].

The amplitudes of the pseudoscalar meson (PM) leptoproduction in the leading twist approximation are sensitive to GPDs $\widetilde{H}$ and $\widetilde{E}$. These contributions were found to be not sufficient to describe spin effects in the PM production at sufficiently low Q^2 [2]. To be consistent with experiment, one needs the essential contributions from the transversity GPDs H_T, $\bar{E}_T$ which go together with a twist-3 pion wave function [5]. We discuss the pion pole and transversity effects in the PM leptoproduction at HERMES and CLAS energies in section 2. Our results are in good agreement with experimental data.

The HERMES data on SDMEs for the ω production indicated essential contributions from unnatural parity exchanges [6]. It was found that the pion pole (PP) contribution [1] is significant in explanation of the large unnatural-parity effects observed by HERMES. The PP contribution to the ρ^0 production is much smaller with respect to the ω case. We discuss the PP effects in ω and ρ production in section 3.

2. Handbag approach. Pseudoscalar meson leptoproduction.

Within the handbag approach the leading contributions to the meson production amplitude off proton at sufficiently high photon virtuality Q^2 can be described in factorized form [4] as a convolution of a hard meson subprocess amplitude off partons with the same helicities $\mathcal{H}^a_{\mu'+,\mu+}$

Content from this work may be used under the terms of the Creative Commons Attribution 3.0 licence. Any further distribution of this work must maintain attribution to the author(s) and the title of the work, journal citation and DOI.
Published under licence by IOP Publishing Ltd

Journal of Physics: Conference Series **678** (2016) 012035 doi:10.1088/1742-6596/678/1/012035

and GPDs as

$$M_{\mu'+,\mu+} \propto \int_{-1}^{1} dx \mathcal{H}^a_{\mu'+,\mu+} F^a(x,\xi,t); \; M_{\mu'-,\mu+} \propto \frac{\sqrt{-t}}{2m} \int_{-1}^{1} d\overline{x}\, \mathcal{H}^a_{\mu'+,\mu+} E^a(\overline{x},\xi,t). \tag{1}$$

Here a is a flavor factor and μ and μ' are helicities of the photon and produced meson.

GPDs contain information on the hadron structure. With the help of sum rules they are connected with hadron form factors, and information on the parton angular momenta can be extracted. In the forward limit $t = 0$ and zero skewness $\xi = 0$ GPDs are equivalent to ordinary Parton Distribution Functions (PDFs). We estimate GPDs by using the double distribution representation [7] which connects GPDs with PDFs. The PDFs parameterizations are obtained, e.g, from the analysis [8], or from the nucleon form factor study [9].

The handbag approach was successfully applied to light meson leptoproduction [3]. The cross sections and spin observables of light vector meson (VM) leptoproduction were found to be in good agreement with HERMES, COMPASS and HERA data.

The leading twist contributions are not sufficient to describe the experimental results on PM electroproduction at low Q^2. This can be found from analysis of $A_{UT}^{\sin(\phi_s)}$ asymmetry which in the leading twist handbag approximation is small. However, HERMES found [10] that this asymmetry is large, about 0.5. This effect can be explained by the large contributions to the $A_{UT}^{\sin(\phi_s)}$ asymmetry from the amplitude $M_{0-,++}$. At low Q^2 the amplitudes $M_{0\pm,++}$ are determined by the transversity GPDs H_T and $\bar{E}_T$ contributions which have the twist-3 nature. Within the handbag approach the transversity GPDs are accompanied by a twist-3 meson wave function in the hard subprocess amplitude $\mathcal{H}$ [5] which is the same for both the $M^{tw-3}_{0\pm,++}$ amplitudes

$$M^{tw-3}_{0-,++} \propto \int_{-1}^{1} d\overline{x} \mathcal{H}_{0-,++}(\overline{x},...)\, H_T; \; M^{tw-3}_{0+,++} \propto \frac{\sqrt{-t'}}{4m} \int_{-1}^{1} d\overline{x} \mathcal{H}_{0-,++}(\overline{x},...)\, \bar{E}_T. \tag{2}$$

The H_T GPDs in the forward limit and $\xi = 0$ are equal to transversity PDFs δ and are parameterized by using the model [11]. The double distribution is used to calculate GPDs as before. For details, see [5].

Information on $\bar{E}_T$ is obtained now only from the lattice QCD [12]. The lower moments of $\bar{E}^u_T$ and $\bar{E}^d_T$ were found to be quite large, have the same sign and a similar size. At the same time H^u_T and H^d_T GPDs have a different sign. These properties of GPDs provide an essential compensation of the $\bar{E}_T$ contribution in the π^+ amplitude, but H_T effects are not small there. For the π^0 production we have large $\bar{E}_T$ contributions and smaller H_T effects.

We present here our results on the PM leptoproduction based on the handbag approach. In calculation, we use the leading contributions together with the transversity effects (2) which are essential at low Q^2.

In Fig.1, we show our results for the π^+ production which are in agreement with HERMES [13]. The dashed line indicates the model results for $H_T = 0$. It is closed to the PP contribution to the π^+ cross section. So in this channel the PP effects are very essential. The pion induced Drell-Yan process $\pi^- p \to \gamma^* n \to l^+ l^- n$ which is a time-like analog of π^+ leptoproduction was investigated in [15]. It was found that the pion pole gives the predominant contribution in the longitudinal cross section of this process. Other effects are rather small in σ_L.

In Fig. 2, we present the model results for the π^0 production cross section [5]. Here the PP contribution is absent but transversity effects are essential. At small momentum transfer the H_T contribution is visible and provides a nonzero cross section. At $-t' \sim 0.2\text{GeV}^2$ the $\bar{E}_T$ contribution becomes predominant and gives a maximum in the cross section. A similar contribution from $\bar{E}_T$ is observed in the interference cross section σ_{TT}. Note that the $\bar{E}_T$ effects in the unseparated cross section σ which is saturated by σ_T is strongly correlated with similar effects in σ_{TT}. The fact that we describe well both the unseparated σ and σ_{TT} cross sections can indicate that transversity effects were probably observed in CLAS [14].

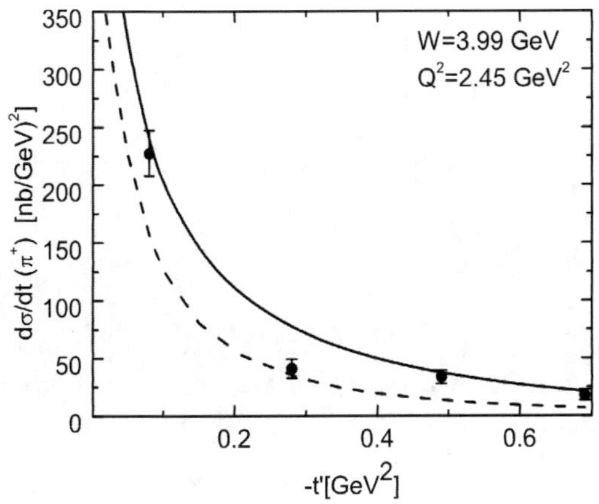

Figure 1. Cross section of π^+ production at HERMES energies with the data [13]. Full line- unseparated cross section. Dashed line- model results for $H_T = 0$.

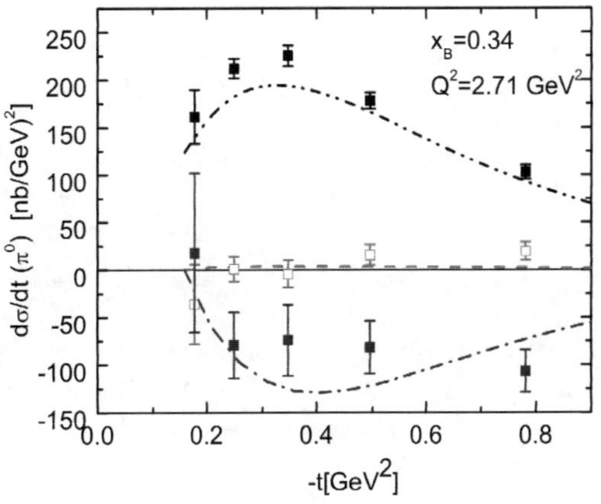

Figure 2. π^0 production in the CLAS energy range together with the data [14]. Dashed-dot-dotted line- $\sigma = \sigma_T + \epsilon\sigma_L$, dashed line-$\sigma_{LT}$, dashed-dotted line- σ_{TT}.

3. Large unnatural parity effects in ω production. Pion pole contribution.

In most reactions the unnatural parity (UP) contributions are small with respect to the natural ones. The HERMES data on the spin density matrix elements (SDMEs) for the ω production indicate the strong contributions from UP effects [6]. It was found that the ratio of the unnatural and natural parity cross section, which was expected to be small, is larger than unity. This can be caused by the large pion pole contribution to this process.

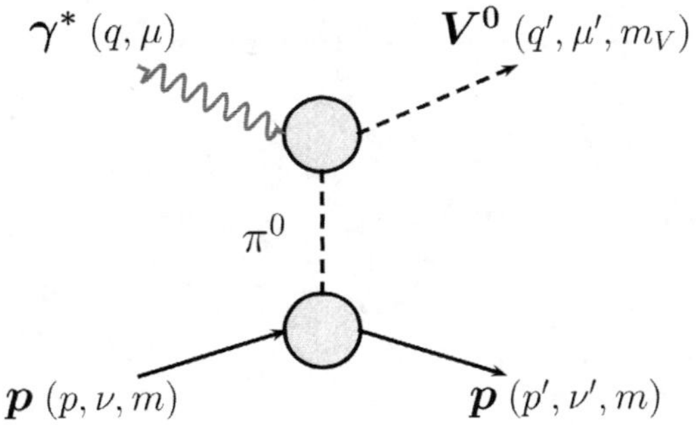

Figure 3. Pion pole contribution to VM production. The UP helicity amplitudes determined by pion pole looks as follows:

$$M^{pole}_{++,++} \sim \frac{\rho_{\pi V}}{t-m_\pi^2}\,\frac{m\,\xi\,Q^2}{\sqrt{1-\xi^2}},$$

$$M^{pole}_{+-,++} \sim -\frac{\rho_{\pi V}}{t-m_\pi^2}\,\frac{\sqrt{-t'}\,Q^2}{2};$$

with

$\rho_{\pi V} \sim g_{\pi V}(Q^2)\, g_{\pi N\, N}\, F_{\pi N\, N}(t)$.

In Fig.3, we show the PP contribution to the VM production together with the helicity amplitudes generated by the pion pole. These amplitudes have the UP nature and are controlled by the πV transition form factor. The transition form factor $g_{\pi V}(0)$ can be determined from the VM radiative decay

$$\Gamma(V \to \pi\gamma) \sim \frac{\alpha_{elm}}{24}\,|g_{\pi V}(0)|^2 M_V^3. \tag{3}$$

We find

$$|g_{\pi\omega}(0)| = 2.3\mathrm{GeV}^{-1}; \qquad |g_{\pi\rho}(0)| = .85\mathrm{GeV}^{-1}. \tag{4}$$

This means that $|g_{\pi\omega}(0)|$ is about 3 times larger with respect to $|g_{\pi\rho}(0)|$ and we should observe large PP effects in ω and small in ρ production. The Q^2 dependence of $g_{\pi V}(Q^2)$ was extracted [1] from the HERMES data [6] on the ratio of the unnatural to the natural parity cross section U_1 at $Q^2 < 4\text{GeV}^2$.

We will discuss our results on PP effects in the ω and ρ production [1] and give a comparison with HERMES data [6]. In calculations we use GPDs from our analysis of hard meson leptoproduction.

The natural and unnatural parity asymmetry P is determined as follows:

$$P = \frac{d\sigma^N(\gamma^*_T \to V_T) - d\sigma^U(\gamma^*_T \to V_T)}{d\sigma^N(\gamma^*_T \to V_T) + d\sigma^U(\gamma^*_T \to V_T)}. \tag{5}$$

If the UP contribution is small, we find $P \sim 1$. If it is large, the value of the P asymmetry will be far from unity. The model results for this asymmetry for ω production are shown in Fig.4. We find that the PP contribution to the ω asymmetry give $P \sim -0.5$ (full line) in agreement with experiment. While neglecting the PP contribution we obtain $P \sim 0.5$ (dashed line). In this figure, we show for comparison the model results for CLAS energy $W = 3.5\text{GeV}$ by the dotted line and for COMPASS energy $W = 8\text{GeV}$ by the dashed-dotted curve. It can be seen that at COMPASS energies PP effects are rather small for ω. For the ρ production PP effects in P asymmetry are shown in Fig. 5. They are small and the asymmetry is close to unity.

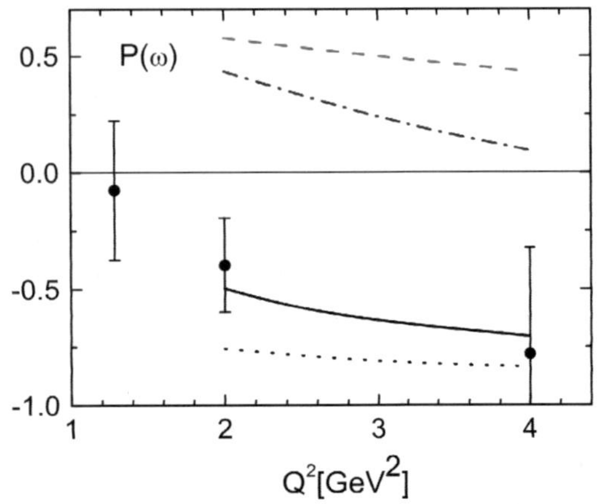

Figure 4. $P(\omega)$ at HERMES. Solid line-with PP contribution, dashed line - without PP. Dotted line-for $W = 3.5\text{GeV}$ (CLAS), dashed-dotted line for $W = 8\text{GeV}$ (COMPASS)

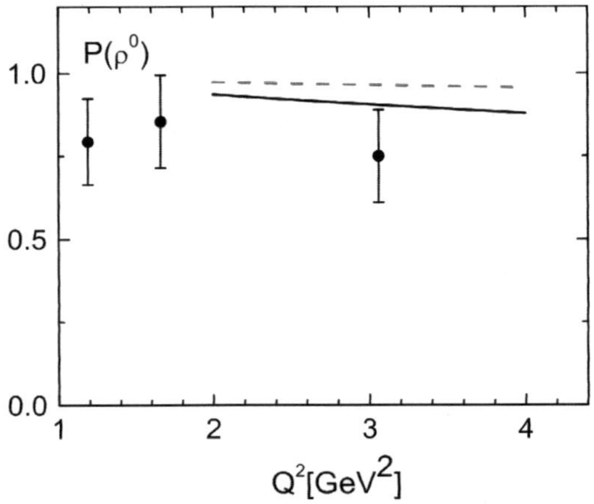

Figure 5. $P(\rho^0)$ at HERMES. Solid line with PP, dashed line -without PP.

The SDME $r^1_{1-1} = -\text{Im}r^2_{1-1}$ shows the difference of the natural and unnatural parity contributions

$$r^1_{1-1} = \frac{d\sigma^N(\gamma^*_T \to V_T) - d\sigma^U(\gamma^*_T \to V_T)}{2d\sigma}. \tag{6}$$

The results for this SDMEs for the ω production are shown in Fig. 6. We see that the PP effects are very strong for ω. With PP contribution we obtain $r^1_{1-1} \sim -0.2$ and if we omit PP, $r^1_{1-1} \sim 0.2$. For the ρ meson production the PP contribution is small and the results are close to the ω case without PP (see Fig.7). We find that HERMES experimental results can not be explained without consideration of the PP contributions.

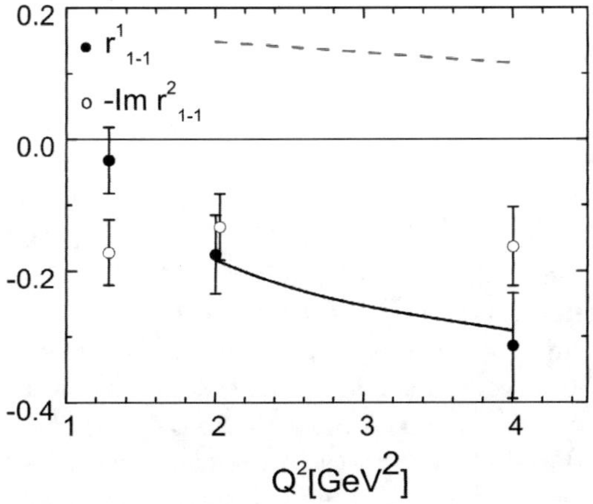

Figure 6. SDMEs for ω production at HERMES.

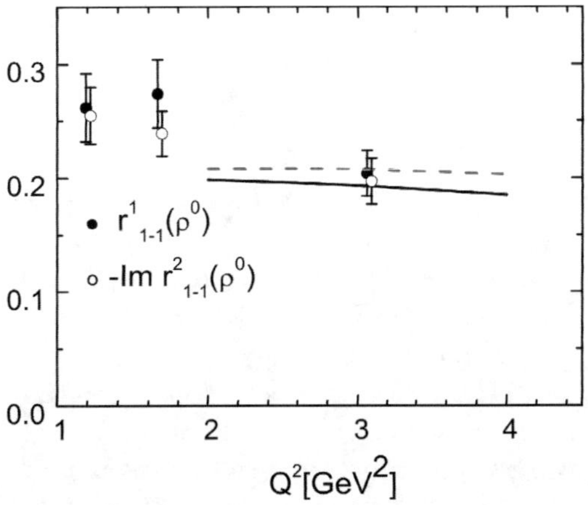

Figure 7. SDMEs for ρ production at HERMES.

Solid line- with PP, dashed -without PP.

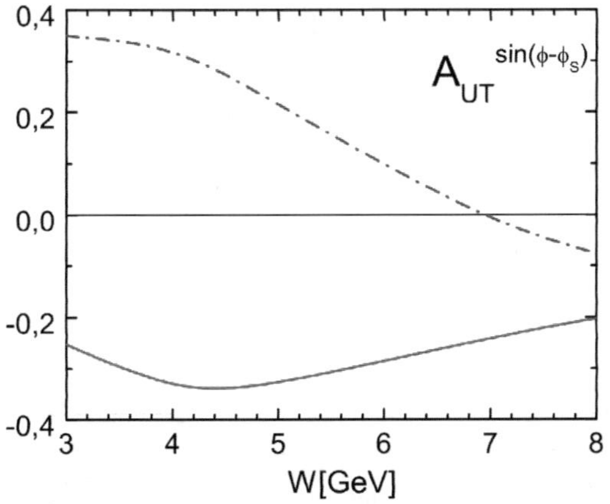

Figure 8. $\sin(\phi - \phi_s)$ modulation of A_{UT} asymmetry.

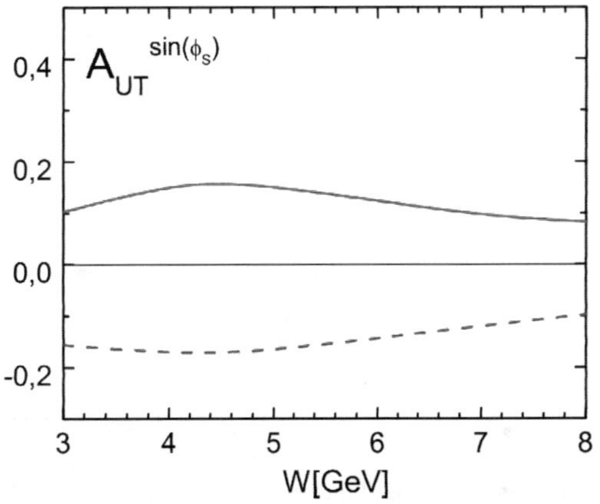

Figure 9. $\sin(\phi_s)$ modulation of A_{UT} asymmetry.

Full line - positive, dashed- dotted line - negative $\pi\omega$ transition form factor.

Pion pole effects can be observed in spin asymmetries. We find two not small interference terms with PP

$$\mathrm{Im} M^{N*}_{++++} M^{pole}_{+-++} \quad \text{and} \quad \mathrm{Im} M^{U*}_{++++} M^{pole}_{+-++} \tag{7}$$

which contribute to spin asymmetries for longitudinal and transverse beam and target polarizations. Note that the M^{U*}_{++++} amplitude in (7) is determined by $\tilde{H}$ effects. The PP contributions to asymmetries change the sign with the transition πV form factor.

Some examples for asymmetries in Figs. 8, 9 show substantially different results for the positive and negative sign of the $\pi\omega$ transition form factor. This means that generally, the sign of this form factor might be determines experimentally. New HERMES data where 5 modulations of A_{UT} asymmetry was analysed [16] are probably consistent with the positive transition $\pi\omega$ form factor. Unfortunately, the large experimental uncertainties prevent a definite conclusion on the sign of this transition form factor.

Journal of Physics: Conference Series **678** (2016) 012035 doi:10.1088/1742-6596/678/1/012035

4. Conclusion

We calculated the light meson leptoproduction reactions within the handbag approach, where the amplitudes factorize into hard subprocesses and GPDs [4]. The results on the cross sections and various spin observables based on this approach are in good agreement with data at HERMES, COMPASS and HERA energies at high Q^2 [3].

We considered pion pole contribution to π^+ leptoproduction. The PP effect in this case is determined by photon interaction with a charged pion. It was shown that PP gave an essential contribution to the leading twist longitudinal cross section [2]. Similar effects were found in the pion-induced Drell-Yan process [15]. The transversity H_T and $\bar{E}_T$ GPDs effects in the PM leptoproduction were investigated in [2, 5]. It was found that the transversity contributions are essential in the PM leptoproduction where they lead to large transverse cross sections which for π^0 production at low Q^2 exceed essentially the leading twist longitudinal cross section.

The pion pole effects in vector meson production have a different nature. They are determined by the πV transition form factor. It is found that the PP contributions are essential in ω and rather small in ρ^0 production. Based on our approach we found that the model results for ω SDMEs [1] are in good agreement with HERMES experimental results [6]. The large unnatural-parity effects observed by HERMES can be explained by the substantial PP contributions there. The PP effects in ω spin asymmetries are sensitive to the sign of the $\pi\omega$ transition form factor [1]. The experimental test of these asymmetries by HERMES [16] does not give a definite conclusion on the sign of the $\pi\omega$ transition form factor because of the large experimental uncertainties.

The work was supported in part by the Heisenberg-Landau program.

References

[1] Goloskokov S V and Kroll P 2014 *Euro. Phys. J.* A **50** 146
[2] Goloskokov S V and Kroll P 2010 *Euro. Phys. J.* C **65** 137
[3] Goloskokov S V and Kroll P 2005 *Euro. Phys. J.* C **42** 281;
Goloskokov S V and Kroll P 2007 *Euro. Phys. J.* C **50** 829;
Goloskokov S V and Kroll P 2008 *Euro. Phys. J.* C **53** 367;
Goloskokov S V and Kroll P 2009 *Euro. Phys. J.* C **59** 809
[4] Ji X 1997 *Phys. Rev.* D **55** 7114;
Radyushkin A V 1996 *Phys. Lett.* B **380** 417;
Collins J C et al. 1997 *Phys. Rev.* D **56** 2982
[5] Goloskokov S V and Kroll P 2011 *Euro. Phys. J.* A **47** 112
[6] Airapetian A et al. 2014 *(HERMES Collab.) Euro. Phys. J.* C **74** 3110
[7] Musatov I V and Radyushkin A V 2000 *Phys. Rev.* D **61** 074027
[8] Pumplin J et al. 2002 *JHEP* **0207** 012
[9] Diehl M, Feldmann T, Jakob R and Kroll P 2005 *Euro. Phys. J.* C **39** 1;
Diehl M and Kroll P 2013 *Euro. Phys. J.* C **73** 2397
[10] Airapetian A et al. *(HERMES Collab.)* 2010 *Phys. Lett.* B **682** 345
[11] Anselmino M, Boglione M, D'Alesio U, Kotzinian A, Murgia F, Prokudin A and Melis S 2009 *Nucl. Phys. Proc. Suppl.* **191** 98
[12] Gockeler M *et al. (QCDSF Collab. and UKQCD Collab.)* 2007 *Phys. Rev. Lett.* **98** 222001
[13] Airapetian A et al. *(HERMES Collab.)* 2008 *Phys. Lett.* B **659** 486
[14] Bedlinskiy I et al. *(CLAS Collab.)* 2012 *Phys. Rev. Lett.* **109** 112001
[15] Goloskokov S V and Kroll P 2015 *Phys. Lett* B **748** 323
[16] Airapetian A et al. *(HERMES Collab.)* Transverse-target-spin asymmetry in exclusive ω -meson electroproduction 2015 ArXiv:1508.07612

Cosmic censorship and stationary states of half-spin particles in the field of Reissner-Nordström naked singularity

M V Gorbatenko[1], V P Neznamov[1,2*], E Yu Popov[1] and I I Safronov[1]

[1]Russian Federal Nuclear Center – All-Russian Research Institute of Experimental Physics, Sarov, Mira 37, Nizhni Novgorod region, Russia, 607188
[2]National Research Nuclear University MEPhI, Moscow, Russia

E-mail: neznamov@vniief.ru

Abstract. The paper explores quantum mechanics of half-spin particle motion in the field of Reissner-Nordström (RN) naked singularity. It is shown that for any quantum mechanical Dirac particle, irrespective of availability and sign of its electrical charge, the RN naked singularity is separated by an infinitely high positive potential barrier. With like charges of a particle and the source of the RN naked singularity, near the origin there exists the second completely impenetrable potential barrier. It has been proved that in the field of the RN naked singularity, bound states of half-spin particles can exist. The conditions for appearance of such states were revealed and computations were performed to find energy eigenvalues and eigenfunctions.

1. The Reissner-Nordström metric

The line element is

$$ds^2 = f_{R-N}dt^2 - \frac{dr^2}{f_{R-N}} - r^2\left(d\theta^2 + \sin^2\theta d\varphi^2\right), \tag{1}$$

where $f_{R-N} = \left(1 - \frac{r_0}{r} + \frac{r_Q^2}{r^2}\right)$, $r_0 = \frac{2GM}{c^2}$ is the gravitational radius of the Schwarzschild field, $r_Q = \frac{\sqrt{G}Q}{c^2}$ is the "charge" radius, G is the gravitational constant, c is the velocity of light.

The case $r_0 < 2r_Q$ corresponds to the naked singularity. In this case, there is always $f_{R-N} > 0$ and the domain of wave functions is the entire region $r \in [0, \infty)$.

Below, we will analyze the behavior of effective potentials of the Dirac equation in the field of RN naked singularity.

For obtaining the Schrödinger-type equation with the self-conjugate Hamiltonian the initial Dirac Hamiltonian must be also self-conjugate.

Earlier, a self-conjugate Hamiltonian of a half-spin particle of mass m and charge e for the Reissner-Nordström metric was derived in [1]

$$\begin{aligned} H_\eta = {} & \sqrt{f_{R-N}}\beta m - i\alpha^1\left(f_{R-N}\frac{\partial}{\partial r} + \frac{1}{r} - \frac{r_0}{2r^2}\right) \\ & - i\sqrt{f_{R-N}}\frac{1}{r}\left[\alpha^2\left(\frac{\partial}{\partial\theta} + \frac{1}{2}\mathrm{ctg}\theta\right) + \alpha^3\frac{1}{\sin\theta}\frac{\partial}{\partial\varphi}\right] + \frac{eQ}{r}. \end{aligned} \tag{2}$$

Content from this work may be used under the terms of the Creative Commons Attribution 3.0 licence. Any further distribution of this work must maintain attribution to the author(s) and the title of the work, journal citation and DOI.
Published under licence by IOP Publishing Ltd

Journal of Physics: Conference Series **678** (2016) 012036 doi:10.1088/1742-6596/678/1/012036

In (2), α^k, β are Dirac matrices.

After separation of variables (see, for instance, [1]), the system of equations for radial functions $F_{R-N}(\rho), G_{R-N}(\rho)$ takes the form of

$$\begin{aligned} f_{R-N}\frac{dF_{R-N}(\rho)}{d\rho} + \left(\frac{1+\kappa\sqrt{f_{R-N}}}{\rho} - \frac{\alpha}{\rho^2}\right) F_{R-N}(\rho) - \left(\varepsilon - \frac{\alpha_{em}}{\rho} + \sqrt{f_{R-N}}\right) G_{R-N}(\rho) = 0, \\ f_{R-N}\frac{dG_{R-N}(\rho)}{d\rho} + \left(\frac{1-\kappa\sqrt{f_{R-N}}}{\rho} - \frac{\alpha}{\rho^2}\right) G_{R-N}(\rho) + \left(\varepsilon - \frac{\alpha_{em}}{\rho} - \sqrt{f_{R-N}}\right) F_{R-N}(\rho) = 0. \end{aligned} \tag{3}$$

In (3), dimensionless variables have been introduced $\rho = \frac{r}{l_c}$; $\varepsilon = \frac{E}{m}$; $\alpha = \frac{r_0}{2l_c} = \frac{GMm}{\hbar c} = \frac{Mm}{M_P^2}$, $\alpha_Q = \frac{r_Q}{l_c} = \frac{\sqrt{G}Qm}{\hbar c} = \frac{\sqrt{\alpha_{fs}}}{M_P} m \frac{Q}{|e|}$; $\alpha_{em} = \frac{eQ}{\hbar c} = \alpha_{fs}\frac{Q}{e}$, $l_c = \frac{\hbar}{mc}$ is the Compton wave-length of a Dirac particle; E is the energy of a Dirac particle; $M_P = \sqrt{\frac{\hbar c}{G}}$ is the Planck mass; α_{fs} is the electromagnetic fine structure constant; α, α_{em} are gravitational and electromagnetic coupling constants; α_Q is the dimensionless constant characterizing source of the electromagnetic field in the Reissner-Nordström metric; $\kappa = \pm\left(j + \frac{1}{2}\right)$; $M_P = 2.2 \cdot 10^{-5} g \left(1.2 \cdot 10^{19} GeV\right)$; $\alpha_{fs} \approx \frac{1}{137}$.

2. Effective potentials for the field of Reissner-Nordström naked singularity

From the system of equations (3), we will derive the second-order equation for the function $\psi(\rho)$ proportional to $F(\rho)$.

$$\frac{d^2\psi(\rho)}{d\rho^2} + 2\left(E_{schr} - U_{eff}(\rho)\right)\psi(\rho) = 0. \tag{4}$$

2.1 Irrespective of signs of charges Q, e, the leading term of the effective potential near the origin is

$$U_{eff} = \frac{3}{8\rho^2} + O\left(\frac{1}{\rho}\right) \; at \; \rho \to 0. \tag{5}$$

2.2 With like signs of charges Q, e near the origin there is the second potential barrier of the form

$$U_{eff}|_{\rho\to\rho_{cl}} \approx \frac{1}{4}\frac{1}{\left(\varepsilon - \frac{\alpha_{em}}{\rho} + \sqrt{1 - \frac{2\alpha}{\rho} + \frac{\alpha_Q^2}{\rho^2}}\right)^2}\left(\frac{\alpha_{em}}{\rho^2} + \frac{\frac{\alpha}{\rho^2} - \frac{\alpha_Q^2}{\rho^3}}{\sqrt{1 - \frac{2\alpha}{\rho} + \frac{\alpha_Q^2}{\rho^2}}}\right)^2. \tag{6}$$

For all the cases considered by the authors, the coefficient at the leading singularity in (6)

$$A = \frac{1}{4}\left(\frac{\alpha_{em}}{\rho_{cl}^2} + \frac{\frac{\alpha}{\rho_{cl}^2} - \frac{\alpha_Q^2}{\rho_{cl}^3}}{\sqrt{1 - \frac{2\alpha}{\rho_{cl}} + \frac{\alpha_Q^2}{\rho_{cl}^2}}}\right)^2 > \frac{3}{8}. \tag{7}$$

It is well-known that such potential barriers are impenetrable to quantum-mechanical particles [3].

Figure 1 presents the typical form of $U_{eff}(\rho)$ for the case of like signs of charges Q, e and $\varepsilon > 1$.

One can see two regions $\rho < \rho_{cl}$ and $\rho > \rho_{cl}$ separated by a potential barrier impenetrable to particles. In the outer region, there are no stationary states of Dirac particles. In the inner region $0 < \rho < \rho_{cl}$, the existence of stationary bound states of half-spin particles is possible.

Journal of Physics: Conference Series **678** (2016) 012036 doi:10.1088/1742-6596/678/1/012036

2.3 For opposite signs of charges Q, e at certain values of initial parameters the typical form of the potential $U_{eff}(\rho)$ is given in figure 2. The form of the potential does not change for the case of an uncharged Dirac particle either. Such $U_{eff}(\rho)$ relations testify to a possibility of existence of stationary bound states of half-spin particles in both cases.

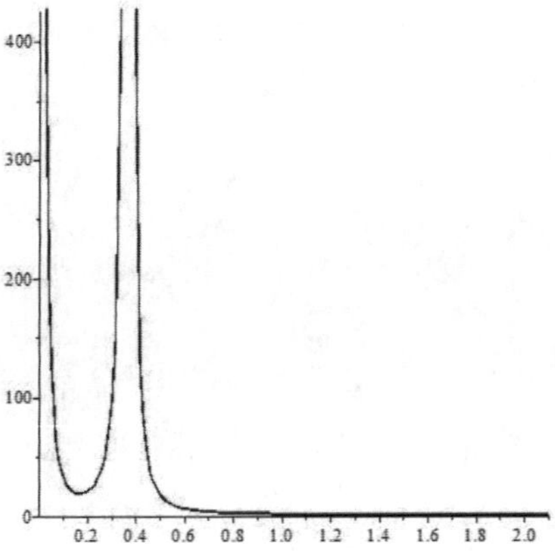

Figure 1. Behavior of effective potentials of the Dirac equation in the field of the RN naked singularity at $\alpha = 0.25,\ \alpha_Q = 0.5,\ \alpha_{em} = 1,$ $\kappa = -1,\ \varepsilon = 1.5,\ \rho_{cl} = 0.3675$.

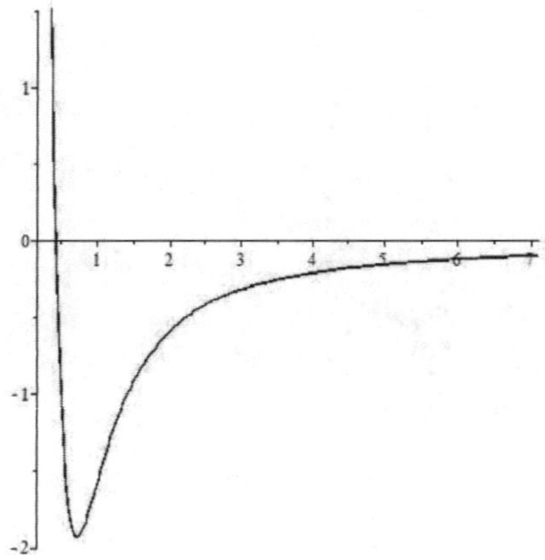

Figure 2. Behavior of effective potentials of the Dirac equation in the field of the RN naked singularity at $\alpha = 0.25,\ \alpha_Q = 0.5,$ $\alpha_{em} = -1,\ \kappa = -1,\ \varepsilon = 0.6$.

3. Determination of stationary bound states of half-spin particles in the field of Reissner-Nordström naked singularity

To solve the problem, we will use the system of equations (3).

Let us determine that

$$\mathrm{tg}\Phi(\rho) = \frac{\mathrm{F}(\rho)}{\mathrm{G}(\rho)},\quad \mathrm{A}^2(\rho) = \mathrm{F}^2(\rho) + \mathrm{G}^2(\rho). \tag{8}$$

Then,

$$F(\rho) = A(\rho)\sin\Phi(\rho),\quad G(\rho) = A(\rho)\cos\Phi(\rho). \tag{9}$$

From the equations (3), one can derive an equation for the phase $\Phi(\rho)$ in Vronsky's form [4] and an equation for $A(\rho)$

$$\frac{d\Phi}{d\rho} = \frac{1}{f_{R-N}}\left(\varepsilon - \frac{\alpha_{em}}{\rho}\right) + \frac{\cos(2\Phi)}{\sqrt{f_{R-N}}} - \frac{\kappa}{\rho\sqrt{f_{R-N}}}\sin(2\Phi), \tag{10}$$

$$\frac{d(\ln A)}{d\rho} = -\frac{1}{\rho f_{R-N}} + \frac{\alpha}{\rho^2 f_{R-N}} + \frac{\kappa}{\rho\sqrt{f_{R-N}}}\cos(2\Phi) + \frac{1}{\sqrt{f_{R-N}}}\sin(2\Phi). \tag{11}$$

3.1 Boundary condition

For $\rho \to \infty$, the leading terms of the asymptotics are

$$F = Ce^{-\rho\sqrt{1-\varepsilon^2}},\quad G = -\sqrt{\frac{1-\varepsilon}{1+\varepsilon}}F,\quad \mathrm{tg}\Phi = -\sqrt{\frac{1+\varepsilon}{1-\varepsilon}}. \tag{12}$$

At $\rho \to 0$, the asymptotic behavior of wave functions was identified in [5]. In our symbols

$$\begin{aligned} F(\rho) &= f_0\rho^{\frac{3}{2}} + f_1\rho^{\frac{5}{2}} + f_2\rho^{\frac{7}{2}} + ... \\ G(\rho) &= g_0\frac{1}{\rho^{\frac{1}{2}}} + g_1\rho^{\frac{1}{2}} + g_2\rho^{\frac{3}{2}} + ... \end{aligned} \tag{13}$$

It follows from (13) that

$$\mathrm{tg}\Phi = \left.\frac{\mathrm{F}(\rho)}{\mathrm{G}(\rho)}\right|_{\rho\to 0} = \frac{|\alpha_Q| - \alpha_{em}}{2\alpha_Q^2}\rho^2 = 0, \quad \Phi(\rho = 0) = k\pi, \quad k = 0, 1, 2... \tag{14}$$

3.2 Determination of the discrete energy spectrum

For three versions of $\alpha, \alpha_Q, \alpha_{em}$ with boundary conditions (12) - (14) discrete spectra of states of Dirac particles in the field of RN naked singularity were determined due to the solution to the equation (10).

4. Discussions

The consideration of the quantum mechanics of the half-spin particle motion in the field of Reissner-Nordström (RN) naked singularity resulted in drawing the following conclusions:

- for any quantum -mechanical half-spin particle, irrespective of availability and sign of the electrical charge, the RN naked singularity is separated by the infinitely high positive potential barrier $\sim \frac{3}{8}r^{-2}$. This agrees with the conclusions [2] concerning the motion of spinless particles in the field of certain singular metrics. As it is noted in [2] the availability of a repulsive barrier shielding the singularity poses no threat to the cosmic censorship.
- with like charges of a Dirac particle and the source of RN naked singularity near the origin there exists the second completely impenetrable positive potential barrier $\sim \frac{A}{(r-r_{cl})^2}$. For elementary charges of a particle and the source of RN naked singularity at $\alpha_{em} \gg \alpha$, $\alpha_{em} \gg \alpha_Q$ and at particle energies $E \sim mc^2$, the value r_{cl} is close to half a classical radius of a charged particle $r_e = \frac{e^2}{mc^2}$. At the value of the particle energy $E >> mc^2$, the radius r_{cl} decreases inversely to E $\left(r_{cl} = \frac{r_e}{E}\right)$.
- the analysis of the effective potentials and the direct numerical solutions to the Dirac equation have shown that the stationary bound states of half-spin particles in the field of RN naked singularity can exist in case of opposite charges of particles and the RN field source. In case of like charges, the bound states of a Dirac particle can exist under the potential barrier in the region $0 \le r \le r_{cl}$. In the region $r > r_{cl}$, there is no bound state of half-spin particles at the explored values of initial parameters.
- the stationary bound states of uncharged Dirac particles implemented due to the forces of gravitational interaction alone were obtained by direct computations of the Dirac equation in the field of RN naked singularity.

Let us pay attention to the difference in the dynamics of classical and quantum half-spin particles. It is well-known that in the classical case there appears the geodesic incompleteness of the RN naked singularity. The geodesics cannot pass across the surface whose radius is equal to half a classical radius of the source of RN naked singularity: $r_{cl} = \frac{Q^2}{2Mc^2}$. However, in quantum mechanics, depending on sings of charges, a Dirac particle can either be reflected from the repulsive barrier (6) or be in one of stationary bound states.

References

[1] Gorbatenko M V, Neznamov V P, *Preprint* gr-qc/1302.2557
[2] Horowitz G T and Marolf D 1995 *Phys. Rev.* D**10** 5670
[3] Dittrich J, Exner P 1985 *J. Math. Phys* **26** (8) 2000-08
[4] Vronsky M A, Gorbatenko M V, Kolesnikov N S, Neznamov V P, Popov E Yu, Safronov I I, *Preprint* gr-qc/1301.7595.
[5] Pekeris C L and Frankowski K 1986 *Proc. Natl. Acad. Sei. USA* **83** 1978-82

Journal of Physics: Conference Series **678** (2016) 012037 doi:10.1088/1742-6596/678/1/012037

Analysis of half-spin particle motion in static Reissner-Nordström and Schwarzschild fields

M V Gorbatenko[1] ,V P Neznamov[1,2]* and E Yu Popov[1]

[1]Russian Federal Nuclear Center – All-Russian Research Institute of Experimental Physics, Sarov, Mira 37, Nizhni Novgorod region, Russia, 607188
[2]National Research Nuclear University MEPhI, Moscow, Russia

E-mail: eugene_1976@mail.ru

Abstract. The paper presents the analysis of effective potentials of Dirac equations in static Schwarzschild and Reissner-Nordström (RN) fields. It is shown that in all the explored cases the condition of a particle "fall" to appropriate event horizons is fulfilled. The exception is one of the solutions for the Reissner-Nordström extremal field, for which the existence of the stationary bound state of half-spin particles is possible inside the event horizon.

1. The Reissner-Nordström metric

In the early 2000s, it was believed that there is no stable bound state of Dirac particles in the gravitational field of Reissner-Nordström black holes [1], [2]. However, in 2013 came a surprising result that stable bound states are yet possible in the field of extremal black hole both inside and outside the event horizon [3]. Approach of the authors [3] was to find a solution to the eigenvalue problem for the Dirac Hamiltonian in the background of an extremal black hole. There is much more simple method enabling us to clarify this issue, the effective potential method. In this paper, we address the question of whether stable bound states realizable in an extremal black hole by applying the efective potential method to the Dirac equation, and arrive at the conclusion that such states are indeed unfeasible in the region outside the event horizon, but a single state (rather than two) can exist inside the extremal black hole.

The line element RN metric is

$$ds^2 = f_{R-N}dt^2 - \frac{dr^2}{f_{R-N}} - r^2\left(d\theta^2 + \sin^2\theta d\varphi^2\right), \tag{1}$$

where $f_{R-N} = \left(1 - \frac{r_0}{r} + \frac{r_Q^2}{r^2}\right)$, $r_0 = \frac{2GM}{c^2}$ is the gravitational radius of the Schwarzschild field, $r_Q = \frac{\sqrt{G}Q}{c^2}$ is the "charge" radius, G is the gravitational constant, c is the velocity of light.

For obtaining the Schrödinger-type equation with the self-conjugate Hamiltonian the initial Dirac Hamiltonian must be also self-conjugate.

The self-conjugate Hamiltonian of a half-spin particle of mass m and charge e for the Reissner-Nordström metric was derived in [4]

Content from this work may be used under the terms of the Creative Commons Attribution 3.0 licence. Any further distribution of this work must maintain attribution to the author(s) and the title of the work, journal citation and DOI.
Published under licence by IOP Publishing Ltd

$$\begin{array}{c} H_\eta = \sqrt{f_{R-N}}\beta m - i\alpha^1\left(f_{R-N}\frac{\partial}{\partial r} + \frac{1}{r} - \frac{r_0}{2r^2}\right) \\ -i\sqrt{f_{R-N}}\frac{1}{r}\left[\alpha^2\left(\frac{\partial}{\partial\theta} + \frac{1}{2}\text{ctg}\theta\right) + \alpha^3\frac{1}{\sin\theta}\frac{\partial}{\partial\varphi}\right] + \frac{eQ}{r}. \end{array} \tag{2}$$

In (2), α^k, β are Dirac matrices.

After separation of variables (see, for instance, [5]), the system of equations for radial functions $F_{R-N}(\rho), G_{R-N}(\rho)$ takes the form

$$\begin{array}{c} f_{R-N}\frac{dF_{R-N}(\rho)}{d\rho} + \left(\frac{1+\kappa\sqrt{f_{R-N}}}{\rho} - \frac{\alpha}{\rho^2}\right)F_{R-N}(\rho) - \left(\varepsilon - \frac{\alpha_{em}}{\rho} + \sqrt{f_{R-N}}\right)G_{R-N}(\rho) = 0, \\ f_{R-N}\frac{dG_{R-N}(\rho)}{d\rho} + \left(\frac{1-\kappa\sqrt{f_{R-N}}}{\rho} - \frac{\alpha}{\rho^2}\right)G_{R-N}(\rho) + \left(\varepsilon - \frac{\alpha_{em}}{\rho} - \sqrt{f_{R-N}}\right)F_{R-N}(\rho) = 0. \end{array} \tag{3}$$

In (3), dimensionless variables have been introduced $\rho = \frac{r}{l_c}$; $\varepsilon = \frac{E}{m}$; $\alpha = \frac{r_0}{2l_c} = \frac{GMm}{\hbar c} = \frac{Mm}{M_P^2}$, $\alpha_Q = \frac{r_Q}{l_c} = \frac{\sqrt{G}Qm}{\hbar c} = \frac{\sqrt{\alpha_{fs}}}{M_P}m\frac{Q}{|e|}$; $\alpha_{em} = \frac{eQ}{\hbar c} = \alpha_{fs}\frac{Q}{e}$, $l_c = \frac{\hbar}{mc}$ is the Compton wavelength of a Dirac particle; E is the energy of a Dirac particle; $M_P = \sqrt{\frac{\hbar c}{G}}$ is the Planck mass; $\alpha_{fs} \approx 1/137$ is the electromagnetic fine structure constant; $\kappa = \pm\left(j + \frac{1}{2}\right)$.

2. The Reissner-Nordström field with two event horizons ($\alpha^2 > \alpha_Q^2$)

From the system of equations (3), we derive the second-order ordinary differential equation for the function $\psi(\rho)$ proportional to $F(\rho)$.

$$\frac{d^2\psi(\rho)}{d\rho^2} + 2\left(E_{schr} - U_{eff}(\rho)\right)\psi(\rho) = 0. \tag{4}$$

The effective potential has a complicate analytical form, but our concern is not with the precise analytical expression for U_{eff}. It is sufficient to know the behavior of the effective potential in the vicinity of the event horizons and in vicinity of the origin.

$$U_{R-N}(\rho)|_{\rho\to\rho_-} = -\frac{1}{(\rho_- - \rho)^2}\left[\frac{1}{8} + \frac{1}{8}\frac{(\varepsilon\rho_- - \alpha_{em})^2\rho_-^2}{(\alpha - \rho_-)^2}\right] + O\left(\frac{1}{\rho_- - \rho}\right), \tag{5}$$

$$U_{R-N}(\rho)|_{\rho\to\rho_+} = -\frac{1}{(\rho_+ - \rho)^2}\left[\frac{1}{8} + \frac{1}{8}\frac{(\varepsilon\rho_+ - \alpha_{em})^2\rho_+^2}{(\rho_+ - \alpha)^2}\right] + O\left(\frac{1}{\rho_+ - \rho}\right). \tag{6}$$

It is well known [6] that quantum-mechanical particle "falls" to the center if the singular effective potential behaves as $-C/\rho^2$ with $C > 1/8$, otherwise it does not "fall" to center. The above expressions for the singular effective potential behaves as $-C/(\rho - \rho_\pm)^2$ in the vicinities of the horizons, with the numerators of these expressions being positive, and no less than 1/8. It follows that Dirac particles "fall" to the horizons (either inner or outer).

For the Schwarzschild field, $\rho_+ \equiv \rho_0 = 2\alpha$, $\rho_- = 0$, $\alpha_{em} = 0$ in the region $\rho > 2\alpha$ the effective potential (6) in the vicinity of event horizon has a form

$$U_S(\rho)|_{\rho\to 2\alpha} = -\frac{1}{(\rho - 2\alpha)^2}\left[1/8 + 2\alpha^2\varepsilon^2\right] + O\left(\frac{1}{\rho - 2\alpha}\right). \tag{7}$$

It follows that in this case half-spin particles "fall" to the event horizon with $\rho_0 = 2\alpha$.

3. The Reissner-Nordström extremal field ($\alpha^2 = \alpha_Q^2$)

Consider the effective potential for an extremal black hole, $M = |Q|$ or $\alpha = |\alpha_Q|$.

In this case, there is the only event horizon with the radius $\rho_+ = \rho_- = \alpha$. The effective potential in the vicinity of horizon is more singular than $C/(\rho-\alpha)^2$ and therefore that the Dirac particle "falls" to the horizon, but if we take into account Dokuchaev and Eroshenko solution $\varepsilon = \frac{\alpha_{em}}{\alpha}$ [3], then the effective potential in the vicinity of horizon becomes

$$U_{R-N}^{extr}(\rho)\Big|_{\rho\to\alpha} = -\frac{\frac{1}{4} - \kappa^2 - \alpha^2 + \alpha_{em}^2}{2(\rho-\alpha)^2}. \tag{8}$$

The eigenvalue $\varepsilon = \frac{\alpha_{em}}{\alpha}$ with normalized wave can exist if condition

$$\kappa^2 + \alpha^2 - \alpha_{em}^2 > \frac{1}{4} \tag{9}$$

imposed. We claim that, in the exterior region of an extremal black hole, a stable bound states described by the Dirac equation do not exist. This is consistent with the results of Refs. [1] and [2]. Indeed, let us consider two possibilities. First, $\alpha > |\alpha_{em}|$. In this case, the found effective potential has no extremal point. The effective potential $U_{R-N}^{extr}(\rho)$ is schematically shown in figure 1. Second, $\alpha < |\alpha_{em}|$. The shape of the potential $U_{R-N}^{extr}(\rho)$ differs qualitatively from that in the first case (see figure 2).

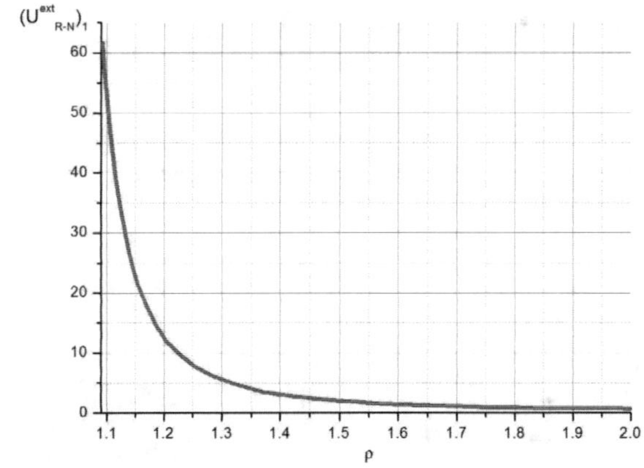

Figure 1. The effective potential in the exterior of the RN extremal black hole for $\alpha > \alpha_{em}$, $\alpha = \alpha_Q = 1, \alpha_{em} = 0.9, \varepsilon = 0.9$.

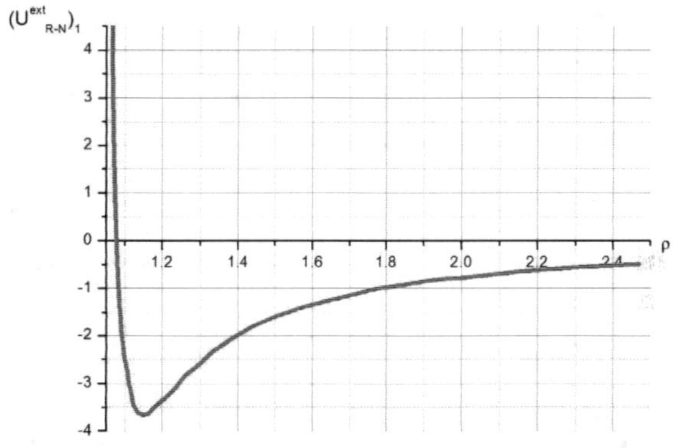

Figure 2. The effective potential in the exterior of the RN extremal black hole for $\alpha < \alpha_{em}$, $\alpha = \alpha_Q = 1, \alpha_{em} = 1.25, \varepsilon = 1.25$.

Let us first assume that the black hole and the particle have like signs of electric charges, $\alpha_{em} = Qe > 0$. Clearly, this energy corresponding to a bound states can not exist because $\varepsilon > 1$ so that this energy level belongs to the continuum part of the spectrum. Let the black hole and the particle have opposite signs of electric charges, $\alpha_{em} = Qe < 0$. It follows that $\varepsilon = -\left|\frac{\alpha_{em}}{\alpha}\right| < -1$, which formally does not prevent to the existence a discrete energy level with $E < -m$. Let us consider the interior region of an extremal Reissner-Nordsröm black hole $(0 < \alpha < \rho)$. The effective potential is positive.

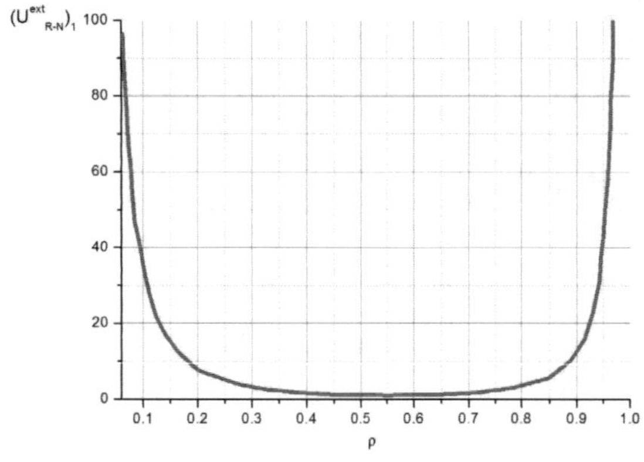

Figure 3. The effective potential in the interior of the RN extremal black hole $0 < \rho < \alpha$, for $\alpha = \alpha_Q = 1, \alpha_{em} = 1.25, \varepsilon = 1.25$.

In the vicinity of the origin, we have $U^{extr}_{R-N}(\rho)\Big|_{\rho\to 0} = \frac{3}{8}\frac{1}{\rho^2} + \mathrm{O}\left(\frac{1}{\rho}\right)$ while in the vicinity of the horizon $U^{extr}_{R-N} \sim C/(\rho-\alpha)^2$. Can a discrete state with $\varepsilon = \frac{\alpha_{em}}{\alpha}$ exist there? The answer is positive provided that the particle and the extremal black hole have the same sign of their electric charges. Otherwise the answer is negative.

4. Discussions

The effective potential method allows us to obtain qualitative results consisting in the following: we show that in all the explored cases, but one, the condition of a particle "fall" to appropriate event horizons is fulfilled. The exception is one of the solutions for the RN extreme field with the single event horizon. For this solution, inside the event horizon the possibility of existence of stationary bound states of hslf-spin particles is shown only at like signs of the particle and the black hole. If $\kappa^2 + \alpha^2 - \alpha_{em}^2 > 1/4$ and $\varepsilon = \frac{\alpha_{em}}{\alpha}$, then the effective potential exhibits the second order pole with a positive coefficient, which implies that the black hole becomes quantum-mechanically impenetrable. This should be compared with the usual situation when a particle "falls" to event horizon.

References

[1] Finster F, Smoller J and Yau S - T. 2000 *J.Math.Phys.* **41** 4 2173-94
[2] Schmid H *Preprint* math-ph/0207039v2
[3] Dokuchaev V I, Yeroshenko Yu N 2013 *JETP* **117** 1 72-77
[4] Gorbatenko M V, Neznamov V P 2010 *Phys. Rev.* D **82** 104056 (*Preprint* gr-qc/1007.4631)
[5] Gorbatenko M V, Neznamov V P 2011 *Phys. Rev.* D **83** 105002 (*Preprint* gr-qc/1102.4067v1)
[6] Landau L D and Lifshitz E M 1965 *Quantum Mechanics. Nonrelativistic Theory* (Pergamon Press, Oxford)

Journal of Physics: Conference Series **678** (2016) 012038 doi:10.1088/1742-6596/678/1/012038

Overview of recent HERMES results

Hrachya Marukyan (On behalf of the HERMES Collaboration)
AANL (Yerevan Physics Institute), Alikhanian Brs. 2, Yerevan 036, Armenia
E-mail: marukyan@mail.desy.de

Abstract. An overview of more recent and important results from the HERMES experiment are presented in this paper. HERMES collected a wealth of data using the 27.6 GeV polarized HERA lepton beam and various polarized and unpolarized gaseous targets. This unique data set opens the door to the measurements of observables sensitive to the multidimensional structure of the nucleon. Among them are semi-inclusive deep-inelastic scattering measurements of azimuthal modulations sensitive to the transverse momentum distributions, such as the leading-twist Sivers and Collins distributions and distributions sensitive to the convolutions of the twist-2 and twist-3 functions. They all provide an information on the three-momentum-dependent quark distributions. Knowledge on the quark distribution as a function of longitudinal momentum and transverse position in impact-parameter space can be accessed, e.g., through exclusive $\omega-$meson leptoproduction, particularly through the measurement of spin density matrix elements and the measurement of azimuthal modulations on transversely polarized proton target. The measurement of Bose-Einstein correlations of hadron pairs in quasi-real leptoproduction are also presented. The transverse polarization of Λ hyperons measured again in quasi-real leptoproduction regime are presented as well. Finally, the new analysis for the search on pentaquark at HERMES are mentioned.

1. Introduction

The HERMES experiment at DESY collected data from 1995 until 2007 using the 27.6 GeV HERA lepton beam. In the experiment, longitudinally polarized electrons or positrons were scattered off longitudinally, transversely or unpolarized hydrogen targets, longitudinally polarized deuterium and helium targets and unpolarized deuterium, helium, or heavier nuclear targets. The scattered lepton and particles produced in the reaction were detected by a forward spectrometer. The lepton-hadron separation was performed using a transition-radiation detector, a scintillator preshower counter, an electromagnetic calorimeter and a threshold gas Cherenkov counter (ring-imaging Cherenkov detector from 1998). The lepton identification efficiency exceeds 98% with a hadron misidentification contamination below 1%. Hadon identification was performed by Cherenkov detectors, allowing the discrimination of pions, kaons and protons.

The analysis of the azimuthal distribution of hadrons in semi-inclusive deep-inelastic (DIS) of leptons off a transversely polarized hydrogen target provides access to the Sivers distribution [1] and Collins fragmentation [2] functions. The former describes the distribution of unpolarized quarks in a transversely polarized nucleon, correlating quark transverse momentum with the nucleon's transverse spin, while the latter the fragmentation of a transversely polarized quark into an unpolarized hadron. The selected results sensitive to these two quantities are presented in section 2. The results from the recent analysis of $\sin\phi_h$ amplitudes of single-spin asymmetry

Content from this work may be used under the terms of the Creative Commons Attribution 3.0 licence. Any further distribution of this work must maintain attribution to the author(s) and the title of the work, journal citation and DOI.
Published under licence by IOP Publishing Ltd

Journal of Physics: Conference Series **678** (2016) 012038 doi:10.1088/1742-6596/678/1/012038

in semi-inclusive DIS leptoproduction of hadrons in case of collision of a longitudinally polarized lepton beam with an unpolarized target are also considered in this section.

Data collected on transversely polarized hydrogen target are used to extract the amplitudes of single-spin asymmetry of the cross section in hard exclusive electroproduction of $\omega-$ mesons. They provide information on the sign of $\pi\omega$ transition form factor. The measurement of amplitudes of azimuthal sine modulations are presented in section 3.

Data from deep-inelastic scattering of longitudinally polarized electrons and positrons off polarized ^{1}H, ^{2}H and ^{3}He targets and unpolarized ^{4}He, N, Ne, Kr and Xe nuclear targets are used to measure the Bose-Einstein correlations of like-sign charged hadrons. The polarized data were summed over the spin orientation. The measurement of these correlation are discussed in section 4.

Measurements of transverse polarization of Λ hyperons from quasi-real photoproduction in the scattering of longitudinally polarized electrons and positrons off longitudinally or transversely polarized hydrogen target and longitudinally polarized deuterium target are discussed in section 5. The direction of the target polarization was reversed in 1-3 min intervals, resulting in a vanishing average target polarization of the data set.

Finally, the renewed analysis for the search of the pentaquark signal again in the regime of quasi-real photoproduction are presented in section 6. In addition to further data from a deuterium target, data from hydrogen targets were used in the analysis and are presented here.

2. Single- and double-spin asymmetries measured in SIDIS on unpolarized and transversely polarized proton target

Semi-inclusive DIS off a transversely polarized target (T) allows to access the single- and double-spin asymmetry amplitudes with characteristic angular modulations. Two angles appear in the description of the cross section of this process: angle ϕ, denoting the angle between the lepton-scattering and the hadron-production planes, and ϕ_S, being the azimuthal angle of the transverse component of the target-spin vector about the virtual photon direction with respect to the lepton scattering plane. Each of the azimuthal amplitudes corresponds to convolutions of different distribution and fragmentation functions. Two of them, proportional respectively to the $\sin(\phi - \phi_S)$ and $\sin(\phi + \phi_S)$ modulations, are interpreted as the convolutions of the Sivers distribution function [1] and the spin-independent fragmentation function and transversity and the Collins fragmentation function [2]. The HERMES results on the Sivers and Collins amplitudes for charged and neutral pions and for charged kaons as a function of x, z and $P_{h\perp}$ were published in Ref. [3] and Ref. [4], respectively. Here, x represents the Bjorken scaling variable, z denotes the fractional hadron energy with respect to the virtual photon energy in the target rest frame and $P_{h\perp}$ the magnitude of the transverse momentum of the final state hadron. In figure 1 the preliminary results of the extracted Collins amplitudes projected in bins of x, z and $P_{h\perp}$ are presented for protons and antiprotons. The statistical (systematic) uncertainties are given as error bars (bands).

Moreover, the fine binned three-dimensional extraction of various azimuthal amplitudes for charged pions, kaons and protons are possible from HERMES data, in projection of above mentioned kinematic parameters, which allows to constrain global fits to the experimental measurements in a more profound way. As an example, in figure 2 and figure 3 are shown respectively the Collins amplitudes for negative pions and Sivers amplitudes for protons vs. x extracted for four z and $P_{h\perp}$ bins. The three-dimensional extraction of all possible single- and double-spin asymmetry amplitudes sensitive to different combination of distribution and fragmentation functions are also available (not shown here).

In case of longitudinal beam (L) and unpolarized target (U) only target spin-independent parts can contribute to the asymmetry. Here, the structure function of interest is related to the asymmetry amplitude $A_{LU}^{\sin\phi_h}$, which is sensitive to convolutions of twist-2 distribution

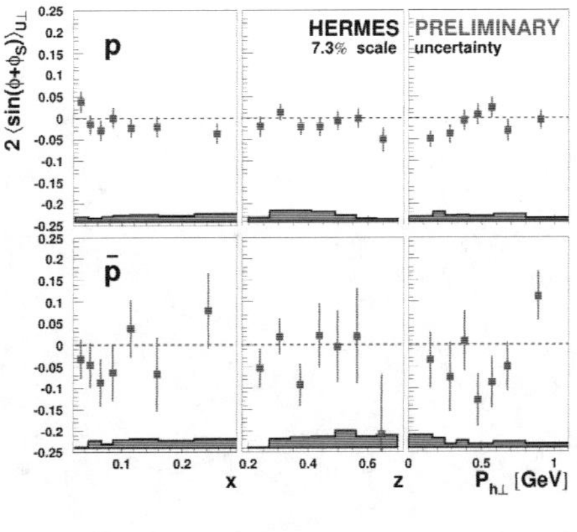

Figure 1. Collins amplitudes for protons and antiprotons as a function of x, z and $P_{h\perp}$.

(fragmentation) functions with twist-3 fragmentation (distribution) functions. HERMES published the results of single-spin asymmetry A_{LU} for charged and neutral pions as a function of x, z and $P_{h\perp}$ [5]. The multidimensional extraction of asymmetry amplitudes $A_{LU}^{\sin\phi_h}$ are shown in figure 4 and figure 5 respectively for negative and positive pions in projections of x, z and $P_{h\perp}$ in four bins for each of the kinematic variable.

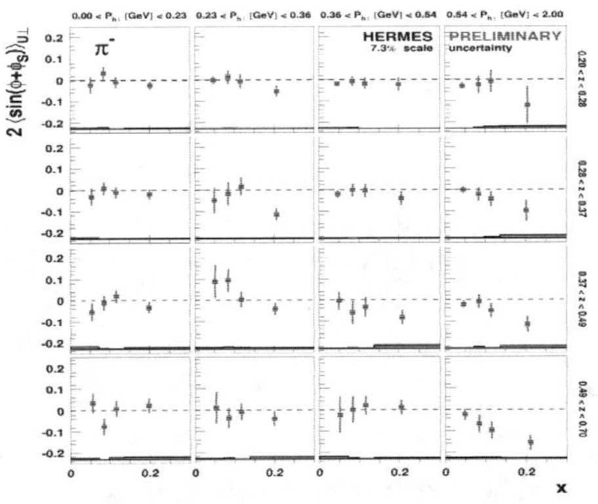

Figure 2. Collins amplitudes for π^- vs. x for four bins in z and $P_{h\perp}$.

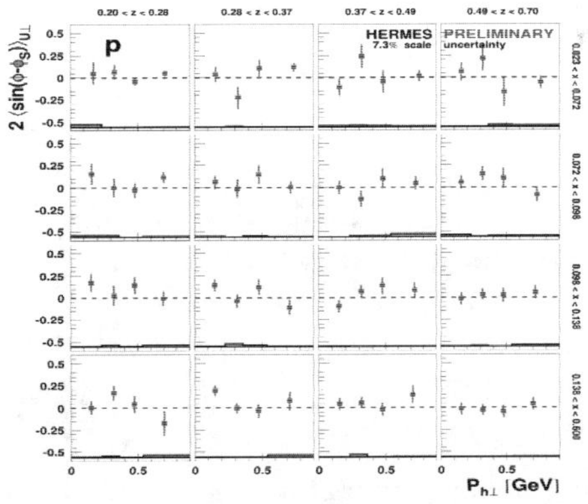

Figure 3. Sivers amplitudes for protons vs. x for four bins in z and $P_{h\perp}$.

3. Exclusive $\omega-$meson leptoproduction

Hard exclusive electroproduction of $\omega-$mesons is studied with the HERMES spectrometer by scattering longitudinally polarized positron and electron beams on unpolarized hydrogen and deuterium targets or on transversely polarized hydrogen target in the kinematic region

Journal of Physics: Conference Series **678** (2016) 012038 doi:10.1088/1742-6596/678/1/012038

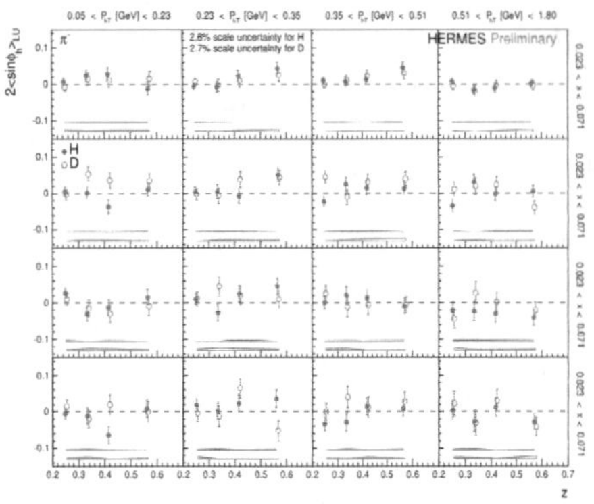

Figure 4. The multidimensional extraction of $A_{LU}^{\sin\phi_h}$ amplitudes for π^- mesons.

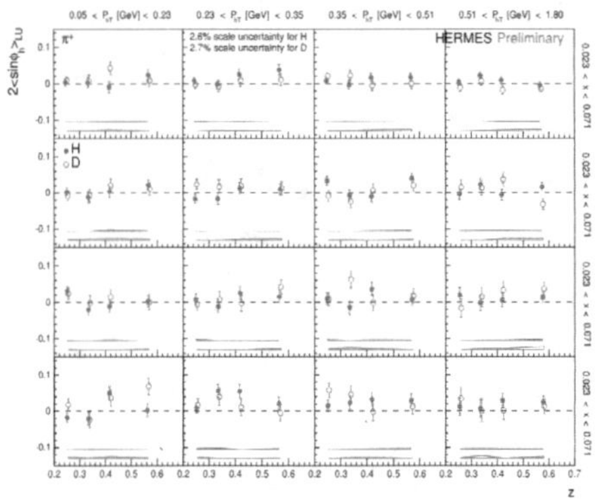

Figure 5. The multidimensional extraction of $A_{LU}^{\sin\phi_h}$ amplitudes for π^+ mesons.

$Q^2 > 1.0\,\mathrm{GeV}^2, 3.0\,\mathrm{GeV} < \mathrm{W} < 6.3\,\mathrm{GeV}, \mathrm{and} - \mathrm{t}' < 0.2\,\mathrm{GeV}^2$. Here, Q^2 represents the negative square of the virtual-photon four-momentum, W is the invariant mass of the photon-nucleon system and $-t'$ is the smallest kinematically allowed value of the momentum transfer to the target. Using an un-binned maximum likelihood method, 15 unpolarized and 8 polarized Spin Density Matrix Elements (SDMEs) are extracted from unpolarized hydrogen and deuterium data [6]. From the analysis of transversely polarized hydrogen data the amplitudes of five sine modulations of the single-spin asymmetry of the cross section with respect to the transverse proton polarization are measured [7], the Q^2 and $-t'$ dependence of which are shown in figure 6. They are determined in the entire kinematic region (open squares) as well as for two bins in photon virtuality and momentum transfer to the nucleon (full circles). The inner error bars represent the statistical uncertainties, while the outer ones indicate the statistical and systematic uncertainties added in quadrature. The results receive an additional 8.2% scale uncertainty corresponding to the target polarization uncertainty. These results are compared to a phenomenological model that includes the pion pole contribution [8]. As can be seen from figure 6, within this model, the data favor a positive $\pi\omega$ transition form factor, the sign of which was not possible to determine from the measurements of SDMEs. Also, a separation of asymmetry amplitudes into longitudinal and transverse components is performed (not shown here).

4. Bose-Einstein correlations in hadron pairs from leptoproduction on nuclei

Bose-Einstein correlations of like-sign charged hadrons produced in deep-inelastic electron and positron scattering are studied in the HERMES experiment using nuclear targets of ^{1}H, ^{2}H, ^{3}He, ^{4}He, N, Ne, Kr, and Xe [9]. A Gaussian approach is used to parameterize a two-particle correlation function determined from events with at least two charged hadrons of the same sign charge. This correlation function is compared to two different empirical distributions that do not include the Bose-Einstein correlations. One distribution is derived from unlike-sign hadron pairs (method MUS), and the second from mixing like-sign pairs from different events (method MEM). The extraction procedure used simulations incorporating the experimental setup in order to correct the results for spectrometer acceptance effects, and was tested using the distribution of unlike-sign hadron pairs. Double ratio correlation function $R(T)$ for like-sign hadron pairs,

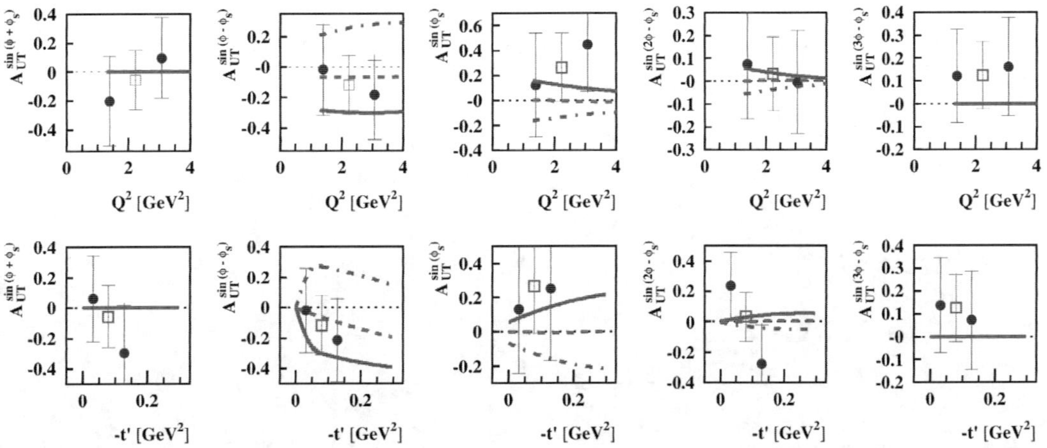

Figure 6. The amplitudes describing the strength of the sine modulations of the cross section for hard exclusive ω−meson production.

obtained with above mentioned two methods from the hydrogen data are shown in figure 7, where T is the square of the difference of the two particle four-momenta. W−dependence of the particle source distribution size r_G (Goldhaber radius) and chaoticity parameter λ are shown respectively on top and bottom panels of figure 8. In figure 9 the world data on measured average value of r_G in lepton nucleon scattering experiments is plotted. No evidence for a dependence on W is found. The A−dependence of the parameters r_G (top panel) and λ (bottom panel) are shown in figure 10. Clear signals of Bose-Einstein correlations for all target nuclei without a significant variation with the nuclear target mass are found.

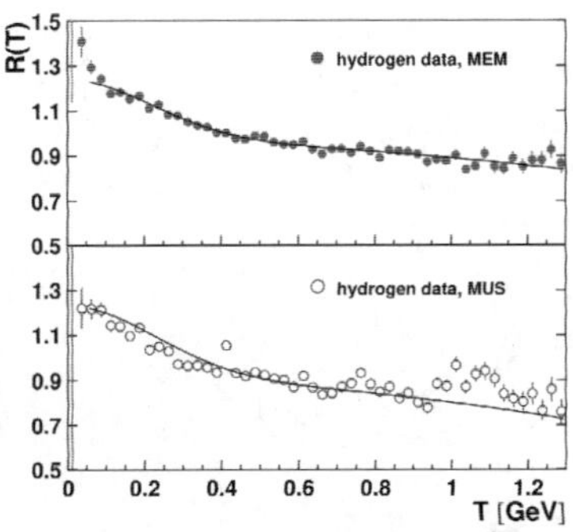

Figure 7. Double ratio correlation function for like-sign hadron pairs obtained with MEM and MUS based on hydrogen target data.

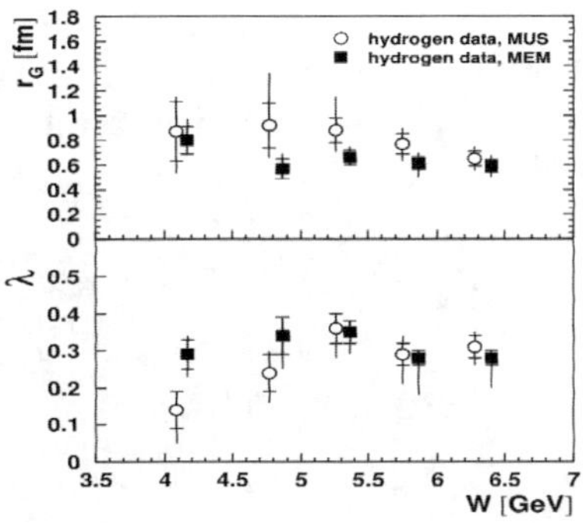

Figure 8. Parameter r_G (top panel) and λ (bottom panel) as a function of W, obtained with MEM and MUS methods on hydrogen.

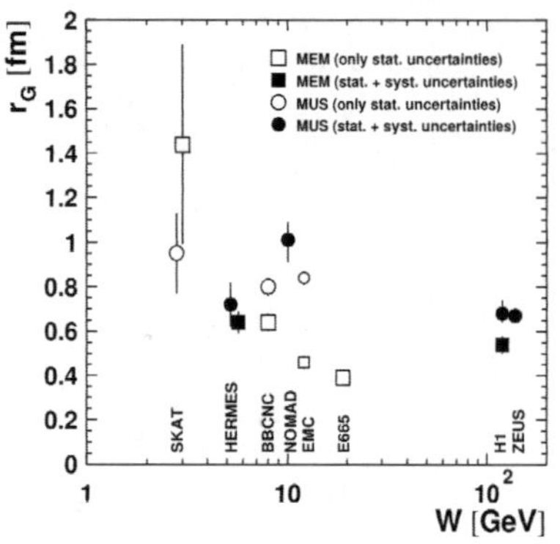

Figure 9. Goldhaber radius r_G, as a function of W, obtained in lepton nucleon scattering experiments.

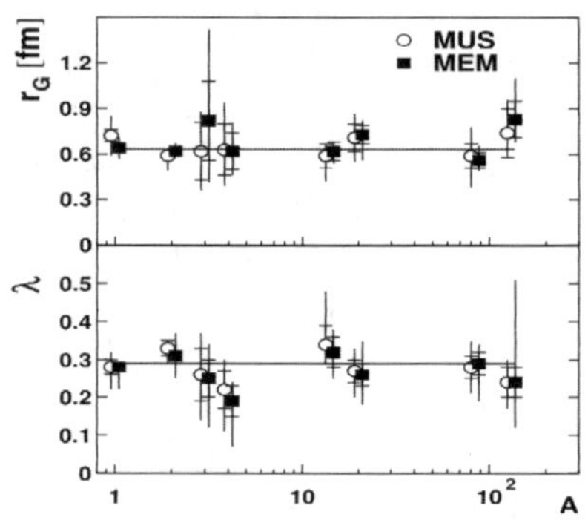

Figure 10. The parameters r_G and λ as a function of the target atomic mass A.

5. Transverse polarization of Λ hyperons from quasi-real photoproduction on nuclei

The transverse polarization of Λ hyperons was measured at HERMES [10] in inclusive quasi-real photoproduction for various target nuclei ranging from hydrogen to xenon. The dependence of the transverse polarization P_n^Λ on the atomic-mass number A of the target nuclei is shown in figure 11. As can be seen from this figure the measured polarization is positive for light target nuclei. It decreases with the atomic-mass number A of the target and is compatible with zero for heavier nuclei krypton and xenon. In figure 12 and figure 13 the dependence of the transverse polarization P_n^Λ is shown respectively as a function of the variable ζ and the transverse momentum p_T of the Λ hyperon for the combined hydrogen and deuterium data (closed symbols) and the combined krypton and xenon data (open symbols). Here, $\zeta = (E^\Lambda + p_L)/(E + k)$ with E^Λ and E being the energies of the produced Λ and the beam lepton, respectively, p_L the Λ's momentum component in the beam direction measured in the target rest frame, and k the lepton momentum. The error bars represent the statistical uncertainty. The values of $\langle p_T \rangle$ for each ζ bin and $\langle \zeta \rangle$ for each p_T bin are shown in the lower panels of corresponding figures.

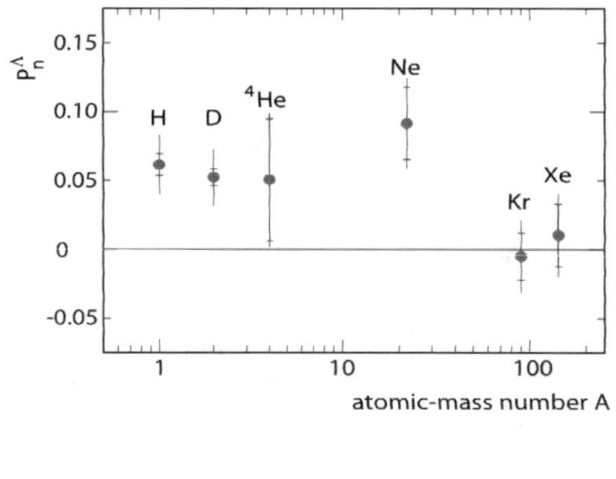

Figure 11. Dependence of the transverse polarization P_n^Λ on the atomic-mass number A of the target nuclei.

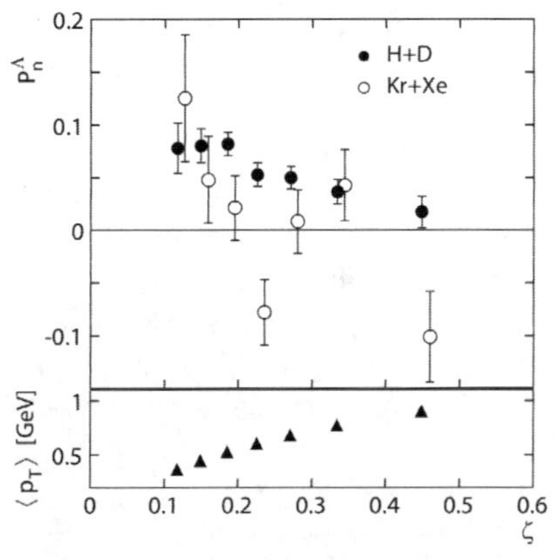

Figure 12. Dependence of the transverse polarization P_n^Λ as a function of the variable ζ.

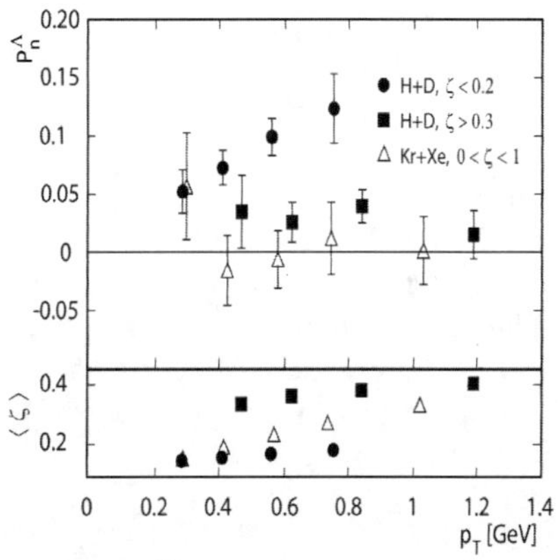

Figure 13. Dependence of the transverse polarization P_n^Λ as a function of the variable p_T.

6. New analysis for the search of pentaquark at HERMES

The earlier search at HERMES for narrow baryon states [11] excited in quasi-real photoproduction, decaying through the channel $pK_S^0 \to p\pi^+\pi^-$, has been extended with improved decay-particle reconstruction (see figure 14), more advanced particle identification, and increased event samples [12]. The invariant mass spectra of two oppositely charged pions plotted in figure 15 show a clear K_S^0 signal peak. For comparison, the standard deviations and mean values of a single Gaussian function fit to the data together with a third-order Chebychev

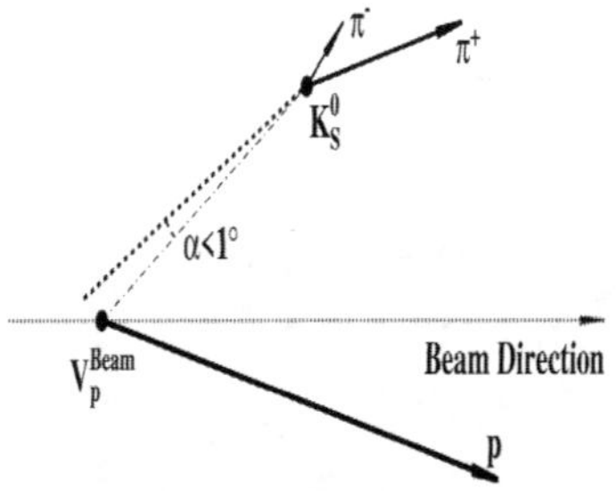

Figure 14. Diagram of the kinematic reconstruction of the decay of a Θ^+. The angle α is the difference in the direction of the K_S^0 momentum (dotted line), as given by the pion momenta, and by the vector connecting the event origin, V_p^{Beam}, with the decay of the K_S^0 (dash-dotted line).

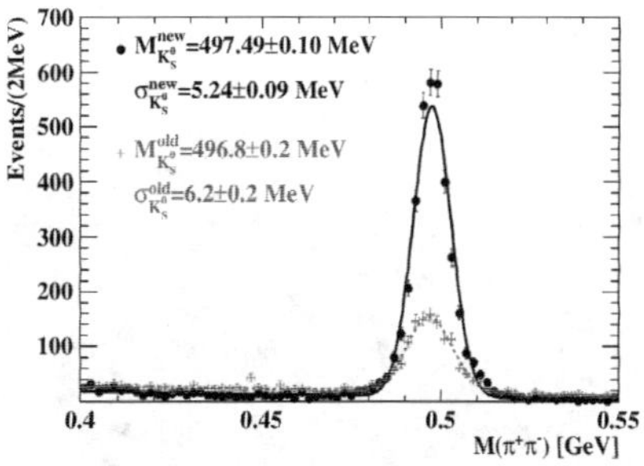

Figure 15. Invariant-mass spectra of two oppositely charged pions. The filled circles denote the analysis with data from 1998-2000 and 2006-2007, while the crosses are from the previously published analysis of the 1998-2000 data [11].

function for the background are given in this figure. As can be seen from figure 15 the new analysis [12] has a much improved mass resolution and signal-to-noise ratio compared to that of the previous HERMES analysis [11]. The $M(pK_S^0)$ spectra for deuterium and hydrogen data taken at the HERMES experiment are shown in figure 16 and figure 17, respectively. A Voigtian (using a Gaussian with a width fixed to 6 MeV) together with two different background hypotheses was fitted to the spectrum in figure 16. The resulting curves are shown separated into signal, background contribution and also their combination. The width Γ of the Breit-Wigner function, the peak position M, and the number of signal events obtained from the fits are given in the panel. The structure observed earlier at an invariant mass of 1528 MeV shifts to 1522 MeV and the statistical significance drops to about 2σ for data taken with a deuterium target. The number of events above background is 68^{+98}_{31}(stat) $\pm$ 13(sys). Note that no such structure is observed in the hydrogen data set.

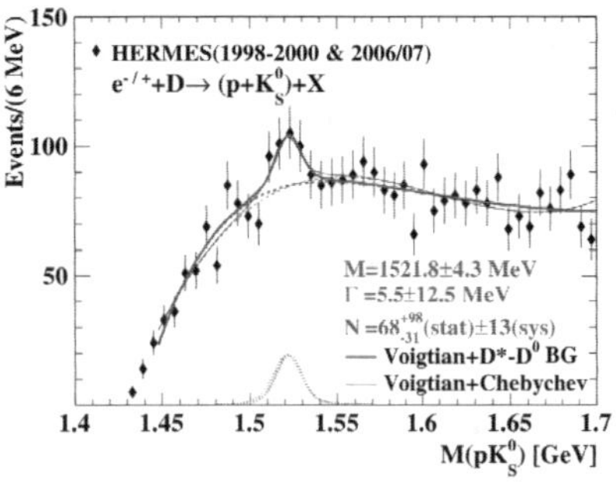

Figure 16. The $M(pK_S^0)$ spectra for deuterium data taken at the HERMES experiment.

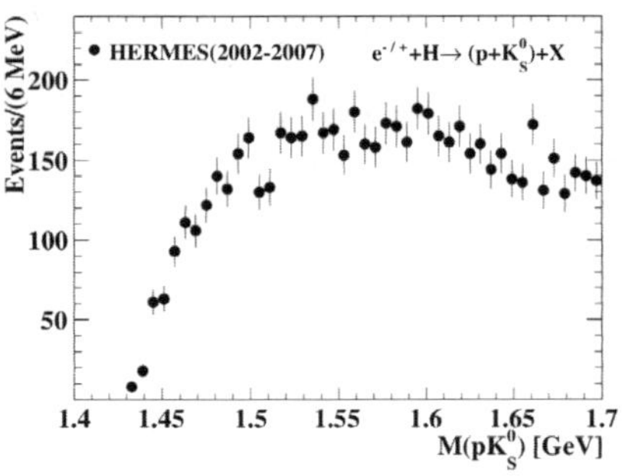

Figure 17. The $M(pK_S^0)$ spectra for hydrogen data taken at the HERMES experiment.

Acknowledgments
I would like to thank the organizers for their support and for a very interesting conference.

References

[1] Sivers D W 1990 *Phys. Rev.* D **41** 83
[2] Collins J C 1993 *Phys. Rev.* B **396** 161
[3] Airapetian A et al. 2009 *Phys. Rev. Lett.* **103** 152002
[4] Airapetian A et al. 2010 *Phys. Lett.* B **693** 11
[5] Airapetian A et al. 2007 *Phys. Lett.* B **648** 164
[6] Airapetian A et al. 2014 *Eur. Phys. J.* C **74** 3110
[7] Airapetian A et al. 2015 *submitted to Eur. Phys. J.* C, arXiv:1508.07612
[8] Goloskokov S V and Kroll P 2014 *Eur. Phys. J.* A **50** 146
[9] Airapetian A et al. 2015 *Eur. Phys. J.* C **75** 361
[10] Airapetian A et al. 2014 *Phys. Rev.* D **90** 072007
[11] Airapetian A et al. 2004 *Phys. Lett.* B **585** 213
[12] Akopov N et al. 2015 *Phys. Rev.* D **91** 057101

Journal of Physics: Conference Series **678** (2016) 012039 doi:10.1088/1742-6596/678/1/012039

Single-spin asymmetry in pp and pA-collisions

V V Abramov

Institute for High Energy Physics National Research Centre "KurchatovInstitute", Protvino 142281, Moscow region, Russia

E-mail: Victor.Abramov@ihep.ru

Abstract. Experimental data on the transverse single-spin asymmetry A_N in the collisions of polarized protons with protons and nuclear targets are analyzed. The existing data are compared with predictions from the chromomagnetic polarization of quarks (CPQ) model for the processes of $\pi^{\pm}$, $K^{\pm}$, p and antiproton production in the inclusive reactions. The results of A_N calculations for the above processes are presented in the following kinematic region: $8.77 \leq \sqrt{s} \leq 500$ GeV, $0 \leq x_F \leq 0.83$, $0 \leq p_T \leq 9$ GeV/c. Predictions of the CPQ model can be used for planning of experiments SPASCHARM(IHEP), SPD(JINR), STAR and PHENIX.

1. Introduction

Studies conducted over the past 35 years have shown that there are significant spin effects in the inclusive processes at energies from a few GeV to hundreds GeV in the c.m.s. In this paper the single-spin asymmetries (SSA), which are measured using a vertically polarised proton beam, are discussed. The observed spin effects tend to be about 10% while the QCD perturbation theory predicts effects of the order of 0.1%. Further experimental research and a global analysis of all available information on single-spin processes are needed to resolve the above contradiction.

2. Chromo-magnetic polarization of quarks model

The main assumptions of the chromo-magnetic polarization of quarks (CPQ) model are:

1) After an initial color charge exchange a longitudinal chromoelectric field $\mathbf{E^a}$ and a circular transverse chromomagnetic field $\mathbf{B^a}$ are generated.
2) Single-spin asymmetry is due to the Stern-Gerlach force acting on a quark in the inhomogeneous transverse circular chromomagnetic field.
3) Quark spin precession in the color field is taken into account.
4) Quark flow diagrams and quark counting rules describe the contributions of quarks and antiquarks to the effective color field. The contributions of quarks and antiquarks are linear functions of their numbers with weights determined by the color factors $C_F(qq)$ and $C_F(q\bar{q})$.

The effective circular transverse chromomagnetic field $\mathbf{B^a}$ is generated by relativistic spectator constituent quarks moving in the forward and backward directions in the c.m. of colliding nucleons (see figure 1). A probe quark from the observed hadron is deflected by the Stern-Gerlach force in the inhomogeneous chromomagnetic field to the left or right depending on the direction of its spin (up or down). Interaction of a chromomagnetic dipole moment of a constituent probe quark with the field $\mathbf{B^a}$ leads to the observed single-spin asymmetries [1, 2, 3].

Content from this work may be used under the terms of the Creative Commons Attribution 3.0 licence. Any further distribution of this work must maintain attribution to the author(s) and the title of the work, journal citation and DOI.
Published under licence by IOP Publishing Ltd

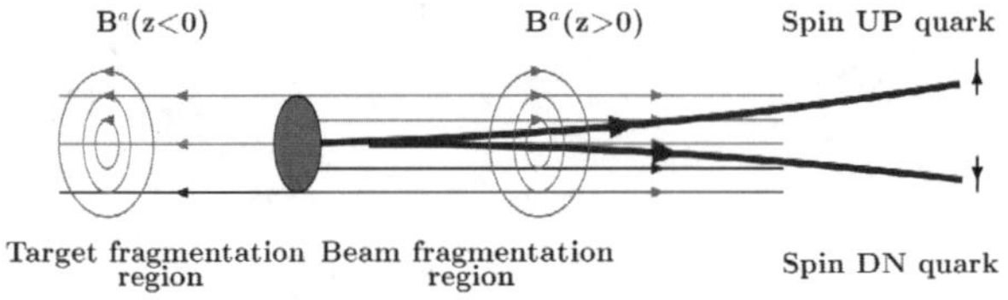

Figure 1. Chromo-magnetic polarization of quarks (microscopic Stern-Gerlach apparatus).

The quark spin precession in the color field changes the direction and magnitude of the Stern-Gerlach force and leads to an oscillation of $A_N(x_F)$ in a strong chromomagnetic field, and to an approximately linear dependence of $A_N(x_F)$ in a weak field. The quark spin precession frequency at high quark energy is proportional to the quark anomalous chromomagnetic moment $\Delta\mu_Q^a$, which is directed along the spin of the quark and has a large negative value [4, 5].

Model-dependent (the CPQ model) estimates of $\Delta\mu_Q^a$ for u, d, s, c and b quarks are obtained from a global analysis of the polarization data [1, 2, 3]. It is found that for the upper u and c quarks $\Delta\mu_Q^a = -0.471 \pm 0.007$, while for the lower d, s, b quarks it is by a factor of $\sqrt{2/3}$ less within experimental uncertainties. The model parameters were estimated from a global analysis of the single-spin data for 80 different reactions with about 3000 data points [1, 2, 3].

The quark counting rules tell us that the integral of the color field along the quark trajectory is a linear function of the number of spectator quarks and antiquarks with different weights. The weight depends on the color factor C_F of a quark pair, where one quark is a spectator and the other one is a probe quark from the observed hadron. It is found that $q\bar{q}$ pair is in a singlet state with the $C_F = 4/3$ and weight equal to 1. A qq pair is in an antitriplet state with the $C_F = 2/3$ and weight $\lambda = -0.1338 \pm 0.0014$ found from the global polarization data fit [1, 2, 3]. The value of λ can be expressed as a function of the ratio $R_F = C_F(qq)/C_F(q\bar{q})$: $\lambda = 1 - exp(R_F^3)$ since the square of the wave function for a pair of quarks is proportional to the factor C_F^3 [6].

3. Comparison of the A_N data and the CPQ model predictions

Below are presented the results of the A_N measurements and the CPQ model predictions. Calculations of A_N in unexplored areas are made for a fixed $p_T = 3$ GeV/c and $\sqrt{s} > 60$

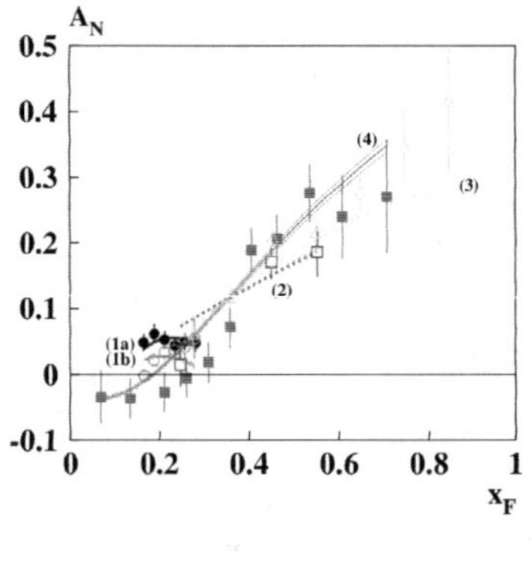

Figure 2. $A_N(x_F)$ data for reaction $p^\uparrow + p(A) \to \pi^+ + X$.

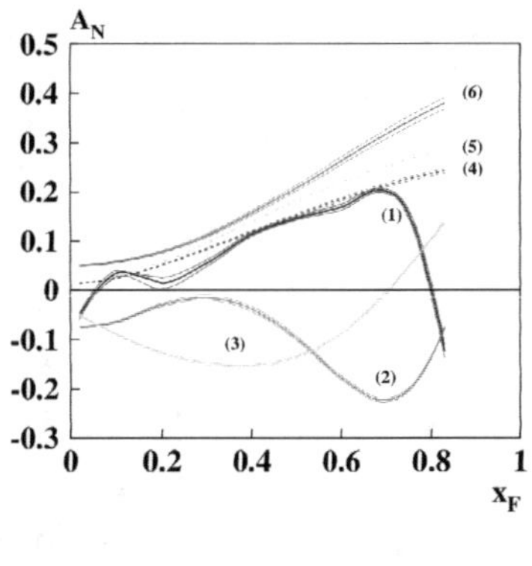

Figure 3. $A_N(x_F)$ predictions for reaction $p^\uparrow + p \to \pi^+ + X$.

GeV and for $p_T = 2$ GeV/c and $\sqrt{s} \leq 60$ GeV. The $A_N(x_F)$ is predicted for $\sqrt{s} = 500$ (1), 200 (2), 130 (3), 64.2 (4), 19.4 (5) and 8.77 (6) GeV, respectively, and is shown in the right panels.

The data for reaction $p^{\uparrow} + p(A) \to \pi^{+} + X$ at $\sqrt{s} = 200$ (1a, 1b) and 64.2 (2) (BRAHMS [7, 8]), 19.4 (3) (E704 [9]) and 8.77 (4) GeV (FODS [10]) are shown in figure 2. These A_N data posess the energy dependence that is consistent with the coresponding CPQ model curves. Predictions of $A_N(x_F)$ in figure 3 show an approximate scaling law for $\sqrt{s} \leq 60$ GeV (curves 4,

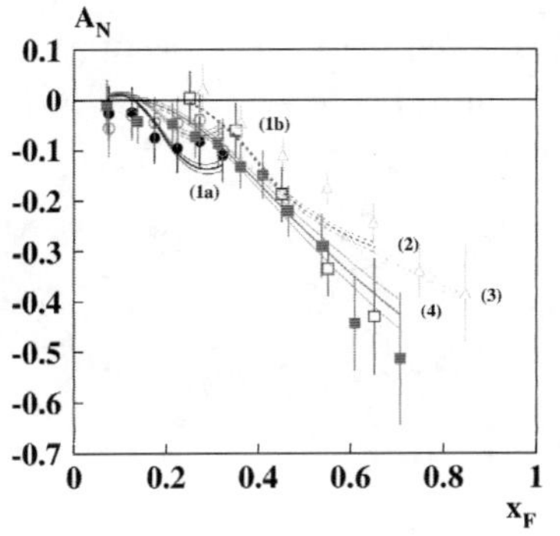

Figure 4. $A_N(x_F)$ data for reaction $p^{\uparrow} + p(A) \to \pi^{-} + X$.

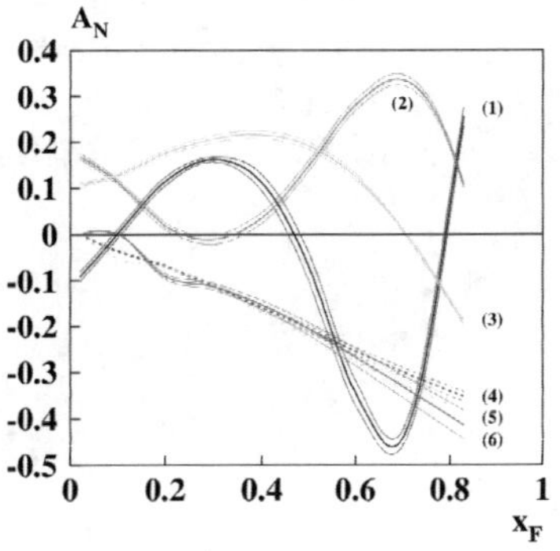

Figure 5. $A_N(x_F)$ predictions for reaction $p^{\uparrow} + p \to \pi^{-} + X$.

5 and 6). At higher energies this scaling is violated due to additional $q\bar{q}$ pair production that changes the number of spectator quarks and, as a result, changes the color field strength and its sign. The second reason for the $A_N(x_F)$ scaling violation is the quark spin precession, which changes the direction of the Stern-Gerlach force and leads to oscillation of $A_N(x_F)$ in a strong chromomagnetic field. The CPQ model predicts $A_N(x_F) < 0$ at $\sqrt{s} = 130$ GeV (curve 3).

The data for reaction $p^{\uparrow} + p(A) \to \pi^{-} + X$ [7, 8, 9, 10] are shown in figure 4. For π^{-} production the SSA scaling is expected to be at $\sqrt{s} \leq 60$ GeV (see figure 5, curves 4, 5 and 6).

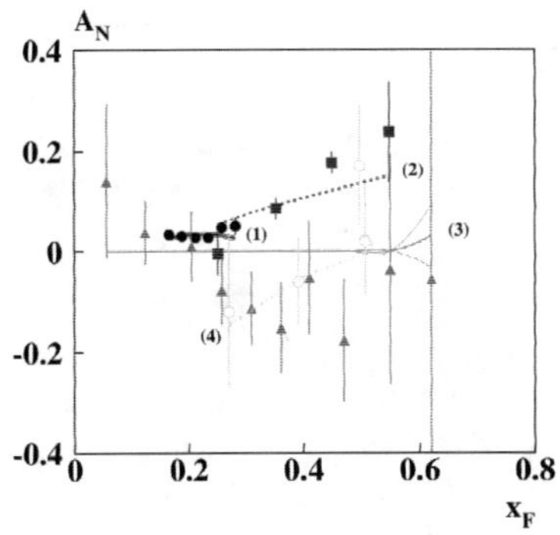

Figure 6. $A_N(x_F)$ data for reaction $p^{\uparrow} + p(A) \to K^{-} + X$.

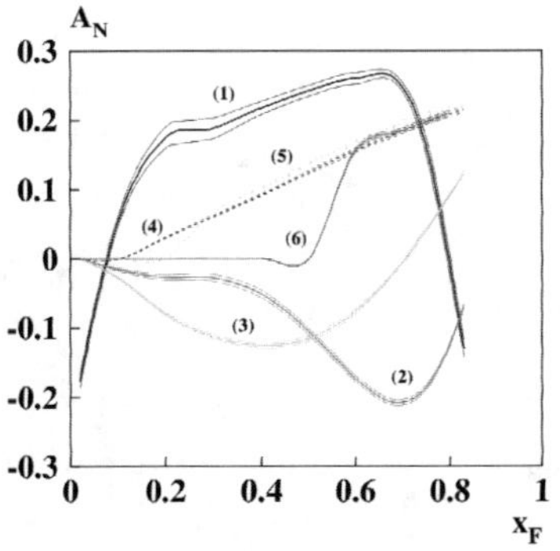

Figure 7. $A_N(x_F)$ predictions for reaction $p^{\uparrow} + p \to K^{-} + X$.

The K^{-} SSA data [7, 8, 10, 12] are compared with the model calculations in figure 6. In a K^{-} meson there is no valence quarks common to the polarized proton but the $A_N(x_F) > 0$ at 64.2 GeV [8]. The positive A_N (solid squares) is explained by the dominating polarized u quark scattering in an effective color field. The positive SSA of the u quark, obtained in the scattering, is then inherited by a K^{-} meson in the process of $u^{\uparrow} \to K^{-} + X$ fragmentation. Scaling law for $A_N(x_F)$ is held only at large $x_F > 0.6$ and $\sqrt{s} \leq 60$ GeV (see figure 7, curves 4, 5 and 6).

Positive SSA was also measured for the antiproton production at $\sqrt{s} = 200$ GeV [7] (not shown). This is due to the dominance of the process of $u^\uparrow \to \bar{p} + X$ fragmentation, analogous to the case of K^- meson production. The dependence on energy of the $A_N(x_F)$ for antiproton production is predicted to be weak in the region of $x_F > 0.6$ and $\sqrt{s} \leq 60$ GeV, in accordance with the CPQ model.

The proton production SSA data are shown in figure 8 [7, 8, 10, 11] and the CPQ model predictions in figure 9. A sizable SSA was measured in FODS experiment [10]. The oscillation of $A_N(x_F)$ (solid squares [10] and curve 3) at 8.77 GeV is due to the spin precession of a polarized u quark. The scaling behavior is not expected for the $A_N(x_F)$ (see figure 9).

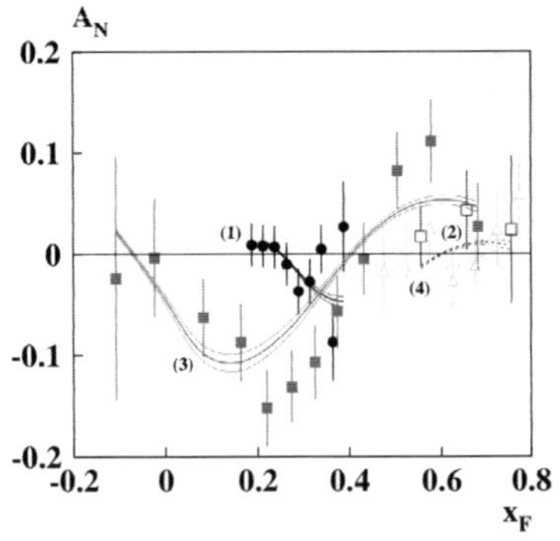

Figure 8. $A_N(x_F)$ data for reaction $p^\uparrow + p(A) \to p + X$.

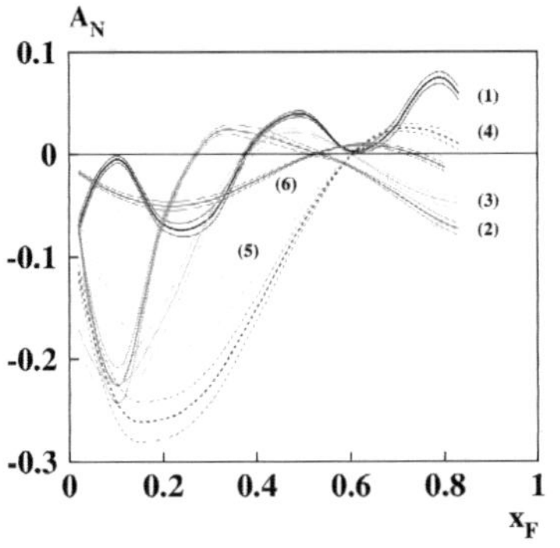

Figure 9. $A_N(x_F)$ predictions for reaction $p^\uparrow + p \to p + X$.

4. Conclusion

1) A scaling law of $A_N(x_F)$ is expected for the production of π^+, π^- and K^+ at $\sqrt{s} < 60$ GeV. The polarized valence quarks fragment into the observed hadrons for these reactions.
2) Generation of new $q\bar{q}$ pairs at $\sqrt{s} > 60$ GeV violates the scaling of $A_N(x_F)$ for the production of π^+, π^- and K^+ mesons. The color field can change its sign and strength at high energies.
3) $A_N(x_F) > 0$ at large $x_F > 0.6$ and $\sqrt{s} < 60$ GeV for K^- and $\bar{p}$, as in the case of $A_N(x_F)$ for π^+ production due to the dominance of polarized u-quark fragmentation to a hadron.
4) A sizable $A_N(x_F)$ and its oscillation is expected for the proton production due to the spin precession of polarized valence quarks in the effective chromomagnetic field.

Acknowledgments

This work was supported in part by the RFBR Contract (Grant No. 15-02-01669).

[1] Abramov V V 2009 *Phys. At. Nucl.* **72** 1872-88
[2] Abramov V V 2011 *J. Phys. Conf. Ser.* **295** 012086
[3] Abramov V V 2014 *Phys. Part. Nucl.* **45** 62-65
[4] Kochelev N I 1998 *Phys. Lett.* B **426** 149-53
[5] Diakonov D 2003 *Prog. Part. Nucl. Phys.* **51** 173-222
[6] Baranov S P 1996 *Phys. Rev.* D **54** 3228-36
[7] Lee J H and Videbaek F 2009 Cross-sections and Single Spin Asymmetries of Identified Hadrons in $p^\uparrow + p$ at $\sqrt{s} = 200$-GeV (*Preprint* arXiv:0908.4551)
[8] Arsene I et al 2008 *Phys. Rev. Lett.* **101** 042001
[9] Adams D L et al 1991 *Phys. Lett.* B **264** 462-66
[10] Abramov V V et al 2007 *Phys. At. Nucl.* **70** 1515-26
[11] Krueger K, Allgower C et al 1999 *Phys. Lett.* B **459** 412-16
[12] Dragoset W H et al 1978 *Phys. Rev.* D **18** 3939-54

Charge exchange $dp \to (pp)n$ reaction study at 1.75 A GeV/c by the STRELA spectrometer

S.N. Basilev[1], Yu.P. Bushuev[1], V.V. Glagolev[1], S.A. Dolgiy[1], D.A. Kirillov[1], N.V. Kostyaeva[1], A.D. Kovalenko[1], A.N. Livanov[1], P.K. Manyakov[1], G. Martinská[2], J. Mušinský[3] N.M. Piskunov[1], A.A. Povtoreiko[1], P.A. Rukoyatkin[1], R.A. Shindin[1], I.M. Sitnik[1], V.M. Slepnev[1], I.V. Slepnev[1] and J. Urbán[2]

[1]Joint Institute for Nuclear Research, 141980 Dubna, Moscow region, Russia
[2]University of P.J. Šafárik, Jesenná. 5, SK-04154 Košice, Slovak Republic
[3]Institute of Experimental Physics, Watsonova 47, SK-04001 Košice, Slovak Republic

E-mail: gabriela.martinska@upjs.sk

Abstract. The differential cross sections of the charge exchange reaction $dp \to (pp)n$ has been measured at 1.75 GeV/c momentum per nucleon for small transferred momenta using the one arm magnetic spectrometer STRELA at the Nuclotron accelerator. The ratio of the differential cross section of the charge exchange reaction $dp \to (pp)n$ to that of the $np \to pn$ elementary process is discussed in order to estimate the spin-dependent part of the $np \to pn$ charge exchange amplitude on the basis of $dp \to (pp)n$ data. The $np \to pn$ amplitude turned out to be predominantly spin-dependent.

1. Introduction

Extracting complex amplitudes of the scattering matrix is a matter of fundamental importance in the theory of nucleon-nucleon scattering. For all amplitudes to be obtained, a complete experiment must be performed, i.e., an experiment with a set of experimentally observed quantities providing a complete and exhaustive description of this process. The complete experiment comprises measurements both with polarized projectile particles and with polarized targets. Many publications emphasized the possibility of using the charge exchange reaction on the unpolarized deuteron $dp \to (pp)n$ to determine the spin-dependent part of elementary $np \to pn$ charge exchange process. The increasing interest in obtaining this information during the past few years may be partly connected with the appearance of accelerated deuterons at the JINR LPHE Nuclotron to energies over 1 A GeV. The basic idea was proposed and mathematically formalized in a number of theoretical works [1–4]. The differential cross section of the elementary $np \to pn$ charge exchange can be represented as sum of the spin-independent (superscript SI) and spin-dependent (superscript SD) parts:

$$(d\sigma/dt)_{np\to pn} = (d\sigma/dt)^{SI}_{np\to pn} + (d\sigma/dt)^{SD}_{np\to pn}\,. \tag{1}$$

Mathematical formalism developed in [3, 4] allows to connect the differential cross section for the deuteron charge exchange break-up and the elementary $np \to pn$ reaction. The differential

Content from this work may be used under the terms of the Creative Commons Attribution 3.0 licence. Any further distribution of this work must maintain attribution to the author(s) and the title of the work, journal citation and DOI.
Published under licence by IOP Publishing Ltd

Journal of Physics: Conference Series **678** (2016) 012040 doi:10.1088/1742-6596/678/1/012040

cross section for $dp \to (pp)n$ charge exchange in the framework of the impulse approximation and at small t is :

$$(d\sigma/dt)_{dp\to(pp)n} = [1 - S(t)]\,(d\sigma/dt)^{SI}_{np\to pn} + [1 - 1/3S(t)]\,(d\sigma/dt)^{SD}_{np\to pn}\,, \tag{2}$$

where $S(t)$ is the deuteron form factor, $t = (P_d - P_1 - P_2)^2$ is the 4-momentum transfer squared from the incoming deuteron to the two fast protons.
This expression implies that at 0° scattering angle $t = 0$, $S(0) = 1$ and the formula reduces to

$$(d\sigma/dt)_{dp\to(pp)n} = 2/3\,(d\sigma/dt)^{SD}_{np\to pn}\,. \tag{3}$$

Thus, the charge exchange break-up reaction of the unpolarized deuteron on the unpolarized proton target $dp \to (pp)n$ in the forward direction ($t = 0$) is determined by the spin-flip part of the $np \to pn$ charge exchange process at 0 scattering angles.To extract the information we compare our experimental data (charge exchange differential cross section at $t = 0$) with the $np \to pn$ (charge exchange cross section at $t = 0$) data at the same energy available in the literature. The first experiment of thus type has been realized at the JINR synchrophasotron irradiating the one-metre hydrogen bubble chamber with deuteron beams of 3.35 GeV/c momenta [5]. The value of the spin-independent part of the elastic $np \to pn$ charge exchange cross sections $R^{ID}_{np} = 0.21 \pm 0.17$ has been obtained. So, the $np \to pn$ amplitude turned out to be entirely spin dependent. Before our investigations, no experiments with a fast deuteron beam were carried out. The aim of present study is to investigate the spin effects in nonpolarized deuteron beams using counter technique for obtaining a statistically justified results in determining the contribution of the spin-dependent components of the elementary $np-$scattering amplitude.

2. The experimental set-up STRELA

Based on the above mentioned ideas and experimental results the experiment STRELA was designed and constructed in the LHEP JINR Dubna with the aim to select and detect charge exchange events in deuteron proton collisions. The experiment demands registration of two protons in a narrow cone from the reaction $dp \to (pp)n$ with momenta approximately equal to the half of the primary deuteron beam momenta. STRELA is a typical one-arm magnetic spectrometer composed of scintillator detectors (S1, S2) used to trigger the setup, block of drift chambers (DC1-DC4) used as coordinate detector and analyzing magnet (M). The recent version of the experimental setup is shown in Figure 1. The sensitive areas of the drift chambers are the following: $125 \times 125\ mm^2$ for $DC1, DC2$ and $250 \times 250\ mm^2$ for $DC3, DC4$. Z axis is chosen in the beam direction. Chambers $DC1, DC3, DC4$ are equiped with xy wires and $DC2$ only with x-wires. $DC1$ and $DC3$ are composed of $8(4y, 4x)$ sensitive planes, $DC4$ is composed of $4(2y, 2x)$ sensitive planes while the $DC2$ contains $4(4x)$ sensitive planes. The intensity of the magnetic field $B = 0.85\ T$ are used to select deuterons and two protons, respectively. The

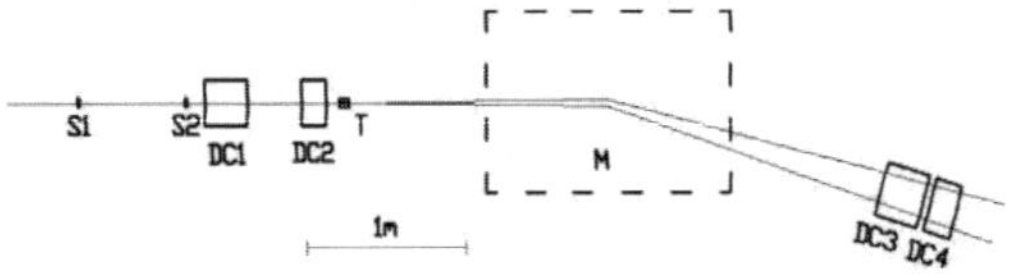

Figure 1. Layout of the experimental setup for determining the spin-dependent part of $np \to pn$ scattering: S1, S2-scintillator counters, (DC1-DC4) drift chambers , M-analyzing magnet, T-target.

drift length for all chambers is $r_{max} = 21\ mm$. The control and read-out electronics were upgraded with new VME crates and all the modules were tested. The basic characteristics of the drift chambers were established from irradiation of a polyethylene target with a deuteron beam of 3.5 GeV/c momentum. For each wire the minimal (t_{min}) and maximal (t_{max}) drift times were determined. The average total drift time was found to be $\approx 450ns$. In the track finding procedure the relation between the measured drift time and the minimal distance (or radius (r)) from the anode wire to the track plays an important role. For determination of the transformation function two methods were applied: the linear one, mainly used for on-line and the cumulative or integral one suitable for off-line purposes. The lateral resolution of the drift chambers which are $(80 \div 90)\ \mu m$ and $(120 \div 130)\ \mu m$ for large and small drift chambers, respectively [6]. The carbon and polyethylene targets were used to extract the dp interaction.

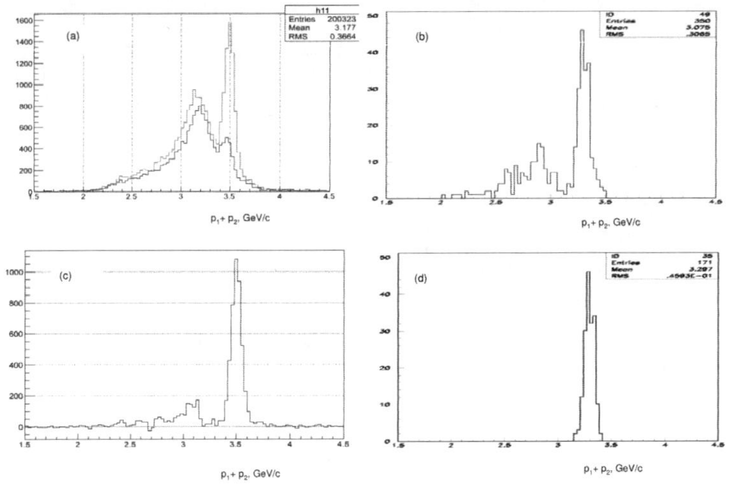

Figure 2. Distributions of the sum of the two proton momenta from dp reaction. Left side- experimental results, right side - simulation: a) targets CH_2-red, C- blue one; c) difference $CH_2 - C$; simulation included all channels *dp*- reaction (b) and only $dp \to ppn$ reaction (d).

The size of the targets were determined by carbon nuclei equivalent. Carbon target was used to account for the background events. The background from other channels of the *dp* reaction and the influence of carbon nuclei were estimated by the use of the GEANT simulation program for transporting the reaction products (taken from the corresponding events of the one meter bubble chamber at the momenta 3.3 GeV/c see [5]) through the experimental setup.

3. Results and discussion

The experimental facility was irradiated in the beam of deuterons with 3.5 GeV/c momenta. In the run of March 2014 one billion triggers were received. The dependences of the sum two proton momenta from dp reaction are presented in Fig. 2 for CH2 and C targets (a) and difference CH2-C (c). The results of simulation include all channels of dp-reaction (b) and channel $dp \to ppn$ only (d) are shown in Fig. 2 b,d. As we can see, the distribution has a characteristic peak near the incoming deuteron momentum kinematically associated with the pair of protons from the reaction $dp \to ppn$ (c). The contribution from the background reactions, other than the studied reaction $dp \to ppn$, which could also produce the two positively charged track in the forward direction, is negligible see Fig.2d). The obtained differential distribution dN/dt of the $dp \to (pp)n$ of the reaction (Fig.3) was extrapolated to $t = 0$ by the expression $dN/dt = (435.6 \pm 6.8) \bullet \exp((440.9 \pm 5.8)t)$ with $\chi^2 = 77.93/78$. Extrapolation to $t = 0$ gives: $dN/dt|_{t=0} = 435.6 \pm 6.8$. This value corresponds to charge exchange differential cross

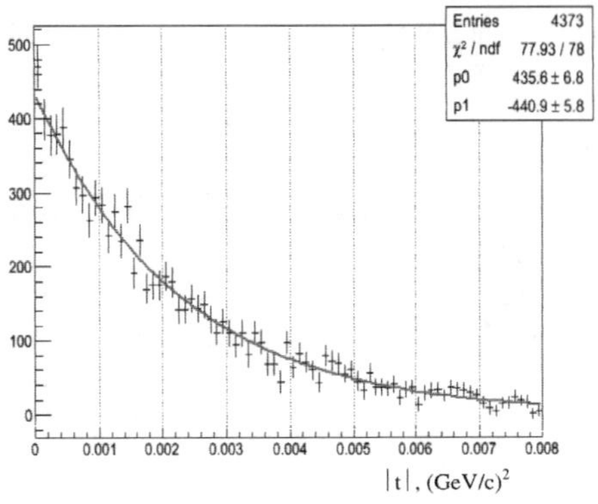

Figure 3. Differential distribution dN/dt of the $dp \to (pp)n$ reaction.

section $d\sigma/dt|_{t=0} = 30.56 \pm 0.48\ mb/(GeV/c)^2$. The obtained charge exchange differential cross section on the deuteron at $t = 0$ was compared with the available data from $np \to pn$ reaction at the same energy. The closest energy data comes from measurements made at the SATURN accelerator [7, 8]. The values of $d\sigma/dt|_{t=0}$ of $np \to pn$ reaction at our incident momentum 1.75 AGeV/c gave the following value of $d\sigma/dt|_{t=0} = 48.0 \pm 0.2 mb/(GeV/c)^2$. The obtained value will be related to the estimated differential cross section of the quasi-elastic $dp \to (pp)n$ charge-exchange at $t = 0$ from our experiment. One would like to stress that the systematic error in the data by Bizard et al. [7] makes 5% . One can introduce the ratio of the differential cross sections at $t = 0$ for the forward scattering (charge exchange) on the deuteron and proton $R = (d\sigma/dt)_{dp} \,/\, (d\sigma/dt)_{np} = 0.637 \pm 0.010$. Under the assumption stated above (see equation (3)) between R and the ratio of the spin-independent part to the spin-dependent one of the elastic $np \to pn$ charge exchange cross sections the following relation holds

$$R_{np}^{ID} = \left(\frac{d\sigma}{dt}\right)_{np}^{SI} \Big/ \left(\frac{d\sigma}{dt}\right)_{np}^{SD} = \frac{2}{3 \times R} - 1 = 0.047 \pm 0.017\ . \qquad (4)$$

4. Conclusion

The value of the charge exchange reaction $dp \to (pp)n$ differential cross section $d\sigma/dt = 30.56 \pm 0.48 mb/(GeV/c)^2$ has been obtained at 1.75 A GeV/c and t=0. The obtained ratio of the charge exchange differential cross sections at $t = 0$ for $dp \to (pp)n$ and $np \to pn$ reactions is $R = 0.637 \pm 0.010$. So, the spin-independent part of the elastic $np \to pn$ charge exchange cross section $R_{np}^{ID} = 0.047 \pm 0.017$ shows onto the predominant contribution of the spin-dependent part of the amplitude $np \to pn$ scattering.

5. Acknowledgment

This work was supported by Slovak Grant Agency VEGA, under number 1/0556/16.

6. References

[1] I.Pomeranchuk 1951 *Dokl. Akad. Nauk SSSR* **78** 249
[2] G.E.Chew 1950 *Phys. Rev.* **80** 196
[3] N.W.Dean 1972 *Phys. Rev.* D bf 5 2832 ; N.W.Dean 1972 *Phys. Rev.* D **5** 1661
[4] D.V.Bugg and C.Wilkin 1987 *Nucl. Phys.* A **167** 575
[5] V.V.Glagolev et al. 2002 *Eur. Phys. J.* A. **15** 471; V.V.Glagolev et al. 2008 *Cent. Eur. J. Phys.* **6** 781
[6] V.V. Glagolev et al. 2013 *Prib. Tekh. Eksp.* **No 4** 20
[7] G. Bizard et al. 1975 *Nuclear Physics* B **85** 14
[8] J. Bystricky and F. Lehar 1978 *Nucleon-Nucleon Scattering data, editors H. Behrens and G. Ebel, Fachinformationszentrum Karlsruhe, 1978 Edition* **N 11-1** 521

Journal of Physics: Conference Series **678** (2016) 012041 doi:10.1088/1742-6596/678/1/012041

Radiative decays of light vector mesons in Poincare invariant quantum mechanics

Viktor Andreev
Francisk Skorina Gomel State University, Theoretical Physics Department, Belarus.
E-mail: vik.andreev@gsu.by

Vadzim Haurysh
Francisk Skorina Gomel State University, Theoretical Physics Department, Belarus.
E-mail: mez0n@inbox.ru

Abstract. This work dedicated to describing two-body compounded system of quarks (meson) in point form of Poincare invariant quantum mechanics with potential, which was offered by so-called Mock-meson model. Authors shows process of calculation the basis parameters of the model by the variational method and followed applications for radiative decay processes. For such calculation authors use the simplest radiative decay scheme. Comparing results with experimental data was shown how to calculate the anomalous part of quark magnetic moment.

1. Introduction

The radiative processes, in particular the decays of vector mesons, has been a convenient tool for studying the structure of hadrons. There are quite a number of approaches for the model to describe radiative transitions mesons(see,[1, 2, 3, 4]). In our work the calculation of the form-factor of the radiative decay conducted within the constituent relativistic quark model, based on the point form of Poincare invariant quantum mechanics (about Poincare invariant quantum mechanics, see, eg [5, 6]).

2. Elements of quark model, based on Poincare invariant quantum mechanics

For the our relativistic quark model, based on the Poincare invariant quantum mechanics, used the potential proposed by the so-called Mock-meson model [7] with a model parameterization running coupling constant of the strong interaction, modified in [8]:

$$\alpha_S(Q^2) = \sum_{k=1}^{7} \alpha_k \exp\left(-Q^2/4\gamma_k^2\right) \,. \tag{1}$$

The interquark potential in coordinate representation from [7] is used, which is considered a sum of Coulomb, linear confinement, and spin-spin parts for pseudoscalar and vector mesons:

$$\hat{V}(\mathrm{r}) = \hat{V}_{\mathrm{Coul}}(\mathrm{r}) + \hat{V}_{\mathrm{conf}}(\mathrm{r}) + \hat{V}_{\mathrm{SS}}(\mathrm{r}) \tag{2}$$

Content from this work may be used under the terms of the Creative Commons Attribution 3.0 licence. Any further distribution of this work must maintain attribution to the author(s) and the title of the work, journal citation and DOI.
Published under licence by IOP Publishing Ltd

Journal of Physics: Conference Series **678** (2016) 012041 doi:10.1088/1742-6596/678/1/012041

where

$$\hat{V}_{\rm Coul}(r) = -\frac{4}{3}\frac{\alpha_S(\rm r)}{\rm r} = -\frac{4}{3\rm r}\sum_{k=1}^{7}\alpha_k {\rm erf}(\tau_k \rm r)\ , \tag{3}$$

$$\hat{V}_{\rm conf}(\rm r) = w_0 + \sigma \rm r\left(\frac{\exp(-b^2 r^2)}{\sqrt{\pi} b r} + \left(1+\frac{1}{2b^2 r^2}\right){\rm erf}(b \rm r)\right) \tag{4}$$

and

$$\hat{V}_{\rm SS}(\rm r) = -\frac{32\left(\mathbf{S}_q \mathbf{S}_{\bar{Q}}\right)}{9\sqrt{\pi} m_q m_{\bar{Q}}}\sum_{k=1}^{7}\alpha_k \tau_k^4 \exp(-\tau_k^2 \rm r^2)\ . \tag{5}$$

Potential has following free parameters: gluon string tension parameter σ, parameter of perturbative part w_0 and masses of quarks $m_{q,\bar{Q}}$. Parameter τ_k determined from relation

$$\tau_k^2 = \frac{\gamma_k^2}{\gamma_k^2 + b} b\ , \tag{6}$$

where b is smearing parameter [7].

Parameter of linear part of the potential in most number of models lies in limits $\sigma = (0.18 \div 0.20)\ {\rm GeV}^2$, that's why we assume, that

$$\sigma = \bar{\sigma} \pm \triangle\sigma = (0.19 \pm 0.01)\ {\rm GeV}^2\ . \tag{7}$$

The parameters α_k, γ_k are fixed on the basis of requirements consistent with experimental data for the difference in the first moments of the proton and neutron spin structure functions (QCD-modified Bjorken sum rule [8, 9]).

In work [10] were calculated integral representation of decay constant for pseudoscalar and vector meson in framework of point form of Poincare invariant quantum mechanics:

$$f_P(m_q, m_{\bar{Q}}, \beta) = \frac{3}{\sqrt{2}\pi}\int_0^\infty {\rm dk}\,{\rm k}^2\,\Phi({\rm k},\beta)\sqrt{\frac{M_0^2 - (m_q - m_{\bar{Q}})^2}{\omega_{m_q}({\rm k})\,\omega_{m_{\bar{Q}}}({\rm k})}\frac{(m_q + m_{\bar{Q}})}{M_0^{3/2}}}\ , \tag{8}$$

$$\begin{aligned} f_V\left(m_q, m_{\bar{Q}}, \beta\right) = \frac{3}{\sqrt{2}\pi}\int_0^\infty {\rm dk}\,{\rm k}^2\,\Phi({\rm k},\beta)\frac{\sqrt{\left(\omega_{m_q}({\rm k}) + m_q\right)\left(\omega_{m_{\bar{Q}}}({\rm k}) + m_{\bar{Q}}\right)}}{\sqrt{\omega_{m_q}({\rm k}) + \omega_{m_{\bar{Q}}}({\rm k})}\sqrt{\omega_{m_q}({\rm k})\,\omega_{m_{\bar{Q}}}({\rm k})}}\times \\ \times\left(1 + \frac{{\rm k}^2}{3\left(\omega_{m_q}({\rm k}) + m_q\right)\left(\omega_{m_{\bar{Q}}}({\rm k}) + m_{\bar{Q}}\right)}\right)\ , \end{aligned} \tag{9}$$

where $M_0 = \omega_{m_q}({\rm k}) + \omega_{m_{\bar{Q}}}({\rm k})$ and $\omega_m({\rm k}) = \sqrt{{\rm k}^2 + m^2}$.

Using equations (2)-(5) and (8),(9) we get the following system of equations [8]:

$$M_P(m_q, m_{\bar{Q}}, w_0, \beta) = M_P^{\rm exp} \pm \Delta M_P\ , \tag{10}$$

$$M_V(m_q, m_{\bar{Q}}, w_0, \beta) = M_V^{\rm exp} \pm \Delta M_V\ , \tag{11}$$

$$M_V(m_q, m_{\bar{Q}}, w_0, \beta) - M_P(m_q, m_{\bar{Q}}, w_0, \beta) = M_V^{\rm exp} - M_P^{\rm exp} \pm \delta M_{V,P}^{\rm exp}\ , \tag{12}$$

$$f_P(m_q, m_{\bar{Q}}, \beta) = f_P^{\rm exp} \pm \Delta f_P^{\rm exp}\ , \tag{13}$$

$$f_V(m_q, m_{\bar{Q}}, \beta) = f_V^{\rm exp} \pm \Delta f_V^{\rm exp}\ , \tag{14}$$

where $M_P^{\exp}, M_V^{\exp}$ - experimental value of pseudoscalar and vector mesons, respectively. The last two equations (13)-(14) express condition of equality lepton constants for the pseudoscalar and vector mesons, calculated in the framework of Poincare covariant model, with experimental values of decay constants. It's should be notice, that during the calculation the wave function was taken in form

$$\Phi(\mathrm{k},\beta) = \frac{2}{\sqrt{3}\pi^{1/4}\beta^{3/2}} \exp\left(-\frac{\mathrm{k}^2}{2\beta^2}\right) . \tag{15}$$

After solving the system of equations (10)-(14) for light mesons we have following values of quark masses and β-parameters of wave function:

$$m_u = (239.9 \pm 2.3)\ \mathrm{MeV}\ ,\quad m_d = (243.8 \pm 2.3)\ \mathrm{MeV}\ ,\quad m_s = (466.6 \pm 28)\ \mathrm{MeV}\ ,$$

$$\beta_{uu} \simeq \beta_{dd} \simeq \beta_{ud} = (328.78 \pm 2.1)\ \mathrm{MeV}\ ,\quad \beta_{us} \simeq (360.3 \pm 12.1)\ \mathrm{MeV}\ . \tag{16}$$

Thus, we have fixed all basic parameters of the model by equation (16).

3. Radiative decay of vector mesons

Matrix element of the radiative decay process $V \to P\gamma^*$ could be parameterized using 4-velocity of the vector and pseudoscalar mesons by expression:

$$_P\langle Q', M_P|\hat{J}^\alpha|Q, M_V\rangle_V = \frac{e}{(2\pi)^3}\ g_{VP\gamma^*}(t)K^\alpha(\mu)\frac{\sqrt{M_V M_P}}{\sqrt{4V_0V_0'}}\ , \tag{17}$$

where $K^\alpha(\mu) = \mathrm{i}\epsilon^{\alpha\nu\rho\sigma}\varepsilon_\nu(\mu)V_\rho V'_\sigma$ and $e = \sqrt{4\pi\alpha_{\mathrm{QED}}}$.

In framework of Poincare invariant quantum mechanics we consider mesons P and V as relativistic constituent quark-antiquark system. In such approach decay caused by the emission of quark a γ^*-quantum. In generalized Breit system it's easy to show, that

$$g_{VP\gamma^*}(t) = \frac{1}{4\pi\sqrt{M_V M_P}}\int \mathrm{dk}\ \mathrm{k}^2\Phi(\mathrm{k},\beta)\sqrt{\frac{1}{\omega_{m_q}(\mathrm{k})\omega_{m_{\bar{Q}}}(\mathrm{k})}}\sqrt{\frac{3+4\nu_1(\lambda-\nu_1)}{2}}\nu_1'\times$$

$$\times\left[\Phi(\mathrm{k}_2,\beta)\sqrt{\frac{\omega_{m_{\bar{Q}}}(\mathrm{k}_2)}{\omega_{m_q}(\mathrm{k}_2)}}\bar{u}_{\nu_1'}(\mathbf{k}_2,m_q)B^{-1}(\boldsymbol{v}_{\mathbf{Q}})\frac{(\Gamma_2\cdot K^*)}{(K\cdot K^*)}u_{\nu_1}(\mathbf{k},m_q)D^{1/2}_{-\nu_1',\lambda-\nu_1}(\mathbf{n}_{W_2}(\mathbf{k},\boldsymbol{v}_Q))+\right.$$

$$\left.\Phi(\mathrm{k}_2,\beta)\sqrt{\frac{\omega_{m_q}(\mathrm{k}_2)}{\omega_{m_{\bar{Q}}}(\mathrm{k}_2)}}\bar{\upsilon}_{\lambda-\nu_1}(\mathbf{k},m_{\bar{Q}})\frac{(\Gamma_1\cdot K^*)}{(K\cdot K^*)}B(\upsilon_{\mathbf{Q}})\upsilon_{-\nu_1'}(\mathbf{k}_2,m_{\bar{Q}})D^{1/2}_{\nu_1',\nu_1}(\mathbf{n}_{W_1}(\mathbf{k},\boldsymbol{v}_Q))\right]\ , \tag{18}$$

where k is relative momentum and

$$\mathbf{n}_{\mathbf{W}_{2,1}} = -\frac{[\mathbf{k},\mathbf{V}]}{\omega_{m_{q,\bar{Q}}} + m_{q,\bar{Q}} - (\mathbf{kV})}\ , \tag{19}$$

$$\Gamma^\mu_{1,2} = F_1(t)\gamma^\mu + \mathrm{i}F_2(t)\frac{\sigma^{\mu\nu}(k_{1,2}-k)}{2m_{q,\bar{Q}}}\ . \tag{20}$$

In relation (20) form-factors $F_1(t)$ and $F_2(t)$ normalized in the natural units magnetic μ_q and anomalous magnetic moments κ_q of quarks :

$$F_1(t=0) + F_2(t=0) = \mu_q,\ \ F_2(t=0) = \kappa_q\ . \tag{21}$$

It's also should be notice, than in (18) relation for $\mathbf{k}_2$ and $\omega_{m_{q,\bar{Q}}}(\mathrm{k}_2)$ given by:

$$\mathbf{k}_2 = \mathbf{k} + \boldsymbol{v}_\mathbf{Q}\left((\varpi+1)\omega_{m_{q,\bar{Q}}} + \sqrt{\varpi^2-1}\ \mathrm{k}\cos\theta_k\right), \tag{22}$$

$$\omega_{m_{q,\bar{Q}}}(\mathrm{k}_2) = \omega_{m_{q,\bar{Q}}}(\mathrm{k})\varpi - \sqrt{\varpi^2-1}\ \mathrm{k}\cos\theta_k, \tag{23}$$

where

$$\varpi = \frac{M_0^2 + M_0'^2 - t}{2M_0M_0'}\ . \tag{24}$$

Using experimental data for radiative decay of light vector mesons $\rho^+, K^{*\pm}$ and K^{*0} [11] and carrying numerical integration from (18) and (21) we obtain following values of magnetic moments of u, d and s quarks in units μ_N (nuclear magneton):

Table 1. Quarks magnetic moments, μ_N

Magnetic moment	This work	[12]	[13]
μ_u	2.080 ± 0.082	2.066	2.08 ± 0.07
μ_d	-1.261 ± 0.015	-1.110	-1.31 ± 0.06
μ_s	-0.621 ± 0.011	-0.633	-0.77 ± 0.06

4. Conclusion and remarks

This work conducted within the framework of the relativistic quark model based on the point form of the Poincare-invariant quantum mechanics to obtain an integral representation for the form factor of the $V \to P\gamma$ transition. From the condition of compliance model calculations with decay width of experimental values found values of the magnetic moments of quarks, which are correlated with the data, obtained using the experimental values of the magnetic moments of baryons.

5. Acknowledgments

We are grateful to the organizational and material support to the organizers of the Conference DSPIN'15. This work was supported by the Belarusian Republican Foundation for Basic Research.

References

[1] Gerasimov S 1989 (*Preprint* JINR-E2-89-837)
[2] Jaus W 1991 *Phys.Rev.* **D44** 2851–2859
[3] Ebert D, Faustov R and Galkin V 2002 *Phys.Lett.* **B537** 241–248 (*Preprint* hep-ph/0204089)
[4] Choi H M 2007 *Phys.Rev.* **D75** 073016 (*Preprint* hep-ph/0701263)
[5] Krutov A and Troitsky V 2009 *Physics of Particles and Nuclei* **40** 136–161 ISSN 1063-7796 URL http://dx.doi.org/10.1134/S1063779609020026
[6] Polyzou W N, Huang Y, Elster C, Glockle W, Golak J, Skibinski R, Witala H and Kamada H 2011 *Few Body Syst.* **49** 129–147 (*Preprint* 1008.5215)
[7] Godfrey S and Isgur N 1985 *Phys.Rev.* **D32** 189–231
[8] Andreev V 2013 (*Preprint* 1305.4266)
[9] Bjorken J 1970 *Phys.Rev.* **D1** 1376–1379
[10] Andreev V 1998 (*Preprint* hep-ph/9912286)
[11] Olive K and Group P D 2014 *Chinese Physics C* **38** 090001
[12] Bijker R, Santopinto E and Santopinto E 2009 *Phys. Rev.* **C80** 065210 (*Preprint* 0912.4494)
[13] Dothan Y 1982 *Physica* **114A** 216–220

Journal of Physics: Conference Series **678** (2016) 012042 doi:10.1088/1742-6596/678/1/012042

Non-singlet Q^2-evolution by inverse Mellin transform in the analytic perturbation theory

A V Sidorov[1], **O P Solovtsova**[1,2]

[1] Bogoliubov Laboratory of Theoretical Physics, JINR, Dubna, 141980, Russia
[2] International Center for Advanced Studies, Technical University, Gomel, 246746, Belarus

E-mail: sidorov@theor.jinr.ru; olsol@theor.jinr.ru

Abstract. We discuss the application of the analytic approach called the fractional Analytic Perturbation Theory (APT) to the QCD analysis of the non-singlet structure function $xF_3(x, Q^2)$. The inverse Mellin transform method applyed for the fit of experimental data and for estimation of the Jacobi polynomial method accuracy in extraction of values of the scale parameter Λ_{QCD} and the form of the xF_3 structure function. Our estimates give the accuracy of the Jacobi polynomials method for the x-shape of the structure function about 10%, and accuracy for the scale parameter Λ_{QCD} 4% .

1. Introduction

Recently application of the APT approach to the QCD analysis of the nucleon structure function data with use of the well known method of the expansion of structure functions in a set of the system of orthogonal Jacobi polynomials (see [1–8]) was done. In this work, using the APT approach [9, 10], we apply the inverse Mellin transform method [11] to the QCD analysis of the xF_3 structure function experimental data. It should be noted that the application of the APT to the QCD analysis of the DIS data required a generalization of the APT on the case of non-integer power of QCD running coupling. Such a generalization was proposed in [12]. The inverse Mellin transform method rather precise and gets an accuracy about five significant digits in our case. We compare the results of both methods in order to estimate the accuracy of the Jacobi polynomial method results. In our analysis, we focus on values of the scale parameter Λ_{QCD} and the form of the $xF_3(x, Q^2)$ structure function.

2. Description of the methods

Let us brifely discuss two methods of the QCD analysis of the structure functions: the inverse Mellin transform method and the method base on the expansion of the structure function on a set of the Jacobi polynomials.

2.1. Jacobi polynomials method

This method based on the expansion of the structure function on a set of the Jacobi polynomials. One can express a shape of the structure function on the x-space to values of the Mellin moments

Content from this work may be used under the terms of the Creative Commons Attribution 3.0 licence. Any further distribution of this work must maintain attribution to the author(s) and the title of the work, journal citation and DOI.
Published under licence by IOP Publishing Ltd

of the structure function. Then the xF_3 structure function can be presented as [2]

$$xF_3^{LT\ N_{max}}(x,Q^2) = x^\alpha(1-x)^\beta \sum_{n=0}^{N_{max}} \Theta_n^{\alpha,\beta}(x) \sum_{j=0}^{n} c_j^{(n)}(\alpha,\beta) M_3\left(j+2,Q^2\right) \quad \text{for PT}, \tag{1}$$

$$= x^\alpha(1-x)^\beta \sum_{n=0}^{N_{max}} \Theta_n^{\alpha,\beta}(x) \sum_{j=0}^{n} c_j^{(n)}(\alpha,\beta) \mathcal{M}_3\left(j+2,Q^2\right) \quad \text{for APT}. \tag{2}$$

Here $\Theta_n^{\alpha,\beta}$ are the Jacobi polynomials, $\alpha = 0.7$ and $\beta = 3.0$ fix the weight function of the Jacobi polynomials.

The perturbative renormalization group Q^2 evolution of the Mellin moments is well known (see, e.g., [13]) and in the leading order reads as

$$M_3^{pQCD}(N,Q^2) = \frac{[\alpha_s(Q^2)]^\nu}{[\alpha_s(Q_0^2)]^\nu} M_3(N,Q_0^2), \quad \nu(N) = \gamma_{NS}^{(0),N}/2\beta_0, \ N = 2,\ 3, \ldots, \tag{3}$$

where $\alpha_s(Q^2)$ is the QCD running coupling, $\gamma_{NS}^{(0),N}$ are the non-singlet one-loop anomalous dimensions, $\beta_0 = 11 - 2n_f/3$ is the first coefficient of the renormalization group β-function, and n_f denotes the number of active flavors ($n_f = 4$ in our analysis).

In the framework of the APT, the expression (3) is converted to:

$$\mathcal{M}_3^{APT}(N,Q^2) = \frac{\mathcal{A}_\nu(Q^2)}{\mathcal{A}_\nu(Q_0^2)} \mathcal{M}_3(N,Q_0^2), \tag{4}$$

where the analytic function $\mathcal{A}_\nu$ is derived from the spectral Källén–Lehmann representation and corresponds to the discontinuity of the ν-th power of the PT running coupling.

In the leading order (LO), the analytic function $\mathcal{A}_\nu$ has a rather simple form (see, e.g., [14])

$$\mathcal{A}_\nu^{LO}(Q^2) = \left[\alpha_{\rm PT}^{LO}\right]^\nu - \left(\frac{4\pi}{\beta_0}\right)^\nu \frac{\mathrm{Li}_\delta(t)}{\Gamma(\nu)}, \tag{5}$$

$$\mathrm{Li}_\delta(t) = \sum_{k=1}^{\infty} \frac{t^k}{k^\delta}, \quad t = \frac{\Lambda^2}{Q^2}, \quad \delta = 1-\nu, \tag{6}$$

where the PT running coupling $\alpha_{\rm PT}^{LO} = 4\pi/[\beta_0 \ln(Q^2/\Lambda_{\rm PT}^2)]$ and Li_δ is the polylogarithm function. Note that the function $\mathcal{A}_{\nu=1}(Q^2)$ defines the APT running coupling, $\alpha_{\rm APT}(Q^2)$ (see [15, 16]).

Unknown quantity $M_3(N,Q_0^2)$ in (3) could be parameterized as the Mellin moments of the structure function xF_3 at some point Q_0^2:

$$M_3(N,Q_0^2) = \int_0^1 dx x^{N-2} xF_3(x,Q_0^2) = \int_0^1 dx x^{N-2} A x^a (1-x)^b (1+\gamma x), \ N = 2,\ 3, \ldots. \tag{7}$$

In our analysis we taken into account the higher twist contribution and therefore

$$xF_3^{exp}(x,Q^2) = xF_3^{LT\ N_{max}}(x,Q^2) + \frac{h(x)}{Q^2}, \tag{8}$$

where $h(x)$ is a shape of the higher twist contribution in the x space.

Table 1. The results of the QCD fit for the scale parameter Λ_{QCD} obtained in the framework of the PT and APT approaches using different methods.

Method of analysis	$\Lambda_{\rm PT}$ (MeV)	$\Lambda_{\rm APT}$ (MeV)
Inverse Mellin transform	378 ± 49	422 ± 77
Jacobi polynomials expansion	363 ± 49	407 ± 74

2.2. Inverse Mellin transform method
One can calculate the structure function at some Q^2-value using the inverse Mellin transform [11]:

$$xF_3(x,Q^2) = \frac{1}{2\pi i}\int_{c-i\infty}^{c+i\infty} dn\, x^{-n}\, M_3(n,Q^2)\,, \qquad \mathrm{PT}\,, \tag{9}$$

$$= \frac{1}{2\pi i}\int_{c-i\infty}^{c+i\infty} dn\, x^{-n}\mathcal{M}_3\left(n,Q^2\right)\,, \ \mathrm{APT}\,. \tag{10}$$

In the PT, it is well known the analytic continuation of the anomalous dimensions on the complex plane of Mellin moments, n. In the APT case, we calculate the function $\mathcal{A}_\nu(Q^2)$ for the complex Mellin moments by the numerical summation of the set in (6). The path of integration for the inverse Mellin transform lies to the right of all singularities of the analytic continuation of moments $M_3(n,Q^2)$ and $\mathcal{M}_3(n,Q^2)$.

3. Results of fit and discussion
The shape of the function $h(x)$ in (8) as well as the parameters A, a, b, γ in (7), and the scale parameter Λ_{QCD} are found by fit of a combined set of the xF_3-data (see details in [7]). The description of the fitting procedure can be found in [17]. The target mass corrections are taken into account up to the terms M_N^2/Q^2 (see [18]).

As seen from Table 1, the accuracy of the determination of the scale parameter Λ_{QCD} by the Jacobi polynomials method is about 15 MeV: $\Delta\Lambda^{M-J} = \Lambda^{M}_{\rm APT/PT} - \Lambda^{J}_{\rm APT/PT} \approx 15$ MeV. Both methods give the same difference for a value of the parameter Λ_{QCD} in the APT and PT approaches: $\Lambda^{M}_{\rm APT} - \Lambda^{J}_{\rm APT} = \Lambda^{M}_{\rm PT} - \Lambda^{J}_{\rm PT} = 44$ MeV.

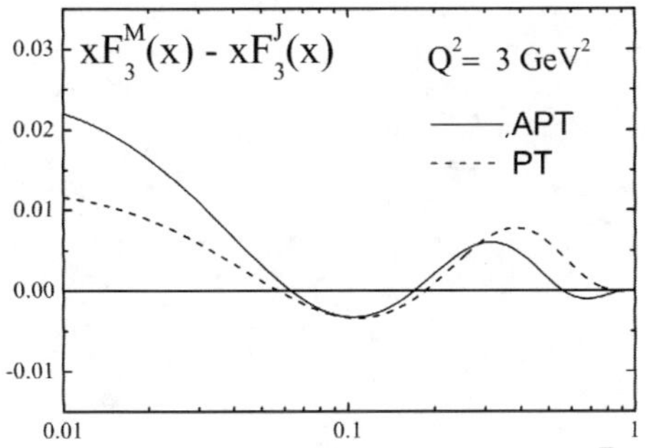

Figure 1. The difference between fitting results for xF_3 structure function data by the inverse Mellin and Jacobi methods presented for the APT and PT approaches.

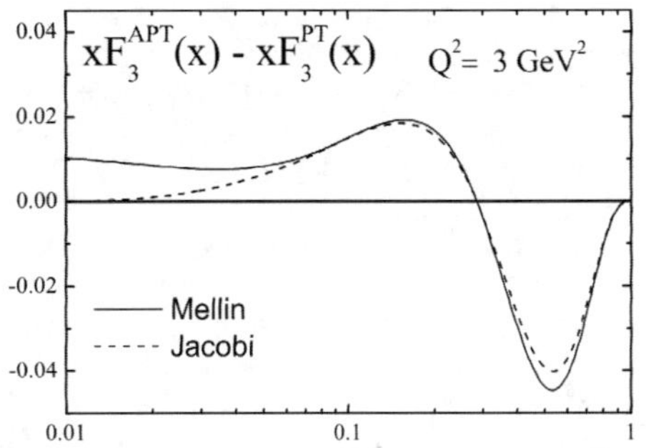

Figure 2. The difference for xF_3 structure function form in the APT and PT approaches for the inverse Mellin and Jacobi methods.

Journal of Physics: Conference Series **678** (2016) 012042 doi:10.1088/1742-6596/678/1/012042

Figure 1 shows the difference in x-space between fitting results for xF_3 structure function data by using the inverse Mellin transform and the Jacobi polynomials expansion methods: $\Delta xF_3^{M-J} = xF_3^M - xF_3^J$ presented for the APT (solid line) and the PT (dashed line) approaches. The number of Jacobi polynomials is $N_{max} = 11$ and $Q_0^2 = 3$ GeV^2. One can see that the value of $\Delta xF_3^{M-J} < 0.025$, which corresponds to the accuracy of the Jacobi method better then 10% for both theoretical approaches. This estimation is in qualitative agreement with the result obtained for the non-singlet xF_2 structure function [19].

Figure 2 shows the difference for the xF_3-shape obtained in the APT (solid line) and the PT (dashed line) approaches using the inverse Mellin transform and the Jacobi polynomials methods: $\Delta xF_3^{\mathrm{APT-PT}} = xF_3^{\mathrm{APT}} - xF_3^{\mathrm{PT}}$. As can be seen from this figure, the Jacobi method gives the same difference with the Mellin one at both theoretical approaches: the APT and the PT. But at small x the Jacobi method is not sensitive to the difference of results in the APT and PT approaches. While the inverse Mellin transform method reveals this difference. We found that at low $Q^2 = 1$ GeV^2 the accuracy of the Jacobi method in the APT is two times better in comparison to the PT.

In conclusion we stress, that the Jacobi polynomials method is fast, but gives about 10% accuracy for x-shape of the structure function and 4% accuracy for the scale parameter Λ_{QCD}.

4. Acknowledgments

It is a pleasure for the authors to thank D. V. Shirkov and O. V. Teryaev for interest in this work and helpful discussions. This research was supported by the JINR-BelRFFR grant F14D-007, JINR-Bulgaria Collaborative Grant and by the RFBR Grants (Nrs 13-02-01005 and 14-01-00647).

References

[1] Parisi G and Sourlas N 1979 *Nucl. Phys.* B **151** 421
[2] Krivokhizhin V G *et al* 1987 *Z. Phys.* C **36** 51
[3] Krivokhizhin V G *et al* 1990 *Z. Phys.* C **48** 347
Benvenuti A *et al* 1987 *Phys. Lett.* B **195** 97
Benvenuti A *et al* 1989 *Phys. Lett.* B **223** 490
Kotikov A V, Parente G and Sanchez-Guillen J 1993 *Z. Phys.* C **58** 465
Kataev A L and Sidorov A V 1994 *Phys. Lett.* B **331** 179
[4] Shaikhatdenov B G *et al* 2010 *Phys. Rev.* D **81** 034008
[5] Kotikov A V, Krivokhizhin V G and Shaikhatdenov B G 2012 *Phys. Atom. Nucl.* **75** 507
[6] Sidorov A V and Solovtsova O P 2013 *Nonlin. Phenom. Complex Syst.* **16** 397
[7] Sidorov A V and Solovtsova O P 2014 *Mod. Phys. Letters* A **29** 1450194
[8] Sidorov A V and Solovtsova O P 2015 *Nonlin. Phenom. Complex Syst.* **18** 222
[9] Shirkov D V and Solovtsov I L 1997 *Phys. Rev. Lett.* **79** 1209
[10] Milton K A and Solovtsov I L 1997 *Phys. Rev.* D **55** 5295
[11] Gluck M and Reya E 1976 *Phys. Rev.* D **14** 3034
Kosower D A 1997 *Nucl. Phys.* B **506** 439
Gluck M, Reya E and Vogt A 1998 *Eur. Phys. J.* C **5** 461
Stratmann M and Vogelsang W 2001 *Phys. Rev.* D **64** 114007
[12] Bakulev A P, Mikhailov S V and Stefanis N G 2005 *Phys. Rev.* D **72** 074014 2005 *Erratum: ibid.* **72** 119908(E)
[13] Buras A J 1980 *Rev. Mod. Phys.* **52** 199
[14] Bakulev A P 2009 *Phys. Part. Nucl.* **40** 715
[15] Milton K A and Solovtsova O P 1998 *Phys. Rev.* **D57** 5402
[16] Shirkov D V and Solovtsov I L 2007 *Theor. Math. Phys.* **150** 132
[17] Kataev A L *et al* 1998 *Phys. Lett.* B **417** 374
[18] Kataev A L, Parente G and Sidorov A V 2000 *Nucl. Phys.* B **573** 405
[19] Ayala C and Mikhailov S V 2015 *Phys. Rev.* D **92** 014028

Journal of Physics: Conference Series **678** (2016) 012043 doi:10.1088/1742-6596/678/1/012043

Study of spin-density matrix in exclusive electroproduction of ω meson at HERMES

S I Manaenkov,

National Research Center "Kurchatov Institute", Petersburg Nuclear Physics Institute, on behalf of the HERMES Collaboration

E-mail: sman@pnpi.spb.ru

Abstract. Exclusive electroproduction of ω mesons on unpolarized hydrogen and deuterium targets is studied in the kinematic region of $Q^2 > 1.0$ GeV^2, 3.0 GeV $< W <$ 6.3 GeV, and $-t' < 0.2$ GeV^2. The data were accumulated with the HERMES forward spectrometer during the 1996-2007 running period using the 27.6 GeV longitudinally polarized electron or positron beam of HERA. Spin-density matrix elements are presented in projections of Q^2 or $-t'$. Violation of s-channel helicity conservation is observed for some of these elements. A sizable contribution from unnatural-parity-exchange amplitudes is established for special combinations of spin-density matrix elements. The determination of the virtual-photon longitudinal-to-transverse cross-section ratio reveals that a dominant part of the cross section arises from transversely polarized photons. Good agreement is found between the HERMES proton data and results of a pQCD-inspired Goloskokov-Kroll model that includes pion-pole contributions.

1. Introduction

Exclusive electroproduction of vector mesons on nucleons gives information both on the reaction mechanisms and on the nucleon structure [1, 2]. The electroproduction at high energies can be considered to consist of three subprocesses: i) the incident lepton emits a virtual photon γ^*, which dissociates into a quark-antiquark pair; ii) this $q\bar{q}$ pair interacts strongly with the nucleon; iii) the observed vector meson is formed from the scattered $q\bar{q}$ pair. In Regge phenomenology, the interaction of the $q\bar{q}$ pair with the nucleon proceeds through the exchanges of a pomeron or/and exchanges of other reggeons (e.g., ρ, f_2, π, a_1, ...). If the quantum numbers of the particle lying on the Regge trajectory are $J^P = 0^+$, 1^-, etc., the process is denoted Natural Parity Exchange (NPE). Alternatively, the case of $J^P = 0^-$, 1^+, etc. corresponds to Unnatural Parity Exchange (UPE). In perturbative quantum chromodynamics (pQCD), the interaction of the $q\bar{q}$ pair with the nucleon can proceed via two-gluon exchange or $q\bar{q}$ exchange, where the former corresponds to the pomeron exchange and the latter to the exchange of other reggeons. In the framework of pQCD valid at large photon virtuality Q^2 and high photon-nucleon center-of-mass (CM) energy W, the nucleon structure can also be studied through hard exclusive meson production as the process amplitude contains Generalized Parton Distributions (GPDs) [3, 4, 5]. The amplitude F_{00} of longitudinal vector meson production by longitudinal virtual photons is proven to factorize rigorously into a perturbative calculable part that describes hard scattering of partons and two soft parts [5, 6]. The soft parts contain GPDs and a meson distribution amplitude. At high Q^2, W and for leading twist, the GPDs H^f and E^f are sufficient to describe exclusive vector-meson production on a nucleon, where f denotes a quark of flavor f or a gluon. These GPDs are related

Content from this work may be used under the terms of the Creative Commons Attribution 3.0 licence. Any further distribution of this work must maintain attribution to the author(s) and the title of the work, journal citation and DOI.
Published under licence by IOP Publishing Ltd

Journal of Physics: Conference Series **678** (2016) 012043 doi:10.1088/1742-6596/678/1/012043

to the total angular momentum carried by the parton in the nucleon [7]. In the Goloskokov-Kroll (GK) model (see [8] and references therein) the validity of factorization is assumed for other amplitudes in addition to F_{00} and this assumption is justified with a good description of the existing data. Recently, Spin-Density Matrix Elements (SDMEs) were studied by CLAS [9] for exclusive ω electroproduction at 1.6 CeV^2 $< Q^2 <$ 5.2 CeV^2 and the contribution of the π-reggeon was found to be dominant. The presented HERMES data can also be well described in the GK model if the pion exchange is included into account.

2. Formalism

The ω meson is produced and decays in the following reactions:

$$e + p \to e + p + \omega, \quad \omega \to \pi^+ + \pi^- + \pi^0, \quad \pi^0 \to 2\gamma \tag{1}$$

with a branching ratio $Br = 89.1\%$ for the ω decay. The angular distribution of the three final-state pions depends on the SDMEs. In the present paper, the formalism proposed in Ref. [10] for SDMEs $r^{\alpha}_{\lambda_V \lambda'_V}$ is used. The vector-meson spin-density matrix is expressed in terms of a set of nine matrices $r^{\alpha}_{\lambda_V \lambda'_V}$ related to various photon polarization states: transversely polarized photon (α=0,...,3) corresponding to the photon helicity $\lambda_\gamma = \pm 1$, longitudinally polarized photon (α=4) for which $\lambda_\gamma = 0$, and terms describing their interference (α=5,...,8) [10]. Here, λ_V and λ'_V denote the helicities of the vector meson. When contributions of transverse and longitudinal photons cannot be separated, the SDMEs $r^{04}_{\lambda_V \lambda'_V}$ being the liner combination of $r^{0}_{\lambda_V \lambda'_V}$ and $r^{4}_{\lambda_V \lambda'_V}$ are customarily used (see details in [10, 11]).

The vector-meson SDMEs $r^{\alpha}_{\lambda_V \lambda'_V}$ can be expressed in terms of helicity amplitudes $F_{\lambda_V \lambda'_N \lambda_\gamma \lambda_N}$. These CM amplitudes of the process $\gamma^* N \to VN$ describe the transition of a virtual photon with helicity λ_γ to a vector meson (V) with helicity λ_V, while λ_N and λ'_N are the helicities of the nucleon (N) in the initial and final states, respectively. Helicity amplitudes depend on W, Q^2, and $t' = t - t_{min}$, where t is the Mandelstam variable and $-t_{min}$ represents the smallest kinematically allowed value of $-t$ at fixed virtual-photon energy and Q^2.

Any helicity amplitude can be decomposed into a sum of a NPE amplitude T and an UPE amplitude U: $F_{\lambda_V \lambda'_N \lambda_\gamma \lambda_N} = T_{\lambda_V \lambda'_N \lambda_\gamma \lambda_N} + U_{\lambda_V \lambda'_N \lambda_\gamma \lambda_N}$, for details see Refs. [10, 11]. In the one reggeon exchange approximation, NPE amplitudes are due to exchanges with pomeron, ρ, f_2, a_2 etc. reggeons while UPE amplitudes are amplitudes of exchanges with π, a_1 etc. reggeons. For brevity, the nucleon helicity indexes will be omitted if this does not lead to misunderstanding.

The SDMEs in exclusive ω electroproduction are determined using the processes in Eqs. (1). They are fitted as parameters of the angular distribution $\mathcal{W}^{U+L}(\Phi, \phi, \cos\Theta)$ to the experimental distribution of the three pions originating from the ω decay. The angular distribution $\mathcal{W}^{U+L}(\Phi, \phi, \cos\Theta)$ is decomposed into $\mathcal{W}^{U}$ containing 15 "unpolarized" SDMEs and $\mathcal{W}^{L}$ determined with 8 "polarized" SDMEs: $\mathcal{W}^{U+L}(\Phi, \phi, \cos\Theta) = \mathcal{W}^{U}(\Phi, \phi, \cos\Theta) + P_b \mathcal{W}^{L}(\Phi, \phi, \cos\Theta)$, where P_b is a longitudinal beam polarization. The "polarized" SDMEs can be extracted only from data collected with a longitudinally polarized beam.

The angle between the ω production plane and the lepton scattering plane in the CM system is denoted Φ. The variables Θ and ϕ are respectively the polar and azimuthal angles of the unit vector $\vec{n}$ normal to the decay plane in the ω-meson rest frame. For more details see [12].

3. Data Analysis

The data were accumulated with the HERMES spectrometer during the running period of 1996 to 2007 using the 27.6 GeV longitudinally polarized electron or positron beam of HERA, and gaseous hydrogen or deuterium targets. The HERMES forward spectrometer is described in Ref. [13]. The spectrometer permitted a precise measurement of charged-particle momenta,

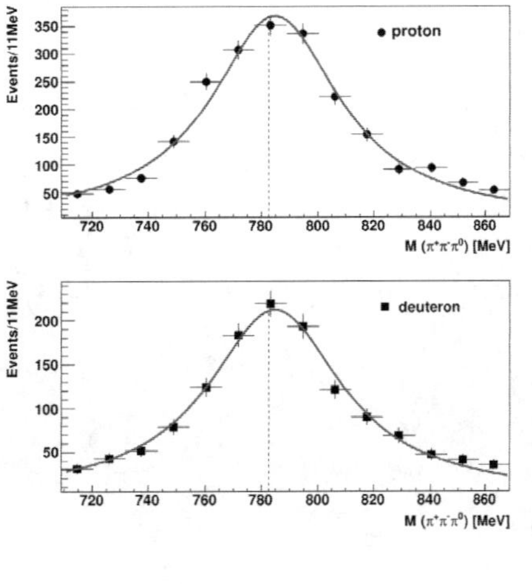

Figure 1. Breit-Wigner fit (solid line) of $\pi^+\pi^-\pi^0$ invariant mass distributions after application of all criteria to select ω mesons produced exclusively from proton (top) and from deuteron (bottom). The dashed line represents the PDG value of the ω mass.

with a resolution of 1.5%. A separation of leptons was achieved with an average efficiency of 98% and a hadron contamination below 1%.

The following requirements were applied for event selection:
i) Exactly two oppositely charged hadrons and one lepton with the same charge as the beam lepton are identified.
ii) A π^0 meson that is reconstructed from two calorimeter clusters is selected requiring the two-photon invariant mass to be in the interval 0.11 GeV $< M(\gamma\gamma) <$ 0.16 GeV.
iii) The three-pion invariant mass is required to obey 0.71 GeV$\leq$ M$(\pi^+\pi^-\pi^0)\leq$ 0.87 GeV.
iv) The scattered-lepton momentum lies above 3.5 GeV.
v) Taking into account the spectrometer resolution, the missing energy ΔE has to lie in the interval -1.0 GeV $< \Delta E <$ 0.8 GeV. Here, $\Delta E = \frac{M_X^2 - M_p^2}{2M_p}$, with M_p being the proton mass and $M_X^2 = (p + q - p_{\pi^+} - p_{\pi^-} - p_{\pi^0})^2$ the missing mass squared, where p, q, p_{π^+}, p_{π^-}, and p_{π^0} are the four-momenta of target nucleon, virtual photon, and each of the three pions respectively.
vi) The constraints $Q^2 > 1.0$ GeV^2, 6.3 GeV $> W >$ 3 GeV, $-t' < 0.2$ GeV^2 are applied.

After application of all these constraints, the proton sample contains 2260 and the deuteron sample 1332 events of exclusively produced ω mesons. These data samples are referred to in the following as data in the "entire kinematic region". The invariant-mass distributions for exclusively produced ω mesons are shown in Fig. 1.

The distributions of missing energy ΔE, shown in Fig. 2, exhibit clearly visible exclusive peaks. The shaded histograms represent semi-inclusive deep-inelastic scattering (SIDIS) background obtained from a PYTHIA Monte Carlo simulation that is normalised to the data in the region of 2 GeV $< \Delta E <$ 20 GeV. The simulation is used to determine the fraction of background under the exclusive peak, which is calculated as the ratio of number of background events to the total number of events. It amounts to about 20% for the entire kinematic region and increases from 16% to 26% with increasing $-t'$.

4. Results

The results on SDMEs are obtained using an unbinned maximum likelihood method. All details of the fit procedure and taking into account the background corrections are described in Ref. [12].

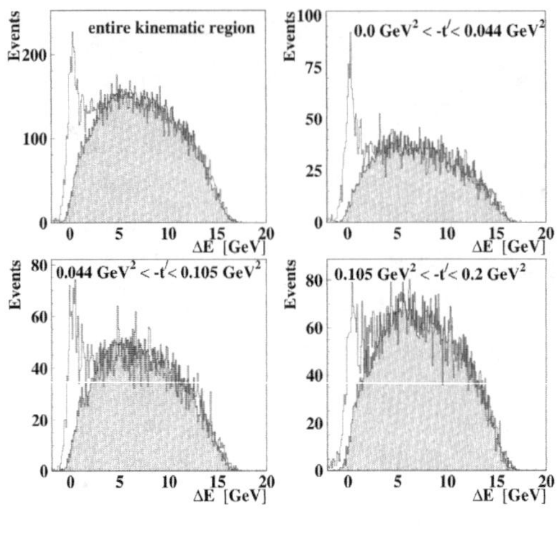

Figure 2. The ΔE distributions of ω mesons produced in the entire kinematic region and three kinematic bins in $-t'$ are compared with SIDIS ΔE distributions from PYTHIA (shaded area). The vertical dashed line denotes the upper limit of the exclusive region.

The SDMEs of the ω and ρ^0 mesons for the entire kinematic region are shown in Fig. 3. These SDMEs are divided into five classes corresponding to various helicity transitions $\lambda_\gamma \to \lambda_V$. The main terms in the expressions of class-A SDMEs correspond to the transitions from longitudinal virtual photons to longitudinal vector mesons, $\gamma_L^* \to V_L$, and from transverse virtual photons to transverse vector mesons, $\gamma_T^* \to V_T$. The dominant terms of class B correspond to the interference of these two transitions. The main terms of class-C, D, and E SDMEs are proportional to small amplitudes of $\gamma_T^* \to V_L$, $\gamma_L^* \to V_T$, and $\gamma_T^* \to V_{-T}$ transitions respectively.

The SDMEs for the proton and deuteron data are found to be consistent with each other within their total uncertainties. In Fig. 3, the uncertainties of the polarized SDMEs are larger than those of the unpolarized SDMEs because the mean polarization $< |P_b| >\approx 40\%$ and in the equation for the angular distribution they are multiplied by the small factor $|P_b|\sqrt{1-\epsilon} \approx 0.2$. Here, ϵ is the ratio of fluxes of the longitudinally and transversely polarized virtual photons.

The linear combination of the class-A and class-B SDMEs which are to be zero, if S-Channel Helicity Conservation (SCHC) approximation is valid, are really zero within experimental uncertainty. Indeed, the proton data yield $r^1_{1-1} + \mathrm{Im}\{r^2_{1-1}\} = -0.004 \pm 0.038 \pm 0.015$, $\mathrm{Re}\{r^5_{10}\} + \mathrm{Im}\{r^6_{10}\} = -0.024 \pm 0.013 \pm 0.004$, $\mathrm{Im}\{r^7_{10}\} - \mathrm{Re}\{r^8_{10}\} = -0.060 \pm 0.100 \pm 0.018$, and the deuteron data give $r^1_{1-1} + \mathrm{Im}\{r^2_{1-1}\} = 0.033 \pm 0.049 \pm 0.016$, $\mathrm{Re}\{r^5_{10}\} + \mathrm{Im}\{r^6_{10}\} = 0.001 \pm 0.016 \pm 0.005$, $\mathrm{Im}\{r^7_{10}\} - \mathrm{Re}\{r^8_{10}\} = 0.104 \pm 0.110 \pm 0.023$. All SDMEs of classes C, D, E are to be zero in the SCHC approximation. The SDME r^5_{00} violates SCHC since it is nonzero as is seen from Fig. 3.

The comparison of ω and ρ^0 [11] SDMEs is shown in Fig. 3. As is seen the SDMEs r^1_{1-1} and $\mathrm{Im}\{r^2_{1-1}\}$ have opposite sign for ω and ρ^0 mesons. The explanation follows from expressions of these SDMEs in terms of the amplitudes given in [10, 11]: $|T_{11}|^2 > |U_{11}|^2$ for the ρ^0 meson, while $|U_{11}|^2 > |T_{11}|^2$ for the ω meson. Moreover, the contribution of the UPE amplitude U_{11} dominates in the cross section of the ω-meson production while its contribution for the ρ^0-meson is small [11]. As the simplest consequence of the UPE dominance, the values of the SDME combinations u_1, u_2, and u_3 are unexpectedly large for the ω meson as is seen from Fig. 4. They are well described in the GK model [8] if the pion exchange is included into consideration as follows from a comparison of solid and dashed curves in Fig. 4. The quantities u_1, u_2, and u_3 are: $u_1 = 1 - r^{04}_{00} + 2r^{04}_{1-1} - 2r^1_{11} - 2r^1_{1-1}$, $u_2 = r^5_{11} + r^5_{1-1}$, and $u_3 = r^8_{11} + r^8_{1-1}$.

Journal of Physics: Conference Series **678** (2016) 012043 doi:10.1088/1742-6596/678/1/012043

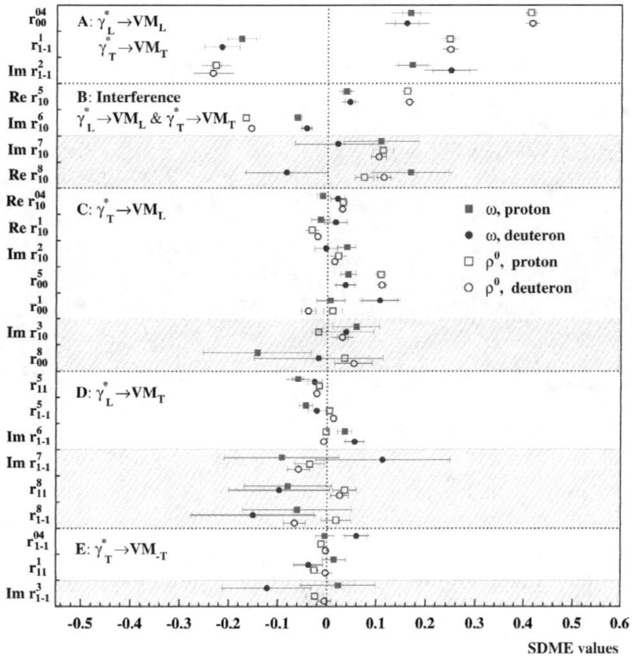

Figure 3. Comparison of 23 SDMEs in exclusive ω and ρ^0 [11] electroproduction at HERMES for the entire kinematic region. The SDMEs for ω are extracted in the region with $\langle Q^2 \rangle = 2.42$ GeV^2, $\langle W \rangle = 4.8$ GeV, $\langle -t' \rangle = 0.080$ GeV^2 while for ρ^0 production $\langle Q^2 \rangle = 1.95$ GeV^2, $\langle W \rangle = 4.8$ GeV, and $\langle -t' \rangle = 0.13$ GeV^2. The inner error bars represent the statistical uncertainties, while the outer ones indicate the statistical and systematic uncertainties added in quadrature. Unpolarized (polarized) SDMEs are displayed in the unshaded (shaded) areas.

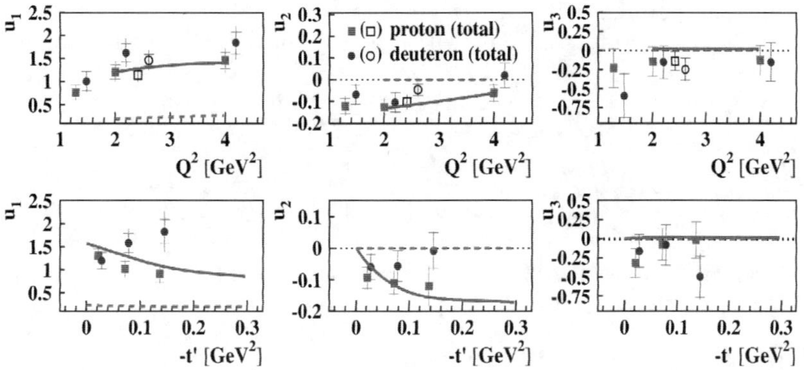

Figure 4. The Q^2 and $-t'$ dependences of u_1, u_2, and u_3. The open symbols represent the values over the entire kinematic region. Solid (dashed) curves are obtained in the GK model when the pion exchange is (not) taken into account. Other notations are as in Fig. 3.

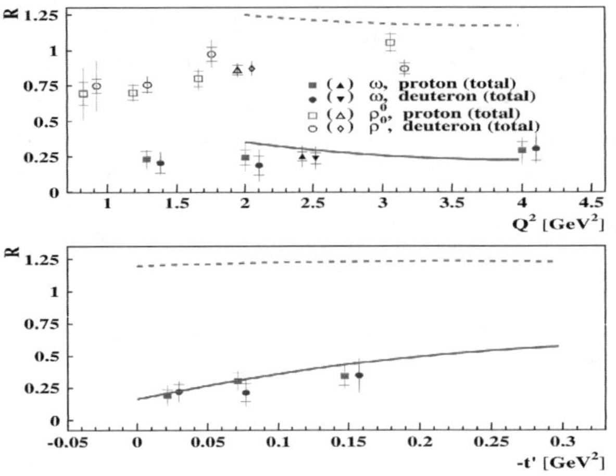

Figure 5. The Q^2 (top) and $-t'$ (bottom) dependence of the longitudinal-to-transverse virtual-photon differential cross-section ratio for exclusive ω and ρ^0 electroproduction at HERMES, where the $-t'$ bin is [0.0-0.2] GeV^2 for ω meson and [0.0-0.4] GeV^2 for ρ^0 meson [11]. The triangle symbols represent the value of R in the entire kinematic region. Otherwise as for Fig. 4.

The differential cross-section ratio, $R = d\sigma_L(\gamma_L^* \to V)/d\sigma_T(\gamma_T^* \to V)$ of longitudinal to transverse photons is experimentally determined from the measured SDME r_{00}^{04} using the equation [11] $R \approx r_{00}^{04}/[\epsilon(1 - r_{00}^{04})]$. This relation is exact in the case of SCHC. A comparison of the ratio for the ω and ρ^0 meson shows that the former is by a factor of about 3 smaller than the latter. This fact is a manifestation of the larger fractional contribution of pion exchange for the case of the ω meson than for the ρ^0 meson. Indeed, the formula $R \approx |T_{00}/T_{11}|^2/(1 + |U_{11}/T_{11}|^2)$ valid for rough estimations shows that since $|U_{11}/T_{11}|^2 > 1$ for ω while $|U_{11}/T_{11}|^2 < 1$ for ρ^0 mesons the ratio R should be smaller for the ω meson if $|T_{00}/T_{11}|^2$ are approximately equal to each other for these two mesons. The crucial importance of the pion pole contribution is illustrated in Fig. 5 by a comparison of calculations in the GK model, in which this contribution is taken into account and ignored. The GK model appears to fully account for the Q^2 and t' dependences of R and shows good agreement with the data only if the pion exchange is taken into account.

References

[1] Frankfurt L et al. 1996 *Phys. Rev.* D **54** 3194
[2] Diehl M 2003 *Physics Reports* **388** 41
[3] Müller D et al. 1994 *Fortschr. Phys.* **42** 101
[4] Ji X 1997 *Phys. Rev.* D **55** 7114
[5] Radyushkin A V 1997 *Phys. Rev.* D **56** 5524
[6] Collins J C Frankfurt L Strikman M S 1997 *Phys. Rev.* D **56** 2982
[7] Ji X 1997 *Phys. Rev. Lett* **74** 610
[8] Goloskokov S Kroll P 2014 arXiv:1407.1141
[9] Morand L at al. (CLAS Collaboration) 2005 *Eur. Phys. J.* A **24** 445
[10] Schilling K Wolf G 1973 *Nucl. Phys.* B **61** 381
[11] Airapetian A et al. (HERMES Collaboration) 2009 *Eur. Phys. J.* C **62** 659
[12] Airapetian A et al. (HERMES Collaboration) 2014 *Eur. Phys. J.* C **74** 3110
[13] Ackerstaff K et al. (HERMES Collaboration) 1998 *Nucl. Instr. and Meth.* A **417** 230

Journal of Physics: Conference Series **678** (2016) 012044 doi:10.1088/1742-6596/678/1/012044

Self-similarity of proton spin and z-scaling

M Tokarev[1], **I Zborovský**[2]

[1] Joint Institute for Nuclear Research, Dubna, Moscow region, Russia
[2] Nuclear Physics Institute, Academy of Sciences of the Czech Republic, Řež, Czech Republic

E-mail: tokarev@jinr.ru

E-mail: zborovsky@ujf.cas.cz

Abstract. The concept of z-scaling previously developed for analysis of inclusive reactions in proton-proton collisions is applied for description of processes with polarized particles. Hypothesis of self-similarity of the proton spin structure is discussed. The possibility of extracting information on spin-dependent fractal dimensions of hadrons and fragmentation process from the cross sections and asymmetries is justified. The double longitudinal spin asymmetry A_{LL} of jet and π^0-meson production and the coefficient of polarization transfer D_{LL} measured in proton-proton collisions at $\sqrt{s} = 200$ GeV at RHIC are analyzed in the framework of z-scaling. The spin-dependent fractal dimension of proton is estimated.

1. Introduction

Spin is one of the most fundamental properties of elementary particles. The proton spin structure is studied for a long time in processes with polarized leptons and protons. The goal is to understand complete picture of the proton spin in terms of quark and gluon degrees of freedom. Microscopic scenario of proton structure implies knowledge of momentum and spin distributions of its constituents at different scales. In the framework of QCD, the non-linear Yang-Mills equations taking into account gauge invariance and Lorentz covariance regulate dynamics of the constituent interactions both at hard and soft regimes.

There is a convincing evidence that inclusive reactions with unpolarized particles reveal self-similarity over a wide scale range [1, 2]. The scaling features of hadron production in $p + p$ and $\bar{p} + p$ collisions (z-scaling) are treated as a manifestation of self-similarity of the structure of colliding protons, interaction mechanism of their constituents, and fragmentation process into real hadrons. The parameters of the scaling, c, δ and ϵ_F, have physical interpretation of a heat capacity of the produced matter, fractal dimension of the proton structure and fractal dimension of the fragmentation process, respectively. Extension of the method for analysis of polarization phenomena and verification of self-similarity of particle production in polarized $p + p$ collisions is an interesting problem which could give new insight into our understanding the origin of proton spin. The spin-dependent fractal dimensions are new parameters of the z-scaling approach. They characterise polarization properties of proton structure, constituent interactions and hadronization process.

In this paper we analyze the double longitudinal spin asymmetry A_{LL} of jet and meson production and the coefficient of polarization transfer D_{LL} measured by the STAR and PHENIX collaborations at RHIC to extract information on spin-dependent fractal dimensions of proton and verify the hypothesis of self-similarity of proton spin in the z-scaling approach.

Content from this work may be used under the terms of the Creative Commons Attribution 3.0 licence. Any further distribution of this work must maintain attribution to the author(s) and the title of the work, journal citation and DOI.
Published under licence by IOP Publishing Ltd

Journal of Physics: Conference Series **678** (2016) 012044 doi:10.1088/1742-6596/678/1/012044

2. z-Scaling

The z-scaling [1, 2] manifests itself in the fact that the inclusive spectra of various types of particles are described with a universal scaling function. The function $\Psi(z)$ depends on single variable z in a wide range of the transverse momentum, registration angles, collision energies, and centralities. The scaling variable has the form:

$$z = z_0 \cdot \Omega^{-1}. \tag{1}$$

The quantity z_0 is ratio of the transverse kinetic energy of the selected binary sub-process required for production of the inclusive particle with its partner (antiparticle) and the charged multiplicity density at midrapidity raised to a power of c. The parameter c is interpreted as a heat capacity of the produced matter. The selected binary sub-process is given by maximum of the function $\Omega(x_1,x_2,y_a,y_b) = (1-x_1)^{\delta_1}(1-x_2)^{\delta_2}(1-y_a)^{\epsilon_a}(1-y_b)^{\epsilon_b}$ with the condition $(x_1P_1 + x_2P_2 - p/y_a)^2 = M_X^2$, where $M_X = x_1M + x_2M + m/y_b$ is the minimal mass of the recoil system. The constraint expresses locality of the interaction at the constituent level. It restricts momentum fractions via kinematics of the constituent interactions and the 4-momenta P_1, P_2 and p of the colliding protons and the inclusive particle with masses M and m, respectively.

The function $\Omega(x_1,x_2,y_a,y_b)$ is proportional to the relative number of the configurations at the constituent level which include the binary sub-processes corresponding to the momentum fractions x_1 and x_2 of colliding protons and to the momentum fractions y_a and y_b of the secondary objects directly produced in the constituent interactions. The parameters δ_1 and δ_2 are fractal dimensions of the protons and ϵ_a and ϵ_b are fractal dimensions of the fragmentation process in the scattered and recoil direction, respectively. For the reactions with no polarizations we have $\delta_1 = \delta_2 = \delta$ and $\epsilon_a = \epsilon_b = \epsilon_F$. The value of ϵ_F depends on type of the inclusive hadron. The parameter dependence of the maximum of $\Omega(x_1,x_2,y_a,y_b)$ used in (1) is briefly indicated as $\Omega \equiv \Omega_{0000} =: \{\delta_1,\delta_2,\epsilon_a,\epsilon_b\}$ in the next. The low index (0000) corresponds to unpolarized particles in the initial and final states. Similar notation applies for processes with polarizations.

The scaling function $\Psi(z)$ is expressed in terms of the inclusive cross section $Ed^3\sigma/dp^3$, multiplicity density $dN/d\eta$, and total inelastic cross section σ_{in} for the inclusive reaction $P_1 + P_2 \to p + X$. The function $\Psi(z)$ is determined by the following expression:

$$\Psi(z) = -\frac{\pi}{(dN/d\eta)\sigma_{in}} J^{-1} E\frac{d^3\sigma}{dp^3}. \tag{2}$$

Here J is Jacobian for the transition from the variables $\{p_T^2, y\}$ to $\{z,\eta\}$. The value of $\Psi(z)$ is probability density to produce the inclusive particle with the corresponding value of the self-similarity parameter z.

3. Self-similarity for processes with polarizations

Below we use the z-scaling approach to formulate self-similarity hypothesis for reactions with polarized particles and discuss a possibility for studying inclusive particle production in polarized $p+p$ collisions to extract information on spin-dependent fractal dimensions of proton [3, 4].

The reaction $\vec{p} + \vec{p} \to h + X$ with two longitudinally polarized protons in the initial state is described by the spin-dependent cross sections $\sigma_{++}, \sigma_{--}, \sigma_{+-}, \sigma_{-+}$. The signs (+) and (−) denote positive and negative helicities of the protons, respectively. The double spin asymmetry A_{LL} of the process is expressed via combination of the cross sections in the form:

$$A_{LL} = \frac{\sigma_{++} + \sigma_{--} - \sigma_{+-} - \sigma_{-+}}{\sigma_{++} + \sigma_{--} + \sigma_{+-} + \sigma_{-+}}. \tag{3}$$

The corresponding quantities Ω are written as follows: $\Omega_{++00} =: \{\delta - \Delta\delta/2, \delta - \Delta\delta/2, \epsilon_F, \epsilon_F\}$, $\Omega_{--00} =: \{\delta - \Delta\delta/2, \delta - \Delta\delta/2, \epsilon_F, \epsilon_F\}$, $\Omega_{+-00} =: \{\delta, \delta + \Delta\delta, \epsilon_F, \epsilon_F\}$, $\Omega_{-+00} =: \{\delta + \Delta\delta, \delta, \epsilon_F, \epsilon_F\}$.

The process $\overrightarrow{p} + p \rightarrow \overrightarrow{h} + X$ with one longitudinally polarized proton in the initial state and one longitudinally polarized particle (e.g. Λ hyperon) in the final state is described by the coefficient of polarization transfer. It is written in the following form:

$$D_{LL} = \frac{\sigma_{++} + \sigma_{--} - \sigma_{+-} - \sigma_{-+}}{\sigma_{++} + \sigma_{--} + \sigma_{+-} + \sigma_{-+}}. \tag{4}$$

The symbols with $(+)$ and $(-)$ denote cross sections corresponding to the parallel and antiparallel spin orientations relative to the respective momenta of the polarized particles (positive and negative helicities). The polarization in the initial state is related to the spin-dependent correction $(\Delta\delta)$ of the proton fractal dimension. Let us consider $(+)$ helicity in the initial state only and denote ϵ_F as the fractal dimension for hadronization of an unpolarized particle (h) in the final state. This corresponds to $\Omega_{+000} =: \{\delta - \Delta\delta/4, \delta + \Delta\delta/4, \epsilon_F, \epsilon_F\}$. If the inclusive particle is polarized, $(\overrightarrow{h})$, the spin-dependent correction $\Delta\epsilon_F$ to the value of ϵ_F is included. The notations for such process are as follows: $\Omega_{+0+0} =: \{\delta - \Delta\delta/4, \delta + \Delta\delta/4, \epsilon_F - \Delta\epsilon_F/2, \epsilon_F\}$, $\Omega_{+0-0} =: \{\delta - \Delta\delta/4, \delta + \Delta\delta/4, \epsilon_F + \Delta\epsilon_F/2, \epsilon_F\}$.

Using experimental information on the spin asymmetries and the unpolarized cross section, the spin-dependent functions $\Psi_{++}, \Psi_{--}, \Psi_{+-}, \Psi_{-+}$ can be constructed [3, 4]. The functions have different arguments which we denote as $z_{++}, z_{--}, z_{+-}, z_{-+}$, respectively. They are expressed (1) by spin-dependent fractal dimensions. Based on existence of the z-scaling in unpolarized $p+p$ collisions, we consider self-similarity of polarization processes at a constituent level in the form:

$$\Psi_{++} = \Psi(z_{++}), \quad \Psi_{+-} = \Psi(z_{+-}), \quad \Psi_{00} = \Psi(z_{00}). \tag{5}$$

This hypothesis assumes universality of $\Psi(z)$ for different spin orientations. The relations include corrections $\Delta\delta$ and $\Delta\epsilon_F$ to the fractal dimensions δ and ϵ_F found for unpolarized reactions. Information on both the polarized and unpolarized cross sections are necessary to extract the spin-dependent fractal dimensions from the polarization characteristics (A_{LL}, D_{LL}) of a given process. Such data allows us to obtain restrictions on the parameters $\Delta\delta$ and $\Delta\epsilon_F$ of the model.

4. z-Scaling and polarized $p+p$ collisions at RHIC

Figure 1 shows the scaling function $\Psi(z)$ for unpolarized $p+p$ collisions. The z-presentation of spectra of jets measured at $\sqrt{s} = 38 - 200$ GeV and π^0 mesons measured at $\sqrt{s} = 23 - 200$ GeV and $\theta_{cms} = 90^0$ is depicted in Figs. 1(a) and 1(b), respectively. The z-scaling of other hadrons is illustrated in Fig. 1(c). The data were obtained at the accelerators ISR, Sp$\bar{\text{p}}$S and RHIC. As

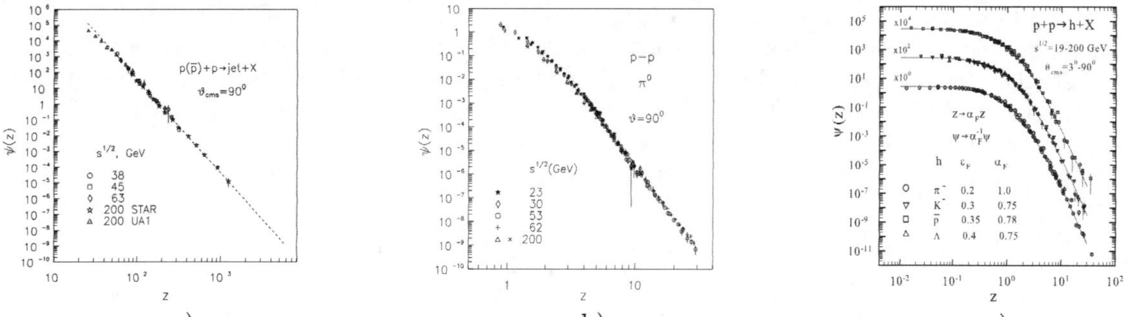

a) b) c)

Figure 1. The data z-presentation of inclusive spectra of jet (a) π^0 meson (b) and hadron (c) production in $p+p$ and $\bar{p}+p$ collisions at ISR, Sp$\bar{\text{p}}$S and RHIC energies.

Journal of Physics: Conference Series **678** (2016) 012044 doi:10.1088/1742-6596/678/1/012044

seen from the figure, similarity of the inclusive cross sections is valid in a wide range of kinematic variables. The function $\Psi(z)$ exhibits a power behavior at high z.

Extension of the method for reactions with polarized particles was tested with data obtained at RHIC. We have constructed the functions Ψ_{++}, Ψ_{+-} and Ψ_{+0} in dependence on corresponding scaling variables to verify the hypothesis (5). Figure 2 demonstrates coincidence of the ratios of spin-dependent and spin-independent functions in dependence on z. The symbols shown in Fig. 2(a) correspond to the inclusive cross section [5] and the spin asymmetry A_{LL} [6] of jet production in proton-proton collisions measured by the STAR collaboration at $\sqrt{s} = 200$ GeV. The data allows us to estimate the spin-dependent fractal dimension of proton from the process.

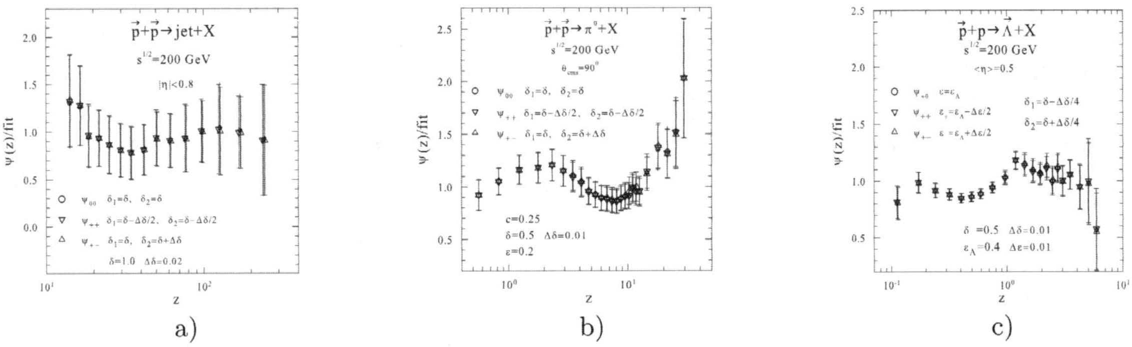

Figure 2. The ratios of scaled spin-dependent and spin-independent functions for $\vec{p} + \vec{p} \to jet + X$ (a) $\vec{p} + \vec{p} \to \pi^0 + X$ (b) and $\vec{p} + p \to \vec{\Lambda} + X$ (c) processes at $\sqrt{s} = 200$ GeV.

Figure 2(b) demonstrates the scaled spin-dependent functions $\Psi(z)$/fit for the reaction $\vec{p} + \vec{p} \to \pi^0 + X$. The double-longitudinal spin asymmetry A_{LL} of π^0 mesons measured by the PHENIX collaboration at $\sqrt{s} = 200$ GeV and $\theta_{cms} = 90^0$ [7] and corresponding unpolarized cross section were used in the analysis. The correction to the fractal dimension δ of unpolarized proton is found to be $\Delta\delta = 0.01$. Figure 2(c) shows ratios of the spin-dependent and spin-independent functions for the process $\vec{p} + p \to \vec{\Lambda} + X$. The comparison exploits data on the longitudinal spin transfer coefficient D_{LL} measured by the STAR collaboration at $\sqrt{s} = 200$ GeV and $< \eta >= 0.5$ [8]. Using the value of $\Delta\delta = 0.01$ obtained from the analysis of π^0-meson production, the spin-dependent correction to ϵ_Λ is found to be $\Delta\epsilon_\Lambda = 0.01$.

Performed analysis justifies application of the z-scaling approach to study the scale properties of spin structure in hadron interactions with polarized particles. The coincidence of functional ratios shown in Fig. 2 indicates self-similarity of spin-dependent processes expressed by (5). This condition was used to estimate corrections $\Delta\delta$ and $\Delta\epsilon_\Lambda$ to the fractal dimensions δ and ϵ_Λ. The suggested procedure of data analysis is applicable to a wide class of polarization reactions. A systematic phenomenological investigation of the processes with polarizations based on the self-similarity principle would contribute to further development of theory and understanding of spin as one of the most important and basic property of particles.

[1] Zborovský I and Tokarev M V 2007 *Phys. Rev.* D **75** 094008
[2] Zborovský I and Tokarev M V 2009 *Int. J. Mod. Phys.* A **24** 1417
[3] Tokarev M V, Zborovský I and Aparin A A 2015 *Part. Nucl. Lett.* **12** 81
[4] Tokarev M V and Zborovský I 2015 *Part. Nucl. Lett.* **12** 214
[5] Li X *Jet measurements in polarized p+p collisions at STAR at RHIC* (*Preprint* hep-ex/1506.06314)
[6] Adamczyk L *et al* 2015 *Phys. Rev. Lett.* **115** 092002
[7] Arschenauer E C *et al* The RHIC Spin Program: Achievements and Future Opportunities (*Preprint* nucl-ex/1304.0079
[8] Xu Q 2013 *Proc. XV Advanced Research Workshop on High Energy Spin Physics (DSPIN-13)* (Dubna: JINR) p 333

Recent results from COMPASS on exclusive muoproduction

Andrzej Sandacz[1]
(on behalf of the COMPASS Collaboration)
Division of High Energy Physics, National Centre for Nuclear Research, PL 00-681 Warsaw, Poland
E-mail: Andrzej.Sandacz@ncbj.gov.pl

Abstract. The 160 GeV polarised muon beam available at CERN, with positive or negative charge, makes COMPASS a unique place for GPD studies. The first GPD related COMPASS results come from exclusive vector meson production on transversely polarised protons and deuterons. The data were taken in 2003-2010 with large solid-state polarised targets, although without detection of recoil particles. Results on various transverse target spin dependent azimuthal asymmetries are presented and their relations to GPDs are discussed. The dedicated COMPASS GPD program started in 2012 with commissioning of a new long liquid hydrogen target and new detectors such as the large recoil proton detector and the large-angle electromagnetic calorimeter. It was followed by a short pilot 'DVCS run'. The performance of the setup and first results on DVCS and exclusive π^0 channels have been demonstrated. The full data taking for the GPD program approved within COMPASS-II proposal is planned for 2016 and 2017.

1. Introduction

General Parton Distributions (GPDs) [1, 2, 3] contain a wealth of information on the partonic structure of the nucleon, which is the one of central problems in hadron physics. In particular, GPDs allow a novel description of the nucleon as an extended object, sometimes referred to as 3-dimensional 'nucleon tomography' [4]. GPDs also allow access to such a fundamental property of the nucleon as the orbital angular momentum of quarks [2]. The mapping of the nucleon GPDs requires comprehensive experimental studies of hard processes such as Deeply Virtual Compton Scattering (DVCS) and Hard Exclusive Meson Production (HEMP) in a broad kinematic range.

2. Exclusive ρ^0 and ω muoproduction on transversely polarised protons

The COMPASS collaboration has analysed exclusive vector meson production on polarised ^{6}LiD (deuterons) and NH_3 (protons) targets using the data from 2003-2010. Here the results for the data taken with polarised protons will be discussed. Although no recoil proton detector was included in the used experimental setup, which is a disadvantage for measurements of exclusive processes, the analysis of these data allows us to obtain first valuable results that are sensitive to 'elusive' GPDs E and chiral-odd GPDs.

[1] Work supported by the Polish NCN Grant DEC-2011/01/M/ST2/02350.

Content from this work may be used under the terms of the Creative Commons Attribution 3.0 licence. Any further distribution of this work must maintain attribution to the author(s) and the title of the work, journal citation and DOI.
Published under licence by IOP Publishing Ltd

2.1. Experimental setup and event selection

COMPASS is a fixed-target experiment situated at the high-intensity M2 beam-line of the CERN SPS. A detailed description can be found in Ref. [5]. The μ^+ beam had a nominal momentum of 160 GeV/c and a longitudinal polarisation of $P_\ell \approx -0.8$. The beam traversed a solid-state ammonia (NH_3) target that provided transversely polarised protons.

To determine the transverse target spin asymmetries for exclusive ρ^0 production the data taken in 2007 and 2010 with polarised protons are used. The details of selection of the sample are given in Ref. [6]. The essential steps of event selection and asymmetry extraction are summarised in the following. The considered events are characterised by an incoming and a scattered muon and two oppositely charged hadrons, h^+h^-, with all four tracks associated to a common vertex in the polarised target. In order to select events in the deep inelastic scattering regime and suppress radiative corrections, the following cuts are used: $Q^2 > 1$ (GeV/c)^2, 0.003 $< x_{\rm Bj} <$ 0.35, $W >$ 5 GeV and 0.1$< y <$ 0.9. The production of ρ^0 mesons is selected in the two-hadron invariant mass range 0.5 GeV/c^2 $< M_{\pi^+\pi^-} <$ 1.1 GeV/c^2, where for each hadron the pion mass hypothesis is assigned. This cut is optimised towards high yield and purity of ρ^0 production, as compared to non-resonant $\pi^+\pi^-$ production. The measurements are performed without detection of the recoiling proton in the final state. Exclusive events are selected by choosing a range in missing energy,

$$E_{\rm miss} = \frac{(p+q-v)^2 - p^2}{2M_{\rm p}}. \tag{1}$$

The four-momenta of proton, photon, and meson, are denoted by p, q, and v respectively, and $M_{\rm p}$ is the proton mass. Although for exclusive events $E_{\rm miss} \approx 0$ holds, the finite experimental resolution is taken into account by selecting events in the 'signal' range $|E_{\rm miss}| <$ 2.5 GeV, which corresponds to $0 \pm 2\sigma$, where σ is the width of the Gaussian signal peak. Non-exclusive background is further suppressed by cuts on the squared transverse momentum of the vector meson with respect to the virtual-photon direction, $p_{\rm T}^2 <$ 0.5 (GeV/c)^2, the energy of the ρ^0 in the laboratory system, $E_{\rho^0} >$ 15 GeV, and the photon virtuality, $Q^2 <$ 10 (GeV/c)^2. An additional cut $p_{\rm T}^2 >$ 0.05 (GeV/c)^2 is used to reduce the contribution of events from coherent production on the target nuclei. The average values of the kinematic variables are $\langle Q^2 \rangle = 2.15$ (GeV/c)^2, $\langle x_{\rm Bj} \rangle = 0.039$, $\langle y \rangle = 0.24$, $\langle W \rangle = 8.13$ GeV, and $\langle p_{\rm T}^2 \rangle = 0.18$ (GeV/c)^2. Details on the method to correct for the contribution of the semi-inclusive background in the sample and on the extraction of asymmetries are given in Ref. [7].

The analysis of exclusive ω production was performed using the data taken in 2010 with transversely polarised protons. An event to be accepted for further analysis was required to have an incident muon track, a scattered muon track, exactly two additional tracks of oppositely charged hadrons, all associated to a vertex in the polarised target material, and a single π^0 meson that is reconstructed using its two decay photons detected in the electromagnetic calorimeters. The photon clusters in the calorimeters have to be time-correlated with the beam track.

Most of the selections of events from exclusive ω production follow those for ρ^0 analysis, except these listed in the following. In order to select π^0 meson, a cut on the range of invariant mass of two photons, $M_{\gamma\gamma}$, was applied, which took into account the energy-dependent resolution of photon energy measurements. The ω resonance was selected by the cut on the invariant mass of $\pi^+\pi^-\pi^0$ system, $|M_{\pi\pi\pi} - M_\omega^{\rm PDG}| < 70\,{\rm MeV}/c^2$, where $M_\omega^{\rm PDG} = 782.65\,{\rm MeV}/c^2$. Due to an additional contribution of calorimetric measurements to the experimental resolution, the exclusivity selection was changed to $|E_{\rm miss}| <$ 3.0 GeV. In the analysis of ω meson the signal and background asymmetries were extracted simultaneously using the unbinned maximum likelihood method. The average values of the kinematic variables are $\langle Q^2 \rangle = 2.2$ (GeV/c)^2, $\langle x_{\rm Bj} \rangle = 0.049$, $\langle y \rangle = 0.18$, $\langle W \rangle = 7.1$ GeV, and $\langle p_{\rm T}^2 \rangle = 0.17$ (GeV/c)^2.

Journal of Physics: Conference Series **678** (2016) 012045 doi:10.1088/1742-6596/678/1/012045

2.2. Results and discussion

For a transversely polarised target five single (UT) and three double (LT) spin asymmetries can be defined. These are $A_{\rm UT}^{\sin(\phi-\phi_s)}$, $A_{\rm UT}^{\sin(\phi+\phi_s)}$, $A_{\rm UT}^{\sin(3\phi-\phi_s)}$, $A_{\rm UT}^{\sin\phi_s}$, $A_{\rm UT}^{\sin(2\phi-\phi_s)}$, $A_{\rm LT}^{\cos(\phi-\phi_s)}$, $A_{\rm LT}^{\cos\phi_s}$, $A_{\rm LT}^{\cos(2\phi-\phi_s)}$. Each asymmetry is related to a modulation of the cross section as a function of ϕ and ϕ_s angles, which is indicated by the superscript. The angle ϕ is the azimuthal angle between the lepton plane, given by the momenta of the incoming and the scattered leptons, and the hadron plane, given by the momenta of the virtual photon and the meson. The angle ϕ_s is the azimuthal angle between the lepton plane and the spin direction of the target nucleon. Average values of measured asymmetries for exclusive ρ^0 production are shown in Fig. 1. The asymmetry $A_{\rm UT}^{\sin\phi_S}$ was found to be $-0.019 \pm 0.008(stat.) \pm 0.003(syst.)$. All other asymmetries were also found to be of small magnitude but consistent with zero within experimental uncertainties. The asymmetries measured as functions of Q^2, $x_{\rm Bj}$ and $p_{\rm T}^2$ can be found in Ref. [7], as well as their comparison to the predictions of the GPD-model by Goloskokov and Kroll [8].

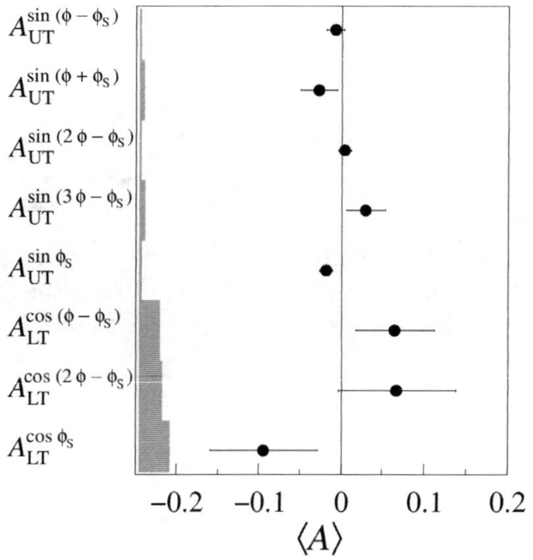

Figure 1. Mean values $\langle A\rangle$ for all transverse target spin symmetries for exclusive ρ^0 productions. The errors bars (left bands) represent the statistical (systematic) uncertainties.

The measured asymmetries are proportional to bilinear combinations of the helicity amplitudes $\mathcal{M}$ for the photoproduction subprocess, $A \propto \sum \mathcal{M}^*_{i'm',im}\mathcal{M}_{i'm',jn}$, where the helicity of the virtual photon is denoted by $i, j = -1, 0, +1$ and the helicity of the initial-state proton by $m, n = -\frac{1}{2}, +\frac{1}{2}$. The sum runs over all combinations of spins given by the spin of the meson $i' = -1, 0, +1$ and the spin of the final-state proton $m' = -\frac{1}{2}, +\frac{1}{2}$. In the following the helicities will be labelled by only their sign or zero.

For an interpretation of results in the framework of the model, the following asymmetries are particularly interesting, for which the dependence on the helicity amplitudes reads

$$\begin{aligned}
\sigma_0 A_{\rm UT}^{\sin(\phi-\phi_s)} &= -2{\rm Im}\left[\epsilon\mathcal{M}^*_{0-,0+}\mathcal{M}_{0+,0+} + \mathcal{M}^*_{+-,++}\mathcal{M}_{++,++} + \tfrac{1}{2}\mathcal{M}^*_{0-,++}\mathcal{M}_{0+,++}\right] , \\
\sigma_0 A_{\rm UT}^{\sin(2\phi-\phi_s)} &= -\ {\rm Im}\left[\mathcal{M}^*_{0+,++}\mathcal{M}_{0-,0+}\right] , \\
\sigma_0 A_{\rm UT}^{\sin\phi_s} &= -\ {\rm Im}\left[\mathcal{M}^*_{0-,++}\mathcal{M}_{0+,0+} - \mathcal{M}^*_{0+,++}\mathcal{M}_{0-,0+}\right] .
\end{aligned} \tag{2}$$

Here σ_0 is the total unpolarised cross section, given by the sum of cross sections for longitudinally, σ_L, and transversely, σ_T, polarised virtual photons,

$$\sigma_0 = \frac{1}{2}\left(\sigma_{++}^{++} + \sigma_{++}^{--}\right) + \epsilon\sigma_{00}^{++} = \sigma_L + \epsilon\sigma_T, \tag{3}$$

and ϵ is the virtual photon polarisation parameter.

The dominant $\gamma^*_L \to \rho^0_L$ transitions are described by helicity amplitudes $\mathcal{M}_{0+,0+}$ and $\mathcal{M}_{0-,0+}$, which are related to chiral-even GPDs H and E, respectively. The subscripts L and T denote the photon and meson helicities 0 and ±1, respectively. The suppressed $\gamma^*_T \to \rho^0_T$ transitions are described by the helicity amplitudes $\mathcal{M}_{++,++}$ and $\mathcal{M}_{+-,++}$, which are likewise related to H

and E. These GPDs are used since several years to describe DVCS and vector meson exclusive production data. By the recent inclusion of transverse, *i.e.* chiral-odd GPDs, it became possible to also describe $\gamma_T^* \to \rho_L^0$ transitions [8]. In their description appear the amplitudes $\mathcal{M}_{0-,++}$ related to chiral-odd GPDs H_T and $\mathcal{M}_{0+,++}$ related to chiral-odd GPDs $\overline{E}_T$.

The small value of $A_{\mathrm{UT}}^{\sin(\phi-\phi_s)}$ asymmetry for ρ^0 is explained as an approximate cancellation of contributions from GPDs E^u and E^d for valence quarks u and d, respectively, which have opposite signs but similar absolute values. The $A_{\mathrm{UT}}^{\sin\phi_s}$ asymmetry represents an imaginary part of two bilinear products of helicity amplitudes. The first product is related to GPDs H and H_T, while the second one is related to GPDs E and $\overline{E}_T$. The asymmetry $A_{\mathrm{UT}}^{\sin(2\phi-\phi_S)}$ represents the same combination of GPDs E and $\overline{E}_T$ as the second term in $A_{\mathrm{UT}}^{\sin\phi_S}$. The observation of a vanishing value for $A_{\mathrm{UT}}^{\sin(2\phi-\phi_S)}$ implies that the non-vanishing value of $A_{\mathrm{UT}}^{\sin\phi_S}$ constitutes the first experimental evidence from hard exclusive ρ^0 leptoproduction for the existence of transverse GPDs H_T.

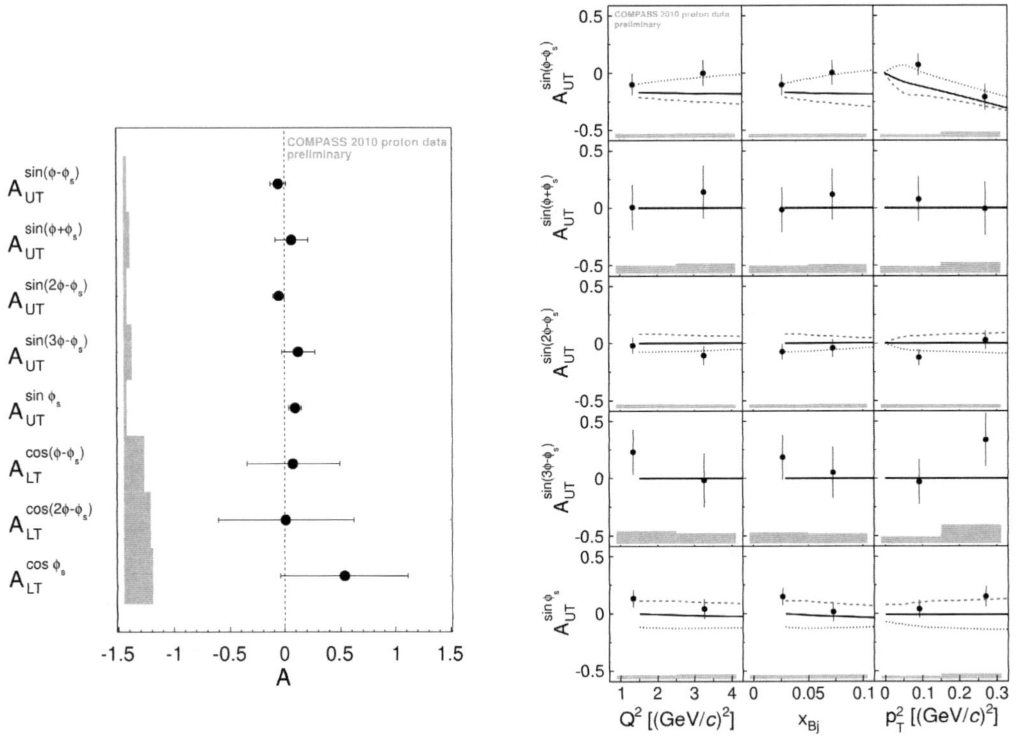

Figure 2. Left: Mean values A for all transverse target spin asymmetries for exclusive ω production. The errors bars (left bands) represent the statistical (systematic) uncertainties. Right: Single transverse target spin asymmetries for exclusive ω production. The curves show the predictions of the GPD model of Ref. [8]. The dashed red and dotted blue lines represent the predictions with the positive and negative $\pi\omega$ form factors, respectively, while the solid black lines represent the predictions without the pion pole exchange.

The preliminary result on average values of all eight transverse target spin asymmetries for exclusive ω production are shown in Fig. 2 (left), while the single spin asymmetries as functions of Q^2, x_{Bj} and p_{T}^2 are presented in Fig. 2 (right) and are compared to the predictions of Goloskokov-Kroll model [9]. In this version of the model the authors have added a contribution from the exchange of π^0 pole. It turned out that this is an important contribution needed to reproduce

Journal of Physics: Conference Series **678** (2016) 012045 doi:10.1088/1742-6596/678/1/012045

HERMES results on SDMEs for exclusive electroproduction of ω mesons [10]. Still, the SDME data do not allow one to distinguish the sign of the $\pi\omega$ transition form factor. Certain azimuthal asymmetries for ω production are sensitive to the pion pole contribution and hence in principle could allow one to determine the sign of the form factor. The effect of the pion pole decreases with the energy of the virtual photon nucleon system, W, but still it could be measurable at COMPASS energies. The authors have provided three sets of predictions: one without the pion pole contribution and two others with the contributions that differ by the sign of $\pi\omega$ form factor. While the negative sign is preferred by the results on $A_{\mathrm{UT}}^{\sin(\phi-\phi_s)}$ and $A_{\mathrm{UT}}^{\sin(2\phi-\phi_s)}$, the results on $A_{\mathrm{UT}}^{\sin\phi_s}$ asymmetry prefer the positive sign. At present this discrepancy is not resolved.

3. The GPD program of COMPASS-II

The GPD part of COMPASS-II proposal [11] is devoted to measurements of both DVCS and HEMP with polarised μ^+ and μ^- beams and a liquid hydrogen target. The new detectors, the 4 m-long recoil proton detector CAMERA and (a central part of) the new large-angle electromagnetic calorimeter ECAL0, which are essential for measurements of exclusive processes, were constructed and incorporated into the COMPASS setup in 2012. The commissioning of these new detectors was done in 2012 and was followed by a short DVCS pilot run. The dedicated data taking for the GPD program, with the complete calorimeter ECAL0, is foreseen in 2016-2017 for a total period of 280 days.

The recoil proton detection is based on the ToF measurement between two barrels of 24 scintillator slats read out at both ends. By installing the CAMERA around the new 2.5 m long LH_2 target COMPASS has been converted into a facility measuring exclusive reactions within a kinematic domain from $x_{\mathrm{Bj}} \sim 0.01$ to ~ 0.1, which cannot be explored at any other existing or planned facility in the near future. The new calorimeter ECAL0 is being constructed. Compared to the existing electromagnetic calorimeters, it will increase the accessible kinematic domain for DVCS and exclusive π^0 production towards higher x_{Bj}, and therefore it will provide an overlap with HERMES and JLAB experiments. ECAL0 will also improve the hermeticity for detection of exclusive events and will contribute to reduce the background to single-photon production, which originates from π^0 and other decays.

An efficient selection of exclusive events, and suppression of the background was possible by using the combined information from the forward COMPASS detectors and the CAMERA. As an example of obtained results we present for exclusive single-photon production the ϕ distributions in bins of x_{Bj}, where ϕ is the angle between the leptonic plane and the $\gamma^*\gamma$ plane. The observed distributions, after applying all cuts and selections and for $Q^2 > 1\ (\mathrm{GeV}/c)^2$, are displayed in Fig. 3 and compared to the predictions from the Monte Carlo (MC) simulations for the BH event yield. The MC is normalised to the data in the small x_{Bj} bin, where only BH contribution is expected. COMPASS offers the advantage to provide various kinematic domains where either BH or DVCS dominates. The collection of almost pure BH events at small x_{Bj} allows one to get an excellent reference yield and to control accurately the global efficiency of the apparatus. In contrast, the collection of a DVCS sample at larger x_{Bj} will allow the measurement of the x_{Bj} dependence of the t-slope of the cross section, which is related to the tomographic partonic image of the nucleon. In the intermediate domain, the DVCS contribution will be boosted by the BH process through the interference term that allows us to investigate DVCS at the amplitude level. The Bethe-Heitler contribution shows a characteristic peak at $\phi \simeq 0$.

The exclusive photon production sample is polluted by π^0 contribution when only one photon of the π^0 decay is observed. In this case, one of the two photons carries most of the energy of the parent meson and the second photon us either absorbed or emitted outside of the ECALs acceptance. We call this a "non-visible" π^0 background, which affects mostly large x_{Bj} events. It comes both from exclusive and semi-inclusive (SIDIS) π^0 productions. The first was studied using HEPGEN MC for hard exclusive π^0 production, while the second one by using LEPTO

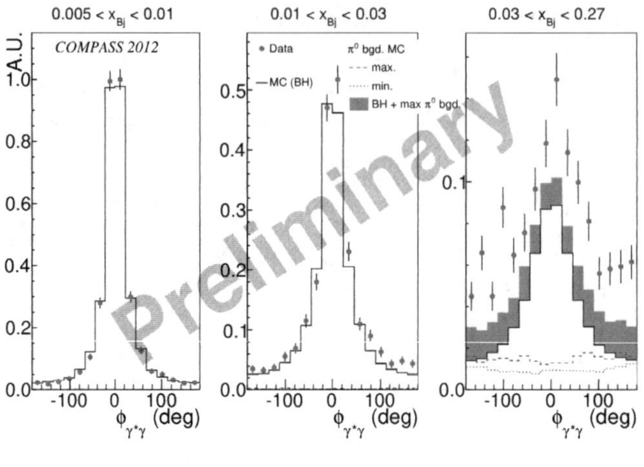

Figure 3. The exclusive single photon events obtained from the 2012 sample as a function of $\phi_{\gamma*\gamma}$ ($\equiv \phi$) compared to the MC estimates of BH contribution (solid line) and the π^0 contamination (dotted and dashed lines). See text for more details.

MC. Presently we use a conservative approach and take the SIDIS π^0 (LEPTO) sample to estimate an upper limit and the exclusive π^0 (HEPGEN) sample for a lower bound on invisible π^0 contribution to the single photon sample. These two estimates are shown in Fig. 3 as dotted (exclusive π^0) and dashed (SIDIS π^0) histograms. The sum of BH contribution (continuous line) and the estimate of the maximal π^0 background is shown as a shaded (blue) histogram (right panel). An excess of the data above this histogram is interpreted as an indication of the DVCS events in the sample and a proof of feasibility to measure DVCS with the present setup.

The projected accuracies for measurements of various observables within the GPD program are presented in the COMPASS-II proposal [11]. Investigation of GPDs with DVCS and HEMP on unpolarised protons using the data from 2016-2017 will allow to determine the x_{Bj}-dependence for t-slopes of the differential cross sections. That is related to the spatial transverse distribution of partons and the 'nucleon tomography'. Measurements of the beam charge and spin sum and difference of single-γ cross sections will give access to the real and imaginary parts of the DVCS amplitude, and will allow to further constrain GPDs H. Studies of exclusive production of vector mesons (ρ, ω, ϕ) will lead to the quark flavour and gluon separation for GPDs H, while that of exclusive π^0 production will provide constrains on the GPD $\widetilde{E}$ and on chiral-odd GPDs. The main goal of future (after 2018) measurements with transversely polarised target is to constrain GPDs E, which are related to the orbital momentum of partons, and also to investigate the role of chiral-odd GPDs in exclusive meson production.

References

[1] D. Mueller *et al*, Fortsch. Phys. **42** (1994) 101.
[2] X. Ji, Phys. Rev. Lett. **78** (1997) 610; Phys. Rev. D **55** (1997) 7114.
[3] A.V. Radyushkin, Phys. Lett. B **385** (1996) 333; Phys. Rev. D **56** (1997) 5524.
[4] M. Burkardt, Phys. Rev. D **62** (2000) 071503; erratum-ibid. D **66** (2002) 119903; Int. J. Mod. Phys. A **18** (2003) 173; Phys. Lett. B **595** (2004) 245.
[5] COMPASS Collaboration, P. Abbon *et al.*, Nucl. Instr. Meth. A **577** (2007) 455.
[6] C. Adolph *et al.* [COMPASS Collaboration], Nucl. Phys. B **865** (2012) 1.
[7] C. Adolph *et al.* [COMPASS Collaboration], Phys. Lett. B **731** (2014) 19.
[8] S.V. Goloskokov and P. Kroll, Eur. Phys. J. C **74** (2014) 2725 and private communication.
[9] S.V. Goloskokov and P. Kroll, Eur. Phys. J. A **50** (2014) 146.
[10] A. Airapetian *et al.* [HERMES Collaboration], Eur. Phys. J. C **74** (2014) 3110.
[11] F. Gautheron *et al.* [COMPASS Collaboration], CERN-SPSC-2010-014, SPSC-P-340, May 2010.

Journal of Physics: Conference Series **678** (2016) 012046 doi:10.1088/1742-6596/678/1/012046

Spin physics and biological membranes

Yury Kiselev
Joint Institute for Nuclear Research, VBLHEP, Dubna, 141980, Russia
E-mail: yury,kiselev@cern.ch

Abstract. Formula for calculating the concentration profile of ions in biological membranes has been obtained. It is assumed that ions are moving in a viscous medium under the action of the electric field and a concentration gradient. The problem is that ions are coated with shells consisting of water dipoles. These dipoles copy the form of the ions and in a strong electric field they can acquire the shape of an ellipsoid which changes the effective ion radius in the membrane. Calculation of the Na^{+1} and K^{+1} profiles leads to a conclusion that active and passive transport of ions is closely associated with the shape of the hydrated shells. The work was performed at the Veksler and Baldin Laboratory of High Energy Physics, JINR, Dubna

1. Introduction

In solid, liquid and gaseous media the force of the electric field acting on ions is balanced by the friction force proportional to the velocity of particle motion. At equality of forces, the drift velocity of charges is to be proportional to the electric field intensity. In a membrane, this velocity also depends on the concentration gradient, and on a specific membrane feature allowing the so-called active or passive transport of ions. Hereafter, we calculate the profiles of the Na^{+1} and K^{+1} concentration in a membrane and show that the active or passive transport is determined by the shape of the hydrate shells. The work was performed at the Laboratory of High Energy Physics, JINR, Dubna.

2. Concentration profile of ions in a membrane

Before considering the ion motion in a biological membrane we recall the classical scheme of these calculations using the electron gas in metal [1]. At the collision with the ion lattice of metals the electrons give up, on the average, a momentum mv during the mean interval of the free motion $\tau/(\mathrm{s})$, where $m/(\mathrm{kg})$ is the mass, and $v/(\mathrm{m{\cdot}s^{-1}})$ is the mean drift velocity of electrons. On equating the momentum transfer velocity $mv/\tau/(\mathrm{N})$ to the electric force $eE/(\mathrm{N})$ of the electric field $E/(\mathrm{V{\cdot}m^{-1}})$ acting on the elementary charge $e/(\mathrm{C})$, results [1] to the equation (1) which is one of the forms of the Ohm law

$$mv/\tau = -eE \;, \qquad (1)$$

from where it follows that the drift velocity v is proportional to the field intensity E.

A physical analogy is also preserved in the case when the ions are moving under the action of the electric field in liquid solutions. For a spherical particle of radius r in a solvent of viscosity $\eta/(\mathrm{kg{\cdot}m^{-1}s^{-1}})$, the friction force is given by Stokes' law $F_S = 6\pi\eta r v$, where v is the velocity of the ion. This time, as in (1), equating of forces gives

$$6\pi\eta r v = zeE \;, \qquad (2)$$

Content from this work may be used under the terms of the Creative Commons Attribution 3.0 licence. Any further distribution of this work must maintain attribution to the author(s) and the title of the work, journal citation and DOI.
Published under licence by IOP Publishing Ltd

where z is the ion valence. The velocity v should be again proportional to the field in a membrane but, in solutions, there is also the thermodynamic force $F_x/(\mathrm{J{\cdot}m^{-1}})$ depending on the gradient of the normalized concentration of ions $C(x)$ [2]

$$F_x = -kT\left(\frac{\partial \ln C}{\partial x}\right) , \tag{3}$$

where $k = 1.38 \cdot 10^{-23}/(\mathrm{J{\cdot}K^{-1}})$ is the Boltzmann constant, $T/(\mathrm{K})$ is the absolute temperature, $x/(\mathrm{m})$ is the ion coordinate along the channel; $0 < x \leq d$, d is the membrane thickness. Adding F_x and $v = u \cdot E$ to equation (2), where $u/(\mathrm{m^2{\cdot}\ s^{-1}{\cdot}\ V^{-1}})$ is the mobility of the ion, we have

$$6\pi\eta ruE = zeE - kT\left(\frac{\partial \ln C}{\partial x}\right) , \tag{4}$$

The electric field is equal to the potential gradient $E = -\partial\varphi/\partial x$. Let $\varphi(x = 0) = 0$ be the electrical potential of the solution, then $\varphi(x = d) = V_r = -(0.05 \div 0.15)$, where $V_r/(\mathrm{V})$ is the rest potential of the membrane, and integrating (4) over x we obtain:

$$\ln C(x) - \ln C(0) = \frac{1}{kT}(ze - 6\pi\eta ru)\int_0^x -d\varphi = \frac{-1}{kT}(ze - 6\pi\eta ru)\left(V_r\frac{x}{d} - 0\right) . \tag{5}$$

The function $V_r \cdot x/d$, describes a linear voltage distribution over x. The mobility u in (5) is conveniently expressed in terms of the diffusion coefficient $D/(\mathrm{m^2{\cdot}s^{-1}})$, since the latter does not depend on the molecular weight of ions. Using the Einstein relation [2]

$$u = \frac{zeD}{kT} , \tag{6}$$

substituting equation (6) into (5), and replacing $(-(-V_r) = |V_r|)$ we obtain the following formula to calculate the ratio of concentrations

$$\frac{C(x)}{C(0)} = exp\{\frac{ze|V_r|}{kT}\frac{x}{d}\left(1 - \frac{6\pi\eta D}{kT}r\right)\} . \tag{7}$$

Let the radius r of the ion at which the expression in the brackets is equal to zero be denoted as r_0. Without a membrane, $C(x) = C(0)$, the ion radius $r = r_0$ is fixed and it is determined by the Stokes - Einstein formulae [2]

$$r_0 = \frac{kT_0}{6\pi\eta_0 D_0} . \tag{8}$$

Hence, the ion profile in a solution is the straight line $C(x)/C(0) = 1$, shown in Fig 1. Combining formulas (7) and (8) we get the concentration profile in the common case:

$$C(x)/C(0) = exp\{\frac{ze|V_r|}{kT}\frac{x}{d}\left(1 - \frac{\eta DT_0}{\eta_0 D_0 T}\frac{r}{r_0}\right)\} = exp\{\frac{ze|V_r|}{kT}\frac{x}{d}\left(1 - \xi\frac{r}{r_0}\right)\} , \tag{9}$$

where r is the actual radius of an ion in the membrane. In a solution of different viscosity and temperature $\xi = \eta DT_0/(\eta_0 D_0 T) = 1$ since the ion radius in (8) is fixed. Assume, for example, that the Na^{+1} ions are in the water at $T_0 = 298/(\mathrm{K})$. In this case $\eta_0 = 0.891 \cdot 10^{-3}$, the diffusion coefficient is $D_0(Na^{+1}) = 1.33 \cdot 10^{-9}$ [2]. At the temperature $T = 309/(\mathrm{K})$, $\eta = 0.699 \cdot 10^{-3}$ [2], and since $\xi = 1$ we have $T_0/\eta_0 D_0 = T/\eta D$ and the diffusion coefficient is changed to $D(Na^{+1}) = 1.76 \cdot 10^{-9}$; however at both temperatures the radius of Na^{+1} ions stays the same: $r = r_0 = 184 \cdot 10^{-12}/(\mathrm{m})$.

Journal of Physics: Conference Series **678** (2016) 012046 doi:10.1088/1742-6596/678/1/012046

Nevertheless, it is known [3] that the calculation by (8) coincides with the experimentally measured radius with the accuracy up to 2. The problem is that the ions are coated with shells consisting of water dipoles. We illustrate the importance of such complication by the following examples. Let us imagine that the radius of $Na^{+1}(z = +1)$ in a membrane is enlarged due to hydration up to $r = 1.5 \cdot r_0 = 1.5 \times 184 \cdot 10^{-12}$. The new profile calculated by (9) is presented in Fig. 1. The ion concentration decays along the membrane. This example demonstrates the concentration profile at the passive transport of ions.

In the next example, we consider the $K^{+1}(z = +1)$ ion having a larger dipole polarizability compared with the Na^{+1} one. Note that dipole and quadrupole interactions of ions with an electric field in membranes are small one to influence any ion motion. The important reason to consider these interactions follows by the fact that the water dipoles copy the shape of the ions. Such complex charge structures are considered in physics of nuclear moments.

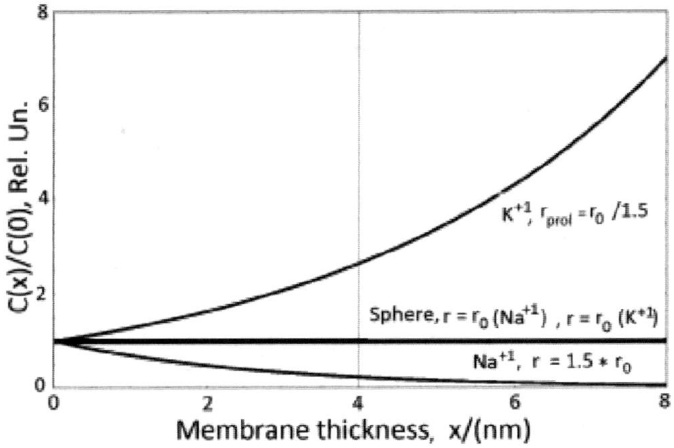

Figure 1. Concentration profiles of ions. The line is a profile in a solution without a membrane at $r = r_0$; the Na^{+1} shows the passive transport at $r = 1.5 \cdot r_0$; the active transport of K^{+1} at $r_p = r_0/1.5$.

In a strong electric field, these objects can acquire an elliptical shape, and the hydrated shell around an ion has to repeat this shape. Because hydrated shells are ingredients of the solvent, the ellipsoidal objects in a solution have a different friction and therefore the different effective radius compared with the spherical one. This means that the concentration profile, in accordance with (9), is changed. Suppose now that $r_0 = (a+b)/2$, where $2a$ are the length of the major and $2b$ of minor ellipse axes, respectively. Let K^{+1} ion be in the water at $T_0 = 298/(K)$ then $\eta_0 = 0.891 \cdot 10^{-3}$, and since $D_0(K^{+1}) = 1.96 \cdot 10^{-9}$ [2] then (8) gives $r_0 = 125 \cdot 10^{-12}/(m)$. Assuming that $a = 195 \cdot 10^{-12}$ and $b = 55 \cdot 10^{-12}$, the effective radius of the prolate ellipsoids in a solution is $r_p \cong (a \cdot b^2)^{1/3} \cong 84 \cdot 10^{-12}$ [2]. Substituting of $r_p/r_0 = 84/125 \cong 0.67$ and $\xi = 1$ in (9) gives the concentration profile of K^{+1} ion, shown in Fig. 1. This profile displays the active transport of the K^{+1} ions. For numeric estimations, it is conveniently to simplify (9) assuming $\xi = 1$, $x = d$, which yields

$$\frac{C(x=d)}{C(x=0)} = exp\{\frac{ze|V_r|}{kT}\left(1 - \frac{r}{r_0}\right)\} \, , \tag{10}$$

where $e/k = 11594/[C/(J{\cdot}K^{-1})]$, T (K). Hence, for the above Na^{+1} ions $r/r_0 = 1.5$ and at $|V_r| = 0.15/(V)$, and (10) gives $C(d) \cong C(0)/25$. Thus, the Na^{+1} intracellular concentration is decreased by 25 times compared with extracellular one; like approximately it does in the cells

Journal of Physics: Conference Series **678** (2016) 012046 doi:10.1088/1742-6596/678/1/012046

of a squid. This is the passive transport of ions. For the K^{+1} ion with the shape of a prolate ellipsoid $r_{prol}/r_0 = 84/125 = 0.67$ and $C(d) \cong 7 \cdot C(0)$; hence, the K^{+1} will increase their concentration on passing the membrane. The case of K^{+1} illustrates a mechanism of the active transport by changing the shape of hydration shells of the ions.

3. Conclusion

Equations (9) and (10) have been obtained to calculate the concentration profile of ions in biological membranes. It is shown that the friction forces and effective ion radius in membranes depends on the shape of shells covering ions. Analysis of the Na^{+1} and K^{+1} profiles lead to a conclusion that a possible mechanism of active and passive transport can be closely associated with changing of hydrated shells of ions by membranes.

3.1. References

[1] Johnk, Carl T.A. 1988 *Engineering electromagnetic fields and waves* (Wiley & Sons, Inc.) 2nd Edition Chapter 3 pp. 111-113.
[2] Atkins P., de Paula J. 2006 *Atkins' Physical Chemistry* (New York: W.H. Freeman and Company) 8th edition Chapter 19 pp. 660, 1017-1020.
[3] Grigoriev I.S. and Melikhov E.Z. 1991 *Physical quantities* Reference book (Moscow: Energoatomizdat) Chapter 17 p. 376.

Polarized antiproton beam at U-70 accelerator of IHEP

S B Nurushev[1,2,3], M A Chetvertkov[3], V A Chetvertkova[3], V I Garkusha[1], A P Meschanin[1], V V Mochalov[1,2], M B Nurusheva[3], V L Rykov[2], P A Semenov[1,2], M N Strikhanov [2], A N Vasiliev[1,2], V N Zapolsky[1]
[1] Institute For High Energy Physics National Research Centre Kurchatov Institute, Protvino, 142280, Russia
[2] National Research Nuclear University (Moscow Engineering Physics Institute), Moscow, 115409, Russia
[3] Federal state budget organization State center Interphysica, Ministry of education and science, Moscow, Russia

E-mail: mochalov@ihep.ru

Abstract. The polarized proton and antiproton beam channel is currently under development at the U-70 accelerator of IHEP, Protvino, Russia. An availability of the both, polarized protons and antiprotons provides an exciting opportunity for the comparative studies of spin effects induced by polarized protons and antiprotons in a variety of hadronic reactions. While the proton and antiproton beams are formed by essentially the same method, there is the specific in the antiproton beam shaping and properties compared to protons. In this report, we address some technical details of forming the polarized antiproton beam and describe its main properties.

1. Introduction

The new experimental program SPASCHARM [1] for the 70 GeV proton synchrotron U-70 is currently under development at the Institute for High Energy Physics (IHEP), Protvino, Russia. The main physics motivation for the SPASCHARM experiment is the systematic study of spin phenomena for a wide range of inclusive and exclusive reactions. Spin asymmetries will be studied in inclusive and exclusive reactions in the beam fragmentation region with polarized beam or target as well as the hyperon and vector mesons polarization and depolarization phenomena. At the first stage, the measurements of single-spin asymmetries are carried out at the existing unpolarized beams, using the polarized target. At the second stage, the special polarized proton and antiproton channel 24A will be constructed. This will create the opportunity for studying the double-spin asymmetries as well as bring the new quality into the measurements of single-spin effects. The ultimate goal is to get the capabilities and reach the sensitivity to the longitudinal double-spin asymmetry A_{LL} in charmonium production in order to explore the gluon polarization $\Delta G/G(x)$ at large x_F. The SPASCHARM physics program and the experimental setup have been presented in details a number of times elsewhere [1–2].

The channel 24A is a part of a pair of two linked new channels, operating from the same slowly extracted to the external target U-70 primary proton beam (see figure 1). The bottom beam-line 24B shown in red in figure 1 delivers the beam to the spectroscopy experiment. The top red beam-line designates the polarized proton and antiproton beam channel 24A. The maximum allowed intensity of

Content from this work may be used under the terms of the Creative Commons Attribution 3.0 licence. Any further distribution of this work must maintain attribution to the author(s) and the title of the work, journal citation and DOI.
Published under licence by IOP Publishing Ltd

Journal of Physics: Conference Series **678** (2016) 012047 doi:10.1088/1742-6596/678/1/012047

the primary beam is $2 \cdot 10^{13}$ protons/cycle, and it is limited by the radiation safety requirements. The secondary beam from the primary target is split between these two channels by the first after the target, so called, 'sweeping' magnet. It directs charged particles to the channel 24B while the neutrals go straight to the channel 24A. The required bending power of this magnet is up to 4.7 Tm. The magnetic field achievable in a conventional 'warm' magnet is ~1.8 T. Thus, the sweeping magnet length must be at least 2.6 m.

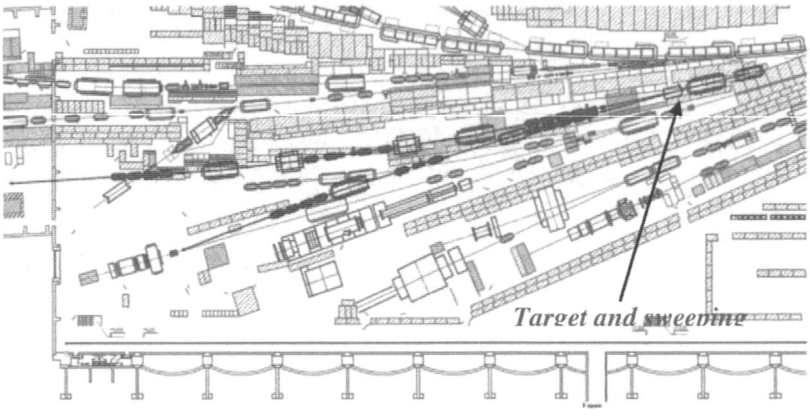

Figure 1. The layout of the new beam channels 24A and 24B in the U70 experimental hall.

2. Polarized beam

The polarization will be created from Λ-decay using standard technique. The idea of such method was suggested almost 50 years ago by O.E. Overseth and J.Sandweiss [3]. It has been successfully used in E-704 (FNAL) [4] and FODS (IHEP, Protvino) [5] experiments. In the Λ- hyperons rest frame, the decay protons or antiprotons are longitudinally polarized. After the Lorenz boost to the laboratory frame, they become transversely polarized, and this polarization depends on their emission angle in Λ-decay. By the appropriately designed channel's magnetic optics, the initial proton distribution over the decay angle is transformed to the spatial distribution at the plane of the so called 'intermediate focus'. There the beam is sorted out into the samples of a certain transverse polarization either by each beam trajectory tagging with fast tracking detectors (hodoscopes) [4], or by a collimation [5]. The beam tagging provides the opportunity for potentially having beam samples of the opposite polarization at the experimental target simultaneously. This is equivalent to potential doubling the beam intensity compared to the collimation with the beam samples of the opposite polarization delivered to the experiment target alternatingly in time. Moreover, the tagging allows the much finer trajectory sorting over the phase space and over the momentum which is measured at the accuracy of better than 1% [7]. This reduces the effects of polarization smearing by the spread of the beam parameters and, as a result, allows to increase the effective average sample polarization. The verification and confirmation of operation of tagging system is done by the independent polarimetry which is discussed elsewhere [8].

The beam optic of the channel 24A is schematically shown in figure 2. It is comprised of 12 quadrupole lenses and four bending magnets. The total beam rotation angle in horizontal plane is equal to ~160 mrad. The total beam-line length from a primary target to the experiment target is 120 m. The high-intensity polarized beams of momenta from ~10 to 45 GeV/c is delivered to the experiment target. The momentum analysis occurs in the horizontal plane, and sorting the beam trajectories over the transverse vertical polarization takes place around the intermediate focus in vertical plane. The one important feature of the channel 24A beam optic is it 'unity'. This means that it transfers the phase space from the primary target plane to the target of experiment without distortions either in the spatial or angular dimension. Such an optic is chosen so as to eliminate the beam depolarization effects in the chain of quadrupole lenses.

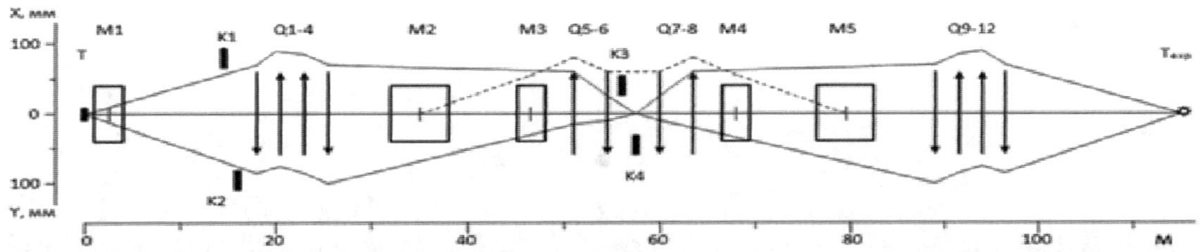

Figure 2. The optics of the polarized beam channel 24A: *T* – external target, *M1* – sweeping magnet, *Q1-Q12* – quadrupole lenses, *M2-M5* – bending magnets, *K1-K4* – collimators, T_{exp} – the target of the experiment.

The intrinsic property of the (anti)proton beam from Λ-decays is that it always includes two approximately equal samples of the same intensity and of equal but of opposite signs transverse polarizations. This property is extremely valuable for reduction the systematic errors by directing to the experiment target the samples of opposite polarizations either simultaneously, using the tagging, or alternatingly in time. The further reduction of systematic errors is achieved by inverting (flipping) the polarization of the entire beam every few tens of minutes. The spin-flip for the entire beam will be performed by the special superconducting helical magnet [6], positions downstream of the last quadrupole lens *Q12*. Spin-flipper is also has the capability to convert the transverse polarization into the longitudinal without modifications of the spatial or angular properties of a trajectory.

The results of simulation of the beam properties are presented in table 1. The achievable intensity of the polarized proton beam satisfies all requirements of the experiment. However, the intensity of the polarized antiproton beam is not as high as desired.

Table 1. Calculation of the polarized beam properties

	proton beam		antiproton beam	
Momentum collimator position	min	max	min	max
Polarized beam energy	45		14	
Momentum spread *Δp/p*, %	±4.5	11.0	±5.5	12.0
	(σ=2.1)	(σ=5.3)	(σ=2.3)	(σ=5.6)
Beam size, $\sigma_x \times \sigma_y$, mm	13 × 11	17 × 14	18 × 19	20 × 19
Beam divergence $\sigma_x \times \sigma_y$, mrad	1.6 × 2.0	1.5 × 1.9	1.6 × 1.9	1.5 × 2.0
Full intensity for 10^{13} protons in the target at 60 GeV	4.9×10^7	1.3×10^8	1.4×10^5	3.9×10^5

3. Polarized antiproton beam

This report is devoted mainly to the issues of forming a polarized antiproton beam.

The dependence of the beam properties on the collimator position (*K4*) in the intermediate focus is shown on figure 2.

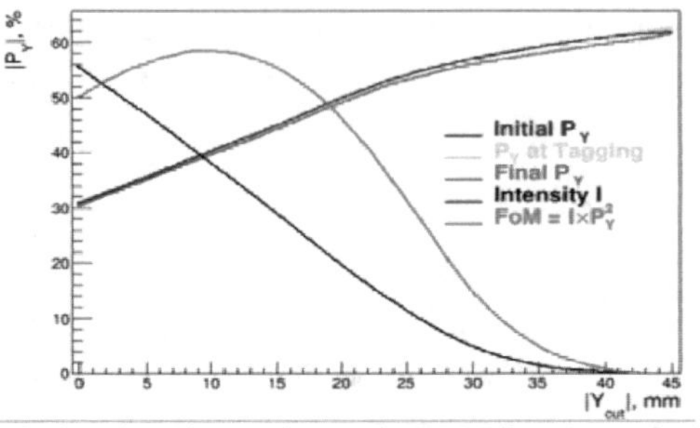

Figure 2. Polarization, intensity and *FOM* for the polarized beam

The coordinate *Y* means the selection of the particles with coordinate greater than |*Y*|. Intensity and Figure of Merit ($FOM=I{\cdot}P^2{}_Y$) are presented in arbitrary units. The maximum value of FOM is achieved at |*Y*|>*10* mm with the average polarization ±39.6%. Beam depolarization is in the range of (0.5-1.3)%. The beam profiles on the target are presented on figure 3.

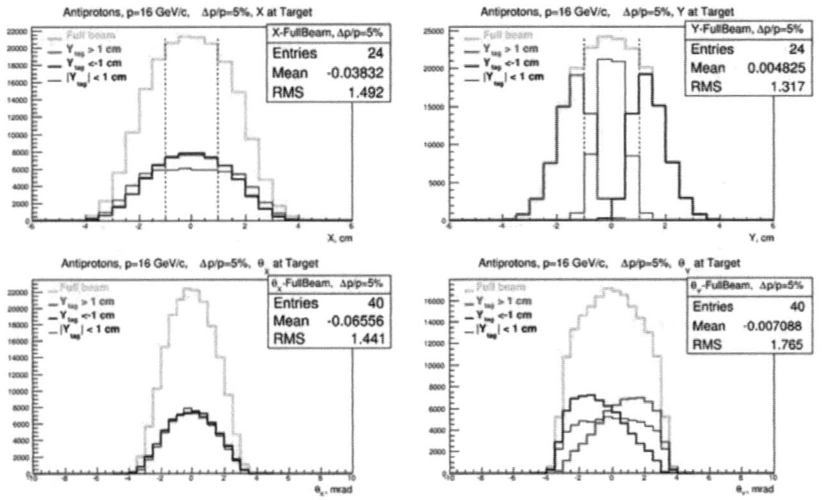

Figure 3. Polarized antiproton Beam at the experiment target.

Dashed line is the size of the polarized frozen target. One can observe that the beam size is larger than the size of available polarized target. This means that the desire of having at the target simultaneously the sample of the both polarizations, sorted by tagging, leads to an ineffective usage of the beam intensity. The solution for the case of this small in diameter target is the alternating in time steering to the target of a sample of only one sign polarization, using the vertical magnet-correctors (not shown in figure 1).

The simulations of the polarized beam and background have been done at the following conditions:

- All numbers for intensities of polarized beams are for the cases when only one part of the polarized beam (positive or negative) has been steered onto the experiment target through the collimator K4 with the slit width of 2 cm.
- For the unpolarized beam is selected the central part of the total beam passed through the collimator K4 of the same slit width 2 cm with the magnets-correctors off.

The intensity of the polarized beams defined above are presented in table 2.

Journal of Physics: Conference Series **678** (2016) 012047 doi:10.1088/1742-6596/678/1/012047

Table 2. Intensities of polarized beams of various energies and of average polarization 40%.

Beam Momentum (GeV/c)	antiproton beam		proton beam	
	In target plane	At 2 cm target[1]	In target plane	At 2 cm target
16	$7{,}34\times10^4$	$2{,}41\times10^4$	$1{,}84\times10^6$	$6{,}90\times10^5$
20	$5{,}70\times10^4$	$2{,}20\times10^4$	$4{,}08\times10^6$	$1{,}56\times10^6$
30	$1{,}60\times10^4$	$7{,}66\times10^3$	$1{,}10\times10^7$	$5{,}63\times10^6$
40	$2{,}71\times10^3$	$1{,}52\times10^3$	$1{,}43\times10^7$	$8{,}04\times10^6$

[1]Number of particles hitting the polarized target of the diameter 2 cm

The results of background simulations from two main sources are presented in the table 3. The comparison of energies 14 and 16 GeV demonstrate the effect of sharp drop of negative pions from Λ-decays which is forbidden by kinematics at and above 16 GeV. This, unfortunately, is not valid for the pion background from K^0-decays.

Table 3. Background estimates

		P=14 GeV/c		P=16 GeV/c	
		In target plane	At 2 cm target	In target plane	At 2 cm target
Polarized part	$\pi^-(K^0)$	$1{,}75\times10^5$	$1{,}81\times10^4$	$1{,}88\times10^5$	$2{,}89\times10^4$
	$\pi^-(\Lambda^0)$	$1{,}35\times10^5$	$6{,}28\times10^4$	0	0
Unpolarized part	$\pi^-(K^0)$	$1{,}50\times10^5$	$1{,}79\times10^4$	$2{,}15\times10^5$	$2{,}76\times10^4$
	$\pi^-(\Lambda^0)$	$4{,}19\times10^5$	$2{,}00\times10^5$	0	0

One can observe that the expected pion background, while usually higher than the antiproton intensity, it does not exceed the capabilities of its significant suppression by the beam Cherenkov threshold counter.

Few approaches have been explored for increasing the antiproton beam intensity within the fiducial area of the experiment target. Among them, was the attempt of compressing the transverse beam size at the target by some departure from the 'unity' channel optic which might potentially be harmful for the polarization conservation. In figure 4 are shown, the result of the respective simulations. It is apparent that the transverse dimensions of the beam are noticeably reduced, the vertical one by a factor of ~2. This came at the price of widening of the angular distributions which was expected of course. It is amazing however, that the departure from the 'unity' optic did not damage much the beam polarization. It dropped down from 39.6% to only 38.9%. The FOM at the 2 cm diameter increased almost twice.

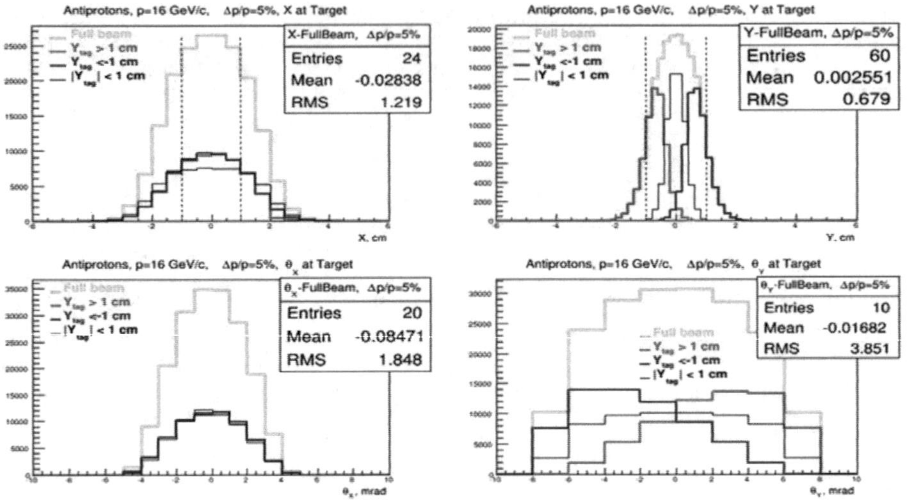

Figure 4. Narrow polarized antiproton Beam at the experiment target with asymmetric optics

Another possibility is to use the whole beam for the asymmetry measurements. The standard method is to select the ensemble of the particle sorted out by either vertical (*Y*) or horizontal (*X*) polarization component and to measure the asymmetry relative to the vertical or horizontal plane. But the SPASCHARM tagging system [7] allows reconstructing of the both components of each Λ- decay and thus assign to each (anti)proton in the beam the both transverse components of polarization. With such a more complete information, the spin asymmetry analysis could be done in respect to the specific plane which rotates from event to event. One can show that such an approach would provide an effective increase of a polarized beam intensity by factor ~1,5.

The very effective way for increasing the beam intensity could be the shortening of the sweeping magnet just after the primary target. However, in order to conserve the bending power 4.7 Tm, the magnetic field must be increased above the achievable level for the conventional warm magnets. Currently, the possibility of using the existing at IHEP 1 m long superconducting magnet with the field of ~5 T is under the study. One more option currently at study is to place the channel 24A primary target inside the conventional sweeping magnet at its middle. The primary target for the channel 24B remains at its current place before the magnet. The expected intensity gain from this option is at about factor 3. However, there is a potential danger of worsening the background situation due to reduced magnet sweeping power downstream the primary target of the channel 24A.

4. Summary

The polarized beam channel for the second stage of the SPASCHARM experiment is under development at IHEP. It is designed so as to create a high-intensity polarized proton beam in a wide momentum range from ~10 to 45 GeV/c.

The polarized antiproton beam would be of an intensity $\sim 4 \cdot 10^5$ antiprotons per each 9 sec long accelerator cycle, and it is adequate for carrying out many single-spin measurements with low background from pions. It will be possible to do the measurements with polarized protons and antiprotons of the same energies in the range of 16-24 GeV with the same detector configurations.

There are under study a few the possibilities for increasing the antiproton beam intensity by factor of 3-5 which would be sufficient for meaningful double-spin asymmetry measurements, including the ones in the production of charmonia.

Acknowledgments

This work has been supported in part by the RFBR Grant No.15-02-1669.

Journal of Physics: Conference Series **678** (2016) 012047 doi:10.1088/1742-6596/678/1/012047

References

[1] Mochalov V V 2013 Phys. Part. Nucl. **44** No 6 930; Mochalov V V et al Systematic Study of Spin Effects at the SPASCHARM Experiment at 70-GeV Accelerator in Protvino 2015 To be published in *Int. J. Mod. Phys.: Conf. Series (Proc. of 21st Int. Symp. on Spin Phys. Spin2014, Beijing, China, 2014)*

[2] Mochalov V V et al 2015 Study of single-spin asymmetries with polarized target at the SPASCHARM experiment To be pub. in J. Phys.: Conf. Series (these proceedings)

[3] O.E. Overseth and J. Sandweiss, A Polarized Proton Beam From Λ^0 Decay; O.E. Overseth, Polarized Protons at the 200-GeV Accelerator, 1969 *Proceedings of the 1969 FNAL Summer Study*, eConf C690609 006.

[4] Grosnick D P et al 1990 *Nucl. Inst. Meth.* A **290** 269

[5] Abramov V V et al 1997 *Nucl. Phys.* B **492** 3

[6] Koop I A et *al Phys.Part.Nucl.* 2014 **45** 279

[7] Nurushev S B et al 2014 *DSPIN-2013 Proceedings* 267

[8] Bogdanov A A et al *DSPIN-2013 Proceedings* 381, Semenov P A et al Polarimetry for the SPACHARM Experiment 2015 To be published in *Int. J. Mod. Phys.: Conf. Series (Proc. of 21st Int. Symp. on Spin Phys. Spin2014, Beijing, China, 2014)*, Bogdanov A A To be pub. in J. Phys.: Conf. Series (these proceedings), Rykov V L et al To be pub. in J. Phys.: Conf. Series (these proceedings)

Journal of Physics: Conference Series **678** (2016) 012048 doi:10.1088/1742-6596/678/1/012048

Study of single-spin asymmetries with polarized target at the SPASCHARM experiment at U70 accelerator

V V Mochalov[1,2†], V V Abramov[1], N A Bazhanov[3], N I Belikov[1], A A Borisov[1], N S Borisov[3], S I Bukreeva[1], Y M Goncharenko[1], A M Davidenko[1], A A Derevschikov[1], R M Fahrutdinov[1], A S Kozhin[1], A B Lazarev[3], Y M Melnik[1], A P Meschanin[1], N G Minaev[1], D A Morozov[1], A B Neganov[3], L V Nogach[1], S B Nurushev[1,2], V S Petrov[1], A F Prudkoglyad[1], A V Ryazantsev[1], V L Rykov[2], S V Ryzhikov[1], P A Semenov[1,2], V A Senko[1], N A Shalanda[1], O N Shchevelev[3], M M Soldatov[1], L F Soloviev[1], M N Strikhanov[2], A N Vasiliev[1,2], Y A Usov[3], V I Yakimchuk[1], A E Yakutin[1]

[1] Institute For High Energy Physics National Research Centre Kurchatov Institute, Protvino, Moscow region, 142280, Russia

[2] National Research Nuclear University (Moscow Engineering Physics Institute), Moscow, 115409, Russia

[3] Joint Institute for Nuclear Research, Dubna, Moscow region, 141980, Russia

[†]Corresponding author: E-mail: mochalov@ihep.ru

Abstract. A new experiment SPASCHARM for systematic study of polarization phenomena in inclusive and exclusive hadronic reactions is currently under commissioning at IHEP. The universal experimental setup will detect dozens of various resonances and stable particles produced in collisions of unpolarized beams with the polarized target, and at the next stage, using polarized beams. At the first stage with polarized target, the final states composed of light quarks (u, d, s) will be reconstructed. Hyperon polarization and spin density matrix elements of the vector mesons will be measured along with the single-spin asymmetries. The 2π-acceptance in azimuth, which is extremely useful for reduction of systematic errors in measurements of spin observables, will be implemented in the experiment. The solid angle acceptance of the setup, $\Delta\theta\approx 250$ mrad vertically and 350 mrad horizontally in the beam fragmentation region, covers a wide range of kinematic variables p_T and x_F. This provides the opportunity for separating dependences on these two variables which is usually not possible in the setups with a small solid angle acceptance. Unlike some previous polarization experiments, the SPASCHARM will be able to simultaneously accumulate and record data on the both, charged and neutral particle production.

1. Introduction

The interest in studying spin dependences in strong interactions arises from the study of hadron's spin structure through interactions of the constituent partons having non-zero spin. New experimental results in this difficult area for theorists are important and relevant to the development of the theory. The new experimental program, the SPASCHARM for 70-GeV proton synchrotron (U-70), is currently under development at the Institute for High Energy Physics (IHEP), Protvino, Russia. The main physics motivation of the SPASCHARM experiment is the systematic study of spin physics for a wide range of inclusive and exclusive reactions in collisions of high-energy polarized hadrons in the QCD non-perturbative region.

Content from this work may be used under the terms of the Creative Commons Attribution 3.0 licence. Any further distribution of this work must maintain attribution to the author(s) and the title of the work, journal citation and DOI.
Published under licence by IOP Publishing Ltd

Journal of Physics: Conference Series **678** (2016) 012048 doi:10.1088/1742-6596/678/1/012048

2. Physics Motivation

In perturbation theory of QCD at leading twist, the transverse single-spin asymmetries are predicted close to zero [1]. Large spin effects found in a number of experiments in the confinement region stimulated the development of theoretical models. The theoretical models by Sivers [2], Collins [3], and the quark-gluon correlators [4] are well known today. These models predict decreasing single-spin asymmetries in the formation of hadrons as the transverse momentum p_T increases. However, these predictions still are not confirmed in the experiments carried out over a wide energy range, $2 < \sqrt{s} < 200$ GeV, at a moderate $p_T < 3$ GeV/c. Few phenomenological models qualitatively explain the polarization of hyperons [5-7]. The coupling of a quark chromomagnetic moment to chromomagnetic field of a QCD string leads to an appearance of additional quark transverse momentum, the direction of which depends on the direction of the quark spin in polarized proton [8,9]. The global analysis of all available data has revealed interesting peculiarities associated with the dependences of polarization effects on kinematic variables, the quark flavor and atomic weights of particles participating in reactions [10]

In order to discriminate the various theoretical approaches, the systematic studies of single-spin transverse asymmetries in a large number of exclusive and inclusive reactions are required. Such a program can be carried out at the first stage of polarization program of the proposed SPASCHARM experiment.

At the first stage of SPASCHARM experiment, the main task is a detailed study of various polarization effects in the production of particles and resonances, consisting mainly of light u, d and s-quarks. Most of the measurements will be performed at the negatively charged unpolarized beam colliding with the polarized target. Composition of the beam $(\pi^-/K^-/\bar{p}) = (97.9/1.8/0.3\%)$ makes it possible to study the processes initiated by antiquarks and strange quarks. Surprisingly, in collisions of mesons with polarized targets, the non-zero single-spin asymmetries have been observed as in central [11, 12] as well as in unpolarized beam fragmentation [13-15] regions.

Unlike in the most experiments, at SPASCHARM it will be implemented 2π-acceptance in azimuth for both charged and neutral secondary particles, which is very useful for reduction of the systematic errors in measurements of spin physics observables. The large acceptance setup allows us to carry out the spin measurements in a wide range of kinematic variables (p_T, x_F) and to separate the asymmetry dependences on these two variables.

The exclusive reactions, such as $\pi^- p\uparrow \to \omega(782)n$, $\pi^- p\uparrow \to \eta'(958)n$, $\pi^- p\uparrow \to f_2(1270)n$, $\pi^- p\uparrow \to a_2(1320)n$ and others, will be investigated in various decay modes into neutral and charged particles. Previously, these reactions have been studied at PROZA experiment only in the mode of their decays into π^0-mesons and γ-quanta. In SPASCHARM, because of its capabilities for detecting the charged decay channels, it is expected to increase the statistics compared to PROZA experiment by about an order of magnitude for reactions $\pi^- p\uparrow \to \omega(782)n$ and $\pi^- p\uparrow \to \eta'(958)n$, as well as, by a factor of 3-4, for the reactions $\pi^- p\uparrow \to f_2(1270)n$ and $\pi^- p\uparrow \to a_2(1320)n$. A significant increase in statistics in SPASCHARM will allow the precise measurement of t-dependence of single-spin asymmetries in these reactions as well as making discrimination against some theoretical approaches for the description of spin effects in these reactions. It worth noting that, for the first time, the asymmetry in reaction $\pi^- p\uparrow \to a_0(980)n$ will be measured in the mode of $a_0(980)$ decaying into $\eta(550)$ and π^0.

Significant effects are expected near the edges of kinematics allowed phase space, as the large spin effects have been seen in the respective exclusive charge-exchange reactions. Thus, it is believed that, for a wide range of reactions available for study at SPASCHARM setup with meson beams, a number of nonzero single-spin transverse asymmetries A_N will be uncovered. The asymmetry A_N will be measured in a variety of inclusive reactions with high precision due to full azimuth coverage within the wide aperture. The expected statistics with pion beam presented earlier [16-17]. Expected statistics in the interaction of antiprotons on polarized target is presented in table 1. The kinematic region to be covered is $0.2 < x_F < 1$ and $0.5 < p_T < 3$ GeV/c.

Journal of Physics: Conference Series **678** (2016) 012048 doi:10.1088/1742-6596/678/1/012048

Table 1. The anticipated numbers of events collected for the inclusive reactions for three months of detector exposition at $\tilde{p}$ beam with the energy of 34 GeV at the first stage of SPASCHARM experiment.

№	частица	N_{EV}	№	частица	N_{EV}
1	π^+	$2.1 \cdot 10^8$	*7*	n	$1.6 \cdot 10^7$
2	π^-	$2.6 \cdot 10^8$	8	$\tilde{n}$	$1.4 \cdot 10^8$
3	K^+	$1.7 \cdot 10^7$	9	$\tilde{\Lambda} \to \tilde{p}\ \pi^+$	$2.1 \cdot 10^6$
4	K^-	$2.2 \cdot 10^7$	10	$\tilde{\Lambda} \to \tilde{n}\ \pi^0$	$1.1 \cdot 10^6$
5	p	$1.6 \cdot 10^7$	11	$\tilde{\Lambda}^{--} \to \tilde{p}\ \pi^-$	$4.2 \cdot 10^7$
6	$\tilde{p}$	$1.8 \cdot 10^8$	12	$\Xi^- \to \Lambda\ \pi^-$	$1.0 \cdot 10^5$

3. Experimental Setup

The experimental setup of the first phase of SPASCHARM experiment is a spectrometer for detection of charged particles, neutrons, K_L^0-mesons and photons in the forward region. The setup has the 2π coverage in azimuth. This is the critical advantage for minimizing the systematic errors in the measurement of spin asymmetries. The layout of the setup is presented in figure 1. It consists of the polarized proton target, tracking system with the excellent spatial resolution, spectrometric large aperture magnet, identification of the secondary particles, electromagnetic and hadron calorimeters, and muon detector. There are also charged particle multiplicity detectors which serve also as a time-of-flight system (three hodoscopes in front of and behind the spectrometer magnet as well as in front of the electromagnetic calorimeter), the guard system for the target, and trigger counters.

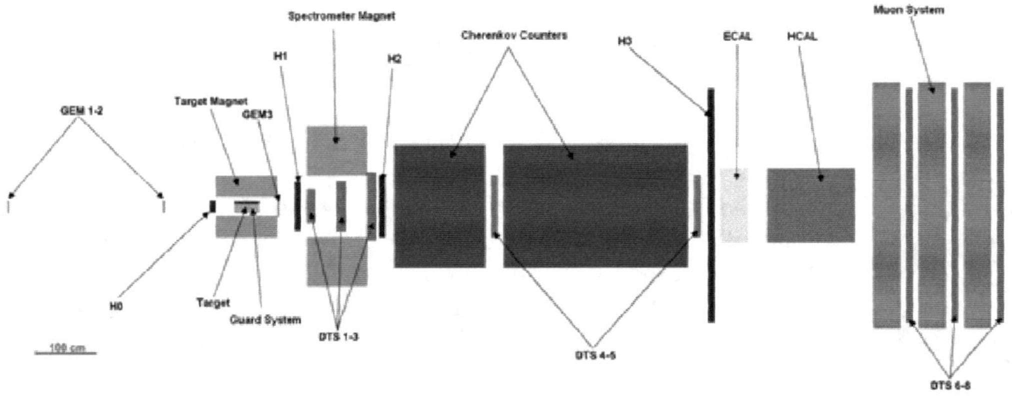

Figure 1. The layout of SPASCHARM experiment.

Special spectrometer magnet of wide aperture ($X \times Y$=200×100 cm^2) has been specially designed and build (see figure 2). The newly developed tracking system is based on 3,000 channels of drift tubes with 15 and 30 mm in diameter and few GEM-detectors. Drift tube system consists of five thin-wall drift tube chamber stations (16 planes, 48 sub-planes). The tracking system has been designed with goal of achievement of the momentum resolution $\Delta P/P$=0.4% at 10 GeV/c with the maximum field.

Three of such stations were commissioned (two of them are shown on figure 3). The on-line profile of the events in X-plane of station DC3 is shown on figure 4 (left). The resolution of the drift chambers has been tested. The result of the test (residual distribution) is presented in the same figure 4 (right). It was found in accord with the specification requirements (<160 μm).

Figure 2. Spectrometric magnet of the SPASCHARM experiment.

Currently, we have the operational lead-glass electromagnetic calorimeter. Later on, we plan to replace it with the thin-segmented shashlyk-type calorimeter with the significantly better energy resolution ($\sigma(E)/E=1.3\oplus2.8/\sqrt{(E)}$ [18-19]) with the cell transverse dimensions of 55×55 mm^2 and the total area of 2×3 m^2.

The compensated lead-scintillator hadron calorimeter has a cell size 10×10 cm^2, thickness 8 nuclear interaction lengths and sampling of 16 mm lead, 4 mm scintillator. The hadronic energy resolution of the calorimeter is $57\%/\sqrt{E}$, the nonuniformity is better than 10% and the measured ratio e/h=1.01 [20].

Two multichannel Cherenkov counters will be used for particle identification. The first one registers π-mesons above 3 GeV/c and K-mesons above 11 GeV/c, while the second one registers pions above 6 GeV/c and kaons above 23 GeV/c. We plan in future to produce also the TOF wall (marked as H3 in figure 1) for p/K-separation up to 2.5 GeV, and K/π separation up to 1.5 GeV. Muon detector will consist of 3 layers of iron and drift tube chambers.

Figure 3. Picture of DC4 and DC3 (inclined plane) stations

New DAQ with parallel reading of electronics has been designed and successfully tested for the SPASCHARM experiment. The information can be read at the speed of about 20 μsec/event from an ADC station and 30-50 μsec/event from a TDC crate. In the test run, the achieved rate for accepted events was 10000 events/0.8 sec. We expect to store about $5\cdot10^4$ events/cycle, while the total event rate expected to be at about 10^5 per cycle.

Journal of Physics: Conference Series **678** (2016) 012048 doi:10.1088/1742-6596/678/1/012048

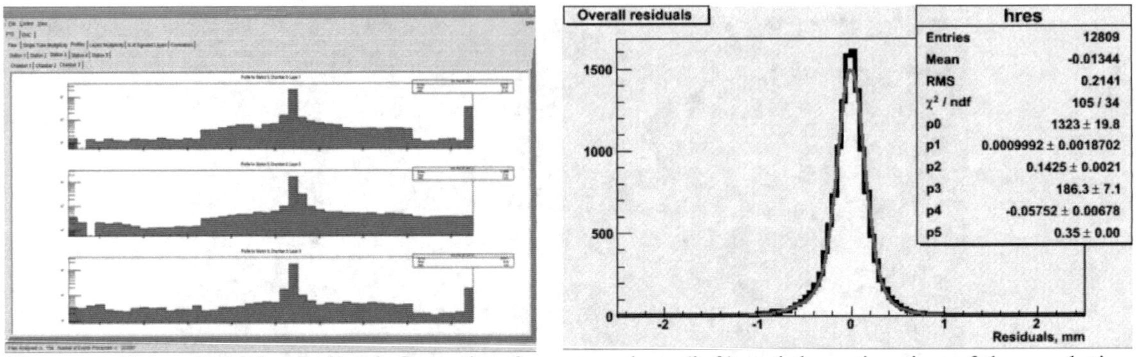

Figure 4. Example of the on-line information from one plane (left) and the estimation of the resolution – parameter *p2* at test measurements (right).

The first experience with the system confirmed its capabilities for measuring the spin asymmetry for charged particles. We successfully reconstructed about the expected number of Δ^{--} and of Δ^{++} in $h^{+(-)}h^{+(-)}$- spectra even without particle identification.

4. The experiment schedule

The current schedule of the experiments looks as follows:

- Autumn 2015 – data taking run about one week long to test the new electronics and DAQ system and tune up their characteristics, commissioning of inclined planes of TDC station, calibration of the electromagnetic calorimeter with the new ADCs.
- 2016 – commissioning of the 4th TDC station, testing and preparation of the new “narrow” TDC station, data taking with “pions” (without Cherenkov counters) to measure a single-spin asymmetry in production of h^+, h^-, $\rho(770)$, $\omega(782)$, $\eta'(958)$, $f_0(980)$, $a_0(980)$, $f_2(1270)$.
- 2017 – commissioning of the full tracking system, consisting of five stations of Drift Tube chambers, installation of the Cherenkov counter with the goal to measure spin effects in production of *K*-mesons, hyperons and especially of ϕ-meson.

In the Spring of 2019, we plan to move to the polarized beam channel 24A [16, 21] which is expected to start operating by the end of the same year. The measurements with the polarized proton and antiproton beams would bring the new quality into the spin physics studies at SPASCHARM. However, the 2nd stage of the SPASCHARM program [16] is beyond the scope of this paper.

5. Summary

The new experiment, SPASCHARM, designed for systematic studies of polarization phenomena in hadron-hadron interactions, is under commissioning now at the 70-GeV accelerator of IHEP, Protvino, Russia. The detection capabilities for the charged and neutral final state particles in a wide acceptance of SPASCHARM spectrometer provide an opportunity for exploring dozens of inclusive and exclusive reactions with a number of beam particles.

It worth mentioning the possibility of simultaneous measurements of different spin-physics observables: SSA, polarization of hyperons, spin density matrix elements for vector mesons, spin-transfer parameters. In the year 2016, we expect to begin taking data at the existing channel with unpolarized beam. The first physics results are expected in about a year later.

An availability of the polarized proton and antiproton beams at the second stage of the experiment will provide an unique opportunity for studying the single-spin effects in the polarized beam fragmentation region for various nuclear targets as well as studying the double-spin asymmetries in charmonium production with the goal to evaluate the gluon contribution into the spin of proton.

The SPASCHARM setup has all the tools for high precision determination of all particle kinematic parameters which is crucial for the charmonium studies as well as for the extraction of resonance signals from combinatorial backgrounds. We plan to start working at the new channel 24 with polarized proton and antiproton beams in the year 2019.

Acknowledgments
We thank the IHEP management and the IHEP Accelerator, Beam and Electronics departments for their invaluable contribution in construction and commissioning of the setup at the beam, and for their support of the SPASCHARM program. This work has been supported in part by the RFBR Grant No.15-02-1669.

References

[1] Kane G L, Pumpkin J and Repko W 1978 *Phys. Rev. Lett.* **42** 1689
[2] Sivers D 1990 *Phys. Rev. D* **41** 83
[3] Collins J 1993 *Nucl. Phys. B* **396** 161
[4] Efremov A V, Korothiyan V and Teryaev O 1995 *Phys. Lett. B* **348** 577
[5] Anderson B, Gustafson G and Ingelman G 1979 *Phys. Lett. B* **85** 417
[6] Thomas L T 1927 *Philos. Mag.* **3** 1
[7] De Grand T and Miettinen H I 1981 *Phys. Rev. D* **23** 1227
[8] Ryskin M G 1988 *Soviet Journal of Nuclear Physics (SJNP)* **48** 708
[9] Nurushev S B and Ryskin M G 2006 *Phys. Atom. Nucl.* **69**, 133
[10] Abramov V V 2009 *Phys. Atom. Nucl.* **72** 1933
[11] Amaglobeli N S *et al* 1989 *Soviet Journal of Nuclear Physics (SJNP)* **50** 432
[12] Apokin V D *et al* 1990 *Phys.Lett. B* **243** 461
[13] Apokin V D *et al* 1989 *Soviet Journal of Nuclear Physics (SJNP)* **49** 97
[14] Mochalov V V *et al* 2010 *Phys.Atom.Nucl.* **73** 2017
[15] Apokin V D *et al* 1989 *Soviet Journal of Nuclear Physics (SJNP) SJNP* **49** 103
[16] Mochalov V V 2013 *Phys. Part. Nucl.* **44** No 6 930
[17] Mochalov V V et al Systematic Study of Spin Effects at the SPASCHARM Experiment at 70-GeV Accelerator in Protvino 2015 To be published in *Int. J. Mod. Phys.: Conf. Series (Proc. of 21st Int. Symp. on Spin Phys. Spin2014, Beijing, China, 2014)*
[18] Morozov D A *et al* 2009 *J.Phys.Conf.Ser.* **160** 012021
[19] Kharlov Yu V et al 2009 *Nucl.Instrum.Meth. A* **606** 432
[20] Alekseev G et al 2001 *Nucl.Instrum.Meth. A* **461** 381
[21] Nurushev S B et al 2015 Polarized antiproton beam at U-70 accelerator of IHEP *To be published in J. Phys.: Conf. Series (these proceedings)*

Journal of Physics: Conference Series **678** (2016) 012049 doi:10.1088/1742-6596/678/1/012049

Fully gauge-invariant maximally path-dependent gluon TMD: Coordinate representation

Igor O. Cherednikov

EDF - Departement Fysica, Universiteit Antwerpen, Belgium

E-mail: igor.cherednikov@uantwerpen.be

Abstract. We propose an approach to obtain gauge-invariant maximally path-dependent operator definition of the gluon transverse-momentum dependent distribution function (gTMD). We demonstrate that the evolution equations for the gTMD can be derived from the shape-variation integral-differential equations formulated in the coordinate space.

1. Introduction

Transverse-momentum dependent (unintegrated) parton density functions (TMD pdfs in what follows) extent the idea of collinear (integrated) pdfs, which absorb mass singularities in any order of perturbative expansion, making it possible to apply the QCD factorisation approach to the inclusive hadron processes (e^+e^- to hadrons, DIS) [1]. In more exclusive cross-sections, which one deals with in the semi-inclusive reactions (semi-inclusive DIS, Drell-Yan, Higgs, vector boson, heavy-flavour production) factorisation often implies *intrinsic* transverse-momentum dependence of non-perturbative pdfs, which must be constructed in such a way to absorb not only mass, but also rapidity singularities [2, 3, 4, 5, 6]. With TMD pdfs one gets access to the three-dimensional structure of the nucleon in the momentum space, which is actively investigated in ongoing and planned experiments at the LHC, JLab + JLab 12 GeV upgrade, RHIC, EIC etc., providing an incentive to the rapid theoretical progress in the field (see, e.g., [7, 8] and Refs. therein).

As compared to the quark TMD case, the gauge-invariant gluon TMD exhibits much more involved structure of the gauge links (Wilson lines), so that the path-dependence brings about extra complications to the understanding of *(non-)universality* of the TMD pdfs. On the other hand, this path-dependence allows us to manipulate freely with the gauge links entering the operator definition of the gTMD. In the present paper we develop a quantum-field theoretic approach to the definition of the gluon TMD (gTMD) making use of the arbitrariness of the trajectories of the Wilson lines to formulate evolution of the gTMD in terms of the equations of motion in the so-called generalized loop space, which elements are the hadronic matrix elements of the arbitrary Wilson loops.

2. Operator structure of TMD pdfs

Following the standard approach one derives the operator definitions of the various TMD pdfs starting with the *factorisation* of a given process in a convenient gauge, which yields an obviously *gauge-dependent* pdf, such as

Content from this work may be used under the terms of the Creative Commons Attribution 3.0 licence. Any further distribution of this work must maintain attribution to the author(s) and the title of the work, journal citation and DOI.
Published under licence by IOP Publishing Ltd

$$\mathbf{G}^{\mu\nu}_{\text{g-d}}(k;P,S) = \int d^4z \ \mathrm{e}^{-ikz} \ \langle h| \ \mathcal{A}^{\mu}(z)\mathcal{A}^{\nu}(0) \ |h\rangle, \tag{1}$$

that is a hadronic matrix element, which corresponds to the distribution of the gluons with momentum k inside the hadron h with momentum P and spin S [2, 9]. Summing up collinear gluon contributions yields a *gauge-invariant* pdf supplied with the Wilson lines, whose path γ is determined by the factorisation scheme:

$$\mathbf{G}^{\mu\nu|\rho\sigma}_{\text{g-i}}(k;P,S) = \int d^4z \ \mathrm{e}^{-ikz} \ \langle h| \ \mathcal{F}^{\mu\nu}(z) \ \mathcal{W}_{\gamma} \ \mathcal{F}^{\rho\sigma}(0) \ |h\rangle, \tag{2}$$

where the field strength is defined in the adjoint representation

$$\mathcal{F}^{\mu\nu} = \partial^{\mu}\mathcal{A}^{\nu} - \partial^{\nu}\mathcal{A}^{\mu} - ig[\mathcal{A}^{\mu},\mathcal{A}^{\nu}], \ \mathcal{F}_{\mu\nu} = F^{a}_{\mu\nu}T^{a}. \tag{3}$$

Generic *gauge-invariant* gTMD c bean obtained by integrating over the minus-light-cone component of the partonic momentum and projecting onto the transverse Lorentz indices:

$$\mathbf{G}^{ij}(x,\boldsymbol{k}_{\perp};P,S) \sim \int dk^{-} \ \mathbf{G}^{+i|+j}(k;P,S) = \tag{4}$$

$$\int dz^{-}d^2z_{\perp} \ \mathrm{e}^{-ikz} \ \langle h| \ \mathcal{F}^{+i}(z) \ \mathcal{W}_{\gamma} \ \mathcal{F}^{+j}(0) \ |h\rangle \tag{5}$$

Let us explore a possibility of an alternative approach to the definition of the TMD pdfs. We construct first a generic *fully gauge-invariant* (by constuction) and *maximally path-dependent* object consisting of the Wilson lines/loops defined on the set of absolutely arbitrary trajectories γ:

$$\mathcal{W}_{\gamma} = \mathcal{P}_{\gamma}\mathrm{exp} \oint_{\gamma} \ d\zeta_{\mu}\mathcal{A}^{\mu}(\zeta). \tag{6}$$

So far we do not assume any factorisation scheme and even do not identify the processes to which those objects are supposed to be applicable. Given the arbitrariness of the paths, one is able to adjust them to a specific factorisation scheme by performing the *evolution in the coordinate space*. Hence we end up with an object which can be further associated with a realistic *gauge-invariant* TMD pdf.

3. Equations of motion in the loop space

Shape variations (or evolution in the coordinate representation) of arbitrary Wilson loops can be consistently formulated in terms of the equations of motion in the loop space, that is the integral-differential equations which the elements of this space (Wilson loops) obey, if the underlying contours γ_i on which the path-ordered exponentials of the gauge fields are defined experience certain infinitesimal local variations. The variations of the contours give rise to the variations of the exponentials themselves, the latter being described by the infinite set of the Makeenko-Migdal loop equations [10, 11, 12, 13, 14, 15]. More specifically, the elements of the loop space are the vacuum matrix elements of products of the Wilson loops, that is

$$\langle 0|\mathcal{W}^{n}_{\gamma_1,\dots\gamma_n}|0\rangle = \langle 0|\mathcal{T}\mathcal{W}_{\gamma_1}\cdots\mathcal{W}_{\gamma_n}|0\rangle \tag{7}$$

These fundamental gauge-invariant (by construction) degrees of freedom obey the Makeenko-Migdal loop equations (in the large-N_c limit)

$$\partial^{\nu} \ \frac{\delta}{\delta\sigma_{\mu\nu}(x)} \ \langle 0|\mathcal{W}^{1}_{\gamma}|0\rangle = N_c g^2 \ \oint_{\gamma} \ dz^{\mu} \ \delta^{(4)}(x-z)\langle 0|\mathcal{W}^{2}_{\gamma_{xz}\gamma_{zx}}|0\rangle, \tag{8}$$

where the area and path differential operators are defined as follows [10, 11, 12]:

Journal of Physics: Conference Series **678** (2016) 012049 doi:10.1088/1742-6596/678/1/012049

- Area derivative describes the behaviour of the Wilson loop under the infinitesimal variations of the area $\delta\sigma_{\mu\nu}$ at a given point

$$\frac{\delta}{\delta\sigma_{\mu\nu}(z)}\langle 0|\mathcal{W}_\gamma|0\rangle = \lim_{|\delta\sigma_{\mu\nu}(z)|\to 0} \frac{\langle 0|\mathcal{W}_{\gamma\delta\gamma_x}|0\rangle - \langle 0|\mathcal{W}_\gamma|0\rangle}{|\delta\sigma_{\mu\nu}(z)|} \tag{9}$$

- Path derivative deals with the infinitesimal path extensions and contractions $\delta z_\mu^{-1}\gamma\delta z_\mu$ at a given point

$$\partial_\mu\langle 0|\mathcal{W}_\gamma|0\rangle = \lim_{|\delta z_\mu|\to 0} \frac{\langle 0|\mathcal{W}_{\delta z_\mu^{-1}\gamma\delta z_\mu}|0\rangle - \langle 0|\mathcal{W}_\gamma|0\rangle}{|\delta z_\mu|} \tag{10}$$

These differential operators in the loop space determine the *evolution* of the Wilson loops in the coordinate representation. In other words, starting from a loop having a given shape, one can come to a loop with another shape by solving the above evolution equations. This is exactly what is needed to adjust a generic Wilson loop to some specific geometrical layout prescribed by a factorisation framework.

4. Stokes-Mandelstam gluon TMD

Let us show how this strategy can be practically implemented. Consider first the Wilson loops themselves, not their matrix elements. This is needed since we want to work with the hadronic matrix elements, not with the vacuum ones. Making use of non-Abelian Stokes' theorem

$$\mathcal{P}_\gamma \exp \oint_\gamma d\zeta_\mu \mathcal{A}^\mu(\zeta) = \mathcal{P}_\gamma \mathcal{P}_\sigma \exp \int_\sigma d\sigma_{\mu\nu}(\zeta)\mathcal{F}^{\mu\nu}(\zeta) \tag{11}$$

and the Mandelstam formula

$$\frac{\delta}{\delta\sigma_{\mu\nu}(x)}\mathcal{P}_\gamma \exp \oint_\gamma d\zeta_\mu \mathcal{A}^\mu(\zeta) = \mathcal{P}_\gamma \mathcal{F}^{\mu\nu}(x) \exp \oint_\gamma d\zeta_\mu \mathcal{A}^\mu(\zeta), \tag{12}$$

we see that the generic correlation function (2) can be represented in the following form

$$\tilde{\mathbf{G}}^{\mu\nu|\rho\sigma}(z; P, S) = \frac{\delta}{\delta\sigma_{\mu\nu}(z)}\frac{\delta}{\delta\sigma_{\mu\nu}(0)}\langle h|\mathcal{W}_{\gamma[z,0]}\;|h\rangle \tag{13}$$

$$= \frac{\delta}{\delta\sigma_{\mu\nu}(z)}\frac{\delta}{\delta\sigma_{\mu\nu}(0)}\sum_X \langle h|\mathcal{W}'_{\gamma[z]}|X\rangle\langle X|\mathcal{W}'_{\gamma[0]}\;|h\rangle. \tag{14}$$

We coin this representation the *Stokes-Mandelstam gTMD* definition. The key feature of this definition (and of the entire approach) is that one first calculates the hadronic matrix element of an arbitrary Wilson loop

$$\langle h|\mathcal{W}_{\gamma[z,0]}\;|h\rangle,$$

choosing its shape the most convenient for the practical purposed. In particular, in the situations where the non-Abelian exponentiation is applicable

$$\langle h|\mathcal{W}_{\gamma[z,0]}|h\rangle = \exp\left[\sum a_n W^{(n)}\right],\; W^{(n)} = \text{hadronic correlators}, \tag{15}$$

one can even obtain an explicit expression for this matrix element and thereafter evaluate the area derivatives in terms of the fundamental hadronic correlation functions. The latter can be taken, for instance, from lattice simulations. Let us illustrate the use of this framework by a simple Abelian example.

4.1. Example: Abelian exponentiation
Wilson loops in the Abelian gauge theory are known to exponentiate

$$\langle h|\mathcal{W}_\gamma|h\rangle = \langle h|\mathcal{P}_\gamma \exp \oint_\gamma d\zeta_\mu \mathcal{A}^\mu(\zeta)|h\rangle \quad (16)$$

$$= \exp\left[-\frac{g^2}{2}\oint_\gamma d\zeta_\mu \oint_\gamma d\zeta'_\nu \, D_{\mu\nu}(\zeta-\zeta')\right], \quad (17)$$

where the basic hadronic correlator reads

$$D_{\mu\nu}(\zeta-\zeta') = \langle h|\mathcal{P}_\gamma A_\mu(\zeta)A_\nu(\zeta')|h\rangle. \quad (18)$$

For the sake of simplicity, let us consider a spinless hadron, which entails the following parameterisation of the basic correlator (18)

$$D_{\mu\nu}(z) = g_{\mu\nu}\, D_1(z,P) + \partial_\mu\partial_\nu\, D_2(z,P) + \{P_\mu\partial_\nu\}\, D_3(z,P) + P_\mu P_\nu\, D_4(z,P). \quad (19)$$

Given this representation and the exponentiation, the area derivative can be evaluated straightforwardly

$$\frac{\delta}{\delta\sigma_{\mu\nu}(z)}\langle h|\mathcal{W}_\gamma|h\rangle = -\frac{g^2}{2}\left[\frac{\delta}{\delta\sigma_{\mu\nu}(z)}\oint_\gamma d\zeta_\mu \oint_\gamma d\zeta'_\nu\, D_{\mu\nu}(\zeta-\zeta')\right]\langle h|\mathcal{W}_\gamma|h\rangle \quad (20)$$

After taking the path derivative ∂_ν, the non-vanishing terms are

- the standard 'Makeenko-Migdal' term:

$$\oint_\gamma d\zeta_\nu\, \partial^2\, D_1(z,P) \quad (21)$$

- the hadron momentum-dependent term, obviously absent in the loop space with vacuum matrix elements

$$\oint_\gamma d\zeta_\nu\, (P\partial)^2\, D_4(z,P) \quad (22)$$

Therefore, the *shape evolution* equation for the hadronic Wilson loops in the Abelian gauge theory is given by

$$\partial^z_\mu \frac{\delta}{\delta\sigma_{\mu\nu}(z)}\langle h|\mathcal{W}_\gamma|h\rangle = \quad (23)$$

$$-\frac{g^2}{2}\left[\oint_\gamma d\zeta^\nu \left(\partial^2\, D_1(z,P) + (P\partial)^2\, D_4(z,P)\right)\right]\langle h|\mathcal{W}_\gamma|h\rangle. \quad (24)$$

Taking into account that for the vacuum matrix elements

$$\partial^2 D_1(z) = -\delta^{(4)}(z),$$

and that $D_4 = 0$ in vacuum, one easily re-obtains the Makeenko-Migdal equation in the leading order.

To conclude, we proposed a new definition of the generic gluon TMD, which allows us to work directly with the entirely gauge-invariant and maximally path-dependent hadronic matrix elements *before* introducing relevant gluon TMD. All necessary information is absorbed in the basic hadronic correlator (18) and the evolution equation (24) generalises the Makeenko-Migdal equation for the Abelian gauge theory to the case of the 'hadronic' loop space.

Further progress will be presented in a separate work [16].

Acknowledgments
The work is supported by the Belgian Federal Science Policy Office. Fruitful discussions with T. Mertens, P. Taels and F. Van der Veken are gratefully appreciated.

References
[1] Dokshitzer Y L, Diakonov D and Troian S I 1980 *Phys. Rept.* **58** 269
[2] Collins J C and Soper D E 1981 *Nucl. Phys.* B **193** 381; *Nucl. Phys.* B **194** (1982) 445
[3] Collins J 2011 *Foundations of Perturbative QCD* (Cambridge University Press)
[4] Collins J C 2003 *Acta Phys. Polon.* B **34** 3103
[5] Cherednikov I O and Stefanis N G (2008) *Phys. Rev. D* **77** 094001; Cherednikov I O and Stefanis N G (2008) *Nucl. Phys. B* **802** 146; Cherednikov I O and Stefanis N G (2009) *Phys. Rev. D* **80** 054008
[6] Cherednikov I O, Karanikas A I and Stefanis N G (2010) *Nucl. Phys.* B **840** 379
[7] Boer D *et al.* 2011 *Preprint* arXiv:1108.1713 [nucl-th]
[8] Angeles-Martinez R *et al.* (2015) *Preprint* arXiv:1507.05267 [hep-ph]
[9] Mulders P J and Rodrigues J (2001) *Phys. Rev.* D **63** 094021
[10] Makeenko Y M and Migdal A A 1980 *Phys. Lett.* B **88** 135, [Erratum-ibid. B **89** 437]; Makeenko Y M and Migdal A A 1981 *Nucl. Phys.* B **188** 269; Makeenko Y 2002 *Methods of Contemporary Gauge Theory* (Cambridge University Press)
[11] Brandt R A, Neri F and Sato M A 1981 *Phys. Rev.* D **24** 879
[12] Brandt R A, Gocksch A, Neri F and Sato M A 1982 *Phys. Rev.* D **26** 3611
[13] Tavares J N (1994) *Int. J. Mod. Phys.* A **9** 4511
[14] Cherednikov I O, Mertens T and Van der Veken F F 2014 *Wilson Lines in Quantum Field Theory* (Berlin: De Gruyter)
[15] Cherednikov I O, Mertens T and Van der Veken F F 2012 *Phys. Rev.* D **86** 085035; Cherednikov I O, Mertens T and Van der Veken F F 2013 *Phys. Part. Nucl.* **44** 250
[16] Cherednikov I O, Mertens T and Taels P 2016 *[in preparation]*

Journal of Physics: Conference Series **678** (2016) 012050 doi:10.1088/1742-6596/678/1/012050

Sivers Asymmetry in J/ψ Production in COMPASS 2010 Proton Data

Jan Matoušek, on Behalf of the COMPASS Collaboration
Charles University in Prague, University of Trieste.
E-mail: jan.matousek@cern.ch

Abstract. The Sivers asymmetry in J/ψ production in scattering of muons off transversely polarized protons $\mu^+ + p^\uparrow \to \mu^+ + J/\psi + X$ is measured in two z-bins (inclusive and exclusive production) in the COMPASS 2010 data. Only J/ψ's decaying into muons are detected. About 2 230 and 4 450 J/ψ events in the lower and the higher z-bins are found. The Sivers asymmetry is extracted using a simple double-ratio method and is found to be compatible with zero in the lower z bin and to have a slight preference for negative values in the higher z bin.

1. Introduction

The gluon Sivers function is related to the orbital motion of gluons in a transversely polarized nucleon. One of the cleanest ways of accessing it is the measurement of the Sivers asymmetry (see Sec. 4) of hadrons, produced in the scattering of leptons off transversely polarized nucleons $l + p^\uparrow \to l + h + X$ via the Photon-Gluon Fusion (PGF) process $\gamma g \to q\bar{q}$. When the $q\bar{q}$ is $c\bar{c}$ (or $b\bar{b}$), the leading process $\gamma q \to q$ as well as the QCD Compton process $\gamma q \to qg$ are suppressed and the PGF dominates. This is best ensured in the open-charm (beauty) production [1, 2]. Another way of enhancing the PGF contribution is to select high-p_T hadron pairs [3]. This method was applied recently on COMPASS to measure the gluon Sivers asymmetry [4, 5]. Finally, the third way to the PGF is charmonium production [6, 7].

In this article results of the third method, applied to the COMPASS data collected in 2010, are discussed. In 2010 COMPASS used 160 GeV/c muon beam and transversely polarized proton target. For that case the process of interest is

$$\mu^+ + p^\uparrow \to \mu^+ + J/\psi + X. \tag{1}$$

The J/ψ is identified when it decays into muons (using a peak in $\mu^+\mu^-$ invariant mass distribution), so $2\mu^+ + \mu^- + X$ is looked for in the final state. Since the dominant mechanism of the J/ψ production can be different for the inclusive and exclusive cases, we study the reaction in two z-intervals[1] [0.3, 0.95] and [0.95, 1.05].

2. Event selection

Good quality of the selected events was ensured using standard COMPASS criteria on tracks and primary vertex reconstruction confidence levels, primary vertex position and beam muon

[1] Fraction of the available energy carried out by the J/ψ: $z \overset{\text{lab}}{=} \frac{E_{J/\psi}}{E-E'}$. Practically no events are found for $z < 0.3$.

Content from this work may be used under the terms of the Creative Commons Attribution 3.0 licence. Any further distribution of this work must maintain attribution to the author(s) and the title of the work, journal citation and DOI.
Published under licence by IOP Publishing Ltd

momentum. Outgoing muon tracks are identified using their ability to penetrate through a thick material. They are required to cross at least 30 radiation lengths in the spectrometer and to have associated hits in a detector behind the first muon filter.

Events with extra outgoing muons apart from $2\mu^+1\mu^-$ are rejected After the quality cuts and this cut there are 480 726 events left. There is no way to efficiently distinguish the μ^+ coming from the J/ψ decay and the scattered beam μ^+, so both possible combinations are examined. For each combination the invariant mass $M_{\mu\mu}$ is checked to be in [2.999, 3.239] GeV/c^2, which we call the signal band.

In the signal band there are 8 026 events (2 744 and 5 282 for the inclusive and exclusive bins respectively). Background level in the signal band was calculated (separately for the two z-bins) fitting an invariant mass histogram with function

$$\frac{A}{w}N(M_{\mu\mu},\mu,\sigma)+B(M_{\mu\mu})^C, \tag{2}$$

where $N(x,\mu,\sigma)$ is the normal distribution, w is the bin width and A, μ, σ, B and C are the fit parameters (the signal band approximately corresponds to $\mu\pm2\sigma$). The signal to background ratios are 4.31 and 5.25, resulting in about 2 230 and 4 450 signal events in the lower (inclusive) and the higher (exclusive) z-bin respectively.

For purposes of background estimation two side-bands were defined as $M_{\mu\mu}\in[2.579,\ 2.939]\cup[3.299,\ 3.659]$ GeV/c^2. There are 1 865 and 2 919 events in the side-bands for the two z-bins. The invariant mass distributions together with the fits and the boundaries of the signal band and of the side-bands are shown on Fig. 1 and 2.

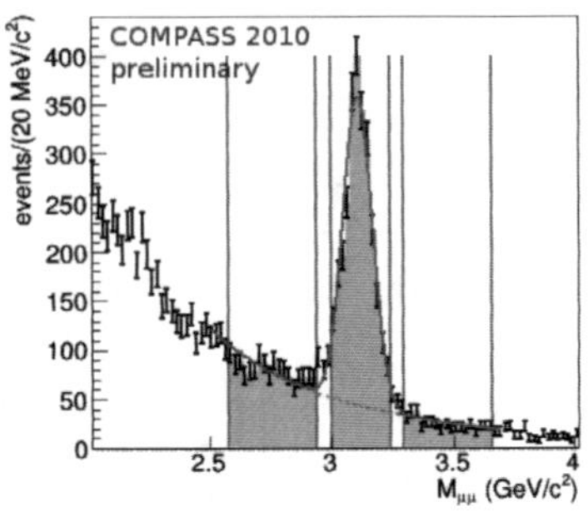

Figure 1. Dimuon invariant mass for $z\in[0.3,\ 0.95]$ (inclusive bin).

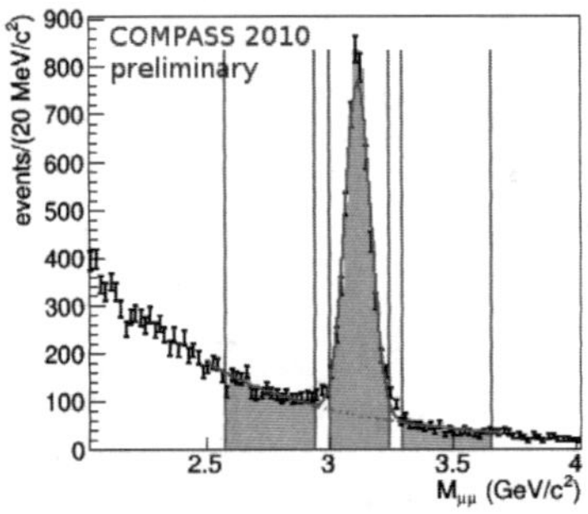

Figure 2. Dimuon invariant mass for $z\in[0.95,\ 1.05]$ (exclusive bin).

3. Kinematic Distributions

In this section we show selected kinematic distributions with all cuts applied (except cuts on the depicted variable, if there are any). The plots contain data from all data-taking periods of 2010 combined. Backgrounds were subtracted using distribution shapes in the side-bands.

The majority of the events lie in the area of low Bjorken x and low momentum transfer Q^2 (Fig. 3). The exclusivity of the produced J/ψ can be seen on Fig. 4, showing the 2 z-bins, and Fig. 5, showing the energy of the undetected system $E_{\text{miss}}=(M_X^2-M_p^2)/(2M_p)$.

The Fig. 6 shows the rapidity y_R of the J/ψ in the μp center of mass system. R. M. Godbole et al. [6] give a formula for the gluon x in their model: $x_g|_{\text{model}}=\frac{M_{J/\psi}}{\sqrt{s}}e^{-y_R}$ (Fig. 7).

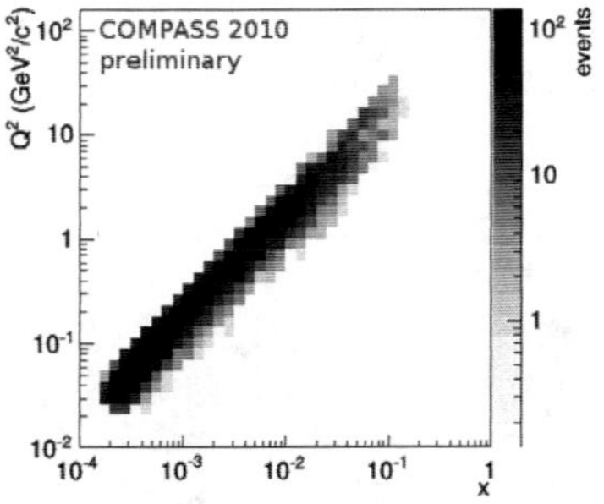

Figure 3. The Bjorken x and the Q^2 are correlated due to the kinematics. Both are low.

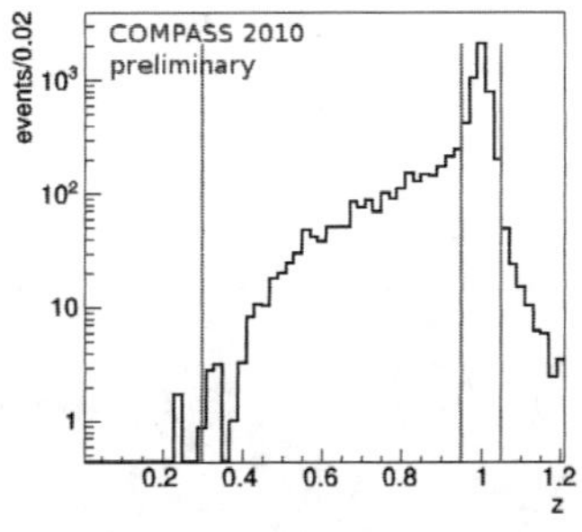

Figure 4. The variable z. The vertical red lines denote the bin boundaries.

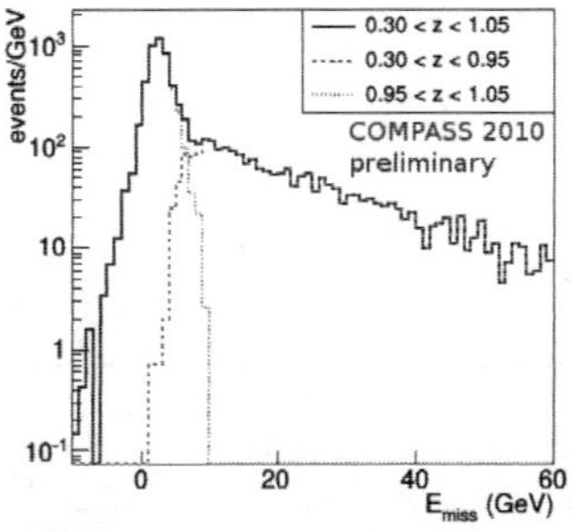

Figure 5. Energy of the undetected system for all events and for the two bins.

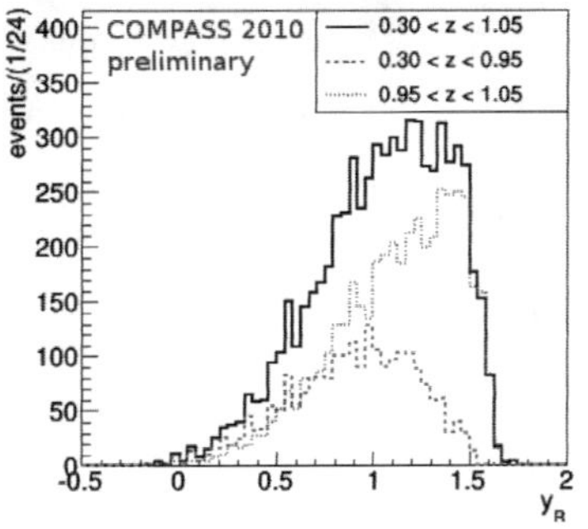

Figure 6. Rapidity of the J/ψ in the μp center of mass system.

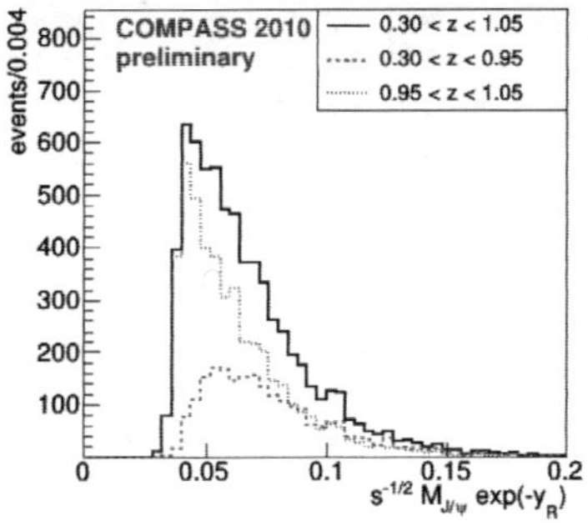

Figure 7. In the model of R. M. Godbole et al. [6] $s^{-1/2}M_{\mathrm{J}/\psi}\,\mathrm{e}^{-y_R}$ is the gluon x.

4. Asymmetry Evaluation

The Sivers asymmetry is an amplitude of a $\sin\phi_{\mathrm{Siv}}$ modulation of the cross-section. The Sivers angle is defined as $\phi_{\mathrm{Siv}} = \phi_h - \phi_S$, that is the difference between the azimuthal angles of the J/ψ momentum and the target polarization.

For this initial study we use a simple double ratio method to evaluate the asymmetry (similar to e. g. [8]). The upstream and downstream cells of the 3-cell target are combined to form an "outer cell". The 2010 run is divided into 12 periods, each consisting of two sub-periods with opposite polarization of the target cells. In each period the double ratio of counts in the 2 cells, 2 sub-periods and in 8 bins in ϕ_{Siv} is calculated

$$A_{\mathrm{Siv}}(\phi_{\mathrm{Siv}}) = \frac{N_{\mathrm{out}}^{\uparrow\downarrow\uparrow}(\phi_{\mathrm{Siv}})N_{\mathrm{cent}}^{\downarrow\uparrow\downarrow}(\phi_{\mathrm{Siv}})}{N_{\mathrm{out}}^{\downarrow\uparrow\downarrow}(\phi_{\mathrm{Siv}})N_{\mathrm{cent}}^{\uparrow\downarrow\uparrow}(\phi_{\mathrm{Siv}})}. \qquad (3)$$

A "raw" Sivers asymmetry $A_{\mathrm{Siv}}^{\mathrm{raw}}$ is obtained by fit $A_{\mathrm{Siv}}(\phi_{\mathrm{Siv}}) = C[1 + 4A_{\mathrm{Siv}}^{\mathrm{raw}}\sin(\phi_{\mathrm{Siv}})]$. It is done period by period to minimize an impact of possible changes in the data-taking conditions. The error-weighted mean of $A_{\mathrm{Siv}}^{\mathrm{raw}}$'s from all subperiods is used in the following.

One can assume that the measured asymmetry $A_{\mathrm{Siv}}^{\mathrm{raw}}$ gets contribution from the real asymmetry $A_{\mathrm{Siv}}^{\mathrm{raw}}|_{\mathrm{sig}}$ and a possible asymmetry of the background $A_{\mathrm{Siv}}^{\mathrm{raw}}|_{\mathrm{bg}}$ weighted by the ratios of counts (determined from the signal to background ratios)

$$A_{\mathrm{Siv}}^{\mathrm{raw}} = \frac{N_{\mathrm{sig}}}{N_{\mathrm{tot}}}A_{\mathrm{Siv}}^{\mathrm{raw}}|_{\mathrm{sig}} + \frac{N_{\mathrm{bg}}}{N_{\mathrm{tot}}}A_{\mathrm{Siv}}^{\mathrm{raw}}|_{\mathrm{bg}}. \qquad (4)$$

Journal of Physics: Conference Series **678** (2016) 012050 doi:10.1088/1742-6596/678/1/012050

If we assume that $A^{\text{raw}}_{\text{Siv}}|_{\text{bg}}$ equals the asymmetry measured in the side-bands (Sec. 2), we can subtract it using the equation (4). The effect of subtraction is small (Tab. 1), because of the large event to background ratio.

Table 1. The raw asymmetries and the final asymmetries.

	$z \in [0.3,\ 0.95]$		$z \in [0.95,\ 1.05]$		$z \in [0.3,\ 0.95]$		$z \in [0.95,\ 1.05]$	
	$A^{\text{raw}}_{\text{Siv}}$	σ_A	$A^{\text{raw}}_{\text{Siv}}$	σ_A	A_{Siv}	σ_A	A_{Siv}	σ_A
Signal band	-0.010	0.031	-0.034	0.018	—	—	—	—
Side bands	-0.025	0.031	-0.030	0.024	—	—	—	—
Background subtracted	-0.006	0.039	-0.034	0.021	-0.05	0.33	-0.28	0.18

The actual Sivers asymmetry is obtained from the raw one by division by the average target polarization $P = 80\ \%$ and the dilution factor $f = 0.15$, which takes into account the fraction of polarizable nuclei in the target material. The final results are shown in Tab. 1 and on Fig. 8.

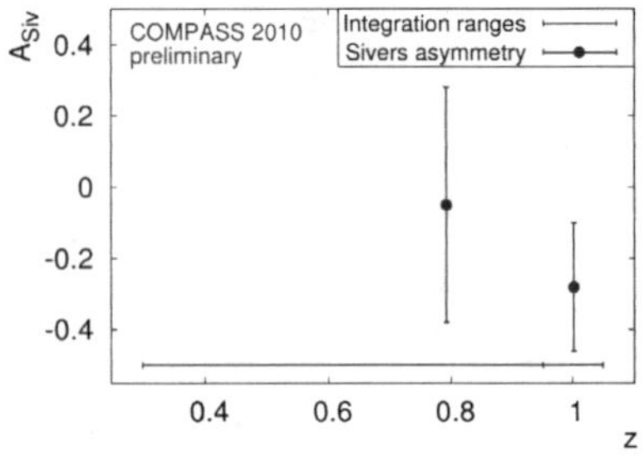

Figure 8. The final Sivers asymmetry.

5. Conclusion and Outlook

In the whole 2010 run data, 2 230 and 4 450 J/ψ events in the lower (inclusive) and the higher (exclusive) z-bins are found. The statistics is small, because of rareness of the process. The Sivers asymmetry, extracted in this initial study by a simple double-ratio method, is found to be compatible with zero in the lower z bin and to have a slight preference for negative values in the higher z bin.

A future improvement could be made utilizing the unbinned maximum likelihood asymmetry extraction method, which is well suited for small data samples. Also, a simple Monte Carlo simulation testing whether the measured $x_g|_{\text{model}}$ distribution is compatible with the model [6] would be an interesting continuation of this study. Finally, looking at all the COMPASS transverse data would increase the statistics, but at most by a factor of 2.

References

[1] Altarelli G and Ross G G 1988 *Phys. Lett. B* **212** 3 391–396.
[2] Glück M and Reya E 1988 *Z. Phys. C* **39** 569–573.
[3] Bravar A, von Harrach D and Kotzinian A 1998 *Phys. Lett. B* **421** 349–359.
[4] Szabelski A (COMPASS Collaboration) 2015 *EPJ Web of Conferences* **85** 02006.
[5] Kurek K (COMPASS Collaboration) 2015 *Presented at the XVI Workshop on High Energy Spin Physics DSPIN-15* (Dubna, Russia).
[6] Godbole R M, Misra A, Mukherjee A and Rawoot V S 2012 *Phys. Rev. D* **85** 094013.
[7] Godbole R M, Misra A, Mukherjee A and Rawoot V S 2013 *Phys. Rev. D* **88** 014029.
[8] Alekseev M et al. (COMPASS Collaboration) 2009 *Phys. Lett. B* **673** 127–135.

XVI Workshop on High Energy Spin Physics (D-SPIN2015) IOP Publishing
Journal of Physics: Conference Series **678** (2016) 012051 doi:10.1088/1742-6596/678/1/012051

Recent results on TMDs from the HERMES experiment

V A Korotkov
(on behalf of the HERMES collaboration)
Institute for High Energy Physics, Protvino 142281, Russia
E-mail: Vladislav.Korotkov@ihep.ru

Abstract. HERMES has taken a wealth of deep-inelastic scattering data using the 27.6 GeV polarized lepton beam at HERA and various pure gas targets, both unpolarized and polarized, which opened the door to several unique results. Among them are the first evidences for the naive-T-odd Sivers and Collins effects. An overview of recent HERMES results on measurement of azimuthal asymmetries in semi-inclusive production of pions, charged kaons and (anti)protons is presented.

1. Introduction

One of the most intriguing puzzles in modern physics is the internal structure of the nucleon. For many years its structure was investigated in deep-inelastic experiments with scattering of leptons by nucleons in inclusive approach. In the last two decades a progress in experimental technique and a rapid theoretical developments in non-perturbative QCD allowed to start more sophisticated study of the nucleon structure.

A formalism of Transverse Momentum Dependent parton distribution functions (TMDs) and of Generalized Parton Distributions (GPDs) was introduced, providing a more comprehensive multi-dimensional description of the nucleon. New distribution functions allow to get complementary descriptions of the nucleon in three dimensions, spanned by the quarks longitudinal momenta and, respectively, by their transverse momenta components and transverse spatial coordinates. They thus may provide, with different approaches, the full phase-space description of the nucleon structure. Experimentally, TMDs and GPDs can be accessed through the analysis of specific azimuthal asymmetries measured, respectively, in semi-inclusive deep-inelastic scattering (SIDIS) and hard exclusive processes.

There are eight leading-twist quark TMDs, which contribute to the SIDIS cross-section in conjunction with a proper fragmentation function (FF). The TMDs depend, in addition to the longitudinal variable x, on the transverse momentum of the quark $\boldsymbol{p}_T$. Three TMDs survive after integrating over $\boldsymbol{p}_T$. They are analogs of usual collinear quark distributions f_1, g_1, and h_1. The other five distributions are the Sivers function $f_{1T}^{\perp}$, the Boer-Mulders function $h_1^{\perp}$, the pretzelosity function $h_{1T}^{\perp}$ and two *worm-gear* functions $h_{1L}^{\perp}$ and $g_{1T}^{\perp}$. They typically describe a correlation between $\boldsymbol{p}_T$ and the polarization of the nucleon and/or the quark.

In the leading order there are two FFs: the familiar polarization-averaged D_1 and the Collins function $H_1^{\perp}$ representing a correlation between the transverse polarization of the fragmenting quark and the transverse momentum of the produced hadron.

Content from this work may be used under the terms of the Creative Commons Attribution 3.0 licence. Any further distribution of this work must maintain attribution to the author(s) and the title of the work, journal citation and DOI.
Published under licence by IOP Publishing Ltd

At the sub-leading order there are many additional TMDs and FFs contributing to the SIDIS cross-section (see e.g. Ref. [1]).

HERMES collaboration has published its results on the one-dimensional Sivers and Collins amplitudes in [2] and [3], respectively. Results on the beam-spin asymmetry (BSA) for pions, obtained with a part of the full statistics, have been presented in [4].

In this talk, a short review of recent HERMES results on azimuthal asymmetries in semi-inclusive processes will be presented.

2. Experiment and extraction of the asymmetry amplitudes

HERMES used the longitudinally polarized lepton beam (e^+ or e^-) of 27.6 GeV scattered off a transversely/longitudinally polarized or unpolarized gas target internal to the HERA storage ring. Scattered leptons and coincident hadrons were detected by the HERMES spectrometer [5]. Leptons were identified with an efficiency exceeding 98% and a hadron contamination of less than 1%. The HERMES dual-radiator ring-imaging Čerenkov detector allows full hadron identification in the momentum range 2–15 GeV. Events were selected subject to the kinematic requirements $Q^2 > 1$ GeV^2, $0.023 < x < 0.6$, $0.1 < y < 0.95$ and $W^2 > 10$ GeV^2, where Q^2, x, y, and W are usual DIS variables. Coincident hadrons were accepted in the range $0.2 < z < 0.7$.

Kinematics of SIDIS on a transversely polarized target is depicted at Fig. 1.

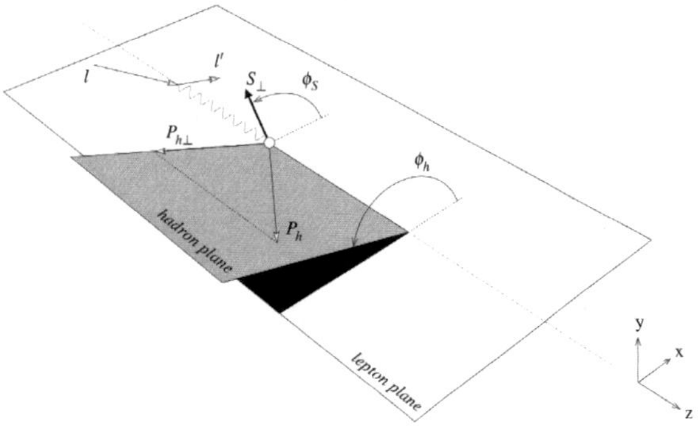

Figure 1. Definition of azimuthal angles ϕ and ϕ_S and the hadron transverse momentum $P_{h\perp}$ for semi-inclusive deep inelastic scattering on a transversely polarized target.

The asymmetry amplitudes were extracted in each particular kinematic bin using an unbinned maximum likelihood fit of the selected SIDIS events to proper probability density function (p.d.f.). Proper p.d.f. should follow to structure of the cross-section. The general structure of SIDIS cross-section was developed and presented in [1]. For a case of a transversely polarized target the probability density function is the following:

$$\begin{aligned} p.d.f. = 1 &+ S_\perp\big(p_1 \cdot \sin(\phi+\phi_S) + p_2 \cdot \sin(\phi-\phi_S) + p_3 \cdot \sin(\phi_S) + p_4 \cdot \sin(2\phi-\phi_S) \\ &+ p_5 \cdot \sin(3\phi-\phi_S) + p_6 \cdot \sin(2\phi+\phi_S)\big) + S_\perp P_L\big(p_7 \cdot \cos(\phi-\phi_S) \\ &+ p_8 \cdot \cos(\phi_S) + p_9 \cdot \cos(2\phi-\phi_S) + p_{10} \cdot \cos(\phi+\phi_S)\big)\,, \end{aligned} \tag{1}$$

where $S_\perp$ is the transverse target polarization and P_L is the beam polarization. Note that two additional $\sin(2\phi+\phi_S)$ and $\cos(\phi+\phi_S)$ modulation terms, with respect to [1], are included here due to existence of the longitudinal component of the target polarization vector [6]. Parameters $p_1, \ldots, p_{10}$ correspond to asymmetry amplitudes $2\langle\sin(\phi+\phi_S)\rangle_{U\perp}, \ldots, 2\langle\cos(\phi+\phi_S)\rangle_{L\perp}$. The first and second subscript of the asymmetry amplitude indicate the respective polarization of beam and target ("U" – unpolarized, "L" – longitudinal, and "⊥" – transverse).

For a case of longitudinal beam and an unpolarized target the probability density function is taken to be

$$p.d.f. = 1 + P_L \cdot p_0 \cdot \sin(\phi)\,, \tag{2}$$

where parameter p_0 corresponds to amplitude $2\langle\sin(\phi)\rangle_{LU}$.

For three-dimensional study, the amplitudes were evaluated in (four-x)×(four-z)×(four-$P_{h\perp}$) bins.

3. Results

The asymmetry amplitudes are known functions [1] of unknown TMDs and FFs.

Leading order amplitudes in a case of the unpolarized beam and the transversely polarized target are the following.

The Sivers amplitude depends on the Sivers distribution function $f_{1T}^{\perp}$ and familiar FF D_1: $2\langle\sin(\phi_h-\phi_S)\rangle_{U\perp} \propto \mathcal{C}\big[-\frac{\hat{\boldsymbol{h}}\boldsymbol{p}_T}{M} f_{1T}^{\perp} D_1\big]$, where $\hat{\boldsymbol{h}} = \boldsymbol{P}_{h\perp}/|\boldsymbol{P}_{h\perp}|$ and $\mathcal{C}\big[wfD\big]$ denotes a convolution integral over the quark transverse momenta $\boldsymbol{p}_T$ and $\boldsymbol{k}_T$ defined for any combination of a TMD $f(x,p_T^2)$ and a FF $D(z,k_T^2)$ multiplied by an arbitrary function $w(\boldsymbol{p}_T,\boldsymbol{k}_T)$ [1].

The Collins amplitude depends on the transversity distribution h_1 and the Collins FF $H_1^{\perp}$: $2\langle\sin(\phi_h+\phi_S)\rangle_{U\perp} \propto \mathcal{C}\big[-\frac{\hat{\boldsymbol{h}}\boldsymbol{k}_T}{M_h} h_1 H_1^{\perp}\big]$

The $2\langle\sin(3\phi_h-\phi_S)\rangle_{U\perp}$ amplitude depends on the pretzelosity distribution function $h_{1T}^{\perp}$ and the Collins FF $H_1^{\perp}$: $2\langle\sin(3\phi_h-\phi_S)\rangle_{U\perp} \propto \mathcal{C}\big[\frac{2\left(\hat{\boldsymbol{h}}\boldsymbol{p}_T\right)\left(\boldsymbol{p}_T\boldsymbol{k}_T\right)+\boldsymbol{p}_T^2\left(\hat{\boldsymbol{h}}\boldsymbol{k}_T\right)-4\left(\hat{\boldsymbol{h}}\boldsymbol{p}_T\right)^2\left(\hat{\boldsymbol{h}}\boldsymbol{k}_T\right)}{2M^2M_h} h_{1T}^{\perp}H_1^{\perp}\big]$

Other possible amplitudes, $2\langle\sin(\phi_S)\rangle_{U\perp}$ and $2\langle\sin(2\phi_h-\phi_S)\rangle_{U\perp}$, are at the twist-3 level.

Leading order amplitude in a case of the longitudinally polarized beam and the transversely polarized target is the following: $2\langle\cos(\phi_h-\phi_S)\rangle_{L\perp} \propto \mathcal{C}\big[\frac{\hat{\boldsymbol{h}}\boldsymbol{p}_T}{M} g_{1T}^{\perp} D_1\big]$. This amplitude depends on the worm-gear function $g_{1T}^{\perp}$ and familiar FF D_1.

Other possible asymmetries, $2\langle\cos(\phi_S)\rangle_{L\perp}$ and $2\langle\cos(2\phi_h-\phi_S)\rangle_{L\perp}$, are at the twist-3 level.

In a case of the longitudinal beam polarization and the unpolarized target the asymmetry amplitude is at the twist-3 level only: $2\langle\sin\phi\rangle_{LU} \propto \frac{2M}{Q}\mathcal{C}\big[-\frac{\hat{\boldsymbol{h}}\boldsymbol{k}_T}{M_h}\big(xe\,H_1^{\perp} + \frac{M_h}{M} f_1 \frac{\tilde{G}^{\perp}}{z}\big) + \frac{\hat{\boldsymbol{h}}\boldsymbol{p}_T}{M}\big(xg^{\perp}D_1 + \frac{M_h}{M} h_1^{\perp}\frac{\tilde{E}}{z}\big)\big]$. It depends on the twist-2 TMDs f_1 and $h_1^{\perp}$, twist-3 TMDs e and $g^{\perp}$, twist-2 FFs D_1 and $H_1^{\perp}$, and twist-3 FFs $\tilde{G}^{\perp}$ and $\tilde{E}$.

3.1. One-dimensional amplitudes for protons and antiprotons

One-dimensional amplitudes for pions and charged pions were published by the HERMES Collaboration a few years ago [2, 3]. In a new study the one-dimensional amplitudes for protons and antiprotons were extracted using (1). The results for the Collins and Sivers amplitudes are presented in Fig. 2. The Collins asymmetry amplitudes for protons show slightly negative values. The Sivers amplitudes for protons are positive. Available statistics of the antiprotons is rather low but the behaviour of the amplitudes have the same tendency as for protons. All other kinds of the asymmetry amplitudes for protons and antiprotons are found to be consistent with zero.

3.2. Three-dimensional amplitudes $A_{U\perp}$

A study of the one-dimensional Sivers amplitudes [2] showed that a magnitude of the K^+ amplitude is larger than a magnitude of the π^+ amplitude. Three-dimensional Sivers amplitudes for π^+ (left panel) and K^+ (right panel) are presented at Fig. 3.

A comparison of the three-dimensional Collins amplitudes for π^+ (left panel) and π^- (right panel) is presented at Fig. 4.

A comparison of the three-dimensional $2\langle\sin(3\phi-\phi_S)\rangle_{U\perp}$ amplitudes for π^+ (left panel) and π^- (right panel) is presented at Fig. 5.

Such multi-dimensional data can be very useful for phenomenolgical analysis of the TMD models.

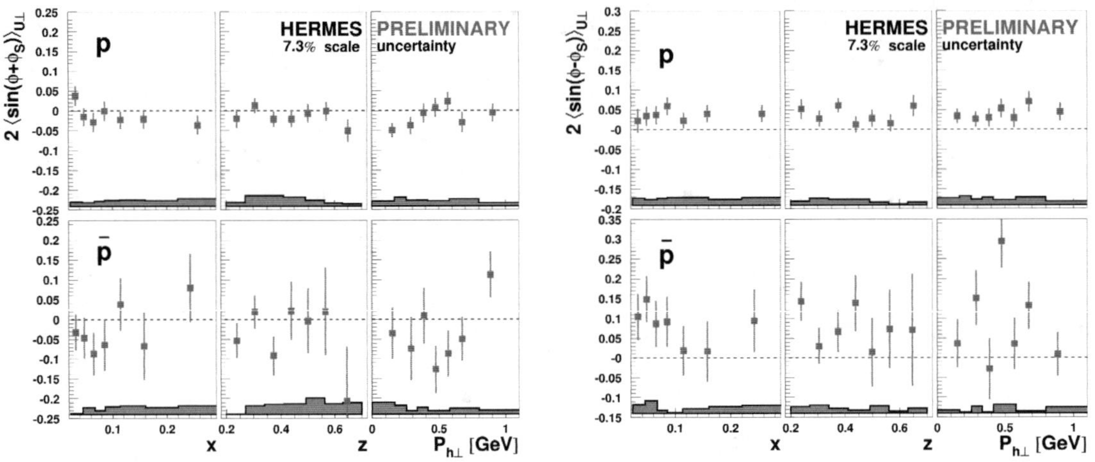

Figure 2. Collins (left panel) and Sivers (right panel) asymmetry amplitudes for protons and antiprotons as a function of x, z or $P_{h\perp}$.

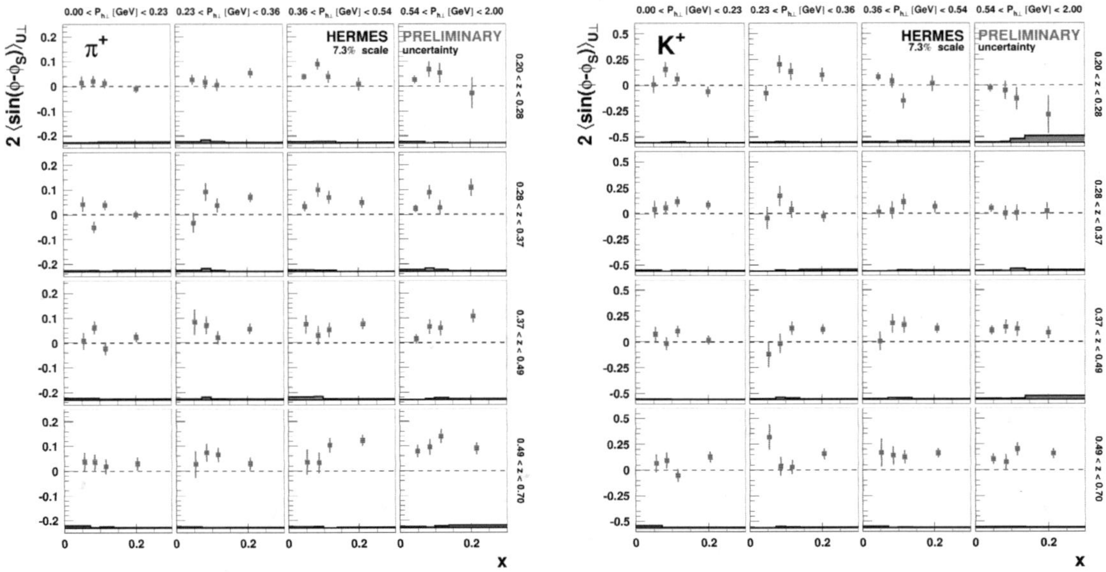

Figure 3. Three-dimensional Sivers asymmetry amplitudes $2\langle\sin(\phi - \phi_S)\rangle_{U\perp}$ for π^+ mesons (left panel) and K^+ mesons (right panel) as a function of x, $P_{h\perp}$, and z.

3.3. Three-dimensional amplitudes $A_{L\perp}$

Three-dimensional $2\langle\cos(\phi - \phi_S)\rangle_{L\perp}$ amplitudes for π^+ (left panel) and π^- (right panel) is presented at Fig. 6.

3.4. Three-dimensional BSAs

Three-dimensional beam-spin asymmetry amplitudes $2\langle\sin\phi\rangle_{LU}$ for π^+ (left panel) and π^- mesons (right panel) are presented at Fig. 7.

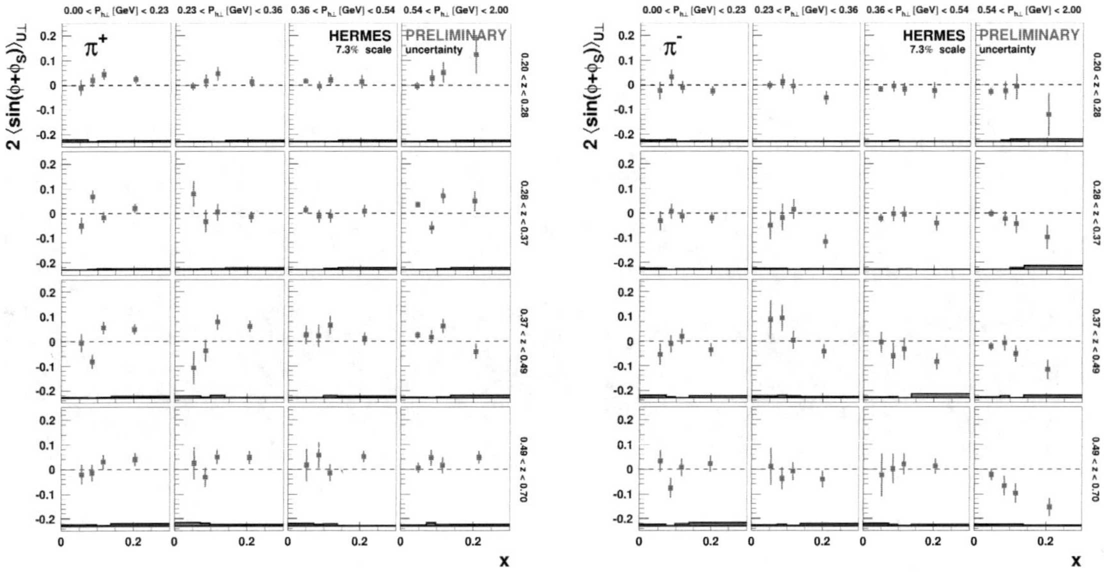

Figure 4. Three-dimensional Collins asymmetry amplitudes $2\langle\sin(\phi+\phi_S)\rangle_{U\perp}$ for π^+ mesons (left panel) and $\pi-$ mesons (right panel) as a function of x, $P_{h\perp}$, and z.

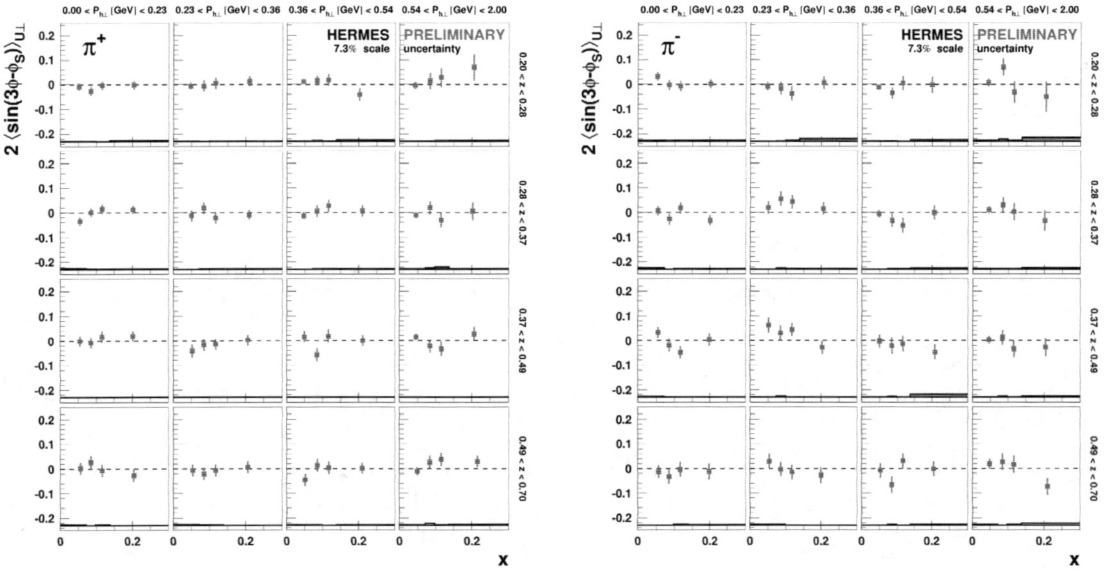

Figure 5. Three-dimensional asymmetry amplitudes $2\langle\sin(3\phi-\phi_S)\rangle_{U\perp}$ for π^+ mesons (left panel) and $\pi-$ mesons (right panel) as a function of x, $P_{h\perp}$, and z.

To conclude, HERMES Collaboration continues its study of the azimuthal asymmetries in SIDIS. Various asymmetries for protons and antiprotons were evaluated. Three-dimensional asymmetry amplitudes as a function of x, z, and $P_{h\perp}$ were evaluated. It is expected that the three-dimensional amplitudes could be much more useful for a phenomenological analysis with a goal to constrain the TMDs.

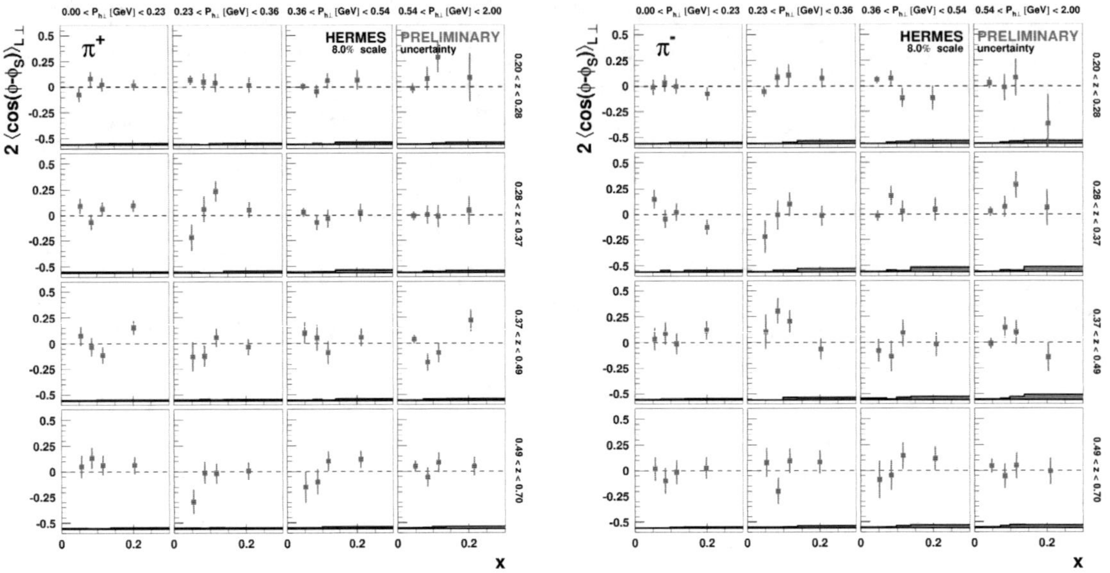

Figure 6. Three-dimensional asymmetry amplitudes $2\langle\cos(\phi - \phi_S)\rangle_{L\perp}$ for π^+ mesons (left panel) and π^- mesons (right panel) as a function of x, $P_{h\perp}$, and z.

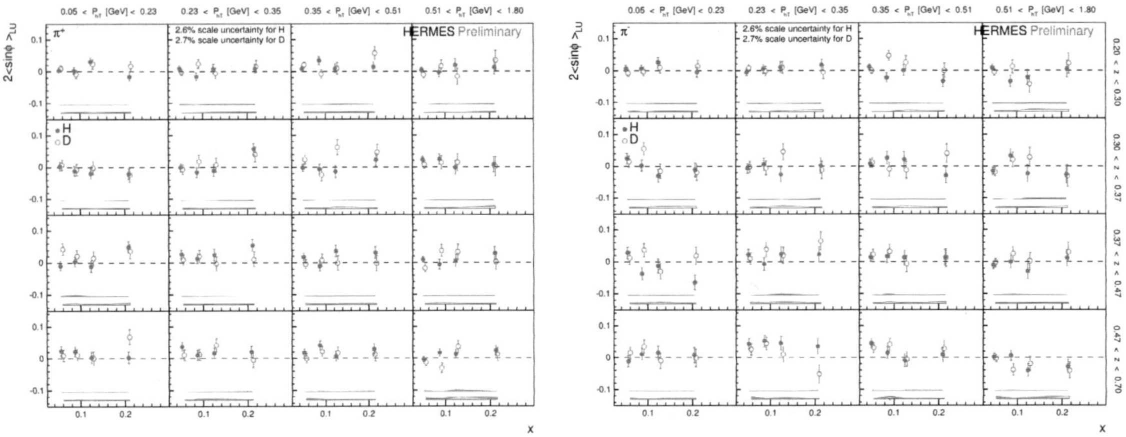

Figure 7. Three-dimensional asymmetry amplitudes $2\langle\sin\phi\rangle_{LU}$ for π^+ mesons (left panel) and π^- mesons (right panel) as a function of x, $P_{h\perp}$, and z. Data obtained with hydrogen (filled circles) and deuterium (empty circles) targets are presented.

References

[1] Bacchetta A et al. 2007 *JHEP* **0702** 093
[2] Airapetian A et al. (HERMES Collaboration) 2009 *Phys. Rev. Lett.* **103** 152002
[3] Airapetian A et al. (HERMES Collaboration) 2010 *Phys. Lett.* B **693** 11
[4] Airapetian A et al. (HERMES Collaboration) 2007 *Phys. Lett.* B **648** 164
[5] Ackerstaff K et al. (HERMES Collaboration) 1998 *Nucl Instr. and Meth.* A **417** 230
[6] Diehl M and Sapeta S 2005 *Eur. Phys. J.* C **41** 515

Journal of Physics: Conference Series **678** (2016) 012052 doi:10.1088/1742-6596/678/1/012052

Constituent quarks in the Standard Model

Sergey Sukhoruchkin
B.P. Konstantinov Petersburg Nuclear Physics Institute, 188300 Gatchina, Russia
E-mail: sergeis@pnpi.spb.ru

Abstract. Tuning effect in particle masses manifests itself in integer relations between masses of leptons, quarks, meson and baryons. It includes also dimensionless relation between such well-known SM-parameters as masses of the muon and Z-boson, m_μ/M_Z=115.9·10^{-5} coinciding with the QED radiative correction $\alpha/2\pi$=115.9·10^{-5} considered for the electron mass m_e by V. Belokurov and D. Shirkov. Integer presentation of particle masses (n=1,13,16,17,18,115) for values m_μ, f_π, m_π, ΔM_Δ, neutron mass and (n=3x16, n=3x18) for constituent quarks M_q''=$m_\rho, m_\omega/2$=780 MeV and M_q=3ΔM_Δ=$m_\Xi/3$=441 MeV were found with the period δ=16m_e. More accurate relations with δ were found from precise ratio m_n/m_e=1838.6836605(11). The shift δm_n=161.65(6) keV of neutron mass from 115δ-m_e accounts integer ratio $\delta m_N/\delta m_n$=8(1.0001(1)) with nucleon mass splitting. With fundamental boson masses the parameters M_q=3ΔM_Δ=$m_\Xi/3$=441 MeV and M_q''=$m_\rho/2$=388.8(2) MeV are in ratios M_Z/M_q=L^Z=206.8 and M_W/M_q''=L^W=207.3 coinciding with lepton ratio L=m_μ/m_e=13·16-1=207.

An integer presentation of the particle masses (n=1,13,16,17,18,115) for the values m_μ, f_π, m_π, ΔM_Δ, neutron mass and (n=3x16, n=3x18) for the constituent quarks M_q''=$m_\rho/2$=389 MeV and M_q=3ΔM_Δ=$m_\Xi/3$=441 MeV were found with the period δ=16m_e [1-3]. Nucleon mass (n=115) is a result of the constituent quark-mass evolution: first it is a mass-reduction for each quark from 3M_q (n=3ΔM_Δ to ΔM_Δ+2f_π (n=18+2·16 equal to 50-51). In Fig.1 [1-3] it corresponds to the shift from 3M_q to Δ. The second step, the spin-flip, is explained in NRCQM [4]. Neutron mass in the nuclear media m_N^{nucl} (circled point) is the end of this sequence.

The neutron mass shift relative to 115δ is equal to $\boxed{\delta m_n\text{=161.65(6) keV}}$ derived from the ratio between the neutron and electron masses evaluated by CODATA [5] $\boxed{m_n/m_e\text{=1838.6836605(4)}}$ and accounts an integer ratio with nucleon mass splitting δm_N $\boxed{\delta m_N/\delta m_n\text{=8(1.0001(1))}}$.

$$m_n = 115 \cdot 16m_e - m_e - \delta m_N/8 \tag{1}$$

$$m_p = 115 \cdot 16m_e - m_e - 9\delta m_N/8 \tag{2}$$

We show here that there are two indirect indications on the reality of mass/energy intervals $m_e/3$=170 keV and 161 keV=$\delta m_N/8$ which are taking part in CODATA relation: values δm_N and $\delta m_N/8$ are observed in nuclear excitations, values Δ – in nuclear binding energies.

Another property of the discussed NRCQM parameters M_q=3ΔM_Δ=$m_\Xi/3$=441 MeV and M_q''=$m_\rho/2$=388.8(2) MeV consists in the empirical observation that their masses and vector boson masses are in ratios M_Z/M_q=L^Z=206.8 and M_W/M_q''=L^W=207.3 coinciding with the lepton ratio L=m_μ/m_e=13×16-1=207 after a small correction $m_e(1-\alpha/2\pi)$.

Content from this work may be used under the terms of the Creative Commons Attribution 3.0 licence. Any further distribution of this work must maintain attribution to the author(s) and the title of the work, journal citation and DOI.
Published under licence by IOP Publishing Ltd

Nucleon interactions within nuclei are commonly considered as a result of one- and many-pion exchange. The role of tensor forces (one-pion exchange) was discussed by T.Otsuka [6] who found out the regions of nuclear chart where tensor-force effects are important. He found that dominant role of one-pion exchange dynamics in nucleon interaction exists in light, near-magic middle-weight and heavy nuclei. Stable intervals in first excitations of the near-magic ^{odd}Sb, ^{116}Sn and ^{101,103}Sn correspond to n=1 and 8 in the periodical systematics with parameters 161 keV=Δ^{TF} close to (1/8) of the nucleon mass splitting δm_N=1293.3 keV. For example a linear trend in excitations E^* of ^{odd}Sb isotopes with a slope of 161 keV was found to be similar to that found in the first excitations of N=21,22 isotopes ^{41}Ca, ^{37,38}S in which E^* values coincide with nδ^{TF} within 2-3 keV.

The stable character of excitation about 170 keV in the near-magic ^{101,103}Sn (n=51,53) was confirmed with stable spacing of ^{97,98}Pd (n=51,52). Here stable intervals D=512 keV (close to 3×170 keV, that is n′=3) and D=648 keV and 1293 keV (n=4,8 in the system with Δ^{TM}) was found. Interval In ^{10}B D=1021.7(4) keV (n′=6) is a result of spin-flip effect (0$^+$-1$^+$) and is close to $2m_e$=1022.0 keV=ε_o. Both systems n·161 keV=(1/8)δm_N and n'·170 keV=(1/6)ε_o coinciding with the parameters of CODATA relation were noticed in many nuclei including magic ^{55}Co [1-3]. This reflects their fundamental origin in particle masses and in nuclear data.

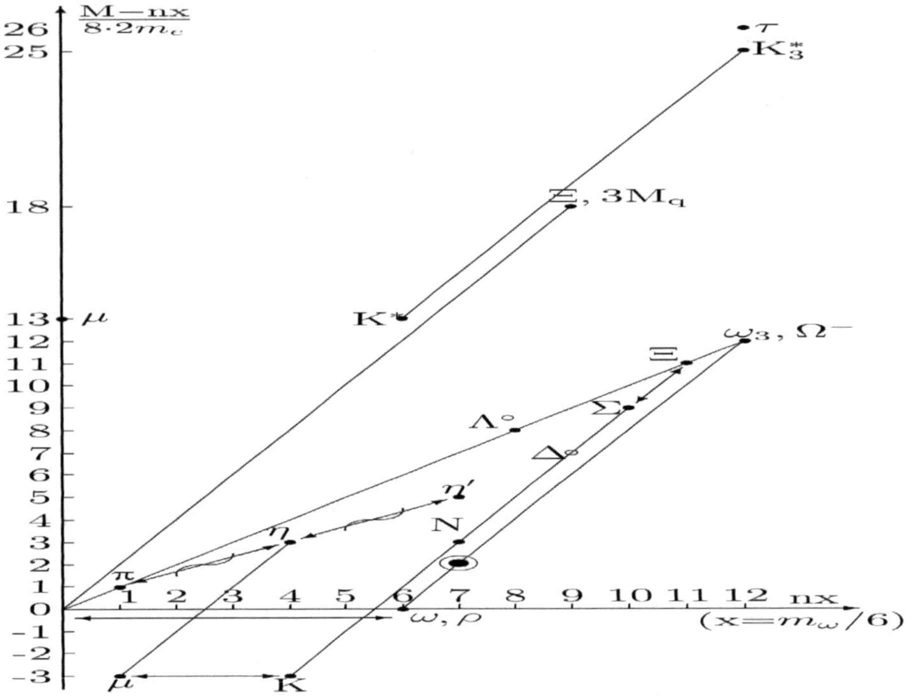

Figure 1. Evolution of the baryon mass N shown in two-dimensional presentation. Values on the horizontal axis are in units $16{\cdot}16m_e = f_\pi$=130.7(4) MeV≈ $m_\rho/6$. The remainder – M_i-kδ=M_i-k16·16m_e) are displayed along vertical axis in units 16m_e. Lines correspond to:
1) the main parameter ΔM_Δ=147 MeV of baryon constituent quark masses from the nucleon Δ-excitation (m_{Δ^o}-m_n)/2=18δ and initial constituent quark mass M_q=3×147 MeV; initial mass $3M_q$=m_Ξ (top);
2) the pion mass 140 MeV=f_π+δ (n=16+1), masses of Λ, Ξ, Ω hyperons are close to $8m_\pi$, $11m_\pi$, $12m_\pi$
3) stable interval in pseudoscalar mesons $m_{\eta'}$-m_η=m_η-$m_\pi^\pm$ (crossed arrows, n=50 in units δ=16m_e.
Nucleon Δ-excitation and two s-quark masses m_s result in a long line from kaon (K, n=4·16-3=61) and the neutron mass (N, n=7·16+3=115) to the Ξ^- hyperon mass (Ξ). Nucleon mass in nuclear medium (m_N^{nucl}, circled point) is close to sum ΔM_Δ+6f_π. It is the last stage of the baryon mass evolution.

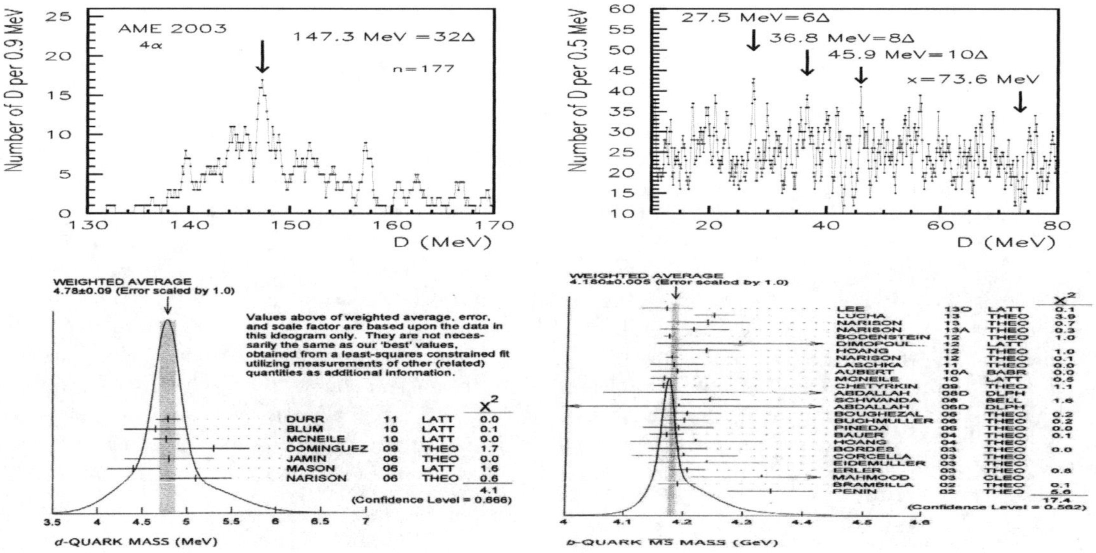

Figure 2. *Top:* Grouping of differences of binding energies ΔE_B at 147.3 MeV=32Δ in light nuclei differing with 4α-clusters (it left) and adjacent ΔE_B distribution for selected x-ΔE_B=73.6 MeV=16Δ in light nuclei (Z≤ 26) [7].
Bottom: Estimations of down- and bottom-quark masses [8] m_d=4.78(9) MeV (close to $\Delta=9m_e$) and m_b=4.180(5) GeV.

In Table 2 [2] the constituent quark masses $M_q = m_{Xi^-}Z/3$=441 MeV and $M''_q = m_\rho/2$ are presented together with the mentioned ratios M_Z/L and M_W/L (lines below M_q and M''_q). In comments to $\Delta M_\Delta = 18\delta = 32\Delta$, $M_q = 54\delta$ and another NRCQM-based estimate $M_q^\Delta = m_\Delta/3 = 50\delta$ three stable intervals in nuclear binding energies ΔE_B=147, 441, 409 MeV are given. Their proximity to intervals in particle masses could be considered as an indirect confirmation of the tuning effect. A special program AIM (Adjacent Interval Method) was used to check ΔE_B distribution: by fixating of x=16Δ=76.3 MeV intervals in the distribution for the adjacent intervals (Fig.2, right [7]) the period of 2Δ=2·4.6 MeV was found (arrows). The down-quark mass m_d=4.78(9) MeV (Fig.2 bottom left) estimation coincides with this value Δ=4.6 MeV.

Table 1. Comparison of particle masses with the period $16m_e=\delta$=8176 keV ($\Delta=9m_e$ is boxed).

Particle	m_i, MeV	k	m_i-k·16m_e	Diff.	Comments (MeV)
μ	105.658367(4)	13	-0.6294		$-m_e$-0.118
f_π	130.7(4) [2]	16		≈0	
$(\Delta^\circ$-n)/2=ΔM_Δ	147.1(2)	18		≈0	ΔE_B = 147.3, Fig.2
M_q NRCQM, $m_{Xi^-}/3$	441	3·18			ΔE_B=441 [1-3]
M_Z/L	440.5	3·18		≈ $-2m_e$	
M_H/18·16	436	3·18-Δ		-5 = -Δ	$m_d \approx \Delta$, Fig.2
η'-η, η-$\pi^\pm$	409	50		≈0	ΔE_B=409, Fig.3 in [7]
M''_q NRCQM, $m_\rho/2$	387.7	48		-4.60 = -Δ	$m_d \approx \Delta$
M_W/L	388.4	3·16		≈ $-2m_e$	$3f_\pi$
n	939.5654(1)	115	-0.6726(1)		$-m_e$-(1/8)δm_N
Σ°	1192.64(2)	146	-1.05(2)		$-2m_e$=-1.022
Ξ°	1314.86(20)	161	-1.47(20)		$-3m_e$=-1.533

The above mentioned new estimates of quark masses (Fig.2 bottom) are presented in the upper left parts of Table 2. Ratio m_d/m_b=114·10^{-5} is coinciding with QED correction $\alpha/2\pi$=115.9·10^{-5}. In Table 2 values differing with the factor $\alpha/2\pi$ are put one under another. The ratio between masses of two well-known SM-parameters m_μ/M_Z=115.9·10^{-5} practically coincides with $\alpha/2\pi$=115.9·10^{-5}. This factor is close to the doubled ratio $m_e/3 : \Delta M_\Delta : M_H$. It was introduced in 70-ties [1] from the empirical stable intervals in nuclear levels $\varepsilon''/\varepsilon'/\varepsilon_\circ$.

Table 2. Presentation of parameters of tuning effect in particle masses (three top sections) and in nuclear data (bottom, sections marked X=-1, 0, 1 at left) by the common expression n·16m$_e$($\alpha/2\pi$)^{X}M with QED radiative correction $\alpha/2\pi$ (α=137^{-1}) [8]. Values m$_\mu$, m$_\pi$-m$_e$, ΔM_Δ=m_s (in baryons), m$_e$/3 and boson masses are boxed. Stable intervals in nuclear binding energies and excitations (E*, D_{ij}, [18-21] X=1) coinciding with nucleon mass splitting δm_N and the neutron mass shift δm_n=115$\delta - m_n - m_e$ (boxed) are considered as a confirmation of relations in particle masses [8]. Unconfirmed mass groupings at M'_H=116 and 58 GeV [13,14] are given in brackets (at X=-1, M=1 and 1/2).

X	M	n = 1	n = 13	n = 16	n = 17	n = 18	Comments
-1	3/2			m$_t$=172.0			
GeV	1	16M_q=δ°	M$_Z$=91.2	M_H=115 [10]		M_H=126	Part.
	1/2	(m_b-M_q)		M^{L3}=58 [9]			masses
0	1	16m_e=2m_d-2m_e	m$_\mu$=106	f$_\pi$=130.7	m$_\pi$-m$_e$	ΔM_Δ=147	NRCQM
MeV	3			M"$_q$=m$_\rho$/2		M$_q$=441=ΔE$_B$	param.
1	1	16m_e=δ=8$\varepsilon_\circ$			kδ-m$_n$-m$_e$=	170 = m$_e$/3	Part.
					=161.651(6)		masses
keV	8				δm_N=1293.3		CODATA
1	1	9.5=δ'=8ε'	123	152	Δ^{TF}=161	170 (Sn)	Nuclear
keV	3				484 (E^*)	512 (Pd)	data
	4		492		648 (Pd)	682(Co)	
	8		984	1212	1293 (E^*)	1360 (Te)	
2	1	11=δ''=8ε''	143	176	749 (Br,Sb)	Neutron	Nuclear
eV	4		570 (Sb)		1500 (Sb,Pd)	reson.	data

Tuning effect in particle masses starts at the large scale. Relations with the constituent quarks masses show an important role of heavy fermions in the properties of physical condensate.

[1] Sukhoruchkin S I 1972 *Stat. Properties of Nuclei* ed. Garg G (Plenum Press) 215
[2] Sukhoruchkin S I 2015 *Nucl. Part. Phys. Proc.* (accepted for publ.)
[3] Landoldt-Boernstein New Series 2012, 2013 **vols. I/25** A,E ed. Schopper H (Springer)
[4] Glozman L 1990 *Nucl. Phys.* A **629** 121c
[5] Mohr P *et al.* 2012 *Rev. Mod. Phys.* **84** 1527
[6] Otsuka T, Suzuku T, Utsino Y 2008 *Nucl. Phys.* A **805** 127c
[7] Sukhoruchkin S I *et al.* 2014 *Proc. ISINN-19* JINR E3-2012-30 284
[8] Olive K A *et al.* (Part. Data Group) *Chin. Phys.* C 2014 **38**
[9] Ting S 1993 *Rept. CERN-PPE/93-34*
[10] Heister A *et al.* ALEPH Collab. 2002 *Phys. Lett.* B **526** 191

Journal of Physics: Conference Series **678** (2016) 012053 doi:10.1088/1742-6596/678/1/012053

Use of the calorimeter to improve analyzing power of the reactions, investigating secondary proton polarization

S N Basilev[1], Yu P Bushuev[1], V V Glagolev[1], D A Kirillov[1], N V Kostyaeva[1], A D Kovalenko[1], A N Livanov[1], D Marchand[7], G Martinska[2], J Musinsky[2], C F Perdrisat[4], N M Piskunov[1], V Punjabi[5] P A Rukoyatkin[1], R A Shindin[1], I M Sitnik[1], V M Slepnev[1], I V Slepnev[1], E Tomasi-Gustafsson[6] J Urban[3] Y Wang[7]

[1] Laboratory of High Energies, JINR, 141980 Dubna, Russia
[2] University of P.J. Safarik, K-04154 Kosice, Slovak Republic
[3] Institute of Experimental Physics, Watsonova 47,SK-04001 Kosice, Slovak Republic
[4] the College of William and Mary, Williamsburg, VA 23187, USA
[5] Norfolk State University, Norfolk, VA 23504, USA
[6]IRFU, SPhn, CEA Saclay and IN2P3/IPN Orsay, France
[7]IPN Orsay, 91406 ORSAY cedex, France

E-mail: sitnik@dubna.ru

Abstract. The reaction $p + CH_2 \rightarrow forward\ charge\ particle + X$ is used for this aim traditionally. Analyzing power of this reaction falls off as $1/p$, where p is the laboratory momentum. At the proton momenta of order 8 GeV/c, which are expected at the JLab experiment, the low analyzing power creates problems for off-line analysis of data. On the other hand, it is well known that the reaction $p+p \rightarrow p+p$ has the much more analyzing power. So, the calorimeter is predestinated for suppression of inelastic events in this reaction. In the report it is shown that the problem is solved quite well.

1. Introduction

The analyzing power of the reaction(AP)

$$p + CH_2 \rightarrow forward\ charge\ particle + X \tag{1}$$

was measured at the Dubna synchrophsotron in range of 1.75 – 5.3 GeV/c [1]. The results are presented in Fig.1.

The $1/p_{lab}$ dependence of AP was found in this work. It turned out that analyzing power of the elastic pp scattering [2], shown in this Fig., follows to the same low. The expected AP at p_{lab} = 7.5 GeV/c, where it is planned to measure secondary proton scattering at the Jlab experiment [3], is too small to provide necessary accuracy for a reasonable run time. As it is seen from Fig.1, the analyzing power of the elastic pp scattering is much higher than AP. That provides the optimistic perspectives to suppress (in part) the contamination of the reaction (1)

Content from this work may be used under the terms of the Creative Commons Attribution 3.0 licence. Any further distribution of this work must maintain attribution to the author(s) and the title of the work, journal citation and DOI.
Published under licence by IOP Publishing Ltd

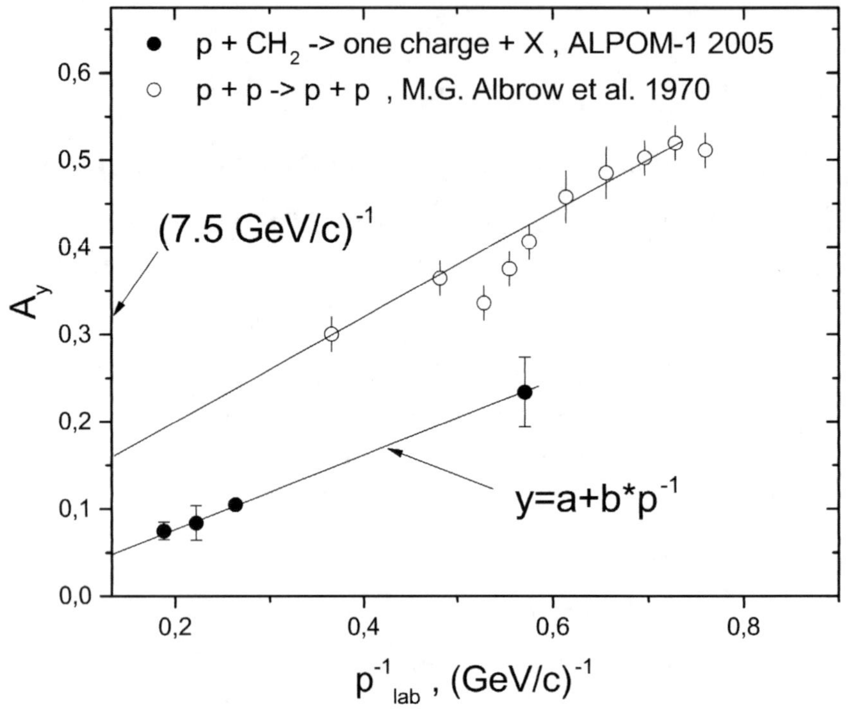

Figure 1. Comparison of the analyzing powers for the reaction (1) and *pp* elastic scattering.

by inelastic scattering events. The process acceptance is too large to use for this aim a dipole. This work is devoted to study how good for this purpose is a calorimeter. The work is fulfilled in the framework of project Alpom-2 [4].

2. Experiment

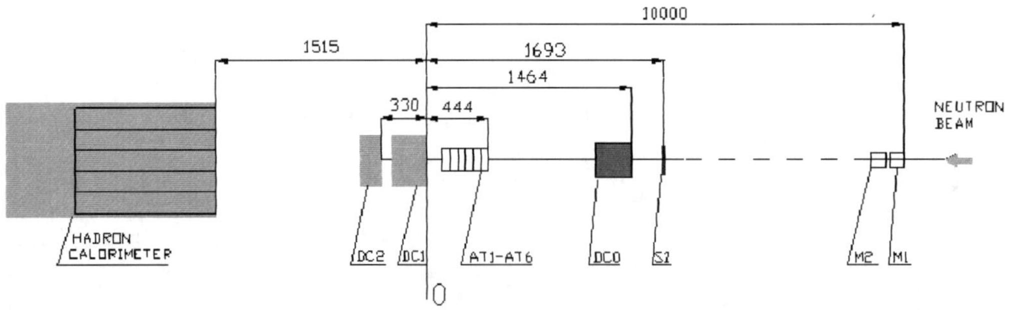

Figure 2. Main scheme of the experiment.
AT_i – target, M – dipole, S_i – scintillator counters, DC_i – drift chambers.

The setup "Alpom-2" (see Fig.2) was situated on the extracted beam line of the Nuclotron. The track parameters were measured by drift chambers. The scintillator counters were used for triggers. The secondary proton ($d \to p$ breakup at zero angle) beam with the momenta of 3.75 and 6.0 GeV/c had the intensity of 10^6, which provided possibility to measure input particle tracks. The space accuracy of the drift chambers was about 0.05 mm. The momentum resolution and efficiency of the setup was of 0.9 %. and $(0.91 \pm 1)\%$, respectively. For the track reconstruction and ofsets adjusting the program package [5] was used.

3. Results

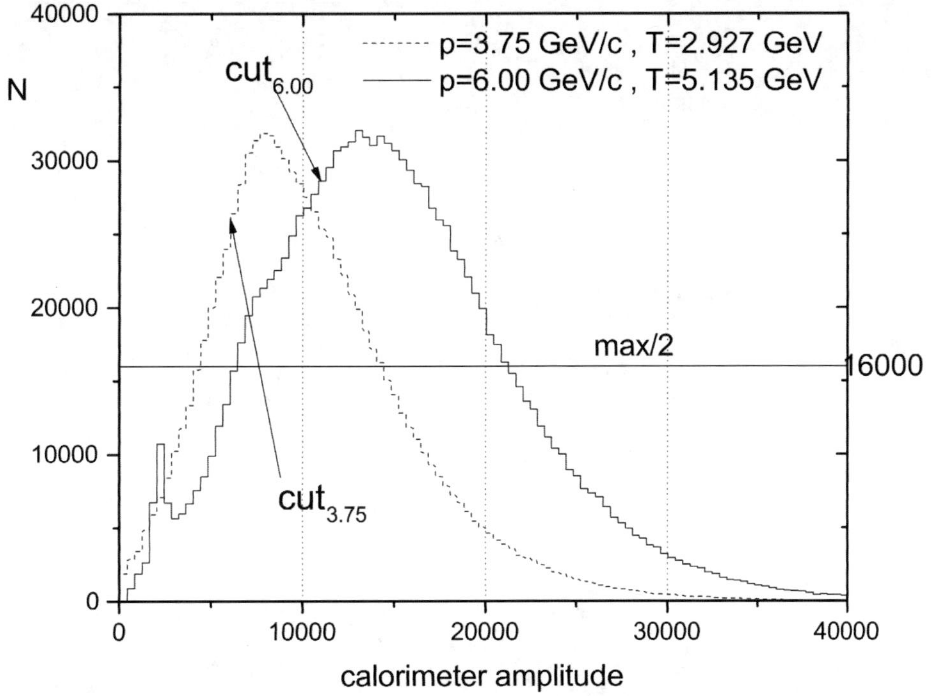

Figure 3. Calorimeter amplitude distributions for unscattered beams.

The distributions of signal amplitudes from the calorimeter is shown in Fig.3 for the unscattered beams of two incident momenta. The amlitude cut for the unscattered beam, $cut_{6.00}$, at 6.00 GeV/c was chosen arbitrary. For the rest events we used $cut = cut_{6.00} * T/T_{6.00}$, where T is measured kinetic energy. The distributions on transversal momentum at 6.00 GeV/c are shown in Fig.4, the ratio of these distribution is shown in Fig.5. One can see that losses of elastic events, which are predominate at small p_t, are of about 20%. At higher p_t the contribution of inelastic events is growing, and we see that suppression of events is also growing and achieves values of about 45%.

4. Conclusions

- The calorimeter is the suitable device for suppression (in part) of inelastic events.
- The test of the setup Alpom-2 on the polarized beam will be carried out soon.

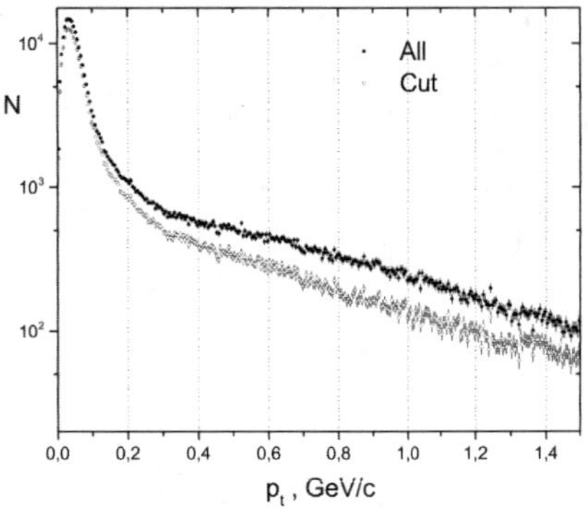

Figure 4. p_t-distributions for all events and events after calorimeter amplitude cut.

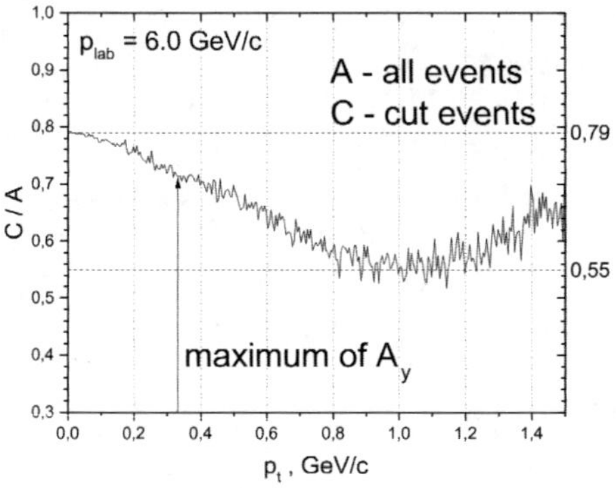

Figure 5. Ratio of these distributions.

5. References

[1] L S Azhgirei et al., Nucl.Instrum.Meth. **A538** (2005) 431
[2] M G Albrow et al., Nucl.Phys. **B23** (1970) 445
[3] E. Brash, E. Cisbani, M. Jones, M. Khandaker, N. Liyanage, L. Pentchev, C.F. Perdrisat, V. Punjabi, B. Wojtsekhowski, Jefferson Lab Experiment 12-07-109, $https : //www.jlab.org/exp_prog/proposals/07/PR12-07-109.pdf$
[4] V. P. Balandin et al., Physics of Particles and Nuclei 45, no. 1 (2014): 330
[5] I.M. Sitnik, *Comput.Phys.Commun.* **185** (2014) 599; JINR Preprint E11-2010-101.

Journal of Physics: Conference Series **678** (2016) 012054 doi:10.1088/1742-6596/678/1/012054

Current status of the muon g-2

A. E. Dorokhov

Bogoliubov Laboratory of Theoretical Physics, JINR, 141980 Dubna, Russia;
N.N.Bogoliubov Institute of Theoretical Problems of Microworld, M.V.Lomonosov Moscow State University, Moscow 119991, Russia;
Department of Physics, Tomsk State University, Lenin ave. 36, 634050 Tomsk, Russia

E-mail: dorokhov@theor.jinr.ru

A. E. Radzhabov

Institute for System Dynamics and Control Theory SB RAS, 664033 Irkutsk, Russia

E-mail: aradzh@icc.ru

A. S. Zhevlakov

Department of Physics, Tomsk State University, Lenin ave. 36, 634050 Tomsk, Russia

E-mail: zhevlakov1@gmail.com

Abstract. The current status of the muon g-2 problem is briefly discussed. We briefly discuss the latest results on the muon g-2 measured in experiment and obtained theoretically within the standard model. Special attention is for the hadronic corrections and in particular the corrections due to the light by light scattering mechanism. For latter we present the results found in the leading in $1/N_c$ approximation with the nonlocal chiral quark model.

1. Introduction

The anomalous magnetic moment (AMM) of charged leptons ($l = e, \mu, \tau$) is defined by

$$a_l = \frac{g_l - 2}{2}, \tag{1}$$

with the gyromagnetic ratio g_l of the lepton magnetic moment to its spin, in Bohr magneton units. For a free pointlike fermion one has $g = 2$ in accordance with the Dirac equation. However, deviations appear when taking into account the interactions leading to fermion substructure and thus to nonzero a_l.

During the first years of the lepton AMM studies the fundamental task was to test the foundations of quantum field theory in general and quantum electrodynamics (QED) in particular. At present, the measurements of the lepton AMM are one of the major low-energy tests of the standard model (SM) and play an important role in the search for new interactions beyond the SM.

Content from this work may be used under the terms of the Creative Commons Attribution 3.0 licence. Any further distribution of this work must maintain attribution to the author(s) and the title of the work, journal citation and DOI.
Published under licence by IOP Publishing Ltd

2. Muon anomalous magnetic moment in the standard model

The nonzero lepton AMMs are induced by radiative corrections due to the coupling of the lepton spin to virtual fields, which in the SM are induced by QED, weak and strong (hadronic) interactions[1]

$$a^{\rm SM} = a^{\rm QED} + a^{\rm weak} + a^{\rm hadr}. \tag{2}$$

The electron and muon AMMs are among the most accurately measured quantities in elementary particle physics. Today, the electron AMM serves as the best quantity to determine the fine structure constant with the highest accuracy. At the same time, for a_μ, there is a deviation at the level of 3-4 σ of the SM prediction from the measured value. Even if this does not give a clear indication for the existence of New Physics, it allows us to provide stringent constraints on the parameters of hypothetical models.

In 2006, there were published the results on a_μ measurements by the E821 collaboration at Brookhaven National Laboratory [5]. The combined result, based on nearly equal samples of positive and negative muons, is

$$a_\mu^{\rm BNL} = 116\ 592\ 08.0\ (6.3) \times 10^{-10} \quad [0.54\ {\rm ppm}]. \tag{3}$$

This exiting result is still limited by the statistical errors and proposals to measure a_μ with a fourthfold improvement in accuracy have been proposed at Fermilab (USA) [6] and J-PARC (Japan) [7]. A future experiments plan to reduce the present experimental error to a precision of 0.14 ppm.

In SM the dominant contribution to the lepton AMM comes from QED. The complete tenth-order QED contribution to a_μ was reported in [8]

$$a_\mu^{\rm QED} = 11\ 658\ 471.8951\ (0.0080) \times 10^{-10}. \tag{4}$$

The accuracy of these calculations is enough for any planed experiments in new future.

In general, the weak contributions are small due to suppressing factor $\alpha/\pi \cdot m_\mu^2/M_w^2 \sim 10^{-9}$, where M_w is a typical mass of heavy $W^\pm, Z$ and H bosons. The one- and two-loop evaluations indicate that they are known with a sufficiently high accuracy [9, 10]

$$a_\mu^{\rm weak} = 15.36\ (0.10) \times 10^{-10}, \tag{5}$$

where the remaining theory error comes from unknown three-loop contributions and dominantly from light hadronic uncertainties in the second-order electroweak diagrams with quark triangle loops. The most important feature of these new estimates, that significantly increase the theoretical precision, is to use the LHC result on the Higgs-boson mass measured by ATLAS and CMS Collaborations.

3. Hadronic contributions to the muon anomalous magnetic moment

Strong (hadronic) interaction produces relatively small contributions to a_μ, however they are known with an accuracy comparable to the experimental uncertainty in (3). In leading in α orders, these contributions can be separated into three terms

$$a_\mu^{\rm hadr} = a_\mu^{\rm HVP} + a_\mu^{\rm ho} + a_\mu^{\rm HLbL}. \tag{6}$$

In (6), $a_\mu^{\rm HVP}$ is the leading in α contribution due to the hadron vacuum polarization (HVP) effect in the internal photon propagator of the one-loop diagram , $a_\mu^{\rm ho}$ is the next-to-leading and next-to-next-to-leading order contributions related to iteration of HVP. The last term is

[1] For comprehensive reviews see [1, 2, 3, 4].

not reduced to HVP iteration and it is due to the hadronic light-by-light (HLbL) scattering mechanism.

Hadronic contributions in (6) are determined by effects dominated by long distance dynamics, the region where the methods of perturbation theory of Quantum Chromodynamics (QCD) do not applicable and one must use less reliable nonperturbative approaches. However, in case of HVP, using analyticity and unitarity (the optical theorem) $a_\mu^{\rm HVP}$ can be expressed as the spectral representation integral

$$a_\mu^{\rm HVP} = \frac{\alpha}{\pi}\int_{4m_\pi^2}^{\infty}\frac{dt}{t}K(t)\rho_{\rm V}^{\rm (H)}(t)\,, \tag{7}$$

which is a convolution of the hadronic spectral function $\rho_{\rm V}^{\rm (H)}$ with the known from QED kinematical factor $K(t)$. The QED factor is sharply peaked at low invariant masses t and decreases monotonically with increasing t. Thus, the integral defining $a_\mu^{\rm HVP}$ is sensitive to the details of the spectral function $\rho_{V}^{\rm (H)}(t)$ at low t, which is related to the total $e^+e^- \to \gamma^* \to$ hadrons cross-section $\sigma(t)$ at center-of-mass energy squared t by

$$\sigma^{e^+e^-\to\rm hadrons}(t) = \frac{4\pi\alpha}{t}\rho_{\rm V}^{\rm (H)}(t)\,. \tag{8}$$

This fact is used to get quite accurate estimate of $a_\mu^{\rm HVP}$. The most precise recent phenomenological evaluations of $a_\mu^{\rm HVP}$, using recent $e^+e^- \to$ hadrons data, provide the results

$$a_\mu^{HVP,LO} = \begin{cases} 692.3\ (4.2)\times 10^{-10}, & [11] \\ 694.91\ (4.27)\times 10^{-10}. & [12] \end{cases} \tag{9}$$

The next-to-leading order corrections are suppressed by the absolute value by extra degree of α. The hadron vacuum polarization contributions in the next-to-leading and next-next-to-leading order are

$$a_\mu^{\rm HVP,NLO} = -9.84(0.06)(0.04)\times 10^{-10} \qquad [12], \tag{10}$$

$$a_\mu^{\rm HVP,NNLO} = 1.24(0.01)\times 10^{-10} \qquad [13], \tag{11}$$

However, one kind of these contributions corresponding to the HLbL is amount to the range from 0.5 to 1.5 ppm and known with accuracy of order 50% It gives an error comparable in magnitude with the uncertainty induced by HVP (9). The problem is that the HLbL scattering contribution can not be calculated from first principles or (unlike to HVP) directly extracted from phenomenological considerations. Instead, it has to be evaluated using various QCD inspired hadronic models that correctly reproduce low- and high- energy properties of strong interaction.

Different approaches to the calculation of the contributions from the HLbL scattering process to a_μ were used. These approaches can be separated in several groups. The first one consists of various extended versions of the vector meson dominance model (VMD) supplemented by ideas of the chiral effective theory, such as the hidden local symmetry model (HLS) [14], the lowest meson dominance (LMD) [15, 16, 17], the resonance chiral theory (RχT) [18, 19]. The second group is based on consideration the effective models of QCD that use the dynamical quarks as effective degrees of freedom. The latter include different versions of the (extended) Nambu–Jona-Lasinio model (E)NJL [20, 21], the Constituent Quark Models with local interaction (CQM) [22, 23, 24, 25], the models based on nonperturbative quark-gluon dynamics, like the non-local chiral quark model (NχQM) [26, 27, 28, 3, 29], the Dyson-Schwinger model [30] (DS), or the holographic models (HM) [31, 32]. More recently, there have been attempts to estimate $a_\mu^{\rm HLbL}$ within the dispersive approach (DA) [33, 34] and the so-called rational approximation (RA) approach [35]. The lattice calculations of HLbL are still at an exploratory stage [36].

4. Results on hadronic light-by-light scattering contributions in the leading in $1/N_c$ approximation

For the numerical estimates, the $SU(2)$- and $SU(3)$- versions of the NχQM model are used by us. In order to check the model dependence of the final results, we also perform calculations for different sets of model parameters.

In the $SU(2)$ model, the scheme of fixing the model parameters was suggested in [27, 28]: fitting the parameters of the nonlocality Λ and the light current quark mass m_c by the physical values of the π^0 mass and the $\pi^0 \to \gamma\gamma$ decay width, and varying the dynamical quark mass m_D in the region $150 - 400$ MeV. For estimation of a_μ^{HLbL} and its error, we use the region for m_D from 200 to 350 MeV.

For the $SU(3)$ version of the model, it is necessary to fix two more parameters: the current and dynamical masses of the strange quark. We suggest to fix them by fitting the K^0 mass and obtaining reasonable values for the η meson mass and the $\eta \to \gamma\gamma$ decay width.

The estimates for the partial contributions to a_μ^{HLbL} (in 10^{-10}) are the π^0 contribution 5.01(0.37) [27], the sum of the contributions from π^0, η and η' mesons 5.85(0.87) [27], the scalar σ, $a_0(980)$ and $f_0(980)$ mesons contribution 0.34(0.48) [3, 28], and the quark loop contribution is 11.0(0.9) [3, 29]. In all cases we estimate the absolute value of the result and its error by calculating $a_\mu^{\mathrm{HLbL,N\chi QM}}$ for the space of model parameters fixed by above mentioned observables, except one, varying m_D. Because in all cases the behavior of the result is quite smooth, it gives to us a credit to point out rather small model errors ($\leq 10\%$) for the intermediate and final results. Thus our claim is that the total contribution obtained in the leading order in the $1/N_c$ expansion within the nonlocal chiral quark model is)

$$a_\mu^{\mathrm{HLbL,N\chi QM}} = 16.8(1.25) \cdot 10^{-10}. \tag{12}$$

This value accounts for the spread of the results depending on reasonable variation of the model parameters and sensitivity to the different choice of the nonlocality shapes.

Summarizing the results of our works [27, 28, 3, 29], we get the total hadronic contribution to a_μ^{HLbL} within the NχQM in the leading order in the $1/N_c$ expansion. The total result is given in Eq. (12). To estimate the uncertainty of this result, we vary some of the model parameters in physically reasonable interval and also study the sensitivity of the result with respect to different model parameterizations.

If we add the result (12) to all other known contributions of the standard model to a_μ, (4)-(11), we get that the difference between experiment (3) and theory is

$$a_\mu^{\mathrm{BNL}} - a_\mu^{\mathrm{SM}} = 18.73 \times 10^{-10}, \tag{13}$$

which corresponds to 2.43σ. The total result becomes more close to the experiment one, however it is still not enough to explain the muon anomaly.

Acknowledgments

Numerical calculations are performed on computing cluster "Ac. V.M. Matrosov". The work is supported by Russian Science Foundation grant (RSCF 15-12-10009).

References

[1] F. Jegerlehner, A. Nyffeler, 2009 *Phys. Rept.* **477**, 1
[2] J.P. Miller, E. de Rafael, B. L. Roberts and D. Stuckinger, 2012 *Ann. Rev. Nucl. Part. Sci.* **62**, 237
[3] A.E. Dorokhov, A.E. Radzhabov and A.S. Zhevlakov, 2014 *JETP Lett.* **100** 2, 133 .
[4] M. Knecht, 2014 arXiv:1412.1228 [hep-ph].
[5] G.W. Bennett, et al. [Muon (g-2) Collaboration], 2006 *Phys.Rev.* **D73**, 072003
[6] G. Venanzoni [Fermilab E989 Collaboration], 2012 *J. Phys. Conf. Ser.* **349**, 012008

Journal of Physics: Conference Series **678** (2016) 012054 doi:10.1088/1742-6596/678/1/012054

[7] N. Saito [J-PARC g-2/EDM Collaboration], 2012 *AIP Conf. Proc.* **1467**, 45
[8] T. Aoyama, M. Hayakawa, T. Kinoshita and M. Nio, 2012 *Phys. Rev. Lett.* **109**, 111808
[9] A. Czarnecki, W. J. Marciano and A. Vainshtein, 2003 *Phys. Rev.* **D67**, 073006 [Erratum-ibid. **D73** , 119901 (2006)].
[10] C. Gnendiger, D. Stockinger and H. Stockinger-Kim, 2013 *Phys. Rev.* **D88**, 053005
[11] M. Davier, A. Hoecker, B. Malaescu, Z. Zhang, 2011 *Eur. Phys. J.* **C71**, 1515 [Erratum-ibid. **C72**, 1874 (2012)].
[12] K. Hagiwara, R. Liao, A. D. Martin, D. Nomura and T. Teubner, 2011 *J. Phys.* **G38**, 085003
[13] A. Kurz, T. Liu, P. Marquard and M. Steinhauser, 2014 *Phys. Lett.* **B734**, 144 .
[14] M. Hayakawa, T. Kinoshita and A.I. Sanda, 1995 *Phys. Rev. Lett.* **75**, 790; M. Hayakawa and T. Kinoshita, 1998 *Phys. Rev.* **D57**, 465 [Erratum-ibid. **D66**, 019902 (2002)].
[15] M. Knecht, A. Nyffeler, 2002 *Phys.Rev.* **D65**, 073034
[16] K. Melnikov, A. Vainshtein, 2004 *Phys.Rev.* **D70**, 113006
[17] A. Nyffeler, 2009 *Phys.Rev.* **D79**, 073012
[18] K. Kampf and J. Novotny, 2011 *Phys. Rev.* **D84**, 014036
[19] P. Roig, A. Guevara and G.L. Castro, 2014 *Phys. Rev.* **D89**, 073016
[20] J. Bijnens, E. Pallante and J. Prades, 1995 Phys. Rev. Lett. **75**, 1447; J. Bijnens, E. Pallante, J. Prades, 1996 *Nucl.Phys.* **B474**, 379; *ibid.* **B626**, 410 (2002).
[21] E. Bartos, A.Z. Dubnickova, S. Dubnicka, E.A. Kuraev, 2002 E. Zemlyanaya, *Nucl. Phys.* **B632**, 330
[22] A. A. Pivovarov, 2003 *Phys. Atom. Nucl.* **66**, 902 [Yad. Fiz. **66**, 934 (2003)]
[23] J. Erler and G. Toledo Sanchez, 2006 *Phys. Rev. Lett.* **97**, 161801
[24] R. Boughezal and K. Melnikov, 2011 *Phys. Lett.* **B704**, 193
[25] D. Greynat and E. de Rafael, 2012 *JHEP* **1207**, 020
[26] A. E. Dorokhov and W. Broniowski, 2008 *Phys. Rev.* **D78** 073011 .
[27] A.E. Dorokhov, A.E. Radzhabov, A.S. Zhevlakov, 2011 *Eur. Phys. J.* **C71**, 1702
[28] A.E. Dorokhov, A.E. Radzhabov, A.S. Zhevlakov, 2012 *Eur. Phys. J.* **C72**, 2227
[29] 2015 A. E. Dorokhov, A. E. Radzhabov and A. S. Zhevlakov, arXiv:1502.04487 [hep-ph].
[30] T. Goecke, C.S. Fischer, R. Williams, 2011 Phys.Rev. **D83**, 094006 [Erratum-ibid. **D86** , 099901 (2012)]; Phys.Rev. **D87**, 034013 (2013).
[31] D.K. Hong, D. Kim, 2009 *Phys.Lett.* **B680**, 480
[32] L. Cappiello, O. Cata, G. D'Ambrosio, 2011 *Phys.Rev.* **D83**, 093006
[33] G. Colangelo, M. Hoferichter, M. Procura and P. Stoffer, 2014 *JHEP* **1409**, 091
[34] V. Pauk and M. Vanderhaeghen, 2014 *Phys.Rev.* **D90**, 113012
[35] P. Masjuan and M. Vanderhaeghen, 2012 arXiv:1212.0357 [hep-ph]; R. Escribano, P. Masjuan and P. Sanchez-Puertas, 2014 *Phys.Rev.* **D89**, 034014
[36] T. Blum, A. Denig, I. Logashenko, E. de Rafael, B. Lee Roberts, T. Teubner, G. Venanzoni, 2013 hep-ph/1311.2198

Journal of Physics: Conference Series **678** (2016) 012055 doi:10.1088/1742-6596/678/1/012055

The Gluon Contribution to the Sivers Effect COMPASS results

Krzysztof Kurek and Adam Szabelski on behalf of the COMPASS collaboration
Adam Szabelski
National Centre for Nuclear Research
Hoza 69, 00-681 Warsaw
Poland

E-mail: adam.szabelski@cern.ch

Abstract. The Sivers effect describes the correlation between the spin of the nucleon and the orbital motion of partons. It can be measured via Semi-Inclusive Deep Inelastic Scattering of lepton on a transversely polarised proton and deuteron targets by determining the azimuthal asymmetry related to the modulation in the Sivers angle ϕ_{Siv}. In the paper a method of obtaining the Sivers asymmetry for gluons is presented. It is based on the model of lepton nucleon interactions via three single-photon-exchange processes: photon-gluon fusion (PGF), QCD Compton (QCDC) and leading process (LP). A method of simultaneous extraction of the Sivers asymmetries of the three processes with the use of Monte Carlo (MC) and neural networks (NN) approach is presented. The method has been applied to COMPASS data taken with 160GeV/c muon beam scattered off transversely polarised deuteron and transversely polarised proton target. For each target a data sample of events containing at least two hadrons with large transverse momentum has been selected. Finally the results for gluon Sivers asymmetry were obtained to be: $A_g^d = -0.14 \pm 0.15(stat.) \pm 0.06(syst.)$ at $\langle x_g \rangle = 0.13$ and $A_g^p = -0.26 \pm 0.09(stat.) \pm 0.08(syst.)$ at $\langle x_g \rangle = 0.15$.

1. Introduction

The transverse momentum dependent structure functions of the nucleon have been studied in semi-inclusive DIS on transversely polarised targets for many years. The strongest emphasis has been put on extracting Collins and Sivers asymmetry (Ref. [1] - deuteron target, Ref.[2] - proton target). In the COMPASS experiment kinematics these measurements are dominated by scattering of the muon on a quark.

To obtain the Sivers asymmetry for gluons from COMPASS data a model of unpolarised muon-nucleon scattering has been assumed. In this paper we use a LEPTO MC model (Ref. [3]) in which three single photon exchange processes: photon-gluon-fusion (PGF) $\gamma^* g \to q\bar{q}$, QCD Compton (QCDC) $\gamma^* q \to qg$ and the leading process (LP) $\gamma^* q \to q$ contibute. The above-mentioned analysis by COMPASS collaboration are dominated by the latter, the absorption of a virtual photon by a quark. To measure the Sivers effect for gluons (PGF Sivers asymmetry) a method of tagging PGF process is needed. The open charm meson production as in Ref. [4] gives a possibility to tag PGF process, however in the case of COMPASS data taken on transversely polarised targets the statistics is too small to extract the asymmetry from reconstructed charmed mesons. The other clean channel is the J/Ψ production which is also not large in statistics.

Content from this work may be used under the terms of the Creative Commons Attribution 3.0 licence. Any further distribution of this work must maintain attribution to the author(s) and the title of the work, journal citation and DOI.
Published under licence by IOP Publishing Ltd

Journal of Physics: Conference Series **678** (2016) 012055 doi:10.1088/1742-6596/678/1/012055

The remaining possibility is the observation of high-p_T hadron pairs in the final state. The analysis presented in this paper is based on a method used for gluon polarisation $\Delta g/g$ determination, introduced in Ref.[5]. This analysis, described in detail in Section 3, uses Monte Carlo simulation and neural network (NN) approach. It enables to simultaneously extract asymmetries of the three contributing processes from a full set of events. The main difference between this analysis and the determination of $\Delta g/g$ is the modulation in the Sivers angle, $\phi_{Siv} \equiv \phi_{2h} - \phi_S$, where ϕ_{2h} is the azimuthal angle of the two high-p_T hadrons observed in the final state and ϕ_S is the azimuthal angle of the spin direction of the nucleon. The angle ϕ_{2h} is measured on the hadron level and it is related to the gluon azimuthal angle ϕ_g, unfortunately distorted by the fragmentation effects. Monte Carlo studies have shown that the strongest correlation between ϕ_g and ϕ_{2h} is when we define the latter as the azimuthal angle of the sum of the leading and next-to-leading hadron momenta. This correlation is further enhanced by the cuts on the transverse momenta of the two hadrons, $p_{T1} > 0.7\mathrm{GeV}/c$, $p_{T2} > 0.4\mathrm{GeV}/c$. The selection has been optimised taking into account the statistics of the sample and the ϕ_g, ϕ_{2h} correlation. Moreover these cuts enhance the PGF fraction in the sample, Ref. [6].

2. Sivers asymmetry in two hadron production

With the above-mentioned selection the scattering of muon on nucleon with the production of at least two hadrons with large transverse momentum is studied:

$$\mu + N \to \mu' + 2h + X \tag{1}$$

The cross section for this reaction is described in details in Ref. [7] in terms of azimuthal angles ϕ_{2h} and ϕ_R corresponding to the vectors $\mathbf{P}_{2h} = \mathbf{P}_1 + \mathbf{P}_2$ and $\mathbf{R} = (\mathbf{P}_1 - \mathbf{P}_2)/2$ respectively. Where $\mathbf{P}_1$ and $\mathbf{P}_2$ are the leading and next-to-leading hadron momenta. In Ref. [8] and [9] have been shown that the Sivers modulation of the cross section, after integrating over ϕ_R, is $\sin(\phi_{2h} - \phi_S)$. Here ϕ_S is the azimuthal angle of the spin of the nucleon.

Let us define the two hadron Sivers asymmetry by:

$$A_T^{2h}(\phi_{Siv}) = \frac{d^8\sigma^{\uparrow}(\phi_{Siv}) - d^8\sigma^{\downarrow}(\phi_{Siv})}{d^8\sigma^{\uparrow}(\phi_{Siv}) + d^8\sigma^{\downarrow}(\phi_{Siv})}. \tag{2}$$

Here $\phi_{Siv} = \phi_{2h} - \phi_S$. Then the number of events in a ϕ_{Siv} bin is given by:

$$N(\vec{x}, \phi_{Siv}) = \alpha(\vec{x}, \phi_{Siv}) \left(1 + f P_T A_T^{2h}(\phi_{Siv})\right), \tag{3}$$

where $\vec{x} = (x_{Bj}, Q^2, p_{T1}, p_{T2}, z_1, z_2)$, f is the dilution factor and P_T is the target polarisation. Here $\alpha = an\phi\sigma_0$, where a is the total spectrometer acceptance, n is the density of scattering centres, ϕ is the beam flux and σ_0 is the azimuthal independent part of the cross section. Throughout this paper only Sivers modulation will be taken into account:

$$N(\vec{x}, \phi_{Siv}) = \alpha(\vec{x}, \phi_{Siv}) \left(1 + f P_T A_{Siv}(\vec{x}) \sin\phi_{Siv}\right). \tag{4}$$

This can be done since azimuthal modulations given in Ref. [7] are orthogonal. The orthogonality has been confirmed during the systematics studies.

To obtain Sivers asymmetry for gluons from two hadron production in SIDIS it is necessary assume that the main contribution to the overall muon-nucleon scattering is due to three processes (Fig. 1) as presented in Ref. [3]. This model is successful in describing the unpolarised data. The leading process is in zero order QCD and is a dominating process. The other two, photon-gluon fusion and QCD Compton, are first order QCD processes and are suppressed. They can be enhanced, however, by the cut on p_T of the produced hadrons. This is due to the

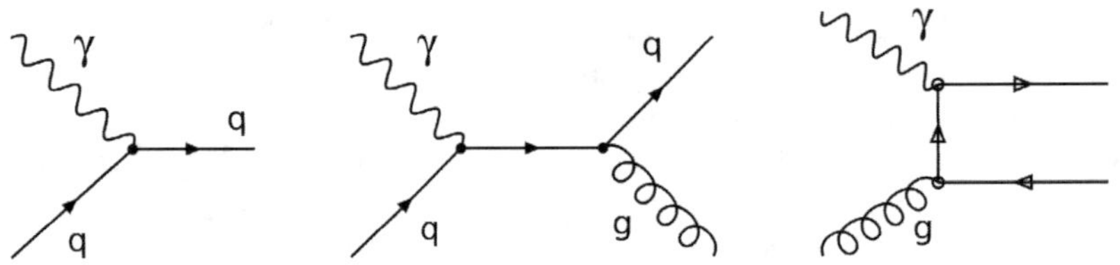

Figure 1: Feynman diagrams considered for $\gamma^* N$ scattering: a) Leading order process (LP), b) gluon radiation (QCD Compton scattering), c) photongluon fusion (PGF).

fact that hadrons in the leading process gain transverse momentum only from intrinsic parton transverse momentum and fragmentation while in the two other processes transverse momentum is also generated at the parton level as discussed in Ref. [6].
It has been checked on Monte Carlo data produced with the LEPTO generator that the fractions R do not depend on ϕ_{Siv}. Hence the Sivers asymmetry given in Eq. 4 can be decomposed into asymmetries of the sub-processes:

$$A_{Siv} = R_{PGF} A_{Siv}^{PGF} + R_{QCDC} A_{Siv}^{QCDC} + R_{LP} A_{Siv}^{LP}. \tag{5}$$

3. The weighted method of asymmetry extraction

The method described in this section has been already applied to longitudinal data for $\Delta g/g$ extraction and featured in Ref [5]. For both, deuteron and proton, targets four target configurations can be introduced; in the case of two-cell target (deuteron): 1 - upstream, 2 - downstream, 3 - upstream′, 4 - downstream′, in the case of three-cell target (proton): 1 - (upstream+downstream), 2 - centre, 3 - (upstream′+downstream′), 4 - centre′. Here upstream′, centre′ and downstream′ denote the cells after the polarisation reversal and configuration 1 has the polarisation pointing upwards in the laboratory frame. Using Eq. 5 and introducing $\beta_t^j(\phi_{Siv}) = R_j f P_T \sin\phi_{Siv}$ one can rewrite Eq. 4:

$$\begin{aligned} N_t(\vec{x}, \phi_{Siv}) = \alpha_t(\vec{x}, \phi_{Siv}) \Big(1 &+ \beta_t^{PGF}(\phi_{Siv}) A_{Siv}^{PGF}(\vec{x}) \\ &+ \beta_t^{QCDC}(\phi_{Siv}) A_{Siv}^{QCDC}(\vec{x}) + \beta_t^{LP}(\phi_{Siv}) A_{Siv}^{LP}(\vec{x}) \Big), \end{aligned} \tag{6}$$

where the target configuration $t = 1, 2, 3, 4$. For each process a statistical weight is introduced chosen to be $\omega^j \equiv \beta^j / P_T$ which optimises the statistical and systematic error.

$$\begin{aligned} p_t^j \equiv \sum_{i=1}^{N_t} \omega_i^j = \tilde{\alpha}_t^j \Big(1 &+ \{\beta_t^G\}_{\omega^j} A_{PGF}^{\phi_{Sivers}}(\langle x_g \rangle) \\ &+ \{\beta_t^{QCDC}\}_{\omega^j} A_{QCDC}^{\phi_{Siv}}(\langle x_C \rangle) + \{\beta_t^{LP}\}_{\omega^j} A_{LP}^{\sin(\phi_{2h} - \phi_s)}(\langle x_{Bj} \rangle) \Big) \end{aligned} \tag{7}$$

where

$$\tilde{\alpha}_t = \int d\vec{x} d\phi_{Siv} \omega(\phi_{Siv}) \alpha_t(\vec{x}) \tag{8}$$

is the integrated acceptance and

$$\{\beta\}_\omega = \frac{\int \beta(\phi_{Siv}) \omega(\phi_{Siv}) \alpha_t(\vec{x}) d\vec{x}}{\int \omega(\phi_{Siv}) \alpha_t(\vec{x}) d\vec{x}} \approx \frac{\sum_i \beta_i \omega_i}{\sum_i \omega_i}. \tag{9}$$

The last approximation assumes that the raw asymmetries are small $\beta A_{Siv} \ll 1$.
In order to avoid approaching to zero for the integrated acceptance defined in Eq. 8 and

Journal of Physics: Conference Series **678** (2016) 012055 doi:10.1088/1742-6596/678/1/012055

present in the denominator in Eq. 9 a binning in ϕ_{Siv} is introduced. Two bins $(\phi^1_{Siv}, \phi^2_{Siv}) = ([0,\pi],[\pi,2\pi])$ have been applied.
In addition it was assumed that A_{Siv} is a linear function of x what allows to write:

$$\left\{A_{Siv}(x)\right\}_{\beta\omega} = A_{Siv}(\langle x\rangle), \tag{10}$$

Here x denotes the momentum fraction carried by parton in given sub-process and it is also assumed that $\{x_j\}_{\omega^k\beta^j_t} \approx \{x_j\}_{\omega^m\beta^j_b} \equiv \langle x_j\rangle \quad j,k,m = PGF, LP, QCDC$ and t,b run over the target configurations $1,2,3,4$, which is true for both data samples.

The given set of 12 equations can be them reduced to three by writing:

$$r^j := \frac{p^j_1 p^j_4}{p^j_2 p^j_3}. \tag{11}$$

Using the assumption that the integrated acceptances ratio is the same before and after the polarisation reversal, $(\tilde{\alpha}^j_1\tilde{\alpha}^j_4)/(\tilde{\alpha}^j_2\tilde{\alpha}^j_3) = 1$ a solvable set of three equations for three unknowns (asymmetries) is obtained. Applying the afore mentioned binning in ϕ_{Siv} and setting a constraint for each asymmetry $A^j(\phi^1_{Siv}) = A^j(\phi^2_{Siv})$ doubles the set of equation which can then be solved by fitting procedure of χ^2 minimisation:

$$\chi^2 = (\vec{R}-\vec{L})^T[prop(12,3)^T Cov(12,12) prop(12,3)]^{-1}(\vec{R}-\vec{L}). \tag{12}$$

Vectors $\vec{R}$ and $\vec{L}$ are defined by right hand side and left hand side of Eqs. 11. The former contains the asymmetries - parameters of the fit, while the latter is given by the measured values of ω^j_i. The $Cov(12,12)$ matrix elements refer to correlations between pairs of p^j_t defined in Eq. 7 and can be approximated by $Cov(p_x,p_y) \approx \sum_{N_t}\omega_x\omega_y$. The propagation matrix $prop(12,3)$ is given by $prop(m,n) = \partial r_n/\partial p_m$.

4. Monte Carlo optimisation and Neural Network training

This section is based on the previous analysis featured in Ref. [6] and Ref. [5]. In this analysis the NN package [10] has been used. The NN has been trained with a Monte Carlo sample with process identification. As an input vector a set of 6 kinematic variables have been chosen, $x_{Bj}, Q^2, p_{T1}, p_{T2}, p_{L1}, p_{L2}$. The latter are the longitudinal part of the hadron momenta. Good agreement of these 6 variables distributions between MC and data are required.

The comparison of the distribution of 6 variables used in NN training is shown in Fig. 2 for the deuteron target case. In case of proton 2010 data the comparison looks similar.
As it was said above the 6 kinematic variables has been used as an input vector for the NN training. The NN output has been parametrised by two values depending on the MC type sub-process. The artificial NN was not able to separate the three processes but for each of the process three probabilities P_{NN} for PGF, LP and QCDC have been assigned. To validate the NN training a MC sample, statistically independent of the MC sample applied for the training, was used. In each bin of P_{NN}, assigned to every MC event by trained NN, true fraction based on process ID was calculated. The results of the comparison for NN trained with MC 2010 are presented in Fig. 3. The results for MC 2004 data are very similar and are omitted here for brevity. In conclusion, the agreement between P_{NN} and P_{MC} for all three processes is very good. As the agreement between MC and real data is also good it is assumed that the fractions in Eq. 5 can be assigned by the trained NN run on real data. In average $R^j = P^j_{NN}$.

5. Systematic uncertainties

The main source of systematic uncertainties in this analysis is the Monte Carlo dependence. To estimate this error different MC settings have been used in the process of Neural

Journal of Physics: Conference Series **678** (2016) 012055 doi:10.1088/1742-6596/678/1/012055

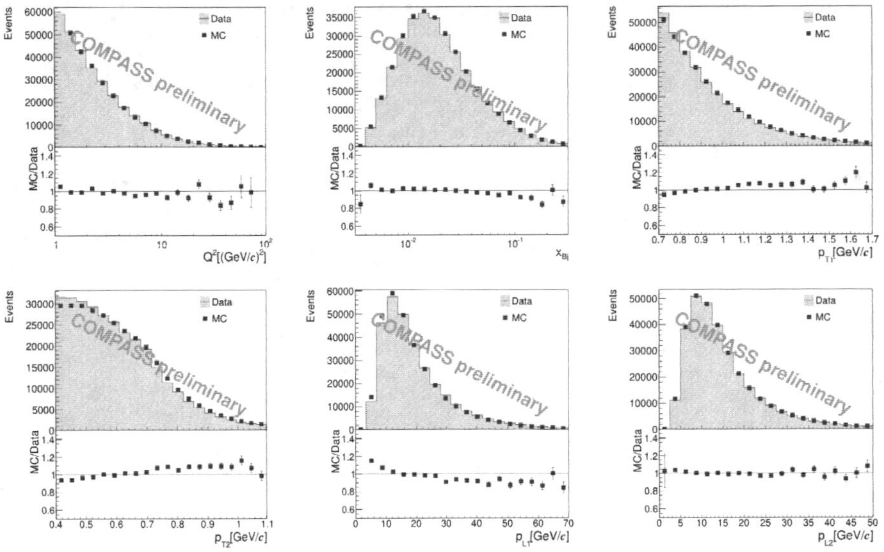

Figure 2: Kinematic variable distributions in MC and Data for high-p_T samples normalised to the same number of events. Deuteron Data and MC 2004. Analogous plots have been obtained for Proton 2010 data

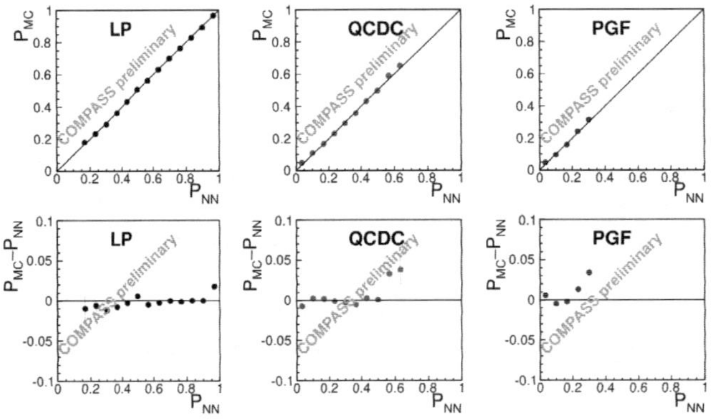

Figure 3: Neural network validation. Here P_{NN} is the fraction of the process given by the NN and P_{MC} is the true fraction of each process from MC in a given P_{NN} bin.

Network training. Seven additional MC samples were prepared with different combinations of fragmentation parameters, tuning (default LEPTO and COMPASS high-p_T tuning), Parton Shower on and off, different choices of the PDFs (MSTW08 or CTEQ5L, Ref.[11]) and F_L from LEPTO and taken from the $R = \sigma_L/\sigma_T$, parametrisation of Ref. [12]. The choosen final Monte Carlo has the best agreement with the data presented in Fig. 2. All additional MCs use GEISHA instead of FLUKA secondary particle generator. For more details the reader is refered to Ref. [5]. Other estimated systematic uncertainties are listed in Table 1.

6. Results

The method presented in Sect. 3 with the use of trained NNs has been applied to the two data sets for deuteron and proton targets. The results are presented in Fig. 4. The gluon

	deuteron		proton	
source	value	% $\sigma_{stat}(=0.15)$	value	% $\sigma_{stat}(=0.085)$
Monte Carlo	0.060	40%	0.054	64%
False asymmetries	0.016	11%	0.032	38%
cut on hadron charges $q_1 \cdot q_2 = -1$	0.05	33%	0.038	45%
radiative corrections	0.018	12%	0.018	21%
large Q^2	-	-	0.014	16%
x_{Bj} binning	0.07	47%	0.011	13%
all asyms vs only Sivers	0.003	2%	0.005	6%
ML vs Weighted	0.008	6%	0.004	5%
target polarisation	0.075	5%	0.0043	5%
dilution factor	0.003	2%	0.0017	2%
total $\sqrt{\sum \sigma_i^2}$	0.13	88%	0.078	92%

Table 1: Systematic studies

Sivers two-hadron asymmetry for deuteron target, $A_g^d = -0.14 \pm 0.15(stat.) \pm 0.06(syst.)$ at $\langle x_g \rangle = 0.13$, is not in contradiction with the zero value while for proton target, $A_g^p = -0.26 \pm 0.09(stat.) \pm 0.08(syst.)$ at $\langle x_g \rangle = 0.15$, it is 3σ below zero. However, the obtained gluon Sivers asymmetries show compatibility between the two data samples which is consistent with the naive expectation that gluons are flavour-independent. The other two extracted asymmetries, for QCDC and LP, differ between the deuteron (isoscalar) target and the proton target.

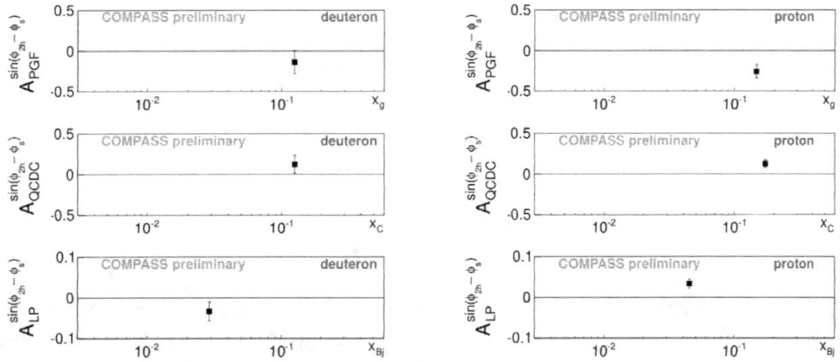

Figure 4: Sivers two-hadron asymmetry of the PGF $A_{PGF}^{sin(\phi_{2h}-\phi_S)}$, QCDC $A_{QCDC}^{sin(\phi_{2h}-\phi_S)}$ and LP $A_{LP}^{sin(\phi_{2h}-\phi_S)}$ for deuteron and proton targets.

References

[1] Alekseev M *et al.* (COMPASS Collaboration) 2009 *Phys.Lett.* **B673** 127–135 (*Preprint* 0802.2160)
[2] Adolph C, Akhunzyanov R, Alexeev M, Alexeev G, Amoroso A *et al.* 2014 (*Preprint* 1408.4405)
[3] Ingelman G, Edin A and Rathsman J 1997 *Comput.Phys.Commun.* **101** 108–134 (*Preprint* hep-ph/9605286)
[4] Adolph C *et al.* (COMPASS Collaboration) 2013 *Phys.Rev.* **D87** 052018 (*Preprint* 1211.6849)
[5] Stolarski M (COMPASS) 2014 *PoS* **DIS2014** 211
[6] Adolph C *et al.* (COMPASS Collaboration) 2013 *Phys.Lett.* **B718** 922–930 (*Preprint* 1202.4064)
[7] Gliske S, Bacchetta A and Radici M 2014 *Phys.Rev.* **D90** 114027 (*Preprint* 1408.5721)
[8] Kotzinian A, Matevosyan H H and Thomas A W 2014 *Phys.Rev.Lett.* **113** 062003 (*Preprint* 1403.5562)
[9] Kotzinian A 2015 *Phys.Rev.* **D91** 054001 (*Preprint* 1408.6674)
[10] Sulej R Netmaker http://www.ire.pw.edu.pl/∼rsulej/NetMaker/
[11] Lai H *et al.* (CTEQ Collaboration) 2000 *Eur.Phys.J.* **C12** 375–392 (*Preprint* hep-ph/9903282)
[12] Abe K *et al.* (E143 Collaboration) 1999 *Phys.Lett.* **B452** 194–200 (*Preprint* hep-ex/9808028)

Journal of Physics: Conference Series **678** (2016) 012056 doi:10.1088/1742-6596/678/1/012056

Novel effects in twist-three SSA

I. V. Anikin

E-mail: anikin@theor.jinr.ru

O. V. Teryaev

E-mail: teryaev@theor.jinr.ru
Bogoliubov Laboratory of Theoretical Physics, JINR, 141980 Dubna, Russia

Abstract. We discuss the role of gluon poles and the gauge invariance for the hadron tensors of Drell-Yan and direct photon production processes with the transversely polarized hadron. These hadron tensors are needed to construct the corresponding single spin asymmetries. For the Drell-Yan process, we perform our analysis within both the Feynman and axial-type (contour) gauges for gluons. In both the Feynman and contour gauges, we demonstrate that the gauge invariance leads to the need of the new (non-standard) diagrams. Moreover, in the Feynman gauge, we argue the absence of gluon poles in the correlators $\langle\bar{\psi}\gamma_\perp A^+\psi\rangle$ related traditionally to $dT(x,x)/dx$. As a result, these terms disappear from the final QED gauge invariant Drell-Yan hadron tensor. For the direct photon production, by using the contour gauge for gluon fields, we find that there are new twist-3 terms present in the hadron tensor of the considering process in addition to the standard twist-3 terms.

1. Introduction

We discuss the direct photon production in two hadron collision where one hadron is transversely polarized. We present the hadron tensor for this process and study the effects which lead to the soft breaking of factorization (or the universality breaking) through the QED and QCD gauge invariance [1]. In a similar manner as in [2], the special role is played by the contour gauge for gluon fields.

Moreover, we present the results of the detailed analysis of hadron tensor in the Feynman gauge with the particular emphasis on the QED gauge invariance [3]. Moreover, the results in the Feynman and contour gauges coincide if the gluon poles in the correlators $\langle\bar{\psi}\gamma_\perp A^+\psi\rangle$ are absent. This is in agreement with the relation between gluon poles and the Sivers function which corresponds to the "leading twist" Dirac matrix γ^+. This also suggests that the gluonic poles appear only in "physical" components of gluon field, coinciding with total ones in the axial gauge.

2. Direct Photon Production

We study the semi-inclusive process where the hadron with the transverse polarization collides with the other unpolarized hadron to produce the direct photon in the final state in:

$$N^{(\uparrow\downarrow)}(p_1) + N(p_2) \to \gamma(q) + q(k) + X(P_X)\,. \tag{1}$$

Content from this work may be used under the terms of the Creative Commons Attribution 3.0 licence. Any further distribution of this work must maintain attribution to the author(s) and the title of the work, journal citation and DOI.
Published under licence by IOP Publishing Ltd

Journal of Physics: Conference Series **678** (2016) 012056 doi:10.1088/1742-6596/678/1/012056

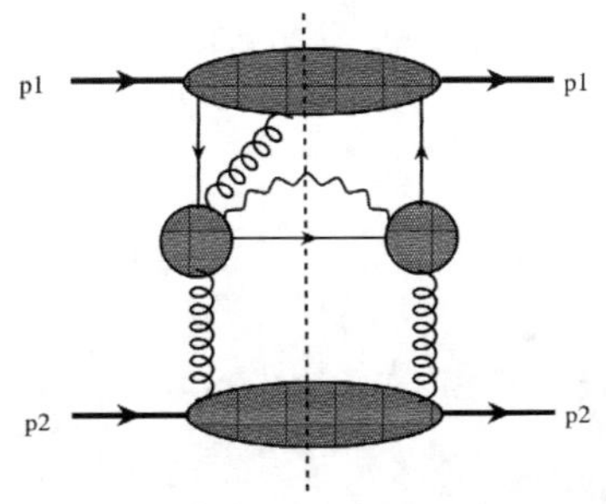

Figure 1. The Feynman diagram describing the hadron tensor of the direct photon production.

For (1) (also for the Drell-Yan process), the gluonic poles manifest. We perform our calculations within a *collinear* factorization and, therefore, it is convenient to fix the dominant light-cone directions as

$$p_1 = \sqrt{\frac{S}{2}}\,n^*\,, \quad p_2 = \sqrt{\frac{S}{2}}\,n\,,$$
$$n^*_\mu = (1/\sqrt{2},\, \mathbf{0}_T,\, 1/\sqrt{2}), \quad n_\mu = (1/\sqrt{2},\, \mathbf{0}_T,\, -1/\sqrt{2})\,. \tag{2}$$

Accordingly, the quark and gluon momenta k_1 and ℓ lie along the plus dominant direction while the gluon momentum k_2 – along the minus direction. The final on-shell photon and quark(anti-quark) momenta can be presented as

$$q = y_B\sqrt{\frac{S}{2}}\,n - \frac{q_\perp^2}{y_B\sqrt{2S}}\,n^* + q_\perp\,, \quad k = x_B\sqrt{\frac{S}{2}}\,n^* - \frac{k_\perp^2}{x_B\sqrt{2S}}\,n + k_\perp\,. \tag{3}$$

And, the Mandelstam variables for the process and subprocess are defined as

$$S = (p_1+p_2)^2, \quad T = (p_1-q)^2, \quad U = (q-p_2)^2,$$
$$\hat{s} = (x_1p_1 + yp_2)^2 = x_1yS, \quad \hat{t} = (x_1p_1 - q)^2 = x_1T, \quad \hat{u} = (q - yp_2)^2 = yU. \tag{4}$$

The amplitude (or the hadron tensor) of (1) constructed by the contributions from (i) the leading (LO) in particle number diagrams: two diagrams with a radiation of the photon before ($\mathcal{A}_1^{\rm LO}$) and after ($\mathcal{A}_2^{\rm LO}$) the quark-gluon vertex with the gluon going to the lower blob, see the right side of Fig. 1; (ii) the next-to-leading order (NLO) diagrams: eight diagrams constructed from the LO diagrams by insertion of all possible radiations of the additional gluon which together with the quark goes to the upper blob, see the left side of Fig. 1. So, we have the hadron tensor related to the corresponding asymmetry:

$$d\sigma^\uparrow - d\sigma^\downarrow \sim \mathcal{W} = \sum_{i=1}^{2}\sum_{j=1}^{8} \mathcal{A}_i^{\rm LO} * \mathcal{B}_j^{\rm NLO}\,. \tag{5}$$

Here, we will mainly discuss the hadron tensor rather than the asymmetry itself. So, the hadron tensor as an interference between the LO and NLO diagrams, $\mathcal{A}_i^{\rm LO} * \mathcal{B}_j^{\rm NLO}$, can be presented by Fig. 1 where the upper blob determines the matrix element of the twist-3 quark-gluon operator while the lower blob – the matrix element of the twist-2 gluon operator related to the unpolarized gluon distribution.

As a result of factorization, we should reach the factorized form for the considered subject:

$$\text{Hadron tensor} = \{\text{Hard part (pQCD)}\} \otimes \{\text{Soft part (npQCD)}\}\,. \tag{6}$$

Journal of Physics: Conference Series **678** (2016) 012056 doi:10.1088/1742-6596/678/1/012056

We now dwell on the QCD gauge invariance of the hadron tensor for the direct photon production (DPP). First of all, let us remind that having used the *contour gauge conception* [1, 4], one can check that the representation

$$B_+^V(x_1, x_2) = \frac{T(x_1, x_2)}{x_1 - x_2 + i\epsilon} \tag{7}$$

belongs to the gauge defined by $[x, -\infty] = 1$, while the representation

$$B_-^V(x_1, x_2) = \frac{T(x_1, x_2)}{x_1 - x_2 - i\epsilon} \tag{8}$$

corresponds to the gauge that defined by $[+\infty, x] = 1$. In both (7) and (8), the function $T(x_1, x_2)$ related to the following prametrization:

$$\begin{aligned} &\langle p_1, S^T|\bar{\psi}(\lambda_1\tilde{n})\,\gamma^+\,\tilde{n}_\nu G_T^{\nu\alpha}(\lambda_2\tilde{n})\,\psi(0)|S^T, p_1\rangle = \\ &\varepsilon^{\alpha + S^T -}\,(p_1 p_2)\int dx_1 dx_2\, e^{ix_1\lambda_1 + i(x_2 - x_1)\lambda_2}\,T(x_1, x_2)\,. \end{aligned} \tag{9}$$

Further, to check the gauge invariance, we have to consider all typical diagrams, depicted in Fig.1, that correspond to the certain ξ-process (see, [5]). Notice that, for the QCD gauge invariance, we have to assume that all charged particles are on its mass-shells. That is, we will deal with only the physical gluons.

As a result of our calculation, the QCD gauge invariance (or the Ward identity) takes place provided only the presence of the different complex prescriptions in gluonic poles dictated by the final or initial state interactions:

$$\left.\begin{array}{l} \textbf{FSI} \;\Rightarrow\; \frac{1}{\ell^+ + i\epsilon} \Rightarrow \text{gauge } [+\infty^-,\, z^-] = 1 \Rightarrow \frac{T(x_1,x_2)}{x_1 - x_2 - i\epsilon} \\ \\ \textbf{ISI} \;\Rightarrow\; \frac{1}{-\ell^+ + i\epsilon} \Rightarrow \text{gauge } [z^-,\, -\infty^-] = 1 \Rightarrow \frac{T(x_1,x_2)}{x_1 - x_2 + i\epsilon} \end{array}\right\} \Rightarrow \textbf{QCD Gauge Invariance}\,, \tag{10}$$

where **FSI** and **ISI** imply the final and initial state interactions, respectively.

We now calculate the full expression for the hadron tensor which involves both the standard and new contributions to the gluon pole terms. After computing the corresponding traces and performing simple algebra within the frame we are choosing, it turns out that the only nonzero contributions to the hadron tensor come from the following terms

$$\begin{aligned} d\mathcal{W}(\mathrm{H1}) \;=\;& \frac{d^3\vec{q}}{(2\pi)^3 2E}\int\frac{d^3\vec{k}}{(2\pi)^3 2\varepsilon}\delta^{(2)}(\vec{\mathbf{k}}_\perp + \vec{\mathbf{q}}_\perp)\,\mathrm{C}_2\int dx_1 dy\delta(x_1 - x_B)\,\delta(y - y_B)\,\mathcal{F}^g(y)\,\times \\ & \int dx_2\,\frac{2S^2\,x_1\,y^2}{[x_2 yS + i\epsilon][x_1 yS + i\epsilon]^2}\,\frac{\varepsilon^{q_\perp + S_\perp -}}{p_1^+}\,B_-^V(x_1, x_2)\,, \end{aligned} \tag{11}$$

$$\begin{aligned} d\mathcal{W}(\mathrm{H7}) \;=\;& \frac{d^3\vec{q}}{(2\pi)^3 2E}\int\frac{d^3\vec{k}}{(2\pi)^3 2\varepsilon}\delta^{(2)}(\vec{\mathbf{k}}_\perp + \vec{\mathbf{q}}_\perp)\,\mathrm{C}_1\int dx_1 dy\delta(x_1 - x_B)\,\delta(y - y_B)\,\mathcal{F}^g(y)\,\times \\ & \int dx_2\,\frac{(-2)S\,T\,x_1\,(y - 3y_B)}{[x_2 T + i\epsilon][x_1 T + i\epsilon]^2}\,\frac{\varepsilon^{q_\perp + S_\perp -}}{p_1^+}\,B_+^V(x_1, x_2)\,, \end{aligned} \tag{12}$$

$$\begin{aligned} d\mathcal{W}(\mathrm{D4}) \;=\;& \frac{d^3\vec{q}}{(2\pi)^3 2E}\int\frac{d^3\vec{k}}{(2\pi)^3 2\varepsilon}\delta^{(1)}(\vec{\mathbf{k}}_\perp + \vec{\mathbf{q}}_\perp)\,\mathrm{C}_1\int dx_1 dy\delta(x_1 - x_B)\,\delta(y - y_B)\,\frac{2}{S}\mathcal{F}^g(y)\,\times \\ & \frac{2S^2\,x_1\,(y - 2y_B)}{[x_1 T + i\epsilon]^2}\,\frac{\varepsilon^{q_\perp + S_\perp -}}{2x_1 p_1^+ + i\epsilon}\int dx_2\,B_+^V(x_1, x_2)\,, \end{aligned} \tag{13}$$

and

$$d\mathcal{W}(\text{H10}) = \frac{d^3\vec{q}}{(2\pi)^3 2E}\int\frac{d^3\vec{k}}{(2\pi)^3 2\varepsilon}\delta^{(2)}(\vec{\mathbf{k}}_\perp+\vec{\mathbf{q}}_\perp)\,\mathrm{C}_3\int dx_1 dy\delta(x_1-x_B)\,\delta(y-y_B)\,\mathcal{F}^g(y)\,\times$$
$$\int dx_2\,\frac{2T(x_1-x_2)(2T+Sy)}{[x_1T+i\epsilon][x_2T+i\epsilon][(x_1-x_2)yS+i\epsilon]}\frac{\varepsilon^{q_\perp+S_\perp-}}{p_1^+}\,B_+^V(x_1,x_2)\,. \qquad (14)$$

Here, $\mathrm{C}_1 = C_F^2 N_c$, $\mathrm{C}_2 = -C_F/2$, $\mathrm{C}_3 = C_F\,N_c C_A/2$. The other diagram contributions disappear owing to the following reasons: (i) the γ-algebra gives $(\gamma^-)^2 = 0$; (ii) the common pre-factor $T + yS$ goes to zero, (iii) cancelation between different diagrams.

Analysing the results for the terms H1, H7, D4 and H10 (see Eqns. (11)–(14)), we can see that

$$d\mathcal{W}(\text{H1}) + d\mathcal{W}(\text{H7}) + d\mathcal{W}(\text{D4}) = d\mathcal{W}(\text{H10})\,. \qquad (15)$$

In other words, as similar to the Drell-Yan process, the new ("non-standard") contributions generated by the terms H1, H7 and D4 result again in the factor of 2 compared to the "standard" term H10 contribution to the corresponding hadron tensor.

3. Drell-Yan process

We now discuss the hadron tensor which contributes to the single spin (left-right) asymmetry measured in the Drell-Yan process with the transversely polarized nucleon (see Fig. 2):

$$N^{(\uparrow\downarrow)}(p_1) + N(p_2) \to \gamma^*(q) + X(P_X) \to \ell(l_1) + \bar{\ell}(l_2) + X(P_X). \qquad (16)$$

Here, the virtual photon producing the lepton pair ($l_1 + l_2 = q$) has a large mass squared ($q^2 = Q^2$) while the transverse momenta are small and integrated out. The left-right asymmetry means that the transverse momenta of the leptons are correlated with the direction $\mathbf{S}\times\mathbf{e}_z$ where S_μ implies the transverse polarization vector of the nucleon while $\mathbf{e}_z$ is a beam direction [6].

Since we perform our calculations within a *collinear* factorization, it is convenient to fix the dominant light-cone directions as

$$p_1 \approx \frac{Q}{x_B\sqrt{2}}\,n^*, \quad p_2 \approx \frac{Q}{y_B\sqrt{2}}\,n, \qquad (17)$$
$$n^*_\mu = (1/\sqrt{2},\,\mathbf{0}_T,\,1/\sqrt{2}), \quad n_\mu = (1/\sqrt{2},\,\mathbf{0}_T,\,-1/\sqrt{2}).$$

So, the hadron momenta p_1 and p_2 have the plus and minus dominant light-cone components, respectively. Accordingly, the quark and gluon momenta k_1 and ℓ lie along the plus direction while the antiquark momentum k_2 – along the minus direction. The photon momentum reads (see Fig. 2)

$$q = l_1 + l_2 = k_1 + k_2 = x_1 p_1 + y p_2 + q_T. \qquad (18)$$

We work within the Feynman gauge for gluons. After calculation of all relevant traces in the factorized hadron tensor and after some algebra, we arrive at the following contributions for the unintegrated hadron tensor (which involves all relevant contributions except the mirror ones): the standard diagram depicted in Fig. 2(a) gives us

$$\overline{\mathcal{W}}^{(\text{Stand.})}_{\mu\nu} + \overline{\mathcal{W}}^{(\text{Stand.},\,\partial_\perp)}_{\mu\nu} = \bar{q}(y)\Bigg\{-\frac{p_{1\,\mu}}{y}\,\varepsilon_{\nu S^T - p_2}\int dx_2\frac{x_1-x_2}{x_1-x_2+i\epsilon}B^{(1)}(x_1,x_2)$$
$$-\Big[\frac{p_{2\,\nu}}{x_1}\varepsilon_{\mu S^T-p_2} + \frac{p_{2\,\mu}}{x_1}\varepsilon_{\nu S^T-p_2}\Big]x_1\int dx_2\frac{B^{(2)}(x_1,x_2)}{x_1-x_2+i\epsilon} + \frac{p_{1\,\mu}}{y}\,\varepsilon_{\nu S^T-p_2}\int dx_2\frac{B^{(\perp)}(x_1,x_2)}{x_1-x_2+i\epsilon}\Bigg\},$$

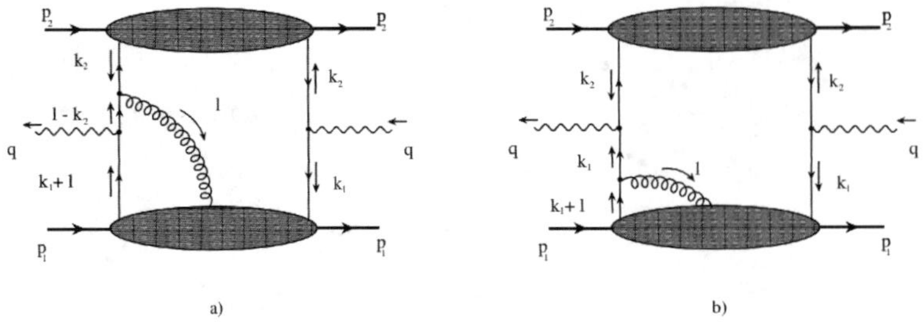

Figure 2. The Feynman diagrams which contribute to the polarized Drell-Yan hadron tensor.

while the non-standard diagram presented in Fig. 2(b) contributes as

$$\overline{\mathcal{W}}^{\text{(Non-stand.)}}_{\mu\nu} = \bar{q}(y)\frac{p_{2\,\mu}}{x_1}\varepsilon_{\nu S^T - p_2}\int dx_2 \Big\{ B^{(1)}(x_1,x_2) + B^{(2)}(x_1,x_2) \Big\}. \tag{19}$$

In Eqns. (19) and (19), all indices have to be treated as contravariant ones independently of the real position in the formulae. We also introduce the shorthand notation: $\varepsilon_{ABCD} = \varepsilon_{\mu_1\mu_2\mu_3\mu_4}A_{\mu_1}B_{\mu_2}C_{\mu_3}D_{\mu_4}$. Moreover, the parametrizing functions are associated with the following correlators:

$$B^{(1)}(x_1,x_2) = \frac{T(x_1,x_2)}{x_1 - x_2 + i\varepsilon} \Longleftarrow \mathcal{F}_2\Big[\langle p_1, S^T|\bar{\psi}(\eta_1)\,\gamma^+\, A^T(z)\,\psi(0)|S^T, p_1\rangle\Big], \tag{20}$$

$$B^{(2)}(x_1,x_2) \Longleftarrow \mathcal{F}_2\Big[\langle p_1, S^T|\bar{\psi}(\eta_1)\,\gamma^\perp\, A^+(z)\,\psi(0)|S^T, p_1\rangle\Big], \tag{21}$$

$$B^{(\perp)}(x_1,x_2) \Longleftarrow \mathcal{F}_2\Big[\langle p_1, S^T|\bar{\psi}(\eta_1)\,\gamma^+\,(\partial^\perp A^+(z))\,\psi(0)|S^T, p_1\rangle\Big], \tag{22}$$

where $\mathcal{F}_2$ denotes the corresponding Fourier transformation. Summing up all contributions from the standard and non-standard diagrams, we finally obtain the expression for the unitegrated hadron tensor. We have

$$\begin{aligned}\overline{\mathcal{W}}_{\mu\nu} &= \overline{\mathcal{W}}^{\text{(Stand.)}}_{\mu\nu} + \overline{\mathcal{W}}^{\text{(Stand.},\,\partial_\perp)}_{\mu\nu} + \overline{\mathcal{W}}^{\text{(Non-stand.)}}_{\mu\nu} = \bar{q}(y)\left\{\left[\frac{p_{2\,\mu}}{x_1} - \frac{p_{1\,\mu}}{y}\right]\varepsilon_{\nu S^T - p_2}\int dx_2 B^{(1)}(x_1,x_2) + \right. \\ &\frac{p_{2\,\mu}}{x_1}\,\varepsilon_{\nu S^T - p_2}\int dx_2 B^{(2)}(x_1,x_2) - \left[\frac{p_{2\,\nu}}{x_1}\varepsilon_{\mu S^T - p_2} + \frac{p_{2\,\mu}}{x_1}\varepsilon_{\nu S^T - p_2}\right] x_1 \int dx_2 \frac{B^{(2)}(x_1,x_2)}{x_1 - x_2 + i\epsilon} + \\ &\left.\frac{p_{1\,\mu}}{y}\,\varepsilon_{\nu S^T - p_2}\int dx_2 \frac{B^{(\perp)}(x_1,x_2)}{x_1 - x_2 + i\epsilon}\right\},\end{aligned} \tag{23}$$

Notice that the first term in Eqn. (23) coincides with the hadron tensor calculated within the light-cone gauge $A^+ = 0$.

Let us now discuss the QED gauge invariance of the hadron tensor. From Eqn. (23), we can see that the QED gauge invariant combination is

$$\mathcal{T}_{\mu\nu} = \Big[p_{2\,\mu}/x_1 - p_{1\,\mu}/y\Big]\,\varepsilon_{\nu S^T - p_2}, \qquad q_\mu \mathcal{T}_{\mu\nu} = q_\nu \mathcal{T}_{\mu\nu} = 0. \tag{24}$$

At the same time, there is a single term with $p_{2\,\nu}$ which does not have a counterpart to construct the gauge-invariant combination $p_{2\,\mu}/x_1 - p_{1\,\mu}/y$. Therefore, the second term in Eqn. (19) should be equal to zero. This also leads to nullification of the second term in Eqn. (19). Now, to get the QED gauge invariant combination (see (24)) one has no the other way rather than to combine the first terms in Eqns. (19)

Journal of Physics: Conference Series **678** (2016) 012056 doi:10.1088/1742-6596/678/1/012056

and (19) which justifies the treatment of gluon pole in $B^{(1)}(x_1, x_2)$ using the complex prescription. In addition, we have to conclude that the only remaining third term in (19) should not contribute to SSA.

Moreover, we can conclude that, in the case with the substantial transverse component of the momentum, there are no any sources for the gluon pole at $x_1 = x_2$. As a result, the function $B^{(2)}(x_1, x_2)$ has no gluon poles and, due to T-invariance, $B^{(2)}(x_1, x_2) = -B^{(2)}(x_2, x_1)$, obeys $B^{(2)}(x, x) = 0$. On the other hand, if we have γ^+ in the correlator, see (20), the transverse components of gluon momentum are not substantial and can be neglected. That ensures the existence of the gluon poles for the function $B^{(1)}(x_1, x_2)$. This corresponds to the fact that the Sivers function, being related to gluon poles, contains the "leading twist" projector γ^+. So, we may conclude that in the Feynman gauge the structure $\gamma^+(\partial^\perp A^+)$ does not produce the imaginary part and SSA as well.

Working within the Feynman gauge, we derive the QED gauge invariant (unitegrated) hadron tensor for the polarized DY process:

$$\overline{\mathcal{W}}^{\rm GI}_{\mu\nu} = \bar{q}(y)\Big[p_{2\,\mu}/x_1 - p_{1\,\mu}/y\Big]\varepsilon_{\nu S^T - p_2}\int dx_2 B^{(1)}(x_1, x_2). \tag{25}$$

Further, after calculation the imaginary part (addition of the mirror contributions), and after integration over x_1 and y, the QED gauge invariant hadron tensor takes the form

$$W^{\rm GI}_{\mu\nu} = \bar{q}(y_B)\Big[p_{2\,\mu}/x_B - p_{1\,\mu}/y_B\Big]\varepsilon_{\nu S^T - p_2} T(x_B, x_B)\,. \tag{26}$$

This expression fully coincides with the hadron tensor which has been derived within the light-cone gauge for gluons. Moreover, the factor of 2 in the hadron tensor, we found within the axial-type gauge [2], is still present in the frame of the Feynman gauge.

4. Conclusions and discussion

the different prescriptions in the gluonic poles. Moreover, the different prescriptions are needed to ensure the QCD gauge invariance. This situation can be treated as a soft breaking of the universality condition resulting in factorization breaking. We find that the "non-standard" new terms, which exist in the case of the complex twist-3 B^V-function with the corresponding prescriptions, do contribute to the hadron tensor exactly as the "standard" term known previously. This is another important result of our work. We also observe that this is exactly similar to the case of Drell-Yan process studied in [2].

We argue the absence of gluon poles in the correlators $\langle\bar{\psi}\gamma_\perp A^+\psi\rangle$ based on the light-cone dynamics. At the same time, we also show that the Lorentz and QED gauge invariance of the hadron tensor calculated within the Feynman gauge requires the function $B^{(2)}(x_1, x_2)$ to be without any gluon poles. This property seems to be natural from the point of view of gluon poles relation [7] to Sivers functions as the latter is related to the projection γ^+. Note also that this may be understood as appearance of gluonic poles due to the physical components [8] of gluonic fields. Another support of this picture is provided by the fact, that imaginary phase in (10) allows to derive [9] Burkardt sum rules from conservation laws.

[1] I. V. Anikin and O. V. Teryaev, Eur. Phys. J. C **75** (2015) 5, 184 [arXiv:1501.04536 [hep-ph]].

[2] I. V. Anikin and O. V. Teryaev, Phys. Lett. B **690** (2010) 519 [arXiv:1003.1482 [hep-ph]]; I. V. Anikin and O. V. Teryaev, J. Phys. Conf. Ser. **295** (2011) 012057 [arXiv:1011.6203 [hep-ph]].

[3] I. V. Anikin and O. V. Teryaev, arXiv:1508.00227 [hep-ph] (in press in PLB).

[4] S. V. Ivanov, G. P. Korchemsky and A. V. Radyushkin, Yad. Fiz. **44**, 230 (1986) [Sov. J. Nucl. Phys. **44**, 145 (1986)]; S. V. Ivanov and G. P. Korchemsky, Phys. Lett. B **154**, 197 (1985); S. V. Ivanov, Fiz. Elem. Chast. Atom. Yadra **21**, 75 (1990).

[5] N. N. Bogolyubov and D. V. Shirkov, "Introduction To The Theory Of Quantized Fields," Intersci. Monogr. Phys. Astron. **3**, 1 (1959).

[6] V. Barone, A. Drago and P. G. Ratcliffe, Phys. Rept. **359**, 1 (2002) [arXiv:hep-ph/0104283].

[7] D. Boer, P. J. Mulders and F. Pijlman, Nucl. Phys. B **667** (2003) 201 [hep-ph/0303034]. X. Ji, J. W. Qiu, W. Vogelsang and F. Yuan, Phys. Rev. Lett. **97** (2006) 082002 [hep-ph/0602239]. P. G. Ratcliffe and O. V. Teryaev, hep-ph/0703293; Mod. Phys. Lett. A **24** (2009) 2984 [arXiv:0910.5348 [hep-ph]].

[8] Y. Hatta, Phys. Lett. B **708** (2014) 186 [arXiv:1111.3547 [hep-ph]].

[9] O. Teryaev, EPJ Web Conf. **85** (2015) 02008.

Journal of Physics: Conference Series **678** (2016) 012057 doi:10.1088/1742-6596/678/1/012057

Measurement of Collins asymmetry at BESIII

Xiao-Rui Lyu
(*on behalf of the BESIII collaboration*)
University of Chinese Academy of Sciences (UCAS), Beijing 100049, China
E-mail: xiaorui@ucas.ac.cn

Abstract. In this talk, I present the recent measurement of the Collins fragmentation function based on a set of 62 pb^{-1} e^+e^- annihilation data at $\sqrt{s} = 3.65$ GeV collected with the BESIII detector at the BEPCII storage rings. In this work, the azimuthal asymmetries of two charged pions in the inclusive process $e^+e^- \to \pi\pi X$ are explored. These asymmetries can be attributed to the Collins fragmentation function, which describes the behavior of a hadron produced from a transversely polarized quark. The corresponding four-momentum transfer of the virtual photon, Q^2, is close to the energy scale of the existing semi-inclusive DIS experimental data. We observe a nonzero asymmetry, which increases with increasing pion momentum and also indicates a larger spin-dependent Collins effect than at higher energy scale. The dependence of the asymmetry on the transverse momentum of hadrons to the reference axis is also investigated. The measured asymmetries are important inputs for the global analysis of extracting the quark transversity distribution inside the nucleon, and are valuable to explore the energy evolution of the spin-dependent fragmentation function.

1. Introduction

The quark-hadron fragmentation process is parameterized with the fragmentation function (FF), which describes the probability that a hadron carrying a fraction of parton energy is found in the hadronization debris of the fragmenting parton. Due to the incorporated non-perturbative physics of the hadronization processes, they are not calculable from the first principle in QCD, but are parameterized with the help of factorization and perturbative calculations, where free parameters can be fixed by fitting to experimental data. The Collins FF, which considers the spin-dependent effects in the fragmentation processes, was firstly discussed by Collins in Ref. [1]. The Collins FF connects transverse quark spin with a measurable azimuthal asymmetry (the so-called Collins effect) in the yields of hadronic fragments along the initial quark's momentum.

The measurement of the Collins FF provides an important test in understanding strong interaction dynamics, and thus, is of fundamental interest in QCD theory. Because it is chiral-odd, it needs to couple to another chiral-odd function to form accessible observables in experiment. The HERMES experiment [2] presented the first non-zero proof of this function in semi-inclusive deep inelastic scattering (SIDIS) process by probing a chiral-even observable of azimuthal asymmetry, where the Collins FF couples to the chiral-odd transversity [3–5]. Transversity corresponds to the tensor charge of the nucleon and is the least known leading-twist quark distribution function. Subsequently, there were more SIDIS measurements on this asymmetry at HERMES [6], COMPASS [7] and JLab [8]. Further exploration of independent measurement of the Collins FF was implemented in the e^+e^- annihilation process, where two Collins FFs couple to construct a chiral-even azimuthal asymmetry, and their measurements from

Content from this work may be used under the terms of the Creative Commons Attribution 3.0 licence. Any further distribution of this work must maintain attribution to the author(s) and the title of the work, journal citation and DOI.
Published under licence by IOP Publishing Ltd

Belle [9–11] and BARBAR [12] give consistent non-zero asymmetries. Base on the assumption of universality of the Collins FF [13], global analyses [14, 15] have been performed to extract the transversity by combing SIDIS and e^+e^- observables. However, the e^+e^- Collins asymmetries, taken from Belle and BABAR, correspond to considerably higher Q^2 ($\sim$100 GeV^2) than the typical energy scale of the existing SIDIS data (mostly 2-15 GeV^2). Thus, the energy evolution of the FF at different Q^2 is a key factor of influencing the uncertainty of the extracted transversity and its theoretical treatment is not well justified [16, 17] due to lack of low-energy e^+e^- data. The BESIII experiment [18] studies the e^+e^- annihilations at a moderate energy scale (Q^2: 4-20 GeV^2) and its measurement of the Collins asymmetry is timely. Ref. [19] predicts that the Collins asymmetry at BESIII will be larger than at B factories. With comparison with the measurement from the B factories, BESIII data will help to understand the Q^2 evolution behavior of the Collins FF. In addition, it holds a unique advantage compared to the B factory that the BESIII data can be connected more reliably to the SIDIS data, as their energy scales are very close.

2. The BESIII experiment

The analysis is based on a data sample with an integrated luminosity of 62 pb^{-1} collected with the BESIII [18] detector at the center-of-mass energy $\sqrt{s} = 3.65$ GeV. The BESIII detector is a general-purpose solenoidal detector at the BEPCII [20] e^+e^- storage rings. The detector has a geometrical acceptance of 93% of the full solid angle. The apparatus relevant to this work includes, from inside to outside, a small-cell main drift chamber (MDC), using a helium-based gas to measure momenta and specific ionizations of charged particles, a time-of-flight (TOF) system based on plastic scintillators that determines the flight times of charged particles, a CsI(Tl) electromagnetic calorimeter (EMC) detects electromagnetic showers. These components are all situated inside a superconducting solenoid magnet, that provides a 1.0 T magnetic field parallel to the beam direction. The momentum resolution for charged tracks in the MDC is 0.5% for a transverse momentum of 1 GeV/c. The energy resolution for showers in the EMC is 2.5% (5.0%) for 1 GeV photons in the barrel (end cap) region. More details on the features and capabilities of the BESIII detector can be found elsewhere [18].

3. Data analysis

Taking into account the spin of the quark, the number density for finding a spinless hadron h with transverse momentum $\mathbf{P}_h^\perp$ produced from a transversely polarized quark q with spin $\mathbf{S}_q$ can be described in terms of the unpolarized FF, D_1^q, and the Collins FF, $H_1^{\perp q}$, at the leading twist [21],

$$D_h^{q\uparrow}(z, \mathbf{P}_h^\perp) = D_1^q(z, \mathbf{P}_h^{\perp 2}) + H_1^{\perp q}(z, \mathbf{P}_h^{\perp 2}) \frac{(\hat{\mathbf{k}} \times \mathbf{P}_h^\perp) \cdot \mathbf{S}_q}{z M_h},$$

where $\hat{\mathbf{k}}$ denotes the motion direction of the initial quark q, $z = 2E_h/Q$ denotes the fractional energy of the hadron relative to half of $Q = \sqrt{s}$, and M_h is the hadron mass. The second term contains the Collins function and depends on the spin orientation of the quark q, which leads to a $\sin(\phi)$ modulation, where ϕ is defined as the angle spanned by the transverse momentum of hadron and the plane normal to the quark spin.

In inclusive hadron production process, $e^+e^- \to q\bar{q} \to hX$, with unpolarized beams, studying on the azimuthal asymmetries of single (anti-)quark fragmentation is impossible due to the absence of the specific (anti-)quark spin. However, the Collins effect can be observed when the fragments of the quark and anti-quark are considered simultaneously. The correlation of quark and anti-quark Collins functions gives a cosine modulation of the distribution of azimuthal angles of the two hadrons, with the azimuthal angle ϕ_0 defined as in Fig. 1.

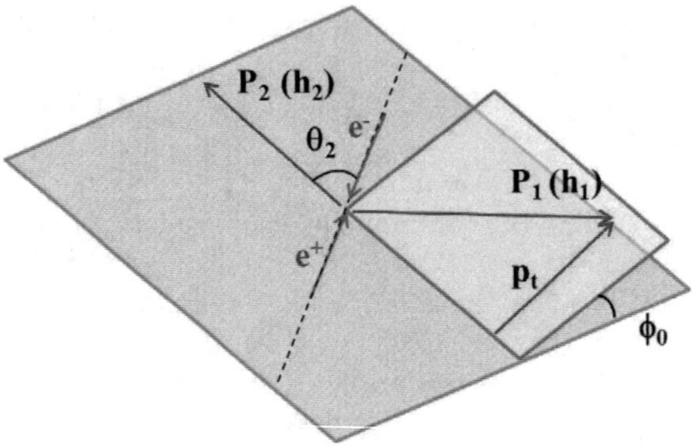

Figure 1. The angle ϕ_0 is defined as the angle between the plane spanned by the beam axis and the momentum of the second hadron (P_2), and the plane spanned by the transverse momentum p_t of the first hadron relative to the second hadron. The θ_2 is the polar angle of the the second hadron.

The normalized dihadron yields is recorded as a function of ϕ_0 and can be parameterized as $a\cos(2\phi_0) + b$ with b referring to the term which is independent on ϕ_0, and [22, 23]

$$a(\theta_2, z_1, z_2) = \frac{\sin^2\theta_2}{1+\cos^2\theta_2} \frac{\mathcal{F}\left(H_1^\perp(z_1)\bar{H}_1^\perp(z_2)/M_1M_2\right)}{D_1(z_1)\bar{D}_1(z_2)}, \tag{1}$$

where $\mathcal{F}$ denotes a convolution over the transverse hadron momenta. M_1 and M_2 are the masses of the two hadrons, z_1 and z_2 are their fraction energies, and θ_2 is the polar angle of the second hadron with respect to the beam axis, $\bar{D}_1$ and $\bar{H}_1^\perp$ denote FFs for anti-quarks.

In this analysis, for each charged π-pair candidate, basic track qualification and particle identification requirements are imposed [24]. The angle between two pions must be larger than 120°. We label the two pions randomly as h_1 and h_2, and we use the momentum direction of h_2 as reference axis. To select the inclusive $e^+e^- \to \pi\pi X$ events of interest, at least three charged tracks are required. To suppress backgrounds from the processes of QED, two photon fusions and ISR returned hadron productions, the visible energy in the detector, which is defined as the total energy of all reconstructed charged and neutral tracks, is required to exceed 1.5 GeV and no electron is present.

We introduce the $2\phi_0$ normalized distribution, $R = \frac{N(2\phi_0)}{\langle N_0 \rangle}$, where $N(2\phi_0)$ is the dipion yields in each ($2\phi_0$) subdivision, and $\langle N_0 \rangle$ is the averaged bin content. The normalized ratios are built for unlike-sign ($\pi^\pm\pi^\mp$), like-sign ($\pi^\pm\pi^\pm$) and charged pion-pairs ($\pi\pi$), as R^{U}, R^{L} and R^{C}, respectively, in which different combinations of favored FFs and disfavored FFs are involved. A favored fragmentation process refers to the fragmentation of a quark into a hadron containing a valence quark of the same flavor, for example $u(\bar{d}) \to \pi^+$, while the corresponding $u(\bar{d}) \to \pi^-$ is a disfavored process. Since the normalized ratio R is strongly affected by detector acceptance effects, to extract the physical asymmetries, double ratios $R^{\mathrm{U}}/R^{\mathrm{L(C)}}$ (UL and UC ratios) are constructed [9, 10]. In the double ratios, instrumental effects, which are charge-independent cancel out, and the QCD radiative effects are negligible to the first order, while the Collins asymmetries which are charge-dependent are kept. The double ratio $R^{\mathrm{U}}/R^{\mathrm{L(C)}}$ follows the expression

$$\frac{R^{\mathrm{U}}}{R^{\mathrm{L(C)}}} = A\cos(2\phi_0) + B, \tag{2}$$

where A and B are free parameters, while B should be consistent with unity, and A mainly contain the Collins effect.

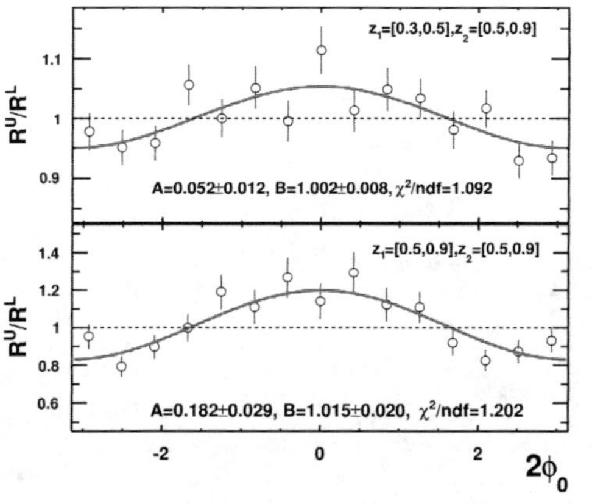

Figure 2. Double ratios $R^{\mathrm{U}}/R^{\mathrm{L}}$ versus $2\phi_0$ in the bin $z_1 \in [0.3, 0.5]$, $z_2 \in [0.5, 0.9]$ (top) and bin $z_1 \in [0.5, 0.9]$, $z_2 \in [0.5, 0.9]$ (bottom). The solid lines show the results of the fit.

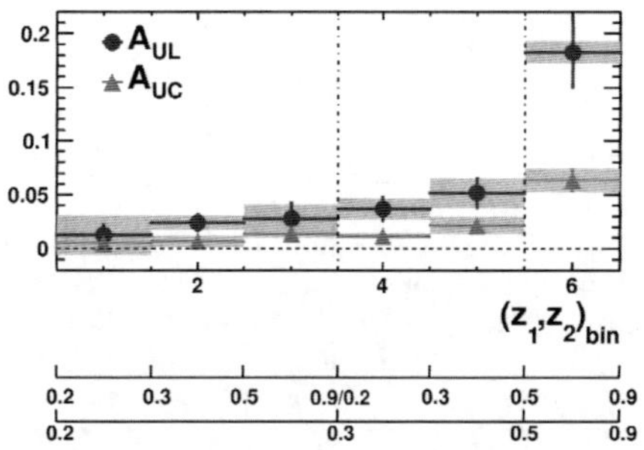

Figure 3. Asymmetries as a function of fractional energies (z_1, z_2) for the UL (dots) and UC (triangles) ratios. Statistical error are represented with sticks and systematic uncertainties represented by the hatched bands. The lower scales show the boundaries of the bins in z_1 and z_2.

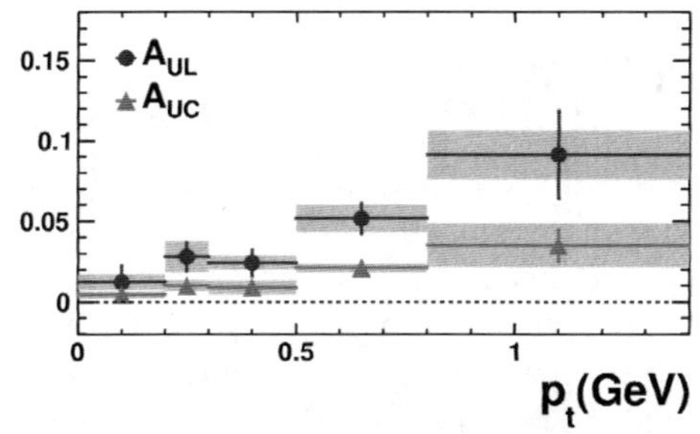

Figure 4. Asymmetries as a function of p_t for UL (dots) and UC (triangles) ratios.

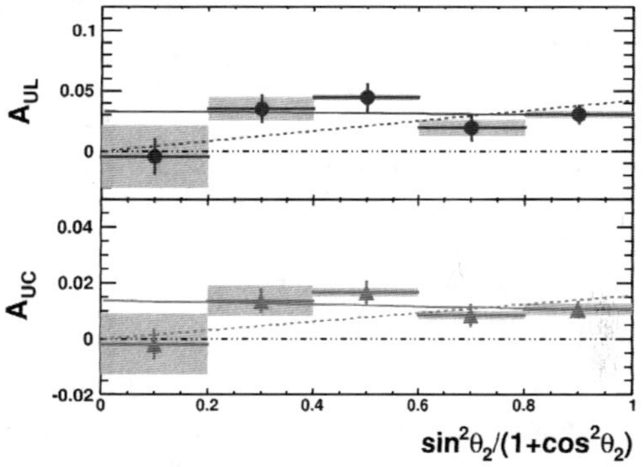

Figure 5. Asymmetries as a function of $\sin^2\theta_2/(1+\cos^2\theta_2)$ for UL (dots) and UC (triangles) ratios. The linear fits are shown.

The analysis is performed in (z_1, z_2) bins with boundaries at z_i= 0.2, 0.3, 0.5 and 0.9 ($i = 1, 2$), where complementary off-diagonal bins (z_1, z_2) and (z_2, z_1) are combined. In each (z_1, z_2) bin, normalized rates $R^{\mathrm{U,L,C}}$ and double ratios $R^{\mathrm{U}}/R^{\mathrm{L,C}}$ are evaluated in 15 bins of constant width in the angles $2\phi_0$. The fits using Eq. (2) for the two highest (z_1, z_2) bins are shown in Fig. 2, while in Fig. 3 and Table 1 are summarized the fit results as a function of six symmetric (z_1,z_2) bins. We see significant nonzero asymmetries [24]. The Collins asymmetries are investigated also in bins of p_t, and in bins of $\sin^2\theta_2/(1+\cos^2\theta_2)$, where definition of p_{t} and θ_2 are illustrated in Fig. 1. The results are shown in Fig. 4 and Fig. 5, and are summarized in Table 1.

As illustrated in Fig. 3, the measured Collins asymmetries rise with fractional energies as expected from theoretical predictions [22]. The moments of Collins asymmetries at the BEPCII energy scale is larger than at B factories, in agreement with the theoretical prediction from Ref. [19]. We observe an increase of the asymmetries with increasing p_t. The expected behavior of Collins asymmetries as a function of $\sin^2\theta_2/(1+\cos^2\theta_2)$ is linear and vanish at $\theta = 0$, as

Table 1. Results of $A_{\rm UL}$ and $A_{\rm UC}$ in each (z_1, z_2) and $p_{\rm t}$ bin. The errors shown are statistical and systematic, respectively. The averages $\langle z_i\rangle$, $\langle p_{\rm t}\rangle$ and $\frac{\langle\sin^2\theta_2\rangle}{\langle 1+\cos^2\theta_2\rangle}$ are also given.

$z_1 \leftrightarrow z_2$	$\langle z_1\rangle$	$\langle z_2\rangle$	$\langle p_{\rm t}\rangle$ (GeV)	$\frac{\langle\sin^2\theta_2\rangle}{\langle 1+\cos^2\theta_2\rangle}$	$A_{\rm UL}(\%)$	$A_{\rm UC}(\%)$
[0.2, 0.3][0.2, 0.3]	[0.245]	[0.245]	0.262	0.589	$1.28 \pm 0.93 \pm 1.84$	$0.50 \pm 0.32 \pm 0.69$
[0.2, 0.3][0.3, 0.5]	[0.311]	[0.311]	0.329	0.576	$2.40 \pm 0.74 \pm 0.68$	$0.67 \pm 0.27 \pm 0.53$
[0.2, 0.3][0.5, 0.9]	[0.428]	[0.426]	0.444	0.572	$2.81 \pm 1.44 \pm 1.26$	$1.36 \pm 0.54 \pm 0.31$
[0.3, 0.5][0.3, 0.5]	[0.379]	[0.379]	0.388	0.563	$3.69 \pm 1.07 \pm 0.97$	$1.17 \pm 0.39 \pm 0.24$
[0.3, 0.5][0.5, 0.9]	[0.498]	[0.499]	0.479	0.564	$5.18 \pm 1.32 \pm 1.41$	$2.17 \pm 0.47 \pm 0.79$
[0.5, 0.9][0.5, 0.9]	[0.625]	[0.628]	0.499	0.570	$18.24 \pm 3.19 \pm 1.05$	$6.37 \pm 0.10 \pm 1.07$
$p_{\rm t}$ (GeV)	$\langle p_{\rm t}\rangle$ (GeV)	$\langle z_1\rangle$	$\langle z_2\rangle$	$\frac{\langle\sin^2\theta_2\rangle}{\langle 1+\cos^2\theta_2\rangle}$	$A_{\rm UL}(\%)$	$A_{\rm UC}(\%)$
[0.00, 0.20]	0.133	0.291	0.348	0.574	$1.22 \pm 1.02 \pm 0.48$	$0.44 \pm 0.36 \pm 0.20$
[0.20, 0.30]	0.253	0.285	0.344	0.579	$2.79 \pm 0.89 \pm 0.93$	$1.00 \pm 0.32 \pm 0.34$
[0.30, 0.45]	0.405	0.327	0.346	0.570	$2.41 \pm 0.79 \pm 0.43$	$0.90 \pm 0.26 \pm 0.43$
[0.45, 0.80]	0.610	0.453	0.349	0.571	$5.16 \pm 0.95 \pm 0.87$	$2.11 \pm 0.41 \pm 0.27$
[0.80, 1.40]	0.923	0.646	0.334	0.584	$9.13 \pm 2.74 \pm 1.52$	$3.50 \pm 0.98 \pm 1.37$

formulated in Eq. (1). Thus, we perform a linear fit (c_0+c_1x) to the points in Fig. 5, with the constant term c_0 set to be zero or float, which gives the goodness of fit to be 1.82 or 1.26 for $A_{\rm UL}$ and 1.70 or 1.09 for $A_{\rm UC}$, respectively, which indicates the results may favor a nonzero constant term. But current data statistics can not provide a conclusive test on whether the constant term is nonzero.

4. Summary and outlook

In summary, we perform a measurement of the azimuthal asymmetry in the inclusive production of charged pion pairs, which can be attributed the the product of a quark and an anti-quark Collins functions. Our results exhibit nonzero asymmetries in the region of large fraction energy z and also indicate larger spin-dependent Collins effect than those at higher energy scale [9–12]. This is the first time the Collins asymmetries are measured at moderate energy scale in e^+e^- collider experiment. The results are of great importance to explore the Q^2 evolution of Collins function, and thus, also to extract the transversity distribution in nucleon.

In 2016, BESIII will take about 0.36 fb^{-1} data at $\sqrt{s} = 3.51$ GeV for studying the Collins fragmentation functions. It will improve the precision of this measurement at least by a factor of two. With larger statistics, we can extend this work to the Collins asymmetries of the charged πK and KK pairs. We also can provide data for the inclusive dipion productions involving natural π^0.

As for higher Q^2 region above charm threshold, BESIII has accumulated more than 5 fb^{-1} data [25]. However, the charmed hadron decay backgrounds need to be well understood before to extract the Collins effect. More efforts will be spent on investigating this issue.

References

[1] Collins J C 1993 *Nucl. Phys. B* **396** 161
[2] Airapetian A *et al.* (HERMES Collaboration) 2005 *Phys. Rev. Lett.* **94** 012002
[3] Ralston J P and Soper D E 1979 *Nucl. Phys. B* **152** 109
[4] Jaffe R and Ji X D 1991 *Phys. Rev. Lett.* **67** 552
[5] Jaffe R and Ji X D 1992 *Nucl. Phys. B* **375** 527
[6] Airapetian A *et al.* (HERMES Collaboration) 2010 *Phys. Lett. B* **693** 11
[7] Adolph C *et al.* (COMPASS Collaboration) 2012 *Phys. Lett. B* **717** 376
[8] Qian X *et al.* (Jefferson Lab Hall A Collaboration) 2011 *Phys. Rev. Lett.* **107** 072003
[9] Seidl R *et al.* (Belle Collaboration) 2006 *Phys. Rev. Lett.* **96** 232002
[10] Seidl R *et al.* (Belle Collaboration) 2008 *Phys. Rev. D* **78** 032011
[11] Seidl R *et al.* (Belle Collaboration) 2012 *Phys. Rev. D* **86** 039905(E)

[12] Lees J *et al.* (BaBar Collaboration) 2014 *Phys. Rev. D* **90** 052003
[13] Metz A 2002 *Phys. Lett. B* **549** 139
[14] Anselmino M, Boglione M, D'Alesio U, Melis S, Murgia F *et al.* 2013 *Phys. Rev. D* **87** 094019
[15] Martin A, Bradamante F and Barone V 2015 *Phys. Rev. D* **91** 014034
[16] Amrath D, Bacchetta A and Metz A 2005 *Phys. Rev. D* **71** 114018
[17] Kang Z B, Prokudin A, Sun P and Yuan F 2015 *Phys. Rev.* **D91** 071501 (*Preprint* 1410.4877)
[18] Ablikim M *et al.* (BESIII Collaboration) 2010 *Nucl. Instrum. Meth. A* **614** 345
[19] Sun P and Yuan F 2013 *Phys. Rev. D* **88** 034016
[20] Zhang C 2010 *Sci. China G* **53** 2084
[21] Bacchetta A, D'Alesio U, Diehl M and Miller C A 2004 *Phys. Rev. D* **70** 117504
[22] Boer D, Jakob R and Mulders P 1997 *Nucl. Phys. B* **504** 345
[23] Boer D 2009 *Nucl. Phys. B* **806** 23
[24] Ablikim M *et al.* (BESIII Collaboration) 2015 *unpublished* (*Preprint* 1507.06824)
[25] Ablikim M *et al.* (BESIII) 2015 *Chin. Phys.* **C39** 093001 (*Preprint* 1503.03408)

Journal of Physics: Conference Series **678** (2016) 012058 doi:10.1088/1742-6596/678/1/012058

Development of polarized ion source for the JINR accelerator complex

V V Fimushkin[1], A D Kovalenko[1], L V Kutuzova[1], Yu V Prokofichev[1], V B Shutov[1], A S Belov[2], V N Zubets[2] and A V Turbabin[2]

[1]Joint Institute for Nuclear Research, Joliot-Curie 6, 141980 Dubna, Moscow region, Russia
[2]Institute for Nuclear Research of RAS, prospect 60letiya Oktyabrya, 7A, 117312, Moscow, Russia

E-Mail: belov@inr.ru

Abstract. Status of the JINR polarized ion source development is described. The source is under tests at the test-bench of LHEP, JINR. A charge-exchange plasma ionizer has been tested initially without a storage cell in the ionization region. An unpolarized deuterium ion beam with peak current of 160 mA, 23 keV energy, pulse duration of 100 μs and repetition rate of 1 Hz has been extracted from the ionizer. With a free polarized atomic hydrogen beam injected into the ionizer a polarized proton beam with peak current of 1.4 mA has been obtained. The nearest plans for the source development include tests of the ionizer with the storage cell and tuning of the high frequency transition units installed in their operating position with a Breit-Rabi polarimeter.

1. Introduction

The high intensity pulsed source of polarized ions (SPI) is developed at JINR in collaboration with INR RAS to ensure injection of polarized deuterons and protons into the Nuclotron and future collider of heavy and light ions NICA [1].

The SPI is an atomic beam-type polarized ion source with nearly resonant charge-exchange plasma ionizer and storage cell in the charge – exchange region [2-4]. Some parts of the polarized ion source CIPIOS from IUCF [5] are used in the project.

2. General description of the source of polarized ions

A schematic diagram of the SPI is shown in figure 1. The source has been described in more detail in [6]. Atomic hydrogen (or deuterium for production of polarized deuterons) is produced in the source using an RF discharge dissociator. The dissociator works in a pulsed mode of operation with repetition rate up to 1 Hz and with RF discharge duration of up to 3 ms. Two electromagnetic valves are used for pulsed gas supply into the dissociator discharge tube: one of the valves is used for injection of gaseous molecular hydrogen (deuterium) and another for injection of molecular oxygen.

The atomic deuterium is cooled during the passage through the pyrex tube with a 5 mm internal diameter (cooling channel) ending by a sonic nozzle 2.3 mm in diameter. The channel walls are cooled to temperature of ~80 K by a cryocooler (model 350 Cryodyne refrigerator system). A skimmer is used to separate the central part of an atomic gas jet and form an atomic beam which is then injected into the sextupole magnets. The skimmer has a cone shape and its central orifice is 6 mm in diameter. The distance between the dissociator nozzle and the skimmer orifice is 30 mm.

Content from this work may be used under the terms of the Creative Commons Attribution 3.0 licence. Any further distribution of this work must maintain attribution to the author(s) and the title of the work, journal citation and DOI.
Published under licence by IOP Publishing Ltd

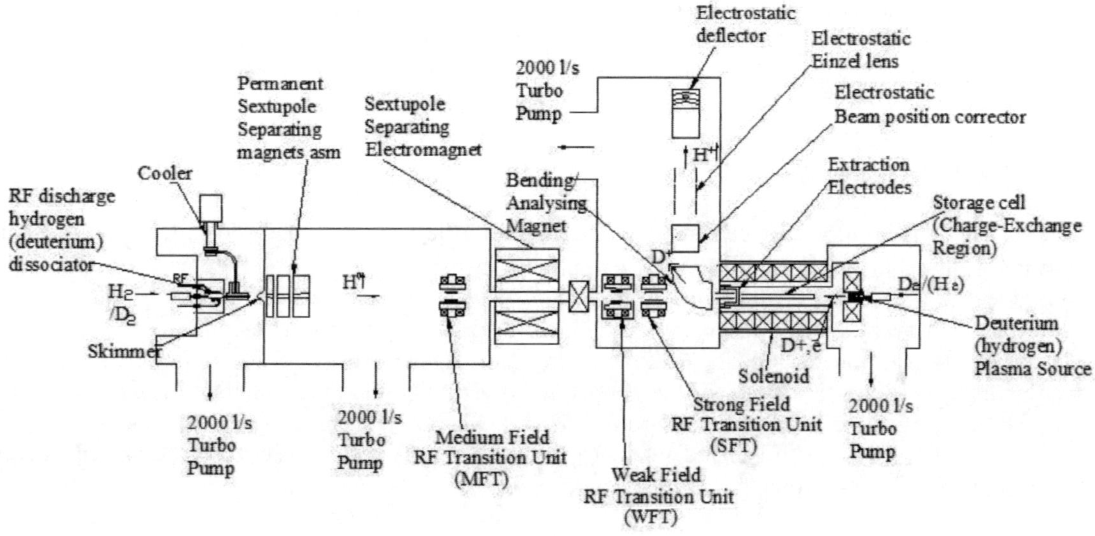

Figure 1. Schematic diagram of the SPI.

The sextupole magnets system consists of assembly of three permanent sextupoles and one electromagnet sextupole. The sextupole magnets parameters are presented in ref. [5].

The system of high frequency transition (HFT) units includes medium field transition unit (MFT), which is installed between the assembly of permanent sextupole magnets and electromagnet sextupole, weak field (WFT) and strong field (SFT) transition units which are installed downstream the electromagnet sextupole. For hydrogen atoms alternate turning on of WFT ($1 \rightarrow 3$) and SFT ($2 \rightarrow 4$) will lead to switching between theoretical values of proton vector polarization -1 and +1. For polarized deuterium use of MFT, WFT and SFT will allow switching deuteron vector polarization between +1 and -1 and tensor polarization between +1 and -2 [5,6].

The polarized atomic beam is injected then into the plasma ionizer where polarized ions are produced via nearly resonant charge-exchange reaction between polarized atoms and unpolarized ions of isotope relative the polarized atoms:

$$H^0\uparrow + D^+ \rightarrow H^+\uparrow + D^0 \quad (1)$$
$$D^0\uparrow + H^+ \rightarrow D^+\uparrow + H^0 \quad (2)$$

The polarized ions move then slowly under influence of weak electric fields in plasma in direction to the extraction electrode system where they are accelerated to energy of up to 25 keV together with unpolarized plasma ions. Radial confinement of the low energy polarized ions is provided by the magnetic field of the ionizer solenoid.

The 25 keV ion beams extracted pass through the 90° bending magnet in which the polarized ion beam is separated from unpolarized ions. The unpolarized ion beam current is recorded using the ion beam collector downstream the bending magnet. The polarized beam comes out the magnet in vertical direction, passes through the electrostatic Einzel lens and then is deflected by the 90° electrostatic deflector into the horizontal plane to x direction. In the deflector spin of ions remains unchanged. At the source exit the polarized ion beam passes through a spin precessor solenoid which rotates the spin to vertical direction. By this way it is possible to ensure at the source exit vertical direction of spin for polarized protons or deuterons.

3. Tests of the source

The atomic beam section (ABS) has been tested at INR RAS [6]. Intensity of the polarized atomic beam 50 cm downstream the sextupole electromagnet was measured with the time-of-flight mass-spectrometer and found to be $1.5\ 10^{17}\ sec^{-1}cm^{-2}$. Efficiency of HFT was determined by placing a HFT between the assembly of permanent sextupole magnets and electromagnet sextupole and measuring the respective decrease of density of polarized atomic beam. Efficiencies of WFT and MFT for deuterium were determined from these measurements close to be 0.95±0.03 [6].

Journal of Physics: Conference Series **678** (2016) 012058 doi:10.1088/1742-6596/678/1/012058

The plasma ionizer has been tested at the JINR test-bench initially without a storage cell.

Efficiency of conversion of polarized atoms into polarized ions is proportional to plasma density in the charge-exchange region and, hence, to current of unpolarized ions extracted from the ionizer. The plasma is produced by an arc-discharge plasma source with cold cathode working in the pulsed mode of operation. The longitudinal magnetic field of ~500 G in the source is produced by the plasma source coil. Molecular deuterium (hydrogen) is injected into the plasma source by a pulsed electromagnetic gas valve similar to one used for the dissociator. An ignition voltage pulse with amplitude of 6 kV and duration of about 40 μs is applied between the cold cathode and the gas valve body to initiate arc-discharge in the source followed by 150 μs voltage pulse applied between the cathode and the anode of the source which produces arc-discharge with current of up to 300 A peak. The deuterium (hydrogen) plasma generated in the source flows out the source into vacuum through the 4 mm diameter hole in the anode and then injected into the ionizer solenoid with the longitudinal magnetic field of ~1 kG created by the ionizer solenoid. The ionizer solenoid coils are at ground potential while the plasma jet inside the solenoid should be at hv positive potential relative ground to make it possible acceleration of ions. Cylindrical screen (made of stainless steel) insulated relative ground is installed inside the solenoid to ensure the plasma injection into region which is at hv potential. The screen is connected electrically to the anode of the plasma source.

Three electrode extraction system has been installed for tests of the ionizer without the storage cell. The first (plasma) electrode and the second electrode are gridded with distance between grids of 4 mm. The extraction electrodes aperture is 17 mm. The plasma electrode is at the plasma source anode potential while the second electrode is at -1 kV potential (DC) relative ground. The third electrode is cylindrical with internal diameter of 20 mm and it is grounded. Potential up to 25 kV is applied to the anode of the plasma source to provide extraction of ions from the plasma region. Negative potential applied to the second electrode produces a trap for secondary electrons which compensate space charge of the intense unpolarized ion beam.

The 25 kV voltage pulse duration can be varied in range of 50-500 μs, rep. rate is up to 1 Hz. The pulse top is stabilized with feed-back system with accuracy of ±0.3%. The power supply has been designed and built at INR RAS.

Figure 2 shows typical oscillograms characterizing the ionizer operation.

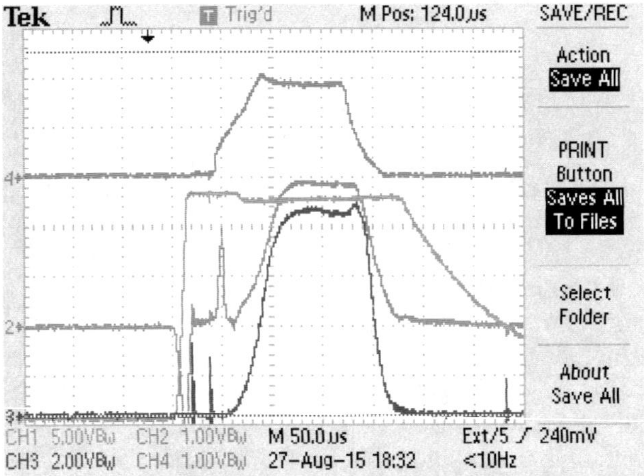

Figure 2. The oscillograms characterizing the ionizer operation: green – the plasma source discharge current pulse with peak current of 190 A; yellow – the extraction voltage pulse with peak value of 23 kV, blue – the current of extraction electrodes with peak value of 560 mA, purple – unpolarized deuterium ion current recorded downstream the analysing magnet (peak value is 160 mA).

With the polarized atomic hydrogen beam "on" a polarized proton beam with peak current of 1.4 mA (difference in the ion beam peak current with atomic beam "on" and "off") has been recorded downstream the bending magnet.

Figure 3 shows photo of the hv power supply for ions extraction from the plasma ionizer during its tests at INR RAS. Photo of the polarized ion source at the test- bench is shown in figure 4.

Figure 3. 25 kV pulsed power supply during its tests at INR RAS.

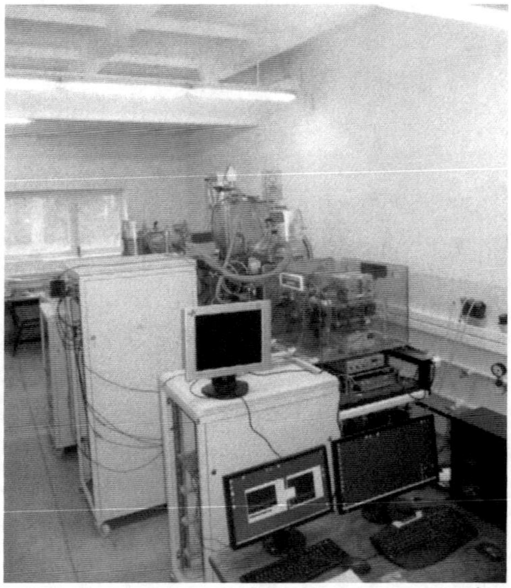

Figure 4. SPI test-bench at JINR.

4. Nearest plans

4.1. Tests of the ionizer with a storage cell

A storage cell is planned to be used in the charge-exchange region to increase density of polarized atoms. This method has been tested at INR RAS [3] and the polarized proton beam with peak current of 11 mA and polarization of 80% has been obtained by this method [4].

A storage cell with the same dimensions (250 mm in length, 15 mm internal diameter) made from aluminum alloy will be installed into the ionizer of the polarized ion source. We plan to start the tests in 2015.

4.2. Measurements of efficiencies of the HFT with a Breit – Rabi polarimeter

HFT determine nuclear polarization of polarized ions but their efficiencies can be impacted by the external magnetic field. For this reason it is important to control efficiencies of HFT directly in their operating position.

To do this we plan use Breit – Rabi type polarimeter consisting of two multipole magnets and TOF mass-spectrometer. A schematic diagram of the ABS components and the polarimeter is shown in figure 5.

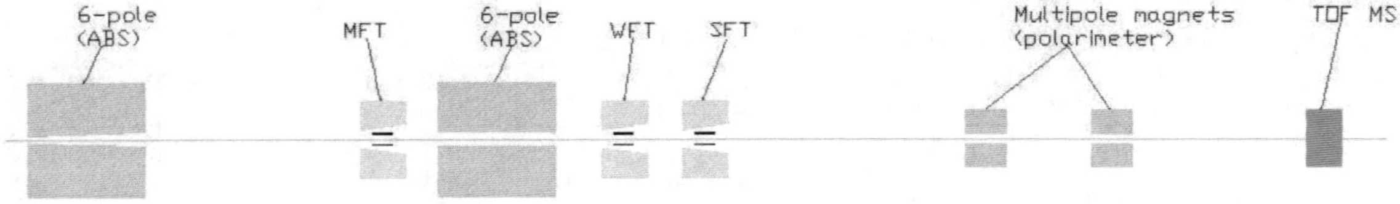

Figure 5. Schematic diagram of the ABS with Breit – Rabi polarimeter.

Two permanent multipole magnets will be used in the polarimeter. Their parameters were determined using race-tracing calculations and are shown in the table 1. The first magnet is a qudrupole with aperture of 17 mm and length of 140 mm. The second magnet is a sextupole with aperture of 18 mm and length of 125 mm.

The maximum magnetic field at the pole tip of both magnets is accepted to be 1.6 T. The magnets are now under construction.

Journal of Physics: Conference Series **678** (2016) 012058 doi:10.1088/1742-6596/678/1/012058

Table 1. Parameters of multipole magnets of the polarimeter

Magnet number	1	2
Poles number	4	6
Aperture (distance between opposite poles in mm)	17	18
Length (mm)	140	125
Magnetic field at pole tip (T)	1.6	1.6

An example of the calculations of atoms trajectories in the source with the polarimeter installed is shown in figure 6.

It is seen from the results of the calculation that atoms focused by the ABS are refocused then by the multipole magnets of the polarimeter if no HFT are "on" in the ABS (figure 6a). If one of HFT is "on", then respective part of the atoms undergoing transition between spin states (with change of their electron spin state in strong magnetic field from $m_j = 1/2$ to $m_j = -1/2$) will be defocused by multipole magnets of the polarimeter (figure 6b). Measurement of relative change of density of the polarized atomic beam by the TOF mass – spectrometer with HFT "off" and "on" will allow one to determine efficiencies of the HFT of the ABS.

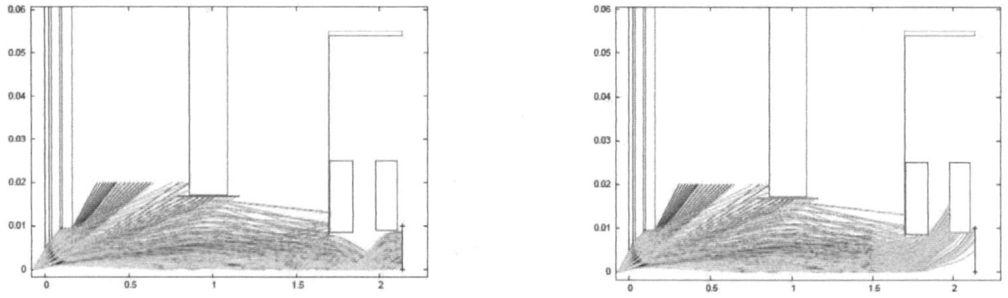

a) b)

Figure 6. Trajectories of atoms in the ABS and the polarimeter: a) HFT are "off" b) one of HFT is "on".

5. Conclusions

Polarized ion source for the JINR accelerator complex is now under tests at the JINR test bench. The ABS part and the plasma ionizer operation have been tested and were found satisfactory. The polarized proton beam of 1.4 mA peak has been produced with free atomic beam in the charge-exchange region of the source. It is planned to continue tests with a storage cell installed into the ionizer. The HFT units installed in their operating position will be retuned with the Breit-Rabi polarimeter.

References

[1] Trubnikov G, Agapov N, Brovko O, Butenko A, Donets E, Eliseev A, Fimushkin V, Gorbachev E, Govorov A, Ivanov E *et al.* 2013 *Proc. of the 4th Int. Particle Accelerator Conf. IPAC2013 (Shanghai, China, 2013)* TUPFI009 p 1343

[2] Belov A S, Esin S K, Kubalov S A, Kuzik V E, Stepanov A A and Yakushev V P 1987 *Nucl. Instrum. and Meth. In Phys. Res.* **A255** 442

[3] Belov A S, Esin S K, Netchaeva L P, Klenov V S, Turbabin A V and Vasil'ev G A 1998 *Proc. Int. Workshop on Polarized Gas Targets and Polarized Beams* (*Urbana IL, USA, 1997)* ed. Roy J. Holt and M. A. Miller, AIP Conf. Proc. **421** p 362

[4] Belov A S, Esin S K, Netchaeva L P, Turbabin A V and Vasil'ev G A 1999 *Proc. Int. Symp. on High Energy Spin Physics (Protvino, Russia*, 1998) ed. N. E. Tyurin *et al*, (Singapore: World Scientific) p 22

[5] Derenchuk V P and Belov A S 2001 *Proc. 2001 Particle Accelerator Conf* (*Chicago, USA)* WPAH008 p 2093

[6] Fimushkin V V, Kovalenko A D, Kutuzova L V, Prokofichev Yu V, Belov A S, Turbabin A V and Zubets V N 2014 *Proc. 20th Int. Symp. on High Energy Spin Physics* (*Dubna, Russia, 2012) PEPAN* **45** part 1 p 297

Journal of Physics: Conference Series **678** (2016) 012059 doi:10.1088/1742-6596/678/1/012059

The RHIC Spin Program Overview

A. Bazilevsky, for the RHIC Spin Collaboration
Brookhaven National Laboratory, Upton, NY 11973, USA

shura@bnl.gov

Abstract. After more than a decade of RHIC running as a polarized proton collider, we summarize recent achievements of the RHIC spin program and their impact on our understanding of the nucleon's spin structure, i.e. the individual parton (quarks and gluons) contributions to the helicity structure of the nucleon, and to understand the origin of the transverse spin phenomena. Open questions are identified and a suite of future measurements with polarized beams at RHIC to address them is laid out.

1. Introduction

This presentation is majorly based on our recently published paper [1], where we summarized the achievements of the RHIC spin physics program for more than a decade of RHIC operation as a polarized proton collider, and outlined the near and longer term plans and experimental upgrades to further advance spin studies at RHIC.

One of the main goals of the RHIC spin program has been the studies of different contributions to the proton spin 1/2, for which it was found from the previous experiments that only small fraction can be attributed to the spins of quarks and anti-quarks. Another set of puzzles is associated with transverse spin effects, which can not be explained by collinear perturbative QCD (pQCD) at leading twist, which requires extension of the theory to higher twist (multi-parton correlations), or introduction of transverse momentum dependencies correlated with nucleon or parton spin. Parton transverse momentum distributions correlated with proton spin are connected with parton orbital momentum in the nucleon – the least known piece in the nucleon spin decomposition at the moment.

2. Helicity Structure of the Proton

The spin structure of the proton has been measured since 1980s in polarized lepton-nucleon deep-inelastic scattering (DIS) experiments, which revealed that only $\sim$1/3 of the proton spin can be attributed to the spins of the quarks and anti-quarks [2] indicating that the proton spin must be largely carried by the spin of the gluons and/or orbital angular momentum of the quarks and gluons. Since then, the main goal of spin physics has been to elucidate the role of gluon spin (Δg) as well as flavor separated contribution of quark and anti-quark spin (Δq and $\Delta \overline{q}$) in the proton spin. Getting access to the orbital angular momentum remains a challenging task from both theoretical and experimental point of view, which will be a major part of the new 12 GeV program at Jefferson Laboratory and at future Electron-Ion Collider. DIS experiments have been providing data to constrain Δg from the scaling violation in inclusive polarized scattering [3] and from semi-inclusive measurements of high transverse momentum (p_T) hadron pairs [4] and heavy flavor production [5] to utilize the photon-gluon

Content from this work may be used under the terms of the Creative Commons Attribution 3.0 licence. Any further distribution of this work must maintain attribution to the author(s) and the title of the work, journal citation and DOI.
Published under licence by IOP Publishing Ltd

Journal of Physics: Conference Series **678** (2016) 012059 doi:10.1088/1742-6596/678/1/012059

process. The flavor separated contribution of quarks and anti-quarks to the proton spin is determined in semi-inclusive DIS using fragmentation processes, which correlate final state hadron with quark flavor.

Polarized proton-proton collisions at RHIC provide complementary approach to study proton spin structure. Gluon-spin contribution is accessed through the hard scattering directly involving gluons. Flavor decomposition of quark and anti-quark polarization can elegantly be obtained from the measurements of W bosons, which couple left-handed quarks and right-handed anti-quarks: $u_L\bar{d}_R \rightarrow W^+$ and $d_L\bar{u}_R \rightarrow W^-$. This approach is free of uncertainties in fragmentation functions (FF), if *W*s are measured through their leptonic decays.

2.1. Polarized gluon contribution

Gluon polarization within the proton is probed through the measurements of double helicity asymmetry, A_{LL}, in different production channels involving gluon dominated hard scattering processes. A_{LL} is defined as the difference between cross sections for colliding bunches with the same helicity and opposite helicity, divided by the sum. The most statistically abundant channels are inclusive pion and jet production. Fig.1 and 2 show the recent updates for π^0 and jet production.

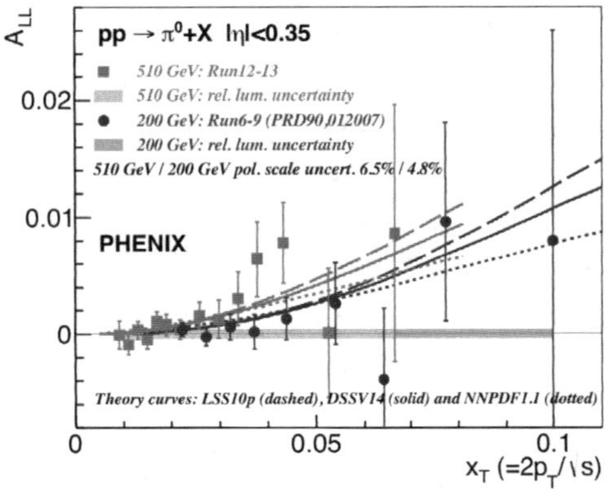

Figure 1. A_{LL} vs. x_T for inclusive π^0-mesons production at mid-rapidity in $p+p$ collisions at √s=200 GeV [9] and 510 GeV [10], compared to predictions from three recent NLO global analyses [11,12,8] (blue curves for √s=200 GeV and red curves for 510 GeV).

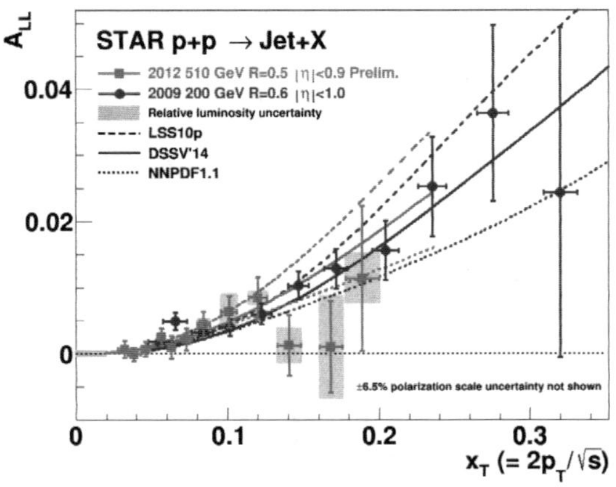

Figure 2. A_{LL} vs. x_T for inclusive jet production at mid-rapidity in $p+p$ collisions at √s=200 GeV [13] and 510 GeV (preliminary), compared to predictions from three recent NLO global analyses [11,12,8] (blue curves for √s=200 GeV and red curves for 510 GeV).

Extraction of Δg is based on a next-to-leading order (NLO) pQCD framework, which was proved to well describe RHIC unpolarized cross section data [6]. The DSSV group has found from their global analysis that the integral of $\Delta g(x,Q^2=10$ GeV$^2)$ in the region $x > 0.05$ is $\mathbf{0.20^{+0.06}_{-0.07}}$ at 90% C.L. [7], consistent with the NNPDFpol1.1 fit [8].

Despite this very important achievement, providing the evidence of sizable gluon polarization within the proton, uncertainties for $\Delta g(x)$ remain significant in the presently unmeasured small x region, see Fig. 3, and prevent a reliable determination of the full integral. Fig.3 also demonstrates the expected improvement of the fit results after inclusion of the data from 2012-2015 RHIC runs (being analyzed), which include data collected in forward rapidity region at √s = 510 GeV. These data will extend the kinematic reach in x towards values of a few times 10^{-3} where Δg is unconstrained so far.

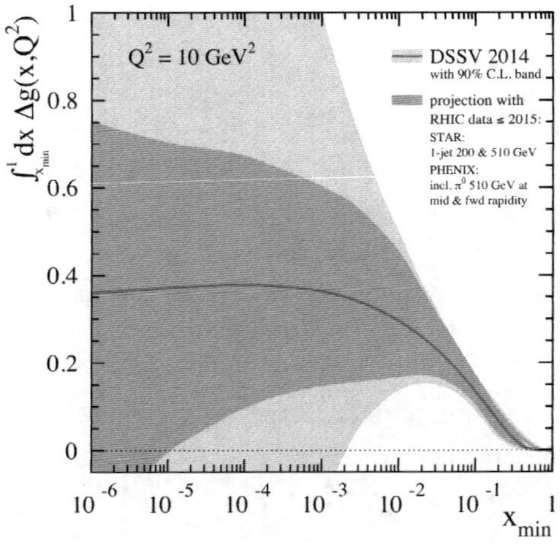

Figure 3. The running integral for Δg as a function of x_{min} at $Q^2 = 10$ GeV^2 as obtained in the DSSV global analysis framework. The inner and outer uncertainty bands at 90% C.L. are estimated with and without including the combined set of projected pseudo-data for preliminary and future RHIC measurements up to Run-2015, respectively.

2.2. Polarized sea quark distribution

The production of $W^{\pm}$ bosons in longitudinally polarized proton-proton collisions serves as an elegant tool to access valence and sea quark helicity distributions at a high scale $Q\sim M_W$ and without the need of FF as in semi-inclusive DIS. W boson production, coupling left-handed quarks and right-handed anti-quarks, provide us with direct access to (anti-)quark preferred helicity state within the longitudinally polarized proton by measuring the parity violating single helicity asymmetry A_L. While the valence quark helicity densities are already well known at intermediate x from DIS, the sea quark helicity PDFs are only poorly constrained. The latter are of special interest due to the differing predictions in various models of nucleon structure (see [14] for a review), which are all able to describe the asymmetry of the unpolarized light quark sea.

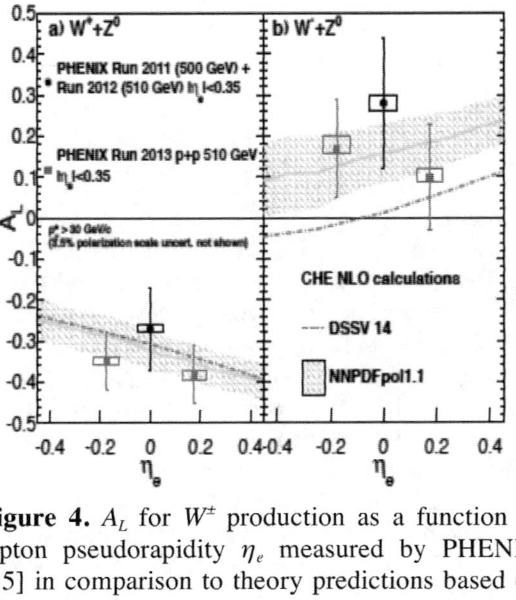

Figure 4. A_L for $W^{\pm}$ production as a function of lepton pseudorapidity η_e measured by PHENIX [15] in comparison to theory predictions based on inclusive and semi-inclusive DIS data.

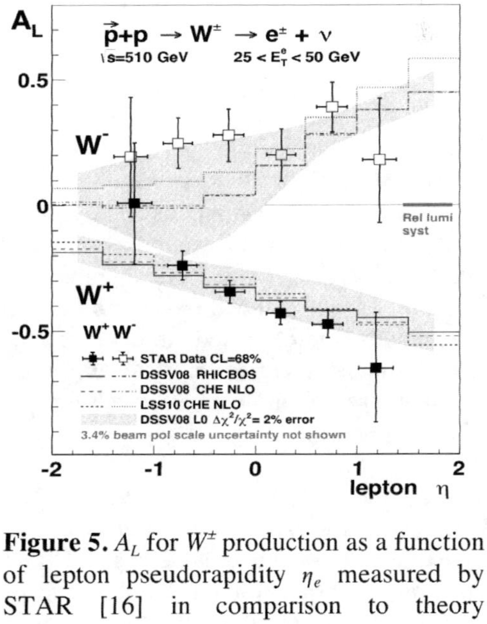

Figure 5. A_L for $W^{\pm}$ production as a function of lepton pseudorapidity η_e measured by STAR [16] in comparison to theory predictions based on inclusive and semi-inclusive DIS data.

Journal of Physics: Conference Series **678** (2016) 012059 doi:10.1088/1742-6596/678/1/012059

Fig. 4 and 5 show the recently released results by PHENIX and STAR collaborations. STAR data covering wider rapidity range prefer W^- decay lepton asymmetries, which are systematically larger than the theory predictions based on inclusive and semi-inclusive DIS data directly indicating more positive $\bar{u}$ helicities, while the W^+ asymmetries are roughly compatible with theory curves. As a consequence these results provide an indication of the flavor symmetry being also broken for the polarized light quark sea with $\Delta\bar{u}$ being positive than $\Delta\bar{d}$ being negative. A positive $\Delta\bar{u}-\Delta\bar{d}$ asymmetry would favor the chiral quark soliton model, the Pauli blocking ansatz and statistical models, while disfavoring any cloud based models. The PHENIX analysis of W production through muon decays in forward region, $1.2<|\eta|<2.4$, as well as STAR analysis of the highest integrated luminosity 2013 RHIC run are ongoing. After inclusion these results in the global analysis of polarized PDFs, the considerable improvement in $\bar{u}$ and $\bar{d}$ helicity distributions are expected in the accessed x range above 0.05. The uncertainty on the flavor asymmetry for the polarized light quark sea $\Delta\bar{u} - \Delta\bar{d}$ will be further reduced and a measurement on the 2σ level will be possible, see Fig. 6.

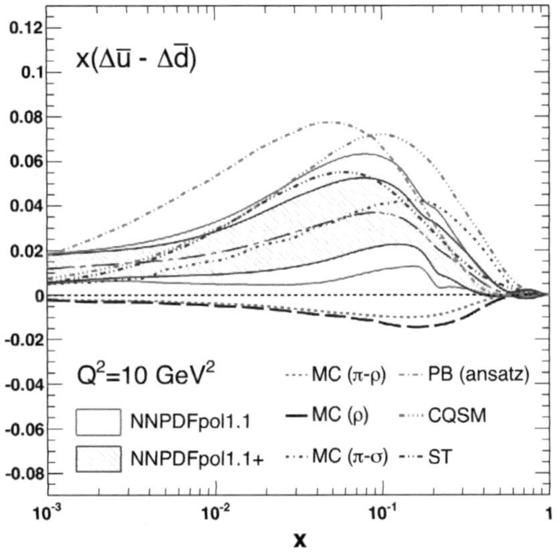

Figure 6. The polarized light sea-quark asymmetry $x(\Delta\bar{u} - \Delta\bar{d})$ computed with NNPDFpol1.1 and NNPDFpol1.1+ PDFs at $Q^2 = 10$ GeV^2, projected to data already collected at RHIC for W program, compared to various models of nucleon structure (see Ref. [14] for a review).

3. Transverse Spin Phenomena

Surprisingly large single-transverse spin asymmetries A_N (SSA) in fixed target $pp \rightarrow \pi X$ pion production have been observed experimentally since more than 30 years ago (see e.g. [17]). A_N relates to left-right asymmetry in particle production in the experiments with one beam transversely polarized colliding on unpolarized target or another beam. Measurements at RHIC have extended the observations from the fixed-target energy range to the collider regime, up to and including the highest center-of-mass energies to date in polarized pp collisions. Fig. 7 summarizes the measured asymmetries in π^0 production from different experiments as function of Feynman-x x_F.

The surprisingly large asymmetries seen are nearly independent of $\sqrt{s}$ over a very wide range. To understand the observed SSAs one has to go beyond the conventional leading twist collinear parton picture in the hard processes. Two theoretical formalisms have been proposed to generate sizable SSAs in the QCD framework: transverse momentum dependent (TMD) parton distribution functions (PDF) and FF, which provide the full transverse momentum information, and the collinear quark-gluon-quark correlations at Twist-3 level in the initial state and the fragmentation process, which provide information about the average transverse momentum. At RHIC the p_T-scale is sufficiently large (see Fig. 8) to make a formalism using collinear quark-gluon-quark correlation functions applicable to calculate the transverse single spin asymmetries. Here, various underlying mechanisms can contribute and need to be disentangled to understand the experimental observations in detail, in

Journal of Physics: Conference Series **678** (2016) 012059 doi:10.1088/1742-6596/678/1/012059

particular the p_T-dependence.

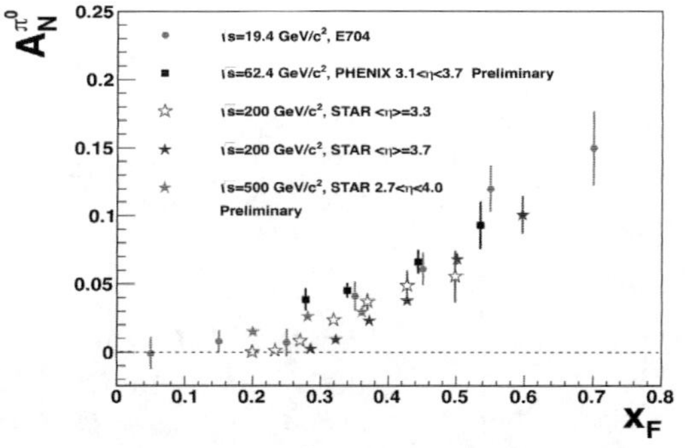

Figure 7. Transverse single spin asymmetry measurements for neutral pions at different center-of-mass energies as function of Feynman-x.

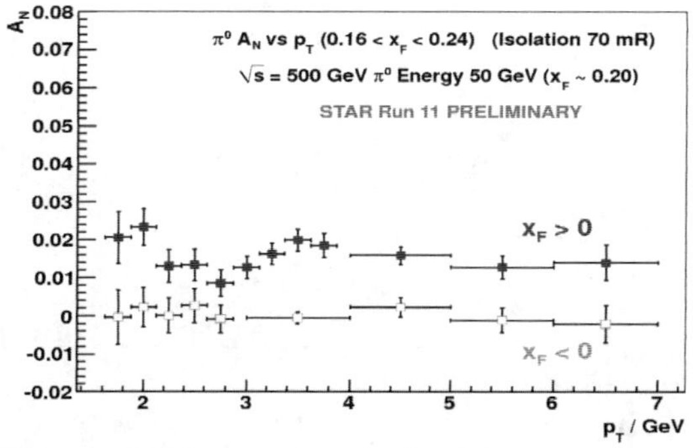

Figure 8. Transverse single spin asymmetry measurements for neutral pions at $\sqrt{s}$ =500 GeV as a function of p_T.

It has been shown that both initial state TMD and multi-parton dynamics (Twist-3 approach) provide equivalent descriptions of transverse asymmetries in an overlap region at intermediate transverse momenta and that the $k_\perp$ moment of a quark's TMD Sivers function is related to the corresponding Twist-3 quark-gluon correlation function $T_{q,F}(x,x)$ [18]. It was also found that the $k_\perp$ moment of the Sivers function, extracted from data for single-spin asymmetries in semi-inclusive deep-inelastic scattering, has opposite sign from that of $T_{q,F}(x,x)$, extracted from single-inclusive hadron production in pp collisions [19]. One of explanations of such a "sign mismatch" problem could be that Sivers effect is not dominantly responsible for the observed single-spin asymmetries in pp collisions. The bigger contribution may come from either fragmentation or some other effects.

Recently, it has been proposed that in the collinear, Twist-3 factorization approach a significant portion of the sizable inclusive pion asymmetries seen at forward pseudorapidity is due to Twist-3 FFs coupled to transversity [20]. This calculation is the first one which showed, similar to the experiment, a flat p_T-dependence for A_N (Fig. 8).

Another interesting observation from RHIC is A_N for "electromagnetic" jets (jets with its energy measured only in the electromagnetic calorimeter) in the pseudo-rapidity range $2.5<\eta<4$, see Fig.9. With increasing number of photons in the "electromagnetic jet" (increasing "jettiness" of the event) the asymmetry becomes smaller. Jets with an isolated π^0 have the largest asymmetry consistent with the asymmetry in inclusive π^0 events. Also, requiring an additional correlated away-side jet the asymmetry for isolated forward π^0 mesons becomes smaller. These observations indicate that the underlying subprocess causing a significant fraction of the large transverse single spin asymmetries in the forward direction may be of diffractive nature.

The main goal of the further RHIC-Spin program is to disentangle the different processes responsible for the large A_N asymmetry in pp collisions. We identify jet, direct photons, heavy flavor, and also W, Z boson and Drell-Yan measurements as ones sensitive to initial state effect and described by Sivers/Twist-3 parton correlation functions. Final state effects can be probed through hadron azimuthal asymmetry within a jet, coupling transversity to TMD Collins FF [21], and hadron pair azimuthal asymmetry coupling transversity to the so-called "interference fragmentation function" (IFF) [22] in the framework of collinear factorization. Diffractive processes will be identified using Roman Pots by tagging one or both scattered protons. The first data with Roman Pots installed for STAR experiment was collected in RHIC 2015 run.

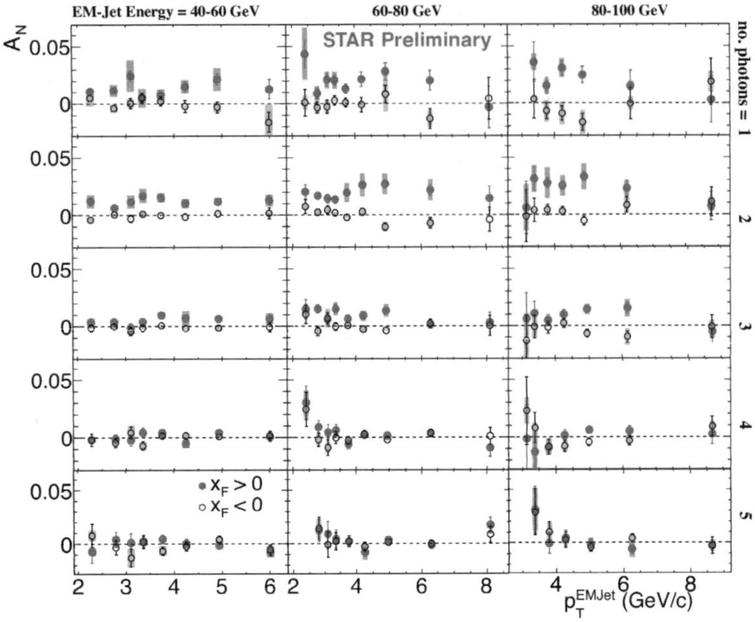

Figure 9. The transverse single spin asymmetry A_N for electromagnetic jets detected in the STAR FMS ($2.5<\eta<4.0$) as function of the jet p_T and the photon multiplicity in the jet in bins of the jet energy.

Studying Siver effect in *pp* collisions is of particular interest due to its fundamental role for testing our understanding of QCD factorization. It was predicted that due to initial or final state effects from gluon exchange, Sivers function is not universal in different processes [23]. Very interesting prediction comes for the comparison of Drell-Yan production and semi-inclusive DIS, for which the Sivers function is predicted to have opposite sign [24]. The other channels like *W* and *Z* boson production also show high sensitivity to the predicted sign change. The first data on *W* and *Z* A_N from the limited data sample was already released by STAR [25]. A factor of 15-30 more data is expected from the dedicated RHIC run in 2017.

Transverse single spin asymmetries in direct photon production provide a different path to access this sign change through the formalism utilizing the Twist-3 parton correlation functions. For the 2015 polarised *pp* run both PHENIX and STAR installed preshowers in front of their forward electromagnetic calorimeters. The collected data in 2015 RHIC run will enable a measurement of the SSA for direct photons.

Measurements of Collins and IFF asymmetries give us access to transversity distribution, which is currently poorly known. Fig. 10 and 11 show the first observation of non-zero Collins and di-hadron asymmetries in *pp* collisions. While the measurements of transversity through the Collins FF need TMD factorization to hold in *pp* scattering, di-hadron asymmetries utilize collinear factorization. Thus, not only can more precise measurements of these effects in *pp* improve our knowledge of transversity, such measurements are/will be invaluable to test the longstanding theoretical questions, such as the magnitude of any existing TMD factorization breaking.

Colliding of transversely polarized protons with nuclei at RHIC opens a new avenue in both spin and heavy ion physics. Kang and Yuan [26] showed that comparison of A_N measurements in *p+A* and *p+p* collisions may be sensitive to gluon saturation effects in nuclei. First data from *p+Au* and *p+Al* collisions were already collected by both PHENIX and STAR collaborations in 2015, and the results are expected soon.

Data from RHIC 2015 run also allowed extending the forward neutron A_N measurements from *p+p* to *p+Al* and *p+Au* collisions. An unexpected A-dependence of the asymmetry was observed [27], which awaits for theoretical interpretation.

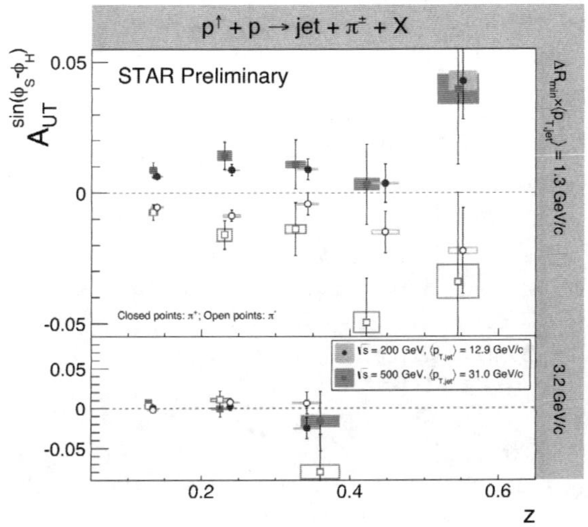

Figure 10. $A_{UT}^{\sin(\phi_s-\phi_h)}$ vs. z for charged pions in jets at $0 < \eta < 1$ from p+p collisions at $\sqrt{s} = 200$ GeV and 500 GeV by STAR.

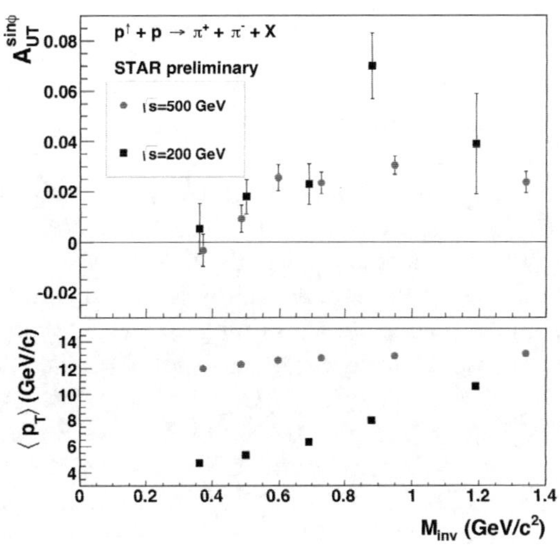

Figure 11. $A_{UT}^{sin\phi}$ as a function of $M_{\pi+\pi-}$ (upper panel) and corresponding $p_{T_{\pi+\pi-}}$ (lower panel), at √s=200 GeV and 500 GeV by STAR for -1 < η < 1.

4. Future upgrades

Both collaborations, PHENIX and STAR, are actively developing their future upgrade programs, with the main focus on forward capabilities. STAR collaboration plans to construct Forward Calorimeter System and Forward Tracking System. PHENIX collaboration initiated even more ambitious project. The current PHENIX detector setup will be decommissioned in 2016, and a new detector system will be built based on superconducting solenoid from the decommissioned BaBar experiment at SLAC, with electromagnetic and hadronic calorimetry and tracking in the barrel region -1<η<1. The open geometry of the detector allows for the addition of a forward spectrometer, covering 1<η<4, and consisting of hadronic calorimeter and tracking system. Part of the current PHENIX detector systems (e.g. forward silicon tracker and muon identification systems) will be re-used in the new detector setup.

The suggested physics program centers on a comprehensive set of measurements of jet, hadron and Drell-Yan lepton pair production in transversely spin-polarized p+p and p+A collisions, with the emphasis on low and high region of Bjorken x, exploiting the unique capability of the RHIC collider to provide beams of protons with high polarization in addition to a variety of unpolarized nuclear beams.

References

[1] E. Aschenauer *et al.*, arXiv:1501.01220.
[2] J. Ashman *et al.* (EMC), Phys. Lett. B **206,** 364 (1988), Nucl. Phys. B **328**, 1 (1989); A. Airapetian *et al.* (HERMES), Phys. Rev. D **75**, 012007 (2007); V. Yu. Alexakhin *et al.* (COMPASS), Phys. Lett. B **647**, 8 (2007).
[3] B. Adeva *et al.*, Phys. Rev. D **58**, 112002 (1998); P. L. Antony *et al.*, Phys. Lett. B **493**, 19 (2000).
[4] B. Adeva *et al.*, Phys. Rev. D **70**, 012002 (2004); A. Airapetian *et al.*, Phys. Rev. Lett. **84**, 2584 (2000).
[5] See e.g. L. Silva (for the COMPASS Collaboration), arXiv:1111.0561.
[6] S. S. Adler *et al.* (PHENIX), Phys. Rev. Lett. **91**, 241803 (2003); J. Adams *et al.* (STAR), Phys.

Journal of Physics: Conference Series **678** (2016) 012059 doi:10.1088/1742-6596/678/1/012059

Rev. Lett. **92**, 171801 (2004); B.I. Abelev *et al.* (STAR), Phys. Rev. Lett. **97**, 252001 (2006); S.S. Adler *et al.* (PHENIX), Phys. Rev. Lett. **98**, 012002 (2007); A. Adare *et al.* (PHENIX), Phys. Rev. D **76**, 051106 (2007).
[7] D. de Florian, R. Sassot, M. Stratmann, and W. Vogelsang, Phys. Rev. Lett. **113** (2014) 012001.
[8] The NNPDF Collaboration: E.R. Nocera, R.D. Ball, St. Forte, G. Ridolfi and J. Rojo, Nucl.Phys. B **887** (2014) 276, arXiv:1406.5539.
[9] A. Adare *et al.* (PHENIX), Phys. Rev. D **90**, 012007 (2014).
[10] A. Adare *et al.* (PHENIX), arXiv:1510.02317.
[11] D. de Florian, R. Sassot, M. Stratmann, and W. Vogelsang, Phys. Rev. Lett. **113** (2014) 012001.
[12] E. Leader, A.V. Sidorov, and D.B. Stamenov, Phys. Rev. D **82**, 114018 (2010).
[13] L. Adamczyk *et al.* (STAR), Phys. Rev. Lett. **115**, 092002 (2015).
[14] W.C. Chang and J.C. Peng, Prog. Part. Nucl. Phys. **79** (2014) 95.
[15] A. Adare *et al.* (PHENIX), arXiv:1504.07451.
[16] L. Adamczyk *et al.* (STAR), Phys. Rev. Lett. **113** (2014) 072301.
[17] R.D. Klem *et al.*, Phys. Rev. Lett. **36**, 929 (1976); C.E. Allgower *et al.*, Phys. Rev. D **65**, 092008 (2002); D.L. Adams *et al.*, Phys. Lett. B **261** (1991) 201; D.L. Adams *et al.*, Phys. Lett. B **264** (1991) 462.
[18] X. Ji, J.-W. Qiu, W. Vogelsang, and F. Yuan, Phys. Rev. Lett. **97** 082002 (2006).
[19] Z.-B Kang, J.-W. Qiu, W. Vogelsang, and F. Yuan, Phys. Rev. D **83** (2011) 094001.
[20] K. Kanazawa, Y. Koike, A. Metz, D. Pitonyak, arXiv:1404.1033.
[21] J. C. Collins, Nucl. Phys. B **396**, 161 (1993).
[22] J. C. Collins, S. F. Heppelmann, and G. A. Ladinsky, Nucl. Phys. B **420**, 565 (1994).
[23] S.J. Brodsky, D.S. Hwang, I. Schmidt, Phys. Lett. B **530**, *99 (2002).*
[24] J. C. Collins, Phys. Lett. B **536** (2002) 43.
[25] L. Adamczyk *et al.* (STAR), arXiv:1511.06003.
[26] Z.-B. Kang and F. Yuan, Phys. Rev. D **84** (2011) 034019.
[27] M. Kim, talk at HESZ-2015 workshop, https://indico.cern.ch/event/407524/contribution/13/attachments/1153405/1656765/150910_HESZ_minjung.pdf

The method for measuring of the characteristics of the cosmic ray dipole anisotropy in the PAMELA experiment

S A Voronov, A V Karelin

National Research Nuclear University MEPhI (Moscow Engineering Physics Institute), Kashirskoe highway 31, Moscow, 115409, Russia

Abstract. The large-scale cosmic ray anisotropy has been studied in the equatorial coordinate system in the PAMELA experiment. The method to obtain dipole anisotropy characteristics as well as some its modernizations is discussed in detail.

1. Introduction

The results of the ground-based measurements for studying of an anisotropy of cosmic rays are well known [1-5]. To observe this phenomenon at an anisotropy intensity level of 10^{-3} there were collected a large number of events $\sim$ tens of millions. Unlike these experiments the satellite PAMELA experiment [6] provides with much lower statistics due to the small size of its aperture and exposure. Finally about 600,000 events were obtained after applying the selection criteria for experimental data of 9 years of observations (Jun. 2006 – Nov. 2014) [7, 8].

All the results of the ground-based experiments agree with each other in the fact that a large-scale anisotropy exists and has a dipole character at the first approximation, while the measurements of the phase and amplitude of this dipole gave inconsistent values. The aim of our study was to measure the phase and amplitude of this dipole using information collected in the PAMELA experiment within the energy range 1 – 20 TeV/n. The results were obtained with the use of the electromagnetic calorimeter [9] and published in [7, 8]. Here the method which has been used in those warks is discussed as well as its variation.

2. The general approach

The measurement of the anisotropy of cosmic rays means a construction of a distribution of primary particle arrival directions. In the PAMELA experiment the electromagnetic calorimeter was used to obtain such distribution. The cosmic ray arriving directions were measured as the axis of the showers initiated by the primary particles inside the calorimeter [10].

The anisotropy of cosmic rays was obtained by using of two distributions: the background or reference and the experimental ones. The reference direction distribution of events corresponds to the case when the cosmic ray arrival direction distribution is perfectly isotropic taking into account a real exposure and a registration efficiency of the apparatus. The reference distribution was simulated with a shuffling technique [11]. These distributions were performed in the equatorial coordinate system. In our case only a single coordinate was considered. It was a right ascension which varied from 0 to 360 degrees. All events along a declination coordinate

Content from this work may be used under the terms of the Creative Commons Attribution 3.0 licence. Any further distribution of this work must maintain attribution to the author(s) and the title of the work, journal citation and DOI.
Published under licence by IOP Publishing Ltd

Journal of Physics: Conference Series **678** (2016) 012060 doi:10.1088/1742-6596/678/1/012060

were summarized. The reference distribution contained the 100 times more events than the real one. The existence of the anisotropy of cosmic rays in the PAMELA data was confirmed by comparing of two distributions: the reference and the experimental ones with the Kolmogorov-Smirnov test.

3. The Kolmogorov-Smirnov test

The first step of the study was to check the fact that basing on the collected statistics we are able to make a conclusion about the difference between the reference and the real distributions. For this reason the Kolmogorov-Smirnov test (see fig. 1) was used [12]. The KolmogorovSmirnov statistics $Dexp$=0.3889 while the critical value is Dc=0.1015. Therefore, the null hypothesis about similarities of experimental and reference distributions was rejected at level 0.05.

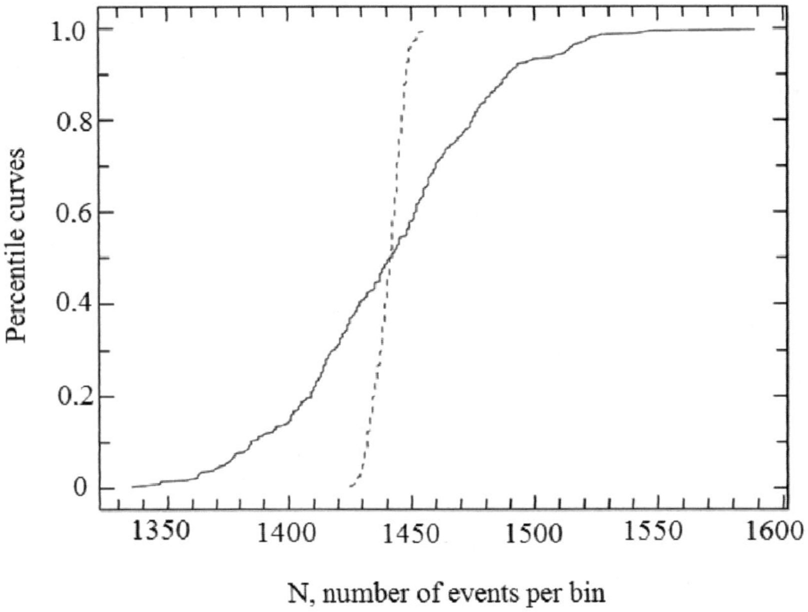

Figure 1. The percentile curves for experimental (solid line) and reference distributions (dashed line).

4. Measuring of the dipole characteristic

Since the existence of the dipole anisotropy based on the ground-based observations is not questioned, so far the task was to obtain its characteristics which are the amplitude and the phase of the dipole. About 600,000 right ascension values represented the arrival directions of the particles in the experimental data while the data in the reference distribution consisted of 100 times more events (as they were simulated their number could be as high as any number).

The basis of the analysis was the experimental fact that the measuring distribution is the dipole one (in the first harmonic). For this reason the size of an integration bin in right ascension equal to 180 degrees was chosen. In our approach this bin is equivalent to an integration radius in the HEALPIX method for the anisotropy study in 3D map [13]. The all events were integrated within this bin. A bin center was beeing shifted throughout the entire range from 0 to 360 degrees with a five degree step. Than bins were averaged to get 12 mean values. The dipole

Journal of Physics: Conference Series **678** (2016) 012060 doi:10.1088/1742-6596/678/1/012060

dependence was obtained in terms of ratio of intensities (Ir/Is 1) where Ir is the experimental number of events per bin and Is the reference one. The number of events in the reference distribution was normalized to the experimental one.

The dipole dependence was approximated by a sine function with the Levenberg-Marquardt algorithm [14]. It is necessary to emphasize a fact that the uncertainties in the figures that has been shown before in the works [7, 8] are not statistical errors and indicate just the average level of the statistics. That is why using of this uncertainties might be questionable. As the obtained numbers of events are correlated to each other the uncertainties do not correspond to the real level of fluctuations. So in the new approach we did not use these uncertainties for the approximation procedure. The scatter of points itself was taking into account. This scatter depends on the statistics level and starts to rise when the statistics goes down.

The amplitude and phase (calculated as a position of the maximum value, differs on $\pi/2$ from a sin phase) were obtained from this new approximation with taking into account factor k=$2/\pi$ (the correction factor between amplitude of the initial dipole and the one obtained with described here method): A=0.0013±0.0003, ϕ =70±10 degrees.

The demonstrated uncertainties as in the previous works [7, 8] were taken from the approximation. It should be noticed that the results in previous publications and these new ones are in agreement within a level of the uncertainties.

5. The result of a reliability test

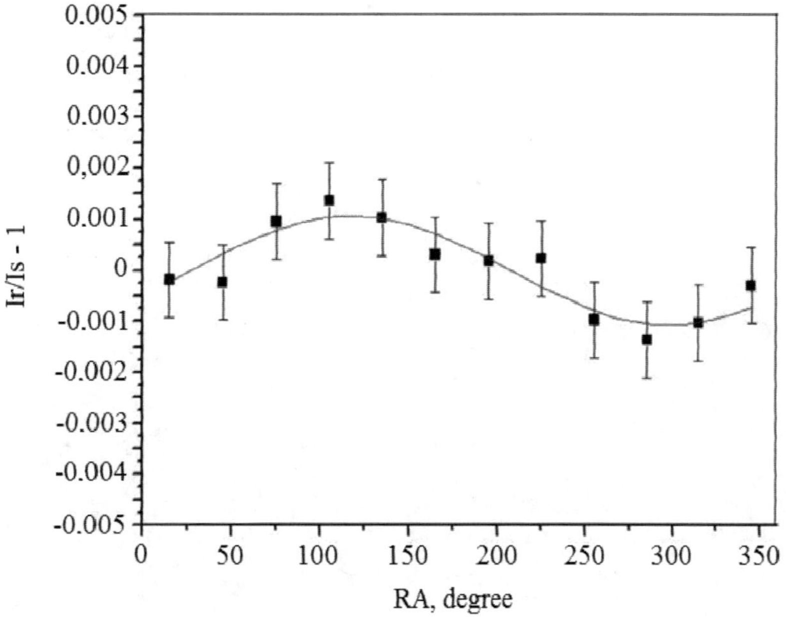

Figure 2. One of the dipoles obtained by using Monte Carlo simulation and the procedure described here.

To check the reliability of the obtained results a numerical simulation test was done. This test included a multiple numerical simulation by the Neuman (rejection) method [15] of events distributed according to a sine function. The simulated dipole dependences had an amplitude value similar to observed one and a phase value equal to 90 degrees. The procedure described above was applied to the simulated dipole distributions to obtain the values similar to the ones

Journal of Physics: Conference Series **678** (2016) 012060 doi:10.1088/1742-6596/678/1/012060

shown in fig. 2. The multiple simulation has given a determination coefficient [16] equal to 0.89 $\pm$ 0.07, while in the actual experimental data its value was 0.88. This coincidence confirmed the reliability of the results obtained at a high probability level (more than 0.9). Furthermore, the level of uncertainties in the approximation for the values of the amplitude and phase in the numerical simulations coincided with the real experimental values.

Acknowledgments

This work has been done with support of the Russian Science Foundation grant N 14-12-00373.

References

[1] Amenomori M, Ayabe S, Cui S W 2005 *Astrophys. J.*. **626** 29
[2] Abdo A A et al., 2009 *Astrophys. J.* **698** 2121
[3] Abbasi R et al., 2011 *Astrophys. J.* **740** 16
[4] BenZvi S 2014 *ASTRA Proc.* **1** 33
[5] Lidvansky A S et al., 2007 *30th ICRC, Merida, Mexico*
[6] Adriani O et al., 2014 *Physics Reports* **544** 4 323
[7] Karelin A V et al., 2015 *JETP Letters* **101** 5 295
[8] Karelin A V et al., 2015 *ASTRA Proc.* **2** 35
[9] Boezio M et al., 2002 *Nucl. Instrum. Meth. A* **487** 407
[10] Borisov S V et al., 2013 *Instruments and Experimental Techniques* **56** 1 1
[11] Karelin A V et al., 2013 *J. Phys.: Conf. Ser.* **409** 012029
[12] Justel A, Pena D, Zamar R 1997 *Statistics and Probability Letters* **35** 3 251
[13] http://healpix.sourceforge.net/documentation.php
[14] Levenberg K, 1944 *Quarterly of Applied Mathematics* **2** 164
[15] http://statweb.stanford.edu/ nzhang/Stat366/lecture6.pdf
[16] Glantz S A, Slinker B K 2000 *Primer of Applied Regression and Analysis of Variance* (New-York: McGraw-Hill) chapter 3 p 72

Journal of Physics: Conference Series **678** (2016) 012061 doi:10.1088/1742-6596/678/1/012061

Measurement conditions in the space experiment MONICA

A.V. Bakaldin, A.V. Karelin, S.V. Koldashov, S.A. Voronov

National Research Nuclear University MEPhI (Moscow Engineering Physics Institute), Kashirskoe highway 31, Moscow, 115409, Russia

Abstract. The present contribution is dedicated to the investigation of the background conditions for cosmic ray ion ionization state measurements in MONICA experiment [1]. The future experiment MONICA is aimed to study the cosmic ray ion fluxes from H till Ni in energy range 10-300 MeV/n by multilayer semiconductor telescope-spectrometer installed onboard satellite. The satellite orbit parameters (circular, altitude is about 600 km, polar) were chosen for the realization of the unique method for the measurement of the charge state of the ions with energies >10 MeV/n by using the Earth magnetic field as a separator of ion charge. The analysis of the background particle fluxes is presented taking into account the recent data of the satellite experiments and known AE-8, AP-8 models and elaborated the recommendations to improve the background conditions for the MONICA experiment.

1. Introduction

New onboard experiment MONICA is aimed for studying the cosmic ray ion fluxes from H to Ni in the energy range 10-300 MeV/n in the near-Earth space [1]. The main scientific objective of the experiment is to measure ion charge states, elemental and isotope composition and energy spectra of SEP fluxes for the individual Solar events as well as to study of evolution of these characteristics over time. At the same time MONICA will investigate ionization states and isotope composition of anomalous and galactic cosmic rays as well.

Ion charge states of the SEP observed in the near-Earth space are considered to be sensitive probes for the conditions in acceleration region such as particle acceleration dynamics and efficiency, solar plasma mean electron density and temperature [2-4]. As it is known the mean ionization state of iron ions with energies more than 10 MeV/n is about $Q_{Fe}\sim+14$ for gradual SEP events and more than $Q_{Fe}\sim+20$ for impulsive events [5], [6]. Thus, Fe mean charge state can be used for identification of SEP event type. The mean charges of more light SEP ions He, O, Mg, Si, Ar, Ca with energies ¿10 MeV/n observed during a few gradual events are the following: $Q_He\sim+2$, $Q_O\sim+7$, $Q_{Mg}\sim+10$, $Q_{Si}\sim+10.5$, $Q_{Ar}\sim+10$, $Q_{Ca}\sim+11.5$ [4]. The He, O, Mg, Si, Ar and Ca ions with energies more 10 MeV/n observed during impulsive solar events are to be practically totally ionized.

Ion charge composition of the cosmic ray fluxes observed during solar quiet times reflects the features of the processes involved in acceleration and propagation of ACRs and GCRs. As it is known today the ACRs are primarily singly charge ions of the elements with high first ionization potential (H, He, C, N, O, Ne, Ar and possibly some others) with energies from several up to 50 MeV/n [7].

Content from this work may be used under the terms of the Creative Commons Attribution 3.0 licence. Any further distribution of this work must maintain attribution to the author(s) and the title of the work, journal citation and DOI.
Published under licence by IOP Publishing Ltd

The investigations will be carried out with the high-acceptance multilayer silicon telescope-spectrometer MONICA installed onboard a satellite which will be launched into the low Earth polar orbit with an altitude about 600 km [1]. The instrument will be pointed to zenith. The spectrometer will detect nuclei in the following energy ranges: H and He from 5 to 70 MeV/n, CNO group from 10 to 150 MeV/n, Fe group from 15 to 300 MeV/n. The instrument will measure nuclear charge (Z), mass, energy, and incident angle for each detected particle. The identification of the registered nuclei will be implemented by modified $\Delta E - E$ method [8]. The spectrometer acceptance is about 100 cm^2sr, the field of view (FOV) is $\pm 45°$.

In the MONICA experiment the method of ion charge (Q) measurement based on the using of the Earth magnetic field as a particle charge separator will be implemented. Electrostatic analyzers [9] and other instrumentation are impossible to use for energies more than 10 MeV/n as the ions will be stripped within several $\mu g/cm^2$ of the material [10].and became nuclei.

This contribution is dedicated to study the background conditions for ion charge measurements in MONICA experiment.

2. The measurement of ion charge [11]

The method for the ion charge state measurement uses of Earth magnetic field as their separator. It is well known that the penetration depth of ions into the magnetosphere depends on their gyro radius. This can be observed as a flux cut-off for the particles with rigidity R at a corresponding magnetic L-shell L_C. This figure 1 shows dependence of registration efficiency $Eff(\Delta E)$ of ions detecting by the instrument in energy range ΔE on L-shell number. This efficiency is defined as a ratio between the counting rate of the ions detected inside the magnetosphere $N_{\mathrm{in}}(\Delta E)$ and the counting rate of the ions detected outside it in interplanetary space $N_{\mathrm{out}}(\Delta E)$.

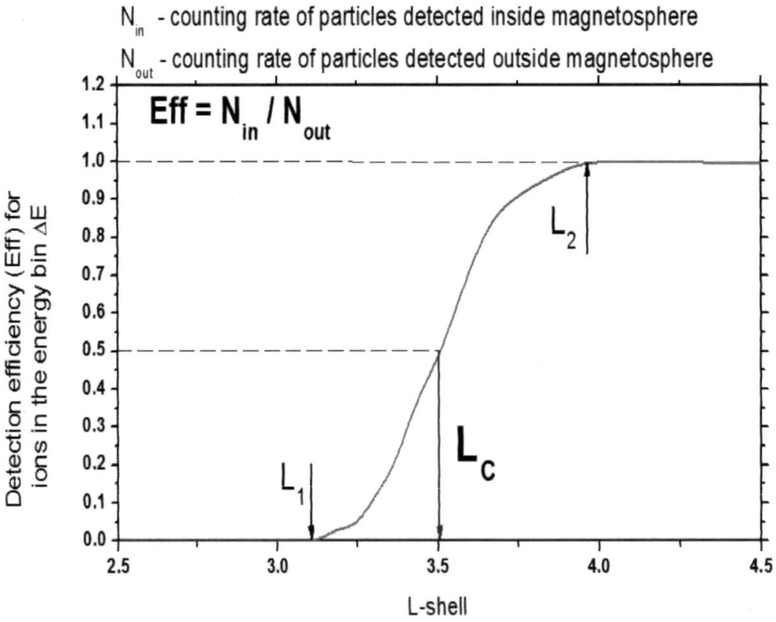

Figure 1. Detection efficiency of ions registered by instrument versus L-shell. $L_1 \div L_2$ is the inflection zone.

Referring to [4] the inverse square of cut-off L-shell value L_C depends on the particle rigidity R linearly:

Journal of Physics: Conference Series **678** (2016) 012061 doi:10.1088/1742-6596/678/1/012061

$1/L_c^2 = aR + b$

This dependence is to be used to obtain the mean charge of the particles Q:

$QR = P = \sqrt{(2M + E)E}$, $Q = a\sqrt{(2M + E)E}/(1/L_c^2 - b)$,

where E, $M = m_0c^2$ and $P = pc$ are the measured ion kinetic energy, the mass at rest and momentum in energy units (eV) correspondingly.

3. Region of the ion charge separation and the sources of the background

The registration of the cosmic ray ions should be carry out at middle and low latitudes where the satellite with instrument onboard will cross the outer and inner Earth radiation belts (RB).

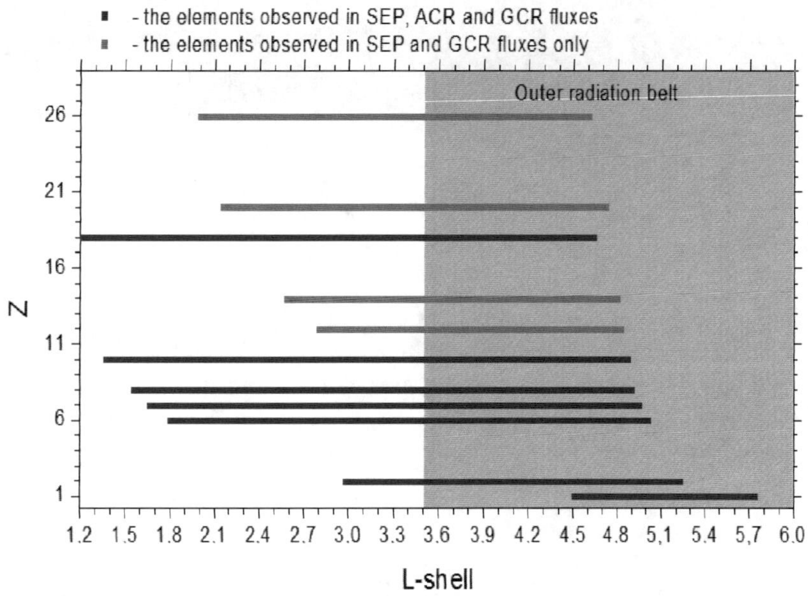

Figure 2. Calculated L-shell intervals $L_{\min} \div L_{\max}$, where the inflection zones of the $Eff(L)$ dependences $L_1 \div L_2$ (see Fig. 1) will be observed in the MONICA experiment, for ions of the different elements from H to Fe.

In Fig. 2 the evaluated L-shell intervals $L_{\min} \div L_{\max}$ are presented where the inflection zones $L_1 \div L_2$ (see Fig. 1) will be observed in MONICA experiment for the different ions from hydrogen to iron. It is seen to realize the geomagnetic separation the observation of the ion fluxes should be carried out in the region where $L < 6$.

In Fig.3 the ion charge separation region is shown in geographic coordinates. It is localized between L=6 lines. In addition, It is shown the counting rate of the top detector of the ARINA spectrometer onboard Resurs-DK1 satellite [12] obtained during the solar quiet period 1 - 30 September 2006. The ARINA top detector detects protons with energy >20 MeV and electrons with energy >1 MeV.

The following main background sources are to be mentioned in the MONICA experiment in the area of the ion charge separation (see Fig. 3): 1) the fluxes of the electrons with energy from some hundred keV to some MeV inside zone of the outer RB ($3.5 < L < 6$), 2) the fluxes of the protons with energy from several MeV to some hundred MeV and electrons with energy from some hundred keV to several tens MeV inside inner zone.

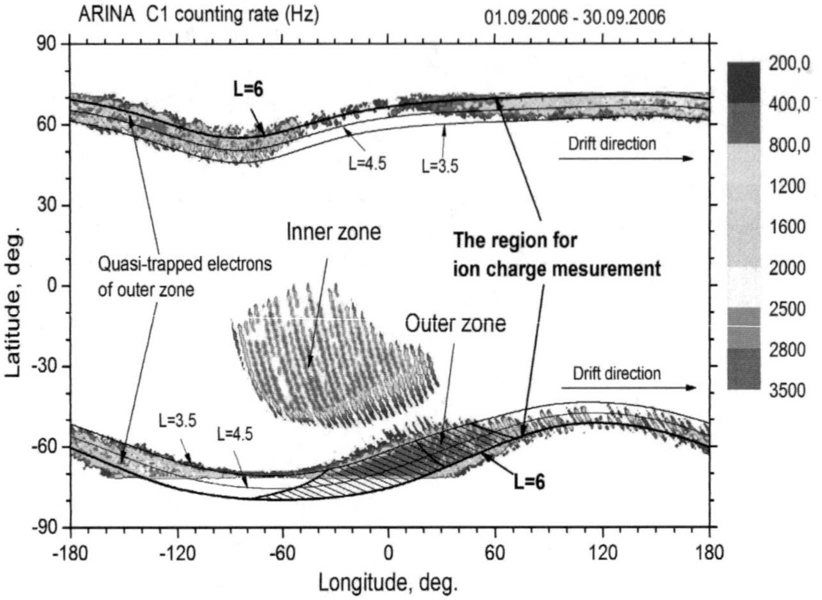

Figure 3. The region of the ion charge measurement in the MONICA experiment and map of the counting rate of the ARINA spectrometer top detector obtained during the period from 01.09.2006 until 30.09.2006.

The intense fluxes of the high energy protons and electrons of the inner RB will penetrate through the MONICA instrument in the South Atlantic Anomaly (SAA) (see Fig. 3) and will cause an overload of all its detectors. As a result the detection of cosmic ray ions will be exceedingly difficult in the SAA region. As the SAA occupies relatively small region in southern hemisphere at altitudes about 600 km the effective detection of ions outside of the SAA can be carried out at all L-shell ($L < 6$). The the SAA zone will lead to a decrease of exposure.

In Fig. 4 the ratio of the calculated exposures outside and inside the SAA are presented for L-shell interval $L_{\min} \div L_{\max}$ ($\eta = T_{L_{\min}-L_{\max}}(outside\ SAA)/T_{L_{\min}-L_{\max}}$) for different ions. The evaluations of η were implemented for Solar minimum and maximum conditions. The position of the SAA was determined with AP-8 model. One can see from Fig. 4 that the rejection of the SAA zone leads to decreasing of the exposure not more than 20%.

Unlike the inner RB the satellite with the MONICA instrument onboard will cross the outer RB for all longitudes in L-shell range $3.5 < L < 6$. As follows from Fig. 2 this range substantially intersects with intervals of the L-shell $L_{min} \div L_{max}$ where the ion charge measurements will be implemented. So if ion detection will be impossible in outer RB zone this will lead to substantial increase of the low boundary of the measured particle rigidities from 93 V (5 MeV proton energy) up to 870 MV (corresponding to vertical cut-off rigidity at L=3.5).

4. The background conditions in outer zone

The zone of the outer RB can be divided into two regions: 1) the region of the stably trapped particles (it is shown by hatching in Fig. 3); 2) the region of the quasi-trapped particles located inside other part of outer RB zone. Inside the first region electrons can be trapped in geomagnetic field the long time doing big number of drift rotations around the Earth. Inside the second region electrons lives short time and can do only one drift rotation after which they are absorbed by

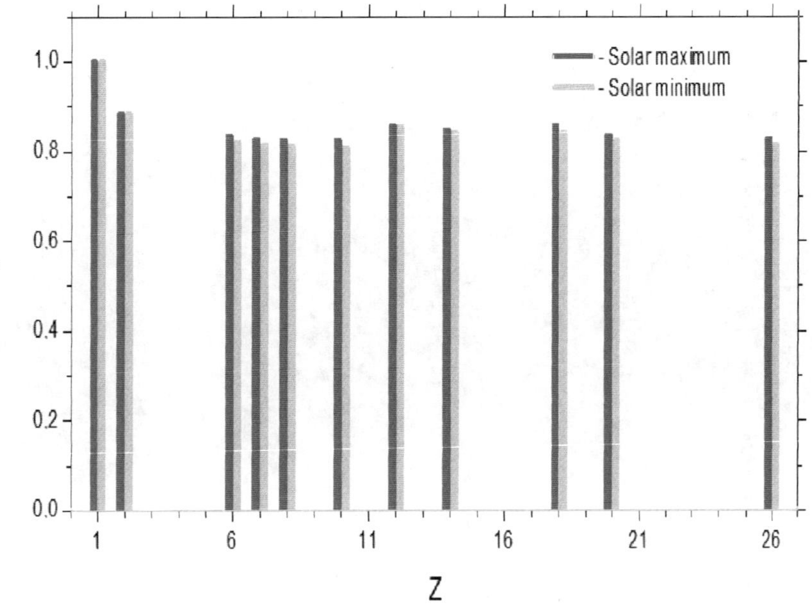

Figure 4. The ratio of the calculated exposures η outside and inside the SAA in L-shell interval $L_{\min} \div L_{\max}$ for ions of the different elements from H to Fe for Solar minimum and maximum. The evaluations were carried out for the polar circular satellite orbit with inclination 82° and altitude about 600 km.

the Earth atmosphere during drift from the East to the West near the west boundary of stable trapped particle area. The quasi-trapped electron fluxes are generated by particle precipitation out of the outer RB. The zone of the quasi-trapped electrons is characterized by high variability of particle fluxes. In the period from 01.09.2006 until 30.09.2006 the changes of these fluxes in several times in the ARINA experiment were observed.

The trapped electron flux in outer RB is not the omnidirectional. For each point of the RB zone there is strictly fixed range of the pitch-angles $\alpha_{\min} \div \alpha_{\max}$ where exists the trapped particle flux. Outside of the range $\alpha_{\min} \div \alpha_{\max}$ this flux is absent (its intensity is exactly equal to zero). Since the energy of the outer RB electrons is quite low (up to several MeV), these particles would be dangerous as a background only in case of their getting into the MONICA spectrometer field of view (FOV) through the input thin window.

The fraction of the spectrometer FOV contamination by outer RB electrons for estimation of the background conditions for MONICA experiment was calculated.

The calculation of the FOV contamination fraction was realized according an algorithm clarified in Fig. 5. At the first stage the pitch-angle range $\alpha_{\min} \div \alpha_{\max}$ was calculated for given point in outer RB zone with coordinates defined by chosen hemisphere (northern or southern), L-shell and geographic longitude. At the beginning the maximum value of the geomagnetic field B_0 was determined in mirror points for trapped particles existing in given point of the satellite orbit. The fact was used that for fixes L-shell a particle can be consider quasi-trapped by geomagnetic field if its mirror points in northern and southern hemispheres (with McIlwain coordinates L, B_0) at the observation longitude are located higher then atmosphere boundary ($h_{\min} \sim 100$ km), i.e. particle can make at least one oscillation between the mirror points. The evaluation of the $B_0(h_{\min} = 100 \text{ km})$ was fulfilled by means of geomagnetic field line

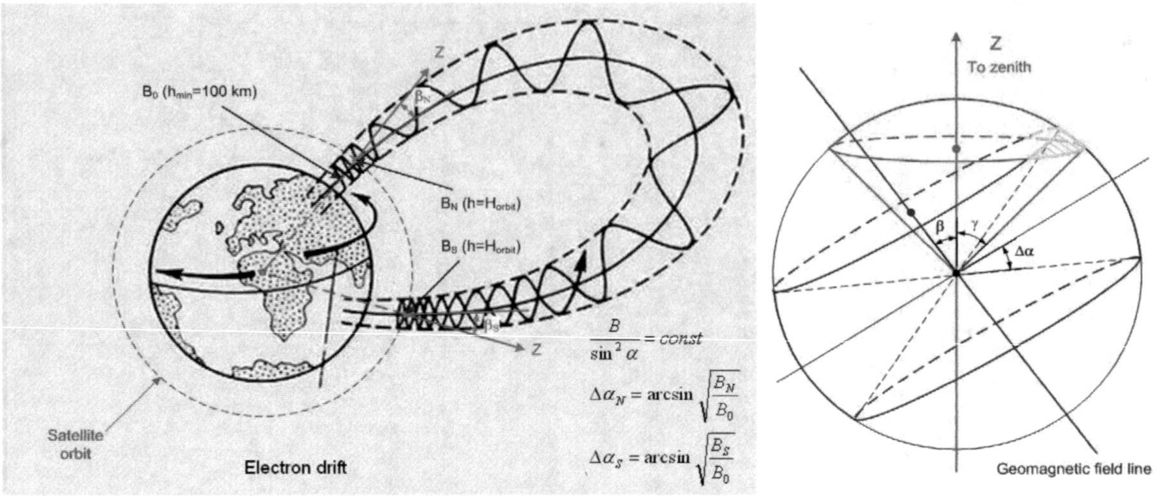

Figure 5. The figure clarifying an algorithm of the evaluation of the instrument FOV contamination fraction inside outer RB.

tracing. Then using adiabatic invariant conservation $B/sin^2\alpha$=const (where B is the value of geomagnetic field and α is pitch-angle) the values α_{min} and α_{max} were determined using the following formulas:

$a_{\text{min}} = \pi/2 - arcsin\sqrt{B(H_{\text{orbit}})/B_0(h_{\text{min}})}$, $a_{\text{max}} = \pi/2 + arcsin\sqrt{B(H_{orbit})/B_0(h_{\text{min}})}$,

where (H_{orbit}) is value of the geomagnetic field in studied point of the satellite orbit (see Fig. 5).

At the second stage the value of the solid angle $\Delta\Omega$ was determined which corresponds to directions of outer RB particles arriving into instrument FOV. Then the fraction of the FOV contamination ξ was calculated as ratio of $\Delta\Omega$ to whole instrument FOV Ω_{FOV}: $\xi = \Delta\Omega/\Omega_{FOV}$. The determining of spatial angle $\Delta\Omega$ is clarified in Fig. 5 where$\Delta\Omega$ is highlighted by green color. The spatial angle $\Delta\Omega$ corresponds to the area of the crossing the FOV (red spherical sector with angle γ=1/2 FOV in Fig. 5) and the area of the feasible trapped particle directions (blue spherical segment with angle $\Delta\alpha$=$\alpha_{\text{min}} - \alpha_{\text{max}}$ in Fig. 5 which axis is the direction of geomagnetic field).

In Fig. 6 the calculated dependences of the instrument FOV contamination fraction ξ by outer RB particles when satellite orbit at altitude 600 km crossing the L-shell L=4.5 (where intensity of the outer RB particles is maximal) on the FOV value and geographical longitude of L=4.5 crossing are presented for northern (left) and southern (right) hemispheres. As it is follows from Fig. 6 the contamination of the MONICA instrument FOV by outer RB electrons is observed in southern hemisphere only in longitude range from $-120°$ to $+50°$ in the region of the stably trapped particles (see Fig. 3). The contamination fraction value ξ does not exceed 6%.

Thus the trapped electron flux in outer RB zone practically will not affect functioning of the MONICA detectors excluding the top semiconductor detector D1 [1].

5. Conclusion

The detailed analysis of the background conditions for cosmic ray ion charge measurements for satellite experiment MONICA was implemented.

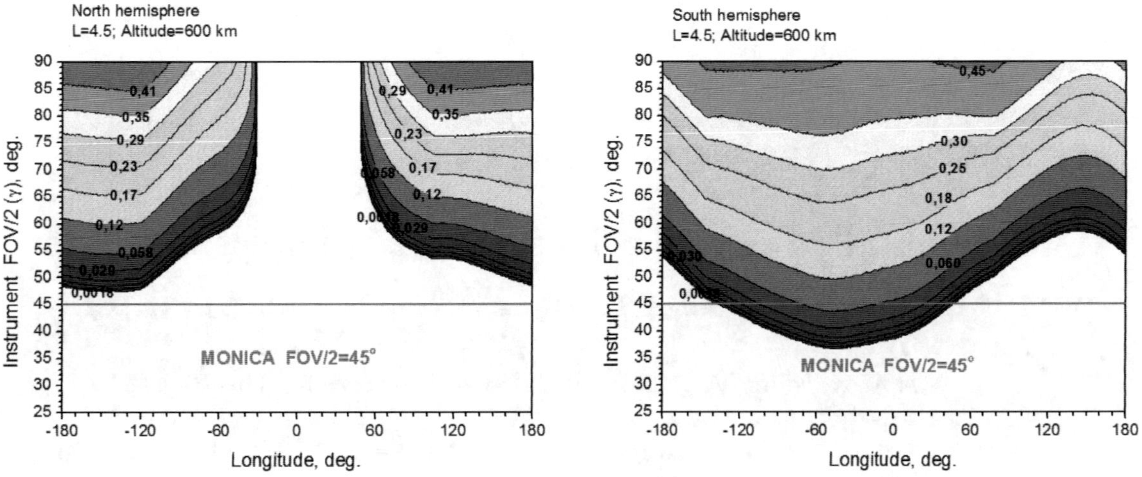

Figure 6. The calculated dependences of the instrument FOV contamination fraction ξ by trapped particles in outer RB when satellite orbit at altitude 600 km crossing the L-shell L=4.5 on the FOV value and geographical longitude of L=4.5 crossing for northern (left) and southern (right) hemispheres.

It was shown that for chosen satellite orbit parameters (circular, polar with inclination $\geq 82°$ and altitude about 600 km) the RB particle background fluxes will not have the significant impact on results of the MONICA experiment. Cosmic ray ion registration will be difficult inside SAA zone especially light ones because of the instrument detectors are overloaded. However the excluding this region out of the observations will lead to decreasing detected particle statistics only inside the regions of the ion charge measurements not more then 20%. The outer RB electron fluxes will not influence on the MONICA instrument functioning. The top semiconductor detector can be shielded against outer RB electrons by additional collimator.

Acknowledgments

This work was supported by the RFBR, grant 13-02-00477.

References

[1] Bakaldin A et al., 2009 *30th ICRC, Lodz, Poland* http://icrc2009.uni.lodz.pl/proc/pdf/icrc0387.pdf
[2] Ostryakov V, Stovpyuk M, 1999 *Solar Physics* **189** 2 357
[3] Kartavykh Y, Ostryakov V, 1999 *Proc. in 26th ICRC* **6** 272
[4] Oetliker M et al., 1997 *Ap.J.* **477** 495
[5] Reames D 1999 *Space Sci. Rev.* **90** 413
[6] Reames D 2013 *Space Sci. Rev.* **175** 53
[7] Mewaldt R A et al., 1996 *Geophysical Monograph* **97** 35
[8] Bakaldin A et al., 2010 *Instruments and Experimental Techniques* **53** 4 490
[9] Mobius E et al., 1998 *Space Sci. Rev.* **86** 447
[10] Shevelko V et al. 2009 *J. Phys. B: At. Mol. Opt. Phys.* **42** 065202
[11] Bakaldin A V, Voronov S A 2015 *Journal of Physics: Conference Series* **632** 012046
[12] Bakaldin A V et al. 2007 *Cosmic Research* **45** 5 445

Journal of Physics: Conference Series **678** (2016) 012062 doi:10.1088/1742-6596/678/1/012062

The method of event selection for nuclei separation with the calorimeter in the PAMELA experiment

A V Karelin, V L Likhacheva, S A Voronov for the PAMELA collaboration

National Research Nuclear University MEPhI (Moscow Engineering Physics Institute), Kashirskoe highway 31, Moscow, 115409, Russia

Abstract. Here we discuss a method for a study of the heavy nuclei cosmic ray composition. The method is based on a charge separation with the PAMELA calorimeter. The work is in progress now. The ability is presented to carry out measurements up to Fe and Ni. As a result we expect to get important data about the abundance of the chemical elements in the cosmic rays.

1. Introduction

The elemental composition of the cosmic rays has been measuring during more than 50 years [1-4]. The relative distribution of elements has both similarities and differences with the composition within Solar system. Moreover it appears to show the composition at the cosmic ray sources.

Being in orbit more than 9 years the space experiment PAMELA [5] is able to carry out the study of this phenomenon as well. A good charge resolution and a wide dynamic range of the instrument detectors give the possibility to obtain the new and very important results in this subject. The obtained results will help to enlarge the understanding of the situation in the sources of the cosmic rays. In turn they can lead to an identification of the optical objects with an injection of the high-energy primary cosmic ray particles.

In this paper the selection criteria are discussed as well as the possibility to study the nuclei energy determinations in the PAMELA experiment.

2. The PAMELA experiment

The PAMELA apparatus [5] was has been working in orbit just 10th year. It is installed on board of the Resurs DK1 satellite and is accumulating the scientific information up to now. The satellite orbit was elliptical and polar with an inclination of 70.4°. The main objective of the PAMELA experiment is to study the composition and energy spectra of cosmic rays over a wide range of energies. The PAMELA apparatus (fig. 1) consists of a number of special detectors: a magnetic spectrometer with position sensitive silicon detectors, a time-of-flight (TOF) system with three double planes of scintillators, an anticoincidence system, a neutron detector, a bottom shower scintillator detector S4, and a multilayer calorimeter. The geometrical factor for high energy particles (>1 GeV) passing through the sensitive volume of the magnetic spectrometer is 21.6 cm^2sr. Its angular aperture is of $19° \times 16°$.

Content from this work may be used under the terms of the Creative Commons Attribution 3.0 licence. Any further distribution of this work must maintain attribution to the author(s) and the title of the work, journal citation and DOI.
Published under licence by IOP Publishing Ltd

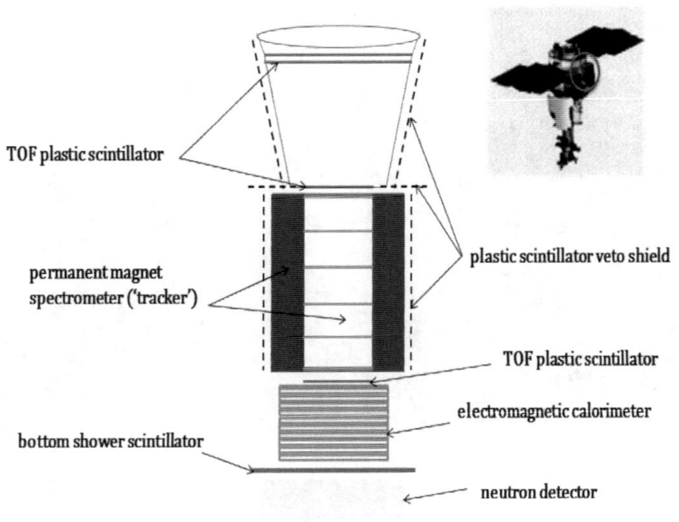

Figure 1. The structure of the PAMELA apparatus and its location in the DK1 satellite.

One of the features of the apparatus is the possibility to extend its angular aperture using as a main detector the position sensitive calorimeter. This makes it possible to increase statistics by two orders. The PAMELA calorimeter consist of 22 tungsten plates with thickness 2.6 mm interleaved by silicon strip detector planes 380 micron thick. Strips of adjacent planes are perpendicular. The strip pitch is 2.2 mm. The detailed description can be found in [6]. Each registered particle gives the information consisting of a set of detector responses formed an event written into instrument memory. As the calorimeter is a position sensitive one the set of event data consist of the amplitudes and coordinates of strips along the particle track. This gives the possibility to restore the particle parameters and to identify a particle type. In this work we discuss the possibility of studying the nuclei with Z more than 6 in the cosmic rays.

3. The track selection

First of all the events were studied for the reconstructed particle tracks in the calorimeter. The events were selected within a wide aperture. This aperture being larger by 100 times than the magnetic spectrometer one is defined by the calorimeter and S4 triggers [7]. This provides a high level of statistics which is especially important from point of view of a heavy nuclei search. The basis of our approach is a track selection procedure for primary particles interacted inside the calorimeter. This procedure (precisely described elsewhere see [8]) was developed to obtain the axis of shower of secondary particles initiated by primary particle within the calorimeter. With some modernizations it was used for a selection of events without any hadron interactions inside the calorimeter or any part of the PAMELA instrument itself before penetration into the calorimeter. The selected events include inter alia the particles which were stopped inside the calorimeter or crossed only several calorimeter planes (see fig. 2). The comparison of events before and after this selection demonstrated the high level of efficiency of this selection. For a primordial analysis only a single track events without any interactions inside the calorimeter were considered.

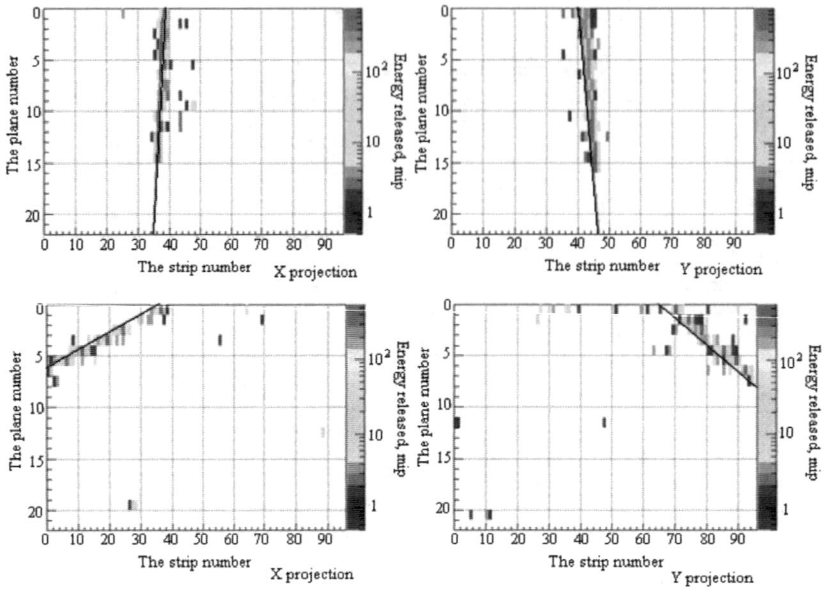

Figure 2. The experimental events. These kinds of events were selected among others for future analysis.

4. The nuclei separation

The next step was to extract the nuclei among the events obtained in the previous stage. The nuclei separation was based on the ionization losses (dE/dx) measurements of the particle along its track in the calorimeter planes. The search was expected to be done for relativistic nuclei with the charge up to heavy elements like Fe and Ni. The selection of nuclei in our method starts from the charge of carbon (Z = 6). This corresponds to a lower threshold for total energy released cut equals to 1500 mip. The (dE/dx) is measured in terms of mip as well, where mip is an energy release of minimum ionization particle. In fig. 3 the visualized events with Ne (Z = 10) Si (Z = 14) and Fe (Z = 26) nuclei are demonstrated. They were selected in accordance with such a relatively simple parameter as an average energy released per single calorimeter plane ~ 100 mip, 200 mip and 700 mip correspondingly. These elements are ones of the most abundant in the cosmic rays [9].

Today our work is in progress and the method is developing and improving. The selection criteria are under the testing. Truncated mean method will be used to obtain the mean value of energy release while the highest values of ionization are not taken into account. This increase the charge resolution as the "tail" values of Landau curve of ionization losses [10] is not used. The other means of data treatment, such as considering the distribution of released energy between strips assembling clusters near the track position in the plane or taking into account the events with interactions, will be consider as well. All these ways should definitely increase statistics of identified nuclei.

5. The future analysis and conclusion

Unfortunately this approach does not allow measuring of nuclei energy because magnetic spectrometer cannot be used. Described separation of particles according their charge with the calorimeter allows only constructing a charge distribution of registered nuclei to study

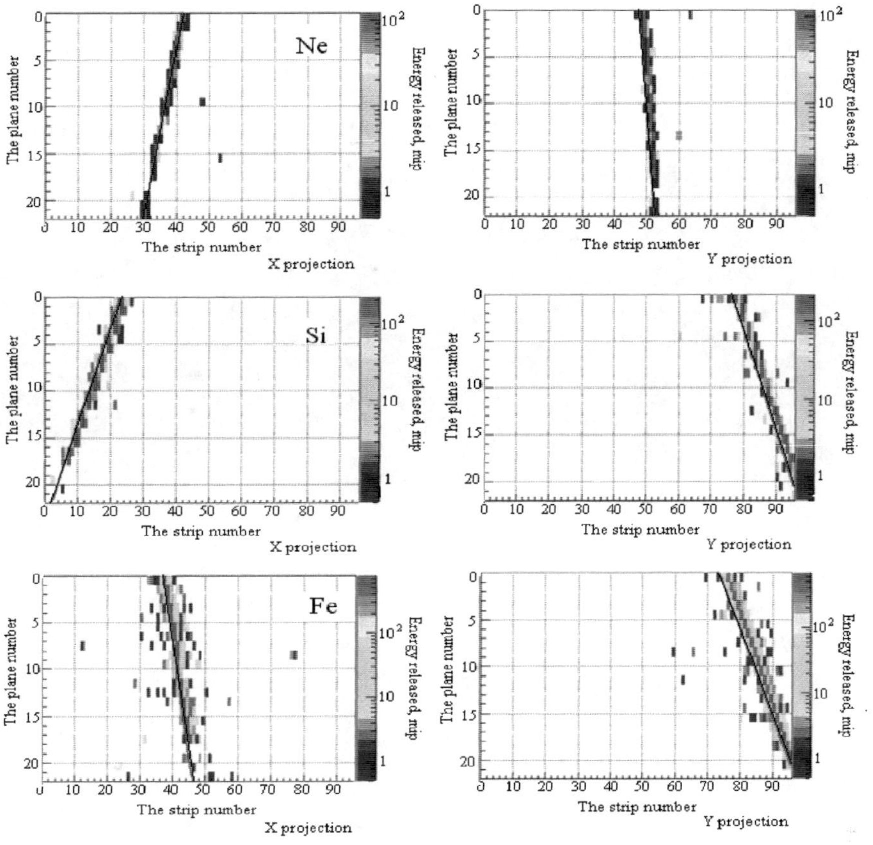

Figure 3. The selected nuclei tracks - Ne, Si, Fe in the experimental data.

an elemental composition of cosmic rays. Nevertheless as the measurements are carrying out in magnetosphere the Earth magnetic field might be used as rigidity separator. The energy thresholds can be installed by taking into account the cut-off rigidity of different areas of the satellite position. By such a way the heavy nuclei charge distribution can be expected to obtain depending on threshold energies.

So the study cosmic ray elemental composition is possible in frame of PAMELA experiment. The events of high Z value nuclei (up to Ni) were found in data set of experiment. The method of nuclei selection is under development and shows a perspective. The energy dependence of Z-distribution can be obtained from cut-off rigidities of measurement positions.

Acknowledgments
This work has been done with support of the Russian Science Foundation grant N 14-12-00373.

References
[1] Simpson J A, 1983 *Ann. Rev. Nucl. And Part. Sci.* **33** 706
[2] Mewaldt R A, 1983 *Rev. Geophys. Space Phys.* **21** 295.
[3] Weel J P, 1988 *Genesis and Propogation of Cosmic Rays* Dordrecht, Holland: D. Reidel and Co.
[4] Longair M S, 1992 High Energy Astrophysics Cambridge Univ. Press.
[5] Adriani O et al., 2014 *Physics Reports* **544** 4 323
[6] Boezio M et al., 2002 *Nucl. Instrum. Meth. A* **487** 407

[7] Karelin A V et al., 2013 *Physics of Atomic Nuclei* **76** 6 737
[8] Borisov S V et al., 2013 *Instruments and Experimental Techniques* **56** 1 1
[9] Panov A D et al., 2006 *Advances in Space Research* **37** 1944
[10] Landau L D, 1944 *J. Phys. USSR* **8** 201

Journal of Physics: Conference Series **678** (2016) 012063 doi:10.1088/1742-6596/678/1/012063

Azimuthal asymmetries of hadrons produced in muon SIDIS off longitudinally polarized deuterons

I. A. Savin, on behalf of the COMPASS Collaboration,
Joint Institute for Nuclear Research, Dubna, Moscow region, Joliot Curie 6, 141980, Russian Federation.

Igor.Savin@cern.ch

Abstract. The azimuthal asymmetries in the cross sections of charged-hadron production in the muon SIDIS reactions off the longitudinally polarised deuterons are determined using the COMPASS data of 2006 and combined data of 2002-2006. The asymmetries are presented as functions of the hadron azimuthal angle ϕ in two ways: first, for hadrons integrated over the kinematic variables and, second, as a function of one of the kinematic variables x, z or p^h_T while integrating over two other. In each case asymmetries were fitted by functions included the ϕ-independent terms and terms amplitude of which are modulated with ϕ as predicted by the theory: sin ϕ, sin 2ϕ, sin 3ϕ and cos ϕ. Results for amplitudes are presented and discussed.

1. Introduction.

Semi-Inclusive Deep Inelastic Scattering (SIDIS)

$$\mu + N \rightarrow \mu' + nh + X, \qquad n = 1, 2 \ldots \qquad (1)$$

of high energy polarised muons μ off polarised nucleons N in the initial state with scattered muons μ', $n = 1, 2 \ldots$ hadrons h and unobserved particles X in the final state give an access to the spin dependent Parton Distribution Functions PDFs of nucleons. The SIDIS cross section depends in particular on the azimuthal angle ϕ of the produced hadron (see e.g. [1]). This dependence leads to the azimuthal asymmetries in hadron production which are related to the Transverse-Momentum-Dependent Parton Distribution Functions (TMD PDFs) and Parton Fragmentation Functions (PFFs) in nucleons.

The TMD PDFs were measured in a number of experiments. The most recent data are obtained by the COMPASS collaboration at CERN using the *unpolarised* [2] and *transversely* polarised [3] ^{6}LiD ("deuteron") and NH_3 ("proton") targets. The data on hadron production off the *longitudinally* polarised deuterons were collected in 2002-2004 and 2006 and off protons in 2007 and 2011. Results on azimuthal asymmetries in hadron production off *longitudinally* polarised deuterons, based on the data of 2002-2004, were published in [4]. The analysis of 2006 and combined 2002-2006 data are presented here.

2. Theoretical framework.

The kinematics of SIDIS reactions are illustrated in figure 1. The 4-momentum of the incident and scattered muons are denoted by l and l', respectively. The 4-momentum of the virtual photon is given by $q = l - l'$ with $Q^2 = -q^2$. The angle of the momentum $\boldsymbol{q}^h$ of the virtual photon with respect to the incident muon in the laboratory frame is denoted by θ_γ. The vectors $\boldsymbol{p}^h$ and $\boldsymbol{P}_\parallel$ are the hadron momentum and the longitudinal target polarisation, respectively. Their transverse components $\boldsymbol{p}_T^h$ and $\boldsymbol{P}_T$ are defined with respect to the virtual photon momentum. The longitudinal component of target polarisation $|\boldsymbol{P}_L| = |\boldsymbol{P}_\parallel| \cos\theta_\gamma$ is approximately equal to $|\boldsymbol{P}_\parallel|$ due to the smallness of the angle θ_γ. Its small transverse component is equal to $|\boldsymbol{P}_T| = |\boldsymbol{P}_\parallel| \sin\theta_\gamma$ with $\sin\theta_\gamma \approx 2(Mx/Q)\cdot(1\text{-}y)^{1/2}$, where M is the nucleon mass, $y = (qp)/(pl)$ is a fractional energy of the virtual photon and p is the 4-momentum of the target nucleon. The angle ϕ is the azimuthal angle between the lepton scattering plane and hadron production plane and $\phi_S = 0, \pi$ is the angle of the target polarisation vector with respect to the lepton scattering plane. The Bjorken variable $x \equiv x_{Bj} = Q^2/(2p\cdot q)$, the fraction of the virtual photon energy taken by hadron $z = (p\cdot p^h)/(p\cdot q)$ and

Content from this work may be used under the terms of the Creative Commons Attribution 3.0 licence. Any further distribution of this work must maintain attribution to the author(s) and the title of the work, journal citation and DOI.
Published under licence by IOP Publishing Ltd

Journal of Physics: Conference Series **678** (2016) 012063 doi:10.1088/1742-6596/678/1/012063

invariant mass of the photon-nucleon system, $W^2 = (p+q)^2$, characterise SIDIS together with Q^2 and y.

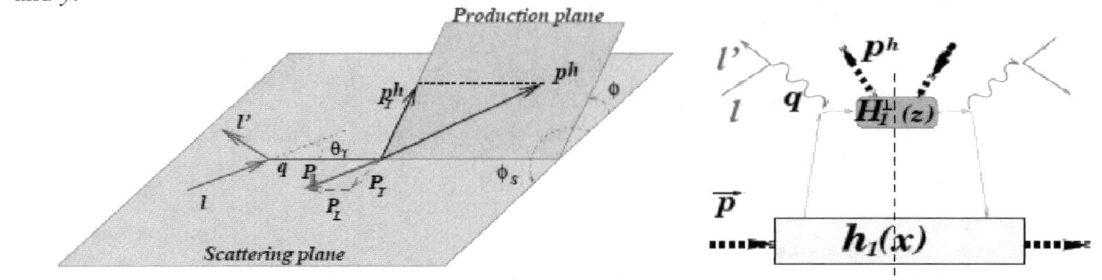

Figure1:The SIDIS kinematics for the antiparallel target polarisation $P_{||}$ with respect to the beam (left) and the diagram accounting a contribution of PDF $h_1(x)$ to the SIDIS cross section (right).

The general expression of the SIDIS cross-section is a linear function of the incident muon polarisation P_μ and of the target polarisation components P_L and P_T. The asymmetry $a\ (\phi)$ for hadron production off a longitudinally polarised target is defined by:

$$a\ (\phi) = (d\sigma^{\leftarrow\rightarrow} - d\sigma^{\leftarrow\leftarrow})/|P_L|(d\sigma^{\leftarrow\rightarrow} + d\sigma^{\leftarrow\leftarrow}) \\ = -\ [d\sigma_{0L} + P_\mu d\sigma_{LL} - \tan\theta_\gamma\ (d\sigma_{0T} + P_\mu d\sigma_{LT})]\ /(d\sigma_{00} + P_\mu d\sigma_{L0}), \qquad (2)$$

where the first superscript ($^{\leftarrow}$) in $d\sigma$ denotes the incident muon polarisation which is opposite to the beam direction in case of positive muon charge. The second superscript ($^{\rightarrow}$) or ($^{\leftarrow}$) in $d\sigma$ denotes the target polarisation, which is parallel (positive) or antiparallel (negative) to the beam. The contributions of the cross section $d\sigma_{0T}$ and $d\sigma_{LT}$ are suppressed by the small value of $\tan\theta_\gamma$.

Following the phenomenological considerations based on the parton model of nucleons and SIDIS in the one-photon exchange approximation (see references in [4]), the matrix element of the SIDIS process is represented by a number of diagrams of the type shown in figure 1. This diagram accounts, via the $d\sigma_{0T}$, a contribution to the SIDIS cross section of the chiral-odd transversity PDF $h_1(x)$ coupled to the chiral-odd Collins PFF $H_1^{\perp}(z)$. Expected contributions of other PDFs to the asymmetry (2) in the hadron production off longitudinally polarised deuterons and their modulations are given in [4]:

$$a_{h\pm}(\phi) = a_{h\pm}{}^{0} + a_{h\pm}{}^{\sin\phi} \sin\phi + a_{h\pm}{}^{\sin2\phi} \sin 2\phi + a_{h\pm}{}^{\sin3\phi} \sin 3\phi + a_{h\pm}{}^{\cos\phi} \cos\phi\ . \qquad (3)$$

None of these modulations, with the exception of $a_{h\pm}{}^{0}$ and $a_{h\pm}{}^{\sin2\phi,}$ which are connected to the twist-2 helicity PDF g_{1L} and the worm-gear-L PDF $h^{\perp}{}_{1L}$, respectively, were definitely observed so far for longitudinally polarised deuterons [6]. COMPASS results [4] showed some indication on the x-dependence of the $a_{h\pm}{}^{\sin2\phi}$ and $a_{h\pm}{}^{\cos\phi}$ amplitudes. The other modulations are either related to twist-3 PDFs or to twist-2 PDFs suppressed by a small factor $\tan\theta_\gamma$, i.e. transversity PDF h_1 and Sivers PDF $f^{\perp}{}_{1T}$ in $a_{h\pm}{}^{\sin\phi}$, pretzelosity PDF $h^{\perp}{}_{1T}$ in $a_{h\pm}{}^{\sin3\phi}$ and worm-gear-T PDF g_{1T} in $a_{h\pm}{}^{\cos\phi}$. In (3) we have disregarded the contribution of the $\cos 2\phi$ term which could appear from the $d\sigma_{00}$ in the denominator of (2). This amplitude is expected [2,7-9] not to exceed a value 0.1 and would enter into (3) with the factor $a_{h\pm}{}^{0}$ that is of order 10^{-3} (see below). This is beyond our experimental accuracy. The same comments are valid for contribution to the $\sin\phi$ and $\cos\phi$ amplitudes from the denominator of (2).

The aim of this study is to continue searches for possible modulations in $a_{h\pm}(\phi)$ as manifestation of TMD PDFs of nucleons in deuterons and to investigate the x, z and $p^h{}_T$ dependence of the corresponding modulation amplitudes using first the 2006 data and second the combination of all COMPASS deuteron data between 2002 and 2006.

3. The 2006 data analysis.

For the 2006 data taking with the ^{6}LiD longitudinally polarised target, a modified COMPASS set up was used that is described elsewhere [5]. Major improvements, influencing the present analysis, were, first, the replacement of two 60-cm long target cells denoted as U and D by three cells (U, M and D) of lengths 30, 60 and 30 cm, respectively, second, replacement of the target solenoid magnet by new one with wider aperture, third, installation of the electromagnetic calorimeter ECAL1 in front of the hadron calorimeter HCAL1. These modifications aimed to further minimizing systematic uncertainties, to increasing an acceptance of the spectrometer and improving its e/γ identification capabilities.

In 2006 the data taking have been performed in two groups of periods. Each group is characterized by an initial setting of the target cell polarisations using different frequencies of the microwave field to polarise the target material. We call one group of periods as Freezing 1 (FR1) and another – as Freezing 2 (FR2). Each period includes certain number of time intervals (runs) of the continuous data taking. The FR1 data taking periods started with initial setting of positive deuteron polarisation in the target cells U and D and negative one in the cell M corresponding to the positive direction of the holding solenoid field f =+ (coincided with the beam direction). After taking some number of runs, the solenoid field was reversed to f =− causing the reversal of the target cell polarisations and taking of data with opposite deuteron polarisations in the cells. The periodic rotation of polarisations continued up to the end of FR1 periods. Within the period the polarisation values, measured for each run, remained constant at the level of ~55 %. If not, polarisations are restored by the microwave field before beginning of the next period. For FR2 periods the procedure was analogous but initial polarisation setting in cells at the same field was opposite to one in FR1.

Events detected by the COMPASS spectrometer [10] were conserved for the analysis under condition $Q^2 > 1$ (GeV/c)^2. The statistic of 2006, excluding runs that not passed the data stability tests (see below), comprise about $64.2\cdot 10^6$ events.

Selection of SIDIS events was similar to that of [4]. For each event a reconstructed vertex with incoming and outgoing muons and one or more additional hadron tracks are required. Applying cuts on the quality of reconstructed tracks and vertices, the target cell effective length (28, 56, 28 cm), effective beam diameter (2.8 cm), the momentum of incoming muons (140 -180 GeV/c), the fractional energy carried by all hadrons from the event ($z < 1$) and the fractional virtual-photon energy ($0.1 < y < 0.9$), about $36.6\cdot 10^6$ SIDIS events selected for further analysis.

The tracks from SIDIS events have been identified as hadron tracks if each of them has $p^h_T >$ 0.05 GeV/c, comes from the current fragmentation region of $x_F > 0$ (x_F is the c.m. Feynman variable), hits one of the calorimeters HCAL1 or HCAL2 and calorimeter has an associated cluster with energy greater than 5 GeV in HCAL1 or greater than 7 GeV in HCAL2. All hadrons from each SIDIS event are included in the analysis. The total number of the hadrons used for 2006 asymmetry calculations is about $15.6\cdot 10^6$ compared to $53\cdot 10^6$ in 2002-2004.

Using advantages of the 3-cell polarised target, the 2006 data stability tests are performed for a number of observables using their ratios, R_i per run from cells U and D (U+D) and M. Requiring that the beam crosses all cells of the target and the acceptance of the spectrometer is stable, one can expect that $R_i = (U_i + D_i)/M_i$, which is luminosity independent, will be stable and close to 1. A number of plots representing R_i was considered. Example of the R_i values per run for SIDIS events is presented in figure 2 for data taking periods FR1 (W32-W37) and FR2 (W40-W46). The R_i values as function of the run number were fitted by a constant and the R_i-distributions are fitted by the Gaussian. All plots have shown that R_i are mostly stable within the $\pm 3\sigma$ limits

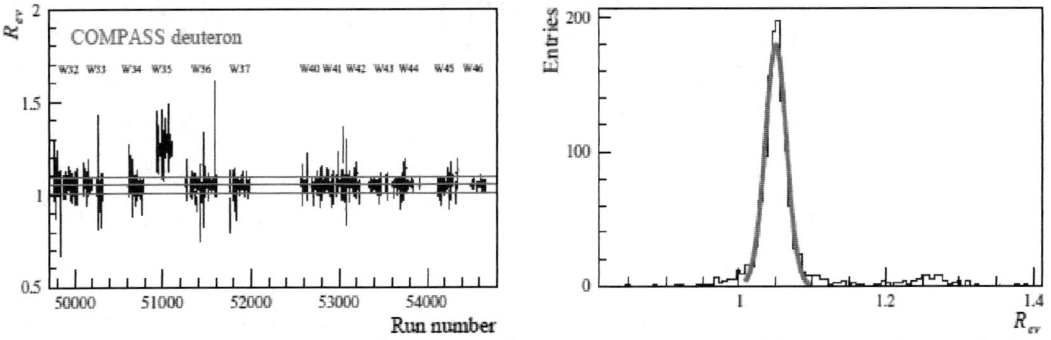

Figure 2. ***Left:*** *the $R_{ev} \equiv R_i = (U_i+D_i)/M_i$ ratios of SIDIS events per run vs. the run number. Its fit by a constant and the $\pm 3\sigma$-corridor are shown lines.* ***Right:*** *the distribution of R_{ev} and its fit by the Gaussian with Mean $\langle R_{ev}\rangle$ and σ values equal to 1.05± 0.001 and 0.015± 0.01, respectively.*

Journal of Physics: Conference Series **678** (2016) 012063 doi:10.1088/1742-6596/678/1/012063

around the average value for all runs, except for separate runs and runs of the period W35.

Stability of measurements of the hadron azimuthal angle ϕ is very important for determination of asymmetries. Distributions of ϕ-values are obtained for each run of data taking and average values per run $< \phi >$ are determined. Tests of $< \phi >$ -values per run as a function of run number have shown that its distribution has the Gaussian shape and for most of runs its values are within the 3σ limits around the mean value.

The method of asymmetry calculations described below requires stability of the spectrometer acceptance. Special tests have confirmed that acceptances are indeed mostly stable.

The acceptance free method of asymmetry calculations have been used in [4]. To cancel acceptances, the method utilizes double ratios of events. For the 3-cell target the method was modified as following. The target cell M was conditionally divided in sub-cells M1 and M2, each of 28 cm long. So, conditionally we have got a pair of targets with the cells (U and M1) and (M2 and D). In these conditional targets the cells have equal lengths and opposite polarisations at the fixed holding solenoid field direction f. For each conditional target one can obtain double ratios of SIDIS events (or selected from them hadrons), N^{i}_{pf}, from the cell i with polarisation p (+ or −) at the field f (+ or −). At $f = +$ these double ratios, DR_1^+ and DR_2^+, are:

$$DR_1^+ = [(N^U_{++} \cdot C^U_+)/(N^{M1}_{-+} \cdot C^{M1}_+)]_{FR1} \cdot [(N^{M1}_{++} \cdot C^{M1}_+)/(N^U_{-+} \cdot C^U_+)]_{FR2} = s_{1+}^2 , \quad (4.1)$$

$$DR_2^+ = [(N^D_{++} \cdot C^D_+)/(N^{M2}_{-+} \cdot C^{M2}_+)]_{FR1} \cdot [(N^{M2}_{++} \cdot C^{M2}_+)/(N^D_{-+} \cdot C^D_+)]_{FR2} = s_{1+}^2 , \quad (4.2)$$

where C^i_+ is the "acceptance factor" (inverse product of a luminosity and acceptance) of the target cell i at the field $f = +$. For the first ratios events are taken from FR1 runs and for the second ones from FR2 runs. If the acceptance factors C^i_+ do not depend on the freezing they cancel in (4.1, 4.2). By definition, each of these double ratios is equal to the squared value of $s = \sigma_+ / \sigma_-$ (or $d\sigma_+ / d\sigma_-$) which is the ratio of the total (or differential) SIDIS cross sections for positive and negative target polarisations. So, within the statistical errors the values s_{1+}^2 and s_{2+}^2 calculated with events obtained from runs with $f = +$ must be equal. Similar expressions for the negative field $f = -$ (with cancel acceptances) are the following:

$$DR_1^- = (N^{M1}_{+-}/N^U_{--})_{FR1} \cdot (N^U_{+-}/N^{M1}_{--})_{FR2} = s_{1-}^2, \quad (4.1a)$$

$$DR_2^- = (N^{M2}_{+-}/N^D_{--})_{FR1} \cdot (N^D_{+-}/N^{M2}_{--})_{FR2} = s_{2-}^2. \quad (4.2a)$$

As one expects, all s_i have the same values within the errors.

Equations $DR_1^\pm$ and $DR_2^\pm$, being combined, include full statistic of measurements. To be combined, all these double ratios should be stable and equal within the statistical errors.

Because determination of the acceptances for SIDIS events and for hadrons involves different criteria, stability tests of each double ratio were performed both for SIDIS events (DR_{ev}) and for hadrons (DR_h) using the corresponding data from consecutive pairs of FR1 and FR2 selected runs. Examples of these tests for DR_1^+ and DR_1^- calculated with SIDIS events are shown in figure 3. The difference in the DR values calculated for full statistics is visible. It is due to different contributions to them of pairs from the period W35. When these pairs are partially excluded by fitting the distributions in the narrow ranges of DR (see the bottom plots), the average values of these double ratios became practically equal to each other within errors. Removal of all pairs of this period brings distributions and their average values in a perfect agreement which is necessary for calculation of asymmetries.

So, tests of the DR stability have shown that most of the DR_{ev} and DR_h are stable over the whole data taking periods and statistically well inside the 3σ-limits except the period W35.

The final selection of the 2006 data was performed rejecting runs for which the acceptance ratios, values of the angle $< \phi >$ and values of the double ratio DR_h and DR_{ev} are outside the 3σ-limits of average, as well as all runs of the W35 period. In total about 10% of 2006 events were rejected.

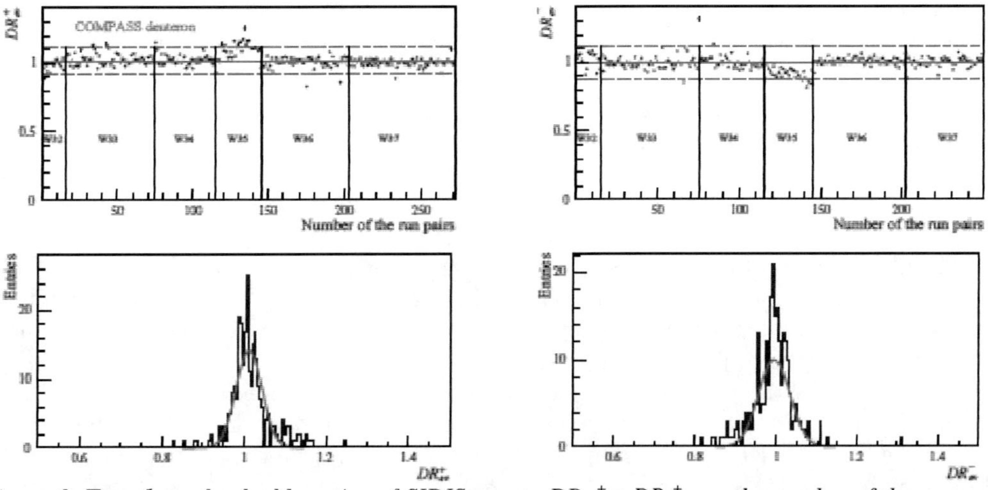

Figure 3. ***Top plots****: the double ratios of SIDIS events, $DR_{ev}^+ \equiv DR_1^+$ vs. the number of the run pair for the positive (left) and $DR_{ev}^- \equiv DR_1^-$ for the negative (right) solenoid field directions. The vertical lines show borders of the selected runs. The DR values calculated for full statistic are equal to 1.017±0.001 and 0.991 ± 0.001, respectively.* ***Bottom plots****: distributions of the DR_{ev}^+ and DR_{ev}^- values fitted by Gaussians in the DR ranges 0.8 – 1.2. Mean values are equal to 1.013± 0.003 and 0.995 ± 0.003, respectively. The σ values are 0.032 ± 0.003 and 0.038 ± 0.05, respectively. The ±3σ-corridors are shown in the upper plots.*

For calculations of azimuthal asymmetries $a_{h\pm}(\phi)$, hadrons, separately h^+ and h^-, have been distributed over 10 ϕ-bins in the range ±180°. These distributions are used for calculations of double ratios $DR_{h\pm}(\phi)$ combining statistically the double ratios $DR_1^\pm$ and $DR_2^\pm$. As it was shown in [4], values $DR_{h\pm}(\phi)$ in the linear approximation are related to $a_{h\pm}(\phi)$ *multiplied by polarisation terms:*

$$DR_{h\pm}(\phi) = [(DR_1^+ \oplus DR_2^+)_{h\pm} \oplus (DR_1^- \oplus DR_2^-)_{h\pm}](\phi) \cong 1 + a_{h\pm}(\phi) \cdot [\Sigma_{i,f}(f^i \cdot P^i_{pf})]_{h\pm}, \quad (5)$$

where symbol ⊕ means statistical sum. The polarisation terms are given by the statistical sum of the $f^i \cdot P^i_{pf}$ values which are expressed via the weighted product of the dilution factor $f^i(x, y)$, calculated as in [4], and absolute value of the polarisations P^i_{pf} (p= + and −) at f = + and f = − in the target cells i = U, M1, M2, D. The values P^i_{pf} are averaged over a number of hadrons (h^+ or h^-) from the cell i. So, the asymmetries in the linear approximation are given by expression:

$$a_{h\pm}(\phi) \cong [DR_{h\pm}(\phi) - 1] / [\Sigma_{i,f}(f^i \cdot P^i_{pf})]_{h\pm} . \quad (6)$$

The asymmetries $a_{h\pm}(\phi)$ are calculated either for hadrons integrated over all kinematic variables (integrated asymmetries) or as functions of one of the variables x, p^h_T or z while integrating over the other two. To be compatible with the data of 2002-2004, the same binning of variables as in [4] has been used for the 2006 data. In each case the data were fitted by (3).

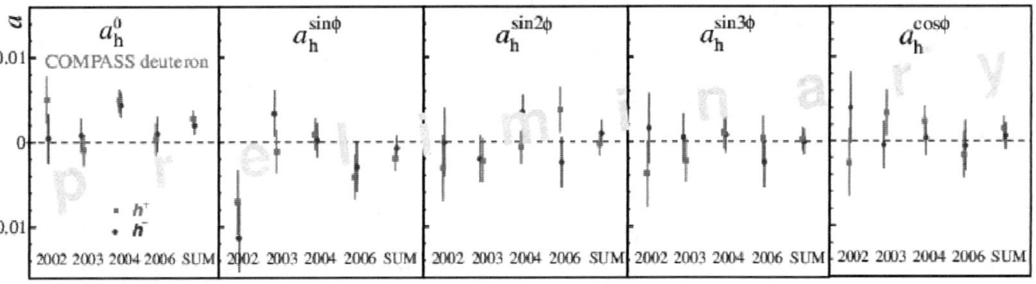

Figure 4. The amplitudes of the integrated azimuthal asymmetries $a \equiv a_{h\pm}(\phi)$ for 2002, 2003, 2004 and 2006 and combined 2002-2006 (SUM).

XVI Workshop on High Energy Spin Physics (D-SPIN2015) IOP Publishing
Journal of Physics: Conference Series **678** (2016) 012063 doi:10.1088/1742-6596/678/1/012063

The modulation amplitudes of the 2006 integrated asymmetries, together with those of the 2002-2004 [4], are shown in figure 4. They are in agreement between themselves. The values of amplitudes combined from the fits of 2002-2004 and 2006 samples (SUM) are also shown. Within the errors, all amplitudes except $a_{h\pm}{}^{0}$ are consistent with zero. Values $a_{h\pm}{}^{0}$ for h^+ and h^-, as expected, are equal to each other within the statistical errors.

Behaviour of the $a_{h\pm}(\phi)$ amplitudes vs. the variables x, z and $p^h{}_T$ for the 2006 data was compared to that of the 2002-4 data and found them to be in agreement within statistical errors.

4. Azimuthal asymmetry for combined deuteron data.

The $a_{h\pm}(\phi)$ modulation amplitudes of the combined 2002-2006 data vs. the variable x, z and $p^h{}_T$ are presented in figure 5. Fits of all distributions by constants, except the $a_{h\pm}{}^{0}(x)$, have produced results consistent with zero. The x-dependence of terms $a_{h\pm}{}^{0}(x)$ is not shown. Instead the x-dependence of terms $a_{h\pm}{}^{0}(x)/D_0(x, y)$, where D_0 is the virtual-photon depolarisation factor calculated as in [11] for each x-bin, is presented. If the main contributions to the terms $a_{h\pm}{}^{0}(x)$ came from the helicity PDF $g_1(x)$ then the values of $a_{h\pm}{}^{0}(x)/D_0(x, y)$, by definition, should be equal to the virtual photon asymmetries $A^h{}_{1d}(x)$: $a_{h\pm}{}^{0}(x)/D_0(x, y) = A^h{}_{1d}(x)$. Figure 5 shows that there is a good agreement between our data on $a_{h\pm}{}^{0}(x)/D_0(x, y)$ and the data on $A^h{}_{1d}(x)$ [11] confirming correctness of the results on asymmetries calculated by the acceptance-free method. The values of $A^h{}_{1d}(x)$ were obtained with the data of 2002-2004. Similar x-behaviour has also asymmetry $A^{\pi}{}_{1d}(x)$ and $A^{K+}{}_{1d}(x)$ for identified pions and Kaons [5] (data of 2002-2006).

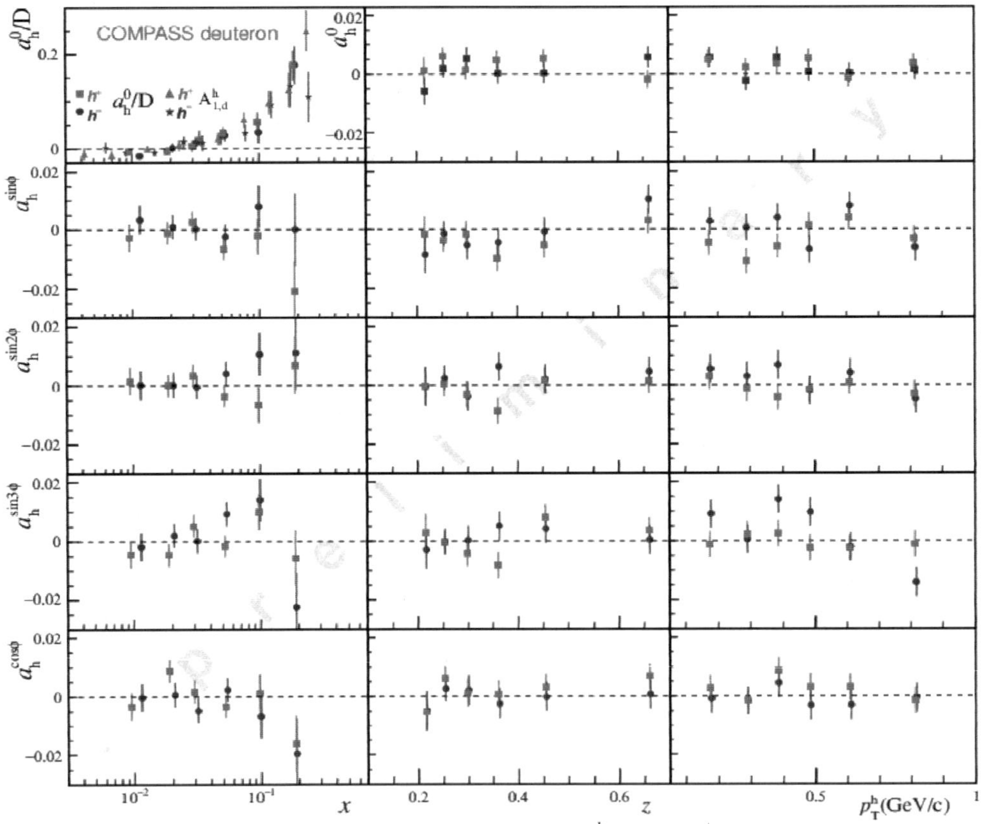

Figure 5. Azimuthal asymmetries as functions of x, z and $p^h{}_T$ for the $h^{\pm}$ combined 2002-2006 data.

5. Estimation of systematic uncertainties.

Possible sources of additive systematic errors in the asymmetry evaluation can be connected with variations of the spectrometer acceptance. The compatibility of data collected during four years and their stability tests indicated absence of large systematic effects. It is therefore assumed that the systematic uncertainties of the asymmetry measurements due to not completely cancel acceptance are smaller than the statistical ones. As a quantitative measure of possible

Journal of Physics: Conference Series **678** (2016) 012063 doi:10.1088/1742-6596/678/1/012063

additive systematic errors in the asymmetry measurements the value $\Delta\ (a_{h\pm}\ (\phi)) = \pm\ 0.003$ is obtained. The uncertainties due to the beam and target polarizations, estimated to be less than 5% each, and uncertainties of the dilution factor, estimated to be ~2%, introduce multiplicative systematic uncertainties in the asymmetry measurements. When combined in quadrature, they give a global systematic multiplicative uncertainty of less than 6 %.

6. Conclusions and discussions.

As one can see from figure 4, the ϕ-independent terms $a_{h\pm}{}^{0}$ of integrated azimuthal asymmetries are about equal for h^{+} and h^{-}. The amplitudes of all ϕ-modulating terms are very small, i.e. of the per-mil-orders, and consisted with zero within statistical errors.

There are some comments concerning the asymmetries $a_{h\pm}(\phi)$ as functions of the kinematic variables x, z and $p^{h}{}_{T}$. The first comment concerns the ϕ-independent terms $a_{h\pm}{}^{0}(x)$. Figure 5 shows that there is a good agreement between the COMPASS data on $a_{h\pm}{}^{0}\ (x)\ /D_{0}(x,\ y)$ and $A^{h}{}_{1d}\ (x)$ [5, 11] confirming expectations that the main contributions to $a_{h\pm}{}^{0}\ (x)$ are coming from the helicity PDF $g_1(x)$.

Fits by the constants of the x-, z- and $p^{h}{}_{T}$ -dependences of the terms modulating in ϕ have not shown any statistically significant peculiarities and produced results consistent with zero. Tendencies in possible x-dependence of the sin 2ϕ and cos ϕ amplitudes are visible. The sin 2ϕ amplitude for h^{-} is mostly positive and rising with increasing x, while for h^{+} it is mostly negative and decreasing with x. This behaviour agrees with those discussed in [6, 12]. The rising with x of absolute values of the cos ϕ amplitudes, connected with the Cahn effect [13] and already visible with the 2002-2004 data [4], persists with all COMPASS data.

Finally one can conclude that contributions of the TMD PDFs to the asymmetries of the hadron production in the SIDIS reactions from the longitudinally polarised deuterons are small either due to possible cancellations of the deuteron constituent quark and anti-quark contributions to the asymmetries, or due to smallness of transverse component of the target polarisation and suppressing factors ~M/Q. Some of these conclusions will be checked studying asymmetries from longitudinally polarised protons.

References

[1] Bacchetta A, Diehl M, Goeke K, Metz A and Mulders P J, Schlegel M 2007 *JHEP* **0702** 093
[2] Adolf C *et al.* [COMPASS Collaboration], 2014 *Nucl. Phys.* B **886** 1046
[3] Adolf C *et al.* [COMPASS Collaboration], 2012 *Phys. Lett.* B **717** 383
Adolf C *et al.* [COMPASS Collaboration], 2012 *Phys. Lett.* B **717** 376
Adolf C *et al.* [COMPASS Collaboration], 2014 *Phys. Lett.* B **736** 124
Adolf C *et al.* [COMPASS Collaboration], 2015 *Phys. Lett.* B **744** 250
[4] Alekseev M G *et al.* [COMPASS Collaboration], 2012 *Eur. Phys. J.* C **70** 39
[5] Alekseev M G *et al.* [COMPASS Collaboration], 2009 *Phys. Lett.* B **680** 217
[6] Airapetian A *et al.* [HERMES Collaboration], 2003 *Phys. Lett.* B **562** 182
[7] Arneodo M *et al.* [European Muon Collaboration], 1987 *Z. Phys.* C **34** 277
[8] Giordano M, Lamb R (On behalf of the HERMES Collaboration), 2009 *AIP Conf. Proc.* 1149 423
[9] Bressan A (on behalf of the COMPASS Collaboration), 17th International Workshop on Deep-Inelastic Scattering and Related Subjects (DIS 2009) [arXiv:0907.5511]
[10] Abbon P *et al.* [COMPASS Collaboration], 2007 *Nucl. Instrum. Meth.* A **577** 455
[11] Alekseev M G *et al.* [COMPASS Collaboration], 2008 *Phys. Lett.* B **660** 458
[12] Avakian H, Efremov A V, Goeke K, Metz A, and Schweitzer P, Teckentrup T 2008 *Phys. Rev.* D **77** 014023
[13] Cahn R N 1978 *Phys. Lett.* B **78** 269; Phys. Rev. D 40 (1989) 3107
Cahn R N 1989 *Phys.* Rev. D **40** 3107

Institute of Physics
Dirac House, Temple Back
Bristol BS1 6BE UK

ISSN: 1742-6588
ISBN 978-1-5108-2089-0

9 781510 820890

CURRAN ASSOCIATES INC.
proceedings.com